FACETS OF DYSLEXIA AND ITS REMEDIATION

STUDIES IN VISUAL INFORMATION PROCESSING 3

Series Editors:

Rudolf Groner
Institute for Psychology
University of Bern
Switzerland

Géry d'Ydewalle
Department of Psychology
University of Leuven
Belgium

NORTH-HOLLAND
AMSTERDAM • LONDON • NEW YORK • TOKYO

FACETS OF DYSLEXIA AND ITS REMEDIATION

Edited by

Sarah F. WRIGHT
Rudolf GRONER
Insitute for Psychology
University of Bern
Switzerland

1993

NORTH-HOLLAND
AMSTERDAM • LONDON • NEW YORK • TOKYO

NORTH-HOLLAND
ELSEVIER SCIENCE PUBLISHERS B.V.
Sara Burgerhartstraat 25
P.O. Box 211, 1000 AE Amsterdam, The Netherlands

ISBN: 0 444 89949 9

This book is printed on acid-free paper.

PRINTED IN THE NETHERLANDS

PREFACE

This volume represents the edited proceeedings of the 18th Rodin Remediation Conference on Reading and Reading Disorders. The conference had as its main theme the impact of fundamental research on new ways of understanding reading disorders, with an emphasis on both the role of basic visual processes in reading and dyslexia, and on developmental issues and methods of early diagnosis. The Conference was held at the University of Bern Children's Hospital, Switzerland, between August 27 and 29th, 1991, preceded by a day of presentations in German. The participants of the main conference were invited to submit a chapter to this book. A selection of those were edited and are included in this volume. A special issue of the Schweizerische Zeitschrift für Psychologie, 1992, Vol. 51, 1, contains papers from the first day.

The conference was organised by Ruth Kaufmann-Hayoz and the editors. It was generously supported by both the Rodin Remediation Academy and the Max and Elsa Beer-Brawand Foundation. The Rodin Academy, based in Kerns in Switzerland, aims at "redressing the adverse personal and social consequences, both congenital and conditioned by development, which retard or impair the learning processes, which in turn are connected with reading, writing and speaking". Composed of researchers and scholars from many scientific fields, one of the Rodin Academy's central activities is the support and organisation of international conferences.

The Max and Elsa Beer-Brawand Foundation supports special events at the University of Bern of cultural and social importance which are beyond the funding capacity of the university. These include meetings and symposia which should be of general interest and open to the public. Additional support was also provided by the Swiss Academy of Humanities and Social Sciences.

The editors would like to express their heartfelt thanks to Ruth Kaufmann-Hayoz for her many ideas and help. Thanks are also due to Björg Jakob and Ruth Parham for their much welcomed assistance and support, both before and during the conference. We would also like to extend our thanks to the members of the programme committee who helped in the selection of papers: the late Günther Baumgartner, Curt von Euler, François Gaillard, Dieter Heller, Giorgio Innocenti and Per Uddén. Finally we would like to thank all those who attended the conference and all who submitted papers.

CONTRIBUTORS

Janette Atkinson, Visual Development Unit, University of Cambridge, 22 Trumpington St., Cambridge CB2 1QA, UK

Monica Biscaldi, Neurology Clinic, University of Freiburg, Hansastr. 9, D-7800 Freiburg, Germany

Bruno Breitmeyer, Department of Psychology, University of Houston, Houston TX 77004-5341, USA

Marc Brysbaert, Laboratory of Experimental Psychology, Catholic University Leuven, Tiensestr. 103, B-3000 Leuven, Belgium

Salome Burri, Psychology Department, University of Bern, Gesellschaftsstr. 49, CH-3012, Bern, Switzerland

Panajotis Cakirpaloglu, Institute of Physiology, Videnska, 1083, CSFR-Prague 4, Czechoslovakia

Evelyne Corcos, Glendon College, University of York, Psychology Department, 2275 Bayview Avenue, Toronto, Ontario M4N 3M6, Canada

Piers Cornelissen, Physiology Department, University of Oxford, Parks Road, Oxford OX1 3PT, UK

Neville Drasdo, University of Aston, Birmingham B4 7ET, UK

Guinevere Eden, Physiology Department, University of Oxford, Parks Road, Oxford OX1 3PT, UK

Janice Edwards, Dyslexia Institute, 133 Gresham Rd., Staines TW18 2AJ, UK

Bruce Evans, Institute of Optometry, 56-62 Newington Causeway, London SE1 6DS, UK

Angela Fawcett, Department of Psychology, University of Sheffield, PO Box 603, Sheffield S10 2TN, UK

Heather Fields, Academic Psychiatry Department, Middlesex Hospital, Mortimer St., London W1N 8AA, UK

Burkhart Fischer, Neurology Clinic, University of Freiburg, Hansastr. 9, D-7800 Freiburg, Germany

François Gaillard, University of Lausanne, BFSH 2, CH-1015 Lausanne, Switzerland

Gad Geiger, Department of Biomedical Engineering, Rutgers University, Piscataway 08855, New Jersey, USA

Evi Graf, Swiss Cerebral Palsy Association, Loretostr., 35, CH-4504 Solothurn, Switzerland

Jane Greene, Basics Plus Education Centers, 4100 General DeGaulle 4, New Orleans LA 70131, USA

Rudolf Groner, Psychology Department, University of Bern, Laupenstr. 4, CH-3008 Bern, Switzerland

Rolf Hänni, Psychology Department, University of Bern, Gesellschaftsstr. 49, CH-3012 Bern, Switzerland

Robyn Hill, Anxiety Disorders Unit, St Vincent's Hospital, 299 Forbes St., Darlinghurst 2010, Australia

Karl-Ludwig Holtz, Pädagogische Hochschule Heidelberg, Keplerstrasse, 87, D-6900 Heidelberg, Germany

Morag Hunter-Carsch, School of Education, University of Leicester, 21 University Rd., Leicester LE1 7RF, UK

Ruth Kaufmann-Hayoz, Interfakultäre Koordinationstelle für Allgemeine Oekologie, Niesenweg, 6, CH-3012 Bern, Switzerland

John Kershner, Ontario Institute for Studies in Education, 252 Bloor St. West, Toronto, Ontario M5S 1V6, Canada

Richard Kruk, Department of Psychology, University of Wollongong, Wollongong, NSW 2500, Australia

Katie LeCluyse, Department of Psychology, University of New Orleans, LA 70148, USA

Jerome Lettvin, Department of Biomedical Engineering, Rutgers University, Piscataway, New Jersey, USA

William Lovegrove, Department of Psychology, University of Wollongong, Wollongong, NSW 2500, Australia

Catherine Meyers, Laboratory of Experimental Psychology, University of Leuven, Tiensestr. 103, B-3000, Leuven, Belgium

David Moseley, School of Education, University of Newcastle-upon-Tyne, Ridley Building, Claremont Place, Newcastle-upon-Tyne NE1 7RU, UK

Peter Müller, Psychology Department, University of Bern, Laupenstr. 4, CH-3008 Bern, Switzerland

Stanton Newman, Academic Psychiatry Department, Middlesex Hospital, Mortimer St., London W1N 8AA, UK

Roderick Nicolson, Department of Psychology, University of Sheffield, Sheffield S10 2TN, UK

Naoyuki Osaka, Department of Psychology, Kyoto University, Kyoto 606, Japan

David Pestalozzi, Ophthalmologist, Solothurnerstr. 19, CH-4600 Olten, Switzerland

Susan Pickering, Department of Psychology, University of Sheffield, Sheffield S10 2TN, UK

Tomas Radil, Institute of Physiology, Videnska, 1083, CSFR-Prague 4, Czechoslovakia

Ian Richards, University of Aston, Birmingham B4 7ET, UK

Alex Richardson, Physiology Department, University of Oxford, Parks Road, Oxford OX1 3PT, UK

Christiane Rieth, Max-Planck Institute for Brain Research, Deutschordenstr. 46, D-6000 Frankfurt, Germany

Anita Rock-Faucheux, Department of Psychology, University of New Orleans, New Orleans, LA 70148, USA

Doris Safra, Ophthalmologist, Myrtenstr. 3, CH-9010 St Gallen, Switzerland

Dieter Schoepf, Neurology Clinic, University of Hamburg, Martinistrasse, 52, D-2000 Hamburg, Germany

Wolf Singer, Max-Planck Institute for Brain Research, Deutschordenstr. 46, D-6000 Frankfurt, Germany

Ruxandra Sireteanu, Max-Planck Institute for Brain Research, Deutschordenstr. 46, D-6000 Frankfurt, Germany

Ronald Stringer, Ontario Institute for Studies in Education, 252 Bloor St. West, Toronto, Ontario M5S 1V6, Canada

John Stein, Physiology Department, University of Oxford, Parks Road, Oxford, OX1 3PT, UK

Fuchuan Sun, Shanghai Institute of Physiology, 320 Yo-yang Rd., 200031 Shanghai, China

Ulrike Tymister, Albert-Ludwigs University of Freiburg, D-7800 Freiburg, Germany

Aryan van der Leij, Department of Special Education, Vrije Universiteit, Van der Boechorststr. 1, NL-1081 BT Amsterdam, The Netherlands

Jean Whyte, Clinical Speech and Language Studies, Trinity College, Dublin 2, Ireland

Mary Williams, Department of Psychology, University of New Orleans, New Orleans, LA 70148, USA

Dale Willows, Ontario Institute for Studies in Education, 252 Bloor St. West, Toronto, Ontario M5S 1V6, Canada

Frank Wood, Bowman Gray School of Medicine, Winston Salem, North Carolina, USA

Sarah Wright, The Barn, Drayton-St-Leonard, Oxford, OX9 8BQ, UK

Regina Yap, Department of Special Education, Vrije Universiteit, Van der Boechorststr. 1, NL-1081 BT Amsterdam, The Netherlands

Wolfgang Zangemeister, Neurology Clinic, University of Hamburg, Martinistrasse 52, D-2000 Hamburg, Germany

CONTENTS

III. LANGUAGE PROCESSING

IV. ATTENTIONAL CORRELATES OF DYSLEXIA

INTRODUCTION

Developmental Dyslexia has been a subject of interest to practitioners for more than a century (e.g. Berlin, 1872; Morgan, 1896; Hinshelwood, 1900). Dyslexia represents a severe and persistence difficulty with the written form of language and affects a substantial proportion of children nationwide. Despite its long research history, dyslexia still provides a challenge for contempory cognitive psychology, education, neurology and physiology. By bringing together contributions from authors working in a wide range of fields and perspectives, it was hoped that the current book would offer a means of considering different facets of dyslexia and enable a greater understanding of reading disorders and their remediation to emerge.

The book is divided into eight major sections. The focus in each section, is on a different facet of dyslexia. Inevitably, given the diversity of the contributions, and the fact that many chapters cover more than one facet, the division is not always optimal. However, the intention was to provide a framework which would enable the reader to assimilate the wide range of pure and applied research, a framework which may even give rise to a new perspective for the understanding of dyslexia. Although the majority of the chapters use - as does the title of the book - the term 'dyslexia', the terms 'specific reading disability', 'reading disability' and 'learning disabled' are also used interchangeably.

In the first chapter, Sireteanu, Singer & Rieth investigate the development of texture segregation in infants and children. They find that while babies at the age of two months can be shown to have a preference for a figure defined by differences in blob size, it is not until the end of the first year that preference is shown for a figure differing from the background by the orientation of its elements. This preference becomes adult-like at an age when children begin to acquire reading and writing skills. Given this finding, an interesting hypothesis might be put forward for a relationship between learning to read and the maturation of cortical mechanisms necessary for the segmentation of figures in a cluttered visual environment.

Sustained and transient channels play an important part in the visual scanning of an extensive display, including textual material. In the second chapter, Breitmeyer argues that from theory and empirical evidence a visual deficit in reading disability based on weak transient channel functioning is likely. The inhibition which transient channels exert on sustained channels is important for the efficient pick-up of pattern information in fixation-saccade sequences and for maintaining constancy and visual stability. A deficit in the transient system would result in less efficient pick-up of pattern information during reading as well as leading to attendant problems with motion smear and visuospatial instabilities.

The discussion of a possible transient system deficit in dyslexics is further explored by Lovegrove. Interestingly, sustained system functioning, which is primarily concerned with extracting information from stationary stimuli, has not been found to differ between dyslexics and non-dyslexics. Addressing the confusion in the literature, Lovegrove argues that one reason some studies have failed to show differences between dyslexics and controls in visual processing may be due to the fact that they have measured sustained system processing while those which did find a difference have measured transient system functioning.

Dyslexics, it seems can foveally identify isolated letters presented tachistoscopically against a blank background, performing as well as non-dyslexic readers. However, when short letter strings or words are presented, dyslexics do considerably worse than non-dyslexics. Geiger & Lettvin investigate this phenonemon further by briefly exposing pairs of letters on a screen with one letter presented at the centre of gaze and the other in the peripheral field along the horizontal axis. In successive presentations the peripheral letters are presented at different eccentricities. The response to each pair is recorded and the plot of the average of correct peripheral letter identification with varying eccentricity determined. Differences in the resulting shapes of these plots between dyslexics and non-dyslexics are attributed to lateral masking. Dyslexics perceive the letters best in the near periphery in the direction of reading but at the same time they perceive a large portion of surrounding text since it is poorly masked.

Studies using computer presentation have found that dyslexics read better when text is presented one word at a time, whereas normal readers perform better when text is presented a whole line at a time. Hill & Lovegrove examine the relationship between one word advantage and transient activity. They suggest that since the transient system is involved with processing the textual information surrounding the fixated word, removing peripheral information (i.e. presenting one word at a time) should result in improved reading in dyslexics who have a transient deficit. The results confirm this suggestion that dyslexics with a weak transient system do read better when potentially disruptive context is removed.

Rock-Faucheux, LeCluyse and Williams expand on Lovegrove's work by studying the perceptual consequences of a transient deficit in disabled readers. Investigating the effect of colour on reading, Williams and co-workers find that blue overlays produce significant improvement in the reading comprehension of disabled and normal readers, while grey overlays produce an improvement for the dyslexics alone. The effect of using blue and grey overlays on reading performance in reading disabled subjects is attributed to the creation of a better temporal separation between transient and sustained responses, which may compensate for the visual deficit of disabled readers.

Corcos and Willows examine the role of visual processing in good and dyslexic readers' utilization of orthographic information in letter strings. The results of their multiple regression analysis shed important light on the types of cognitive processing factors that contribute to the variance in the processing of orthographic information. Their results suggest that Grade 2 and 4 students display an increasing reliance on visual processes when they processed pseudowords, a reliance which decreased in Grade 6 when speed of retrieval becomes more important.

Previous studies on the optimal viewing position effect have shown that the processing time of isolated words depends on the letter initially fixated. Shortest latencies are found when subjects are fixating between the first and the middle letter. Brysbaert & Meyers in the next chapter, compare the response patterns of normal adults, children with reading disability and their matched controls. They found a more pronounced optimal viewing position effect in both samples of children compared to adults. Manipulation of the children's initial fixation location seemed to have a larger effect than manipulation of the adults' initial fixation location, despite the fact that the children's naming latencies were much longer. However, considerable interindividual differences between children can be found.

Atkinson highlights the similarities between sight reading music and reading text and investigates to what extent the same problems encountered by dyslexics in reading text might be found in sight reading music. In one experiment, the children are required to sight read music both by singing and by playing their instruments. A significant difference is found between dyslexics and controls - with a number of the dyslexic children unable to sight read at all despite nine months musical tuition.

In the next chapter, Cornelissen poses the question 'Do children's visual problems affect *how* they read?'. He finds that children with unstable fixation, i.e. those who have failed the Dunlop Test, read differently when print size or viewing conditions are changed, but that the same changes have no effect on children with normal vision. The demands made on children's language processing skills were 'clamped' at a fixed level when they read. Therefore, because the change in children's reading errors is brought about by changing *only* the visual component of the reading task, these experiments suggest a direct link between the efficiency of visual processing and reading and lead to the conclusion that unstable binocular fixation does indeed affect how children read.

Although there is growing evidence that disabled and normal readers do differ in performance on tasks designed to assess visual processing, the extent to which various perceptual, cognitive and linguistic processes may contribute to that performance has seldom been directly assessed. Willows, Corcos and Kershner use factor analysis and multiple regression to relate performance on a large battery of tests assessing visual, phonological, linguistic and memory processes to performance on tasks assessing perception of and memory for unfamiliar visual symbols. They conclude that although memory and linguistic factors clearly account for the largest portion of variance in the perception of visual symbols, the amount of variance accounted for by visual processing is large enough to warrant further study.

Despite the large amount of research and a number of elegant studies, the role of different visual problems in dyslexia continues to be controversial. In an attempt to study the relationship between the different visual factors implicated in dyslexia, Evans, Drasdo and Richards investigate the relationship between different visual factors found to be present in dyslexia. They discover that while some measures of binocular function were reduced in the dyslexic population studied, there is no evidence to suggest that the mild binocular dysfunction was a major cause of reading disability. Their findings provide some evidence for a link between the transient deficit and binocular dysfunction in dyslexia.

Kruk and Willows, in the next chapter, argue that much of the research demonstrating transient system deficits among disabled readers has involved the use of basic psychophysical paradigms and stimuli, such as spatial frequency gratings, that have very limited similarity to text characteristics. Attempting to embrace a more 'ecologically valid' analysis of visual processes in dyslexic readers, further possible effects of a transient system deficit are reviewed and explored by the authors. In particular, studies which have examined differences in visual processes under conditions which approximate 'natural' reading are discussed. Although such studies suggest the presence of visual processing differences between normal and disabled readers, the authors argue that work carried out thus far cannot give unequivocal support for any single explanation of visual processing deficits in disabled readers.

An important aspect of any psychophysical measurement with dyslexic children is motivation. Subjects must be able to concentrate for relatively long periods of time and be able to translate reliably their sensation or perception into a form which can be recorded by the experimenter. In the final chapter in this section, Müller discusses the particular difficulties of running psychophysical experiments with children and concludes that the mathematically optimal procedures, e.g. those based on maximum likelihood estimates are not necessarily the methods that are most suited for investigations with children.

In the second section, the extent to which eye movements hold a possible key to dyslexia is examined. In a task which requires saccades from a central fixation point to a single target, occurring randomly to the right or left, Biscaldi and Fischer analyse various spatio-temporal oculomotor parameters, including express saccades. Significant differences between dyslexics and a control group of readers of the same age are found. The results show that in addition to differences in reaction times and inconsistency in the generation of saccades, fixation stability is also decreased in the dyslexic groups. Thus, while the authors do not conclude that abnormal eye movements cause dyslexia, the data from their paper suggests that the inconsistency in the generation of saccades decreases the performance rate in any task which requires visual attention and timing of saccadic eye movements.

In contrast, Fields, Wright and Newman report that when saccadic eye movements of dyslexics, low achievers and competent readers are compared, there is little which is distinctive about the saccadic eye movements of dyslexics during reading and tracking. The reading eye movements of the dyslexic group are distinguishable from those of the other two groups only in that the dyslexics make significantly more saccades. Furthermore, not only are the group means similar for most of the tracking eye movement variables, but the distributions are highly overlapping, suggesting that inaccurate tracking and dysmetric return sweeps are found in all groups in roughly equal proportions. However, as the authors caution, these results must be viewed in the light of the small number of subjects studied.

In the following chapter, Sun describes and compares eye movements in reading Chinese to those in reading English. He finds that the eye movement patterns for reading passages in Chinese and English are very similar. This, he argues, indicates that reading eye movements are controlled by the high level centre of the brain and not by peripheral visual feature detectors and that recognition spans are determined by linguistic information, not by the visual geometric form of the reading text.

The importance of foveal and parafoveal information for reading is explored by Osaka. To what extent do readers acquire different types of information from the two regions? Subjects are asked to read a Japanese text with a visual patch which obliterates the foveal visual field and which moves together with the eye during reading. As the mask size increases mean reading rate decreases significantly with a marked drop off in performance found beyond the 4-character mask-window. Thus, it seems from these results that a mask of between 4 and 6 characters appears to correspond to the effective visual field and that this does not differ for horizontal and vertical reading.

In the final chapter in this section, Schoepf and Zangemeister, investigate the oculor motor strategies - in particular ocular motor scanning- used in reading by hemianopic patients who only have visual information available in one hemifield. Subjects are required to read aloud 4 texts as accurately and quickly as possible in two conditions: firstly, where the head is fixed and secondly, in a more natural head-free-to-move condition. Schoepf and Zangemeister report that hemianopic patients initially have the intention of compensating for their hemifield loss with active and forced head employment to the affected side. However, in a complex task such as reading, this proves to be an inefficient strategy and they successively reduce their head movements and rely on eye movements instead employing a consistent set of adaptive ocular motor strategies to compensate for their visual handicap.

In the third section, the main focus moves from the visual processing facets of dyslexia to the language processing facets. In the first chapter in this section, Graf undertakes to provide a new theoretical framework for understanding reading and spelling. The 'new' theoretical framework combines the descriptive developmental models for reading with the processing models of cognitive neuropsychology. Graf argues that working within such a framework enables the researcher or practitioner to build up a greater knowledge of the weaknesses and strengths of a specific individual with respect to selecting the most appropriate training or remediation. Testing should, she argues, be broad ranged and include tests which assess visual and language processing.

Continuing on the theme of models, the need for a theory based model of reading is stressed by Holtz in the next chapter. He proposes a model based on Sternberg's triarchic theory of intelligence which enables a synthesis between intelligence, information processing and learning difficulties to be made.

Visuospatial and phonological abilities of dyslexics and controls are compared in the next chapter by Eden, Stein and Wood. Both abilities - phonological and visual are observed to be weaker in the dyslexic group. A regression model incorporating age, IQ, phonological and visual tests, in particular, visual localisation and binocular stability test scores, accounts for 65% of the variance in reading ability. These findings argue against studying dyslexia from a single perspective as it appears that it is likely that dyslexia results from a shifting synthesis of both phonological and visuospatial difficulties.

The possible phonological deficit in dyslexics is further explored in the work of Yap and van der Leij. Dyslexics are compared with a reading age matched group on several kinds of elementary symbol processing tasks. Performances are measured as a function of speed of processing, temporal separation between stimuli and demand on name retrieval ability. The results show that dyslexics have a deficit in rapid processing of successive stimuli but no strong evidence is found for a deficit on the naming task. The dyslexics do perform slower than their peers but faster than the younger readers (matched on reading level) which suggests that naming is at least partly subject to developmental trends.

The final paper in this section moves away from a direct consideration of dyslexia and instead looks at the development of symbolic motor performance in minimal brain dysfunction (MBD) boys - a group in which dyslexia, dysgraphia and dyscalculia are reported to be the most frequent problems. Cakirpaloglu and Radil investigate symbolic motor processes in MBD boys and healthy control children of different ages and cultural and social backgrounds.

Stringer and Kershner start the next section with a chapter which attempts to reexamine and rephrase one of the erstwhile central questions in dyslexia: do children with dyslexia differ from normal readers in the physical lateralisation of their language functions? They ask instead: what are the causes of the situational variability in attentional lateralisation, or more specifically, how might task variations in attentional lateralisation produce the symptoms of dyslexia? Their research suggests that dyslexic children may fail to activate left-lateralised word processing mechanisms when reading connected text and are vulnerable to attentional interference with word decoding when single word processing mechanisms are engaged. While it can be argued that the first problem - failure to activate left lateralised word processing mechanisms is probably a correlated symptom due to lack of exposure to print, the interference effect, appears to be a primary causal factor. The consequences of such an interference effect with regard to phonological and visual word decoding deficits are discussed.

Experiments carried out by Nicolson & Fawcett and reported in the next chapter, test gross motor skills in groups of dyslexics and matched controls. Dyslexic children are found to show significant deficits in balance when required to undertake a further task at the same time, whereas control children are unaffected. It appears that the control children balance automatically whereas the dyslexic children do not. Further experiments led Nicolson and Fawcett to conclude that the established phonological and working memory deficits commonly found in dyslexics can be seen as symptoms of an underlying deficit, and that an automatisation deficit is consistent with most of the known problems of dyslexic children. This deficit is in its turn speculated to derive from 'noisy' neural networks, a hypothesis capable of linking established cytoarchitectonic findings with the cognitive analysis reported in this paper.

The extent to which dyslexics are able to develop autonomous processing at the single word level is explored by van der Leij in a longitudinal study. It is assumed that within the autonomous lexicon, all relevant aspects of words are 'bonded' which facilitates accurate processing, indicated by repeated correct responses over different testing situations. In the present study, van der Leij reports that the reading disabled students are more affected than their reading age controls by increasing task demands and seem to show a very limited ability to develop an autonomous lexicon. Thus, it seems as if a deficit in speed of processing might also be implicated in dyslexia, and although it would be premature to draw too many conclusions from this study, it does seem to offer at least some support for the general automatisation deficit proposed by Nicolson and Fawcett in the previous chapter.

Albert Einstein and Isaac Newton were both reported to have been severely dyslexic and there is some evidence that dyslexics frequently report unusual perceptions and an attentional and cognitive style which may be conducive to original and creative thinking.

The nature of the link between creativity, superior visuo-spatial or mathematical skills and dyslexia is investigated by Richardson and Stein in the first of two chapters on the emotional correlates of dyslexia. In an attempt to measure the personality characteristics which might relate to the unusual perceptual and cognitive styles associated with the dyslexic profile, the authors gave a large sample of adult dyslexics and controls the Eysenck and Claridge questionnaire measures of personality. The main finding to emerge from this study is of higher STA scores (Schizotypal Personality) for dyslexics. They are not found to differ on Eysenck's measures of Extraversion, Neuroticism, Psychotism or the Lie-Scale. The only link between dyslexia and increased emotional problems is found for dyslexic males who report more frustration, mood swings and emotional conflict than others.

In the next chapter Edwards reports on interviews with dyslexic boys and their parents on the emotional effects they perceive to have been a direct result of dyslexia. Four reactions were registered: violence from teachers, unfair treatment, inadequate help and humiliation. From an impressive quantity of descriptive reports, a number of associative reaction to dyslexic problems at school are recorded. These include: truancy, psychosomatic pain, isolation or alienation from peers, lack of communication to family, lack of confidence, self-doubt, competitiveness disorders, sensitivity to criticism and behaviour problems. These are striking findings, particularly as all attended a special school for dyslexics, although the sample size is small which makes generalisations problematic.

Definition and diagnosis, despite the long tradition of dyslexia research, continue to be central issues. The strengths and weaknesses of existing definitions, in addition to the influence of definition on the number and kind of children identified as dyslexic, are discussed by Wright and Groner in the first chapter of this penultimate section. The view that dyslexia may not represent a single disorder but a number of subtypes is also considered.

One of the major groups of ways of determining whether an individual is dyslexic is by establishing whether or not he or she is reading or spelling at a level predicted by his or her measured intelligence, age or grade placement. However, as Whyte argues, the 'normal' rate of progress for reading ability has, in the main, been determined from norm groups based on cross-sectional surveys. Whether this 'normal' rate of progress is equal in every year for groups has rarely been established. Whyte, in a longitudinal study charts a pattern of progress in reading for a normally distributed, but socio-economically and culturally deprived group and for subgroups within that group. In this way, a baseline is provided to which the progress of other children with and without problems can be compared.

Sensitive and accurate pre-screening for dyslexia remains a desired but as yet unattained goal for applied dyslexia research. Fawcett, Pickering and Nicolson argue that recent theoretical developments combined with recent technological developments make the search for a Dyslexia Early Screening Test increasingly more feasible. The authors review earlier studies and outline the reasoning behind their efforts to construct such a test.

The final section in the book continues with the focus on the applied areas of dyslexia research and investigates the principles and techniques of remediation. Gaillard begins by outlining the principles of remediation, in particular the contrasting approaches to therapy made by the researcher, teacher and therapist. The need for cooperation between professionals is also discussed by Hunter-Carsch. This chapter outlines the kinds of professional knowledge and skills which should be shared between special education specialists and generalist class teachers in order that informed individualised teaching of dyslexic children can take place within the classroom.

Greene stresses that the key element in the design and implementation of efficient programs for learning to read is phonology. She argues that methods which simply attempt to present traditional materials and linguistic concepts more slowly will fail unless a systematic approach to the presentation of phonological material is adopted. Reviewing gains in reading made by a group of children following such a structured, sequential phonology program, it is found that they make more progress than children not participating in the program.

Nicolson and Fawcett focus on spelling, given that this is usually an even more severe problem than reading. The chapter describes the SelfSpell environment for dyslexic children which helps them to learn to spell problem words. SelfSpell is written in Hypercard on the Apple Mac using hypermedia techniques to encourage child-centred active learning. Evaluations of the effectiveness of the software indicated substantial improvements in the spelling and motivation of two groups of dyslexic children, including those with severe difficulties in spelling.

The possibility of using computers as a primary means of remediation is also discussed in a study described by Moseley in which computers with a synthesised speech system are used as a means of providing intensive reading practice in the classroom. The visual presentation of the text is designed to minimise problems arising from image instability, control of eye movements and from possible malfunctioning of the transient visual system. Repeated readings of relatively difficult text, with phrase and sentence structure made visually explicit, leads to increasingly accurate and fluent performance together with improvements in comprehension. The progress of seven dyslexic pupils is outlined.

In the next chapters, Pestalozzi and Safra present possibilities for the treatment of dyslexia from the perspective of Opthalmology. Working with dyslexics also suffering from heterophoria (hidden squint) and restoring full binocular vision through the use of prisms, Pestalozzi found that in the majority of cases there was a dramatic improvement in reading and writing ability. Safra, also reports considerable improvement in reading following treatment with prisms and discusses the advantages and drawbacks of such a method of treatment.

The final two chapters in the book look at learning to read in the context of adult illiteracy. Burri and Hänni evaluate the effect of two tutorial programs offered to adults focussing not only on how beneficial these are, and whether they lead to an improvement in the participants' reading and writing skills, but also, given that not all the participants make similar progress, what variables would have the largest impact on progress. They find that adult illiteracy can be successfully remediated but that the degree of remediation is not independent of intelligence.

'The Freiburg Integral Approach' which aims at individually supporting and activating the participants of adult literacy courses is presented by Tymister. Taking into account the inevitable heterogeneity of this group of learners, she presents 4 levels of learning which enable the teacher to assess individual levels of knowledge and thereby concentrate on deficits and weaknesses for a given individual. The possibility of using computers to facilitate such differentiated teaching is also mentioned.

Given the results presented from the various studies in the book, it is clear that a single deficit hypothesis cannot account for the full range of findings related to dyslexia. Recurring evidence suggests that dyslexic children suffer problems not only in skills related to visual and phonological processing, e.g. a transient system or phonological deficit, but also in areas not directly related to reading, e.g. problems in motor skill. Furthermore, dyslexia is a developmental disorder which means that its nature, by definition, will change with time. Nevertheless, the diverse, often confusing, facets of dyslexia are, it seems, gradually beginning to fall into place to give a more cohesive picture. The greater cohesiveness, in combination with an increased communication between researchers and practitioners, has led to better remedial practices in which the success or failure of a given technique can provide a further test of the empirical hypotheses put foward by basic research.

REFERENCES

Berlin, R. (1872). Eine besondere Art der Wortblindheit (Dyslexia). Wiesbaden.
Hinshelwood, J. (1900). Congenital word-blindness. *Lancet*, 1, 1506-1508.
Morgan, W.P. (1896). A case of congenital word-blindness. *British Medical Journal*, 11, 378.

I. VISUAL PROCESSING

Facets of Dyslexia and its Remediation
S.F. Wright and R. Groner (Editors)

3

TEXTURE SEGREGATION BASED ON LINE ORIENTATION DEVELOPS LATE IN CHILDHOOD

R. Sireteanu, W. Singer & C. Rieth

Max-Planck-Institute for Brain Research

Frankfurt/Main, Germany

Segmentation of a figure from its surround is in most cases a preattentive process operating automatically, effortlessly, and in parallel across the visual field (Treisman & Gelade, 1980; Julesz & Bergen, 1983). Segmentation may be based upon differences in contrast, colour, disparity, motion and surface texture. In this study, we investigated the development of texture segregation in infants and children. Segregation based on differences in line orientation and blob size was tested with a forced-choice preferential looking technique (Teller, Morse & Borton, 1974). We found that 2-month-old infants already have a preference for a figure defined by differences in blob size. In contrast, preference for a figure differing from the background by the orientation of its elements emerges at the end of the first year of life and becomes adult-like by school age.

When presented with pairs of equiluminant stimuli, one containing contours and the other not, infants tend to look at the stimulus with contours (Fantz, 1964). Measurements of preferential looking can thus be used to assess the subjects' ability to resolve patterns (Teller et al., 1974). Preferences are also seen when two different patterns, both containing contours, are presented (Fantz, 1964). A preference for one pattern is assumed to indicate that the infant discriminates the two patterns.

The goal of the present study was, first, to determine whether subjects show a spontaneous preference for a pattern containing a figure defined by textural differences, and second, if there are such preferences, to investigate their development.

Subjects were presented with a pair of stimuli. One stimulus contained a figure embedded in a background; the other showed the background alone (Figs. 1 & 2). We used two stimulus configurations: a figure composed of lines of identical orientation embedded in a surround composed of lines of an orthogonal orientation (Fig. 2, upper part), and a figure composed of blobs of identical size, embedded in a surround of blobs of a different (smaller) size (Fig. 2, lower part). Both patterns have been suggested to consist of elementary features ("textons") and both are spontaneously segregated by normal adult observers (Julesz & Bergen, 1983; Nothdurft, 1990a). The line stimulus does not contain global luminance differences, whereas the blob stimulus shows such differences (Nothdurft, 1990a).

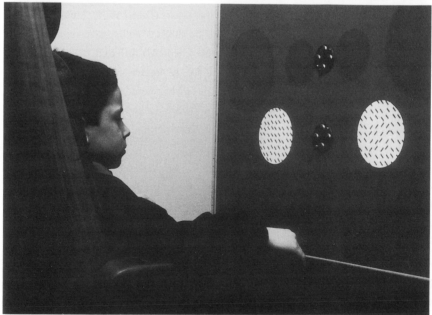

FIG. 1. on opposite page. Set-up for testing visual preferences. Apparatus and procedure were adapted from (Sireteanu et al., 1984; Sireteanu & Fronius, 1987). Subjects sit in front of a large wooden screen containing 2 circular apertures in an otherwise darkened room. Four small red blinking lamps arranged around a peephole serve as a centering stimulus. Two of these lamps can be lit at a time (see upper panel). On each trial, a pair of slides is projected from behind onto the 2 apertures (see lower panel). A video camera is positioned behind the peephole. An adult observer, naive to the side of presentation of the stimulus containing the figure, looked at the subject's face on a video screen. The observer had to decide on the basis of the subject's looking behaviour (side and duration of the first fixation, longest fixation, interested scanning etc.) which side the figure stimulus had been on (forced-choice preferential looking). There was no time limit for the observer's decision. Each session consisted of 20 trials. At the end of the session, percent correct responses was calculated. No preference corresponds to 50% correct. All subjects were naive to the purpose of the experiments.

To test whether segregation can be detected by the forced-choice preferential looking technique, we first applied this method to naive adult observers. Using the orientation stimulus, the adults showed a highly significant preference (91%, p< 0,0005) for the stimulus containing the figure (Fig.3). Figure 3 also shows the percentage correct responses averaged over all children in each age group. Infants aged 3-6 and 6-9 months do not show a significant preference for any stimulus. Only infants from the age of 9-12 months significantly prefer the stimulus containing the figure. This preference increases slowly until the age of 6-7 years, and more rapidly afterwards. At 8-13 years, children show a strong, consistent preference, comparable to that of naive adult observers (Fig.3).

To check whether the subjects were able to see the stimuli, in each session, four additional stimuli (drawings of faces, animals and toys, paired with a blank slide) were included as controls. The contours of these drawings were comparable in width to the individual lines in the main experiment. The responses to these control slides averaged 97% for children younger than 4 years, and 100% afterwards.

In a second experiment, we found that a very different developmental course is followed by segmentation based on blob size. Infants aged only 1-3 months showed a very clear preference (83%, p<0.0005) for the stimulus containing the large blobs. Except for a decrease between 6 months and 2 years of age (a notoriously difficult age range; Mayer and Dobson, 1982), the percentages of correct responses remained high up to the age of 10 years (Fig. 4).

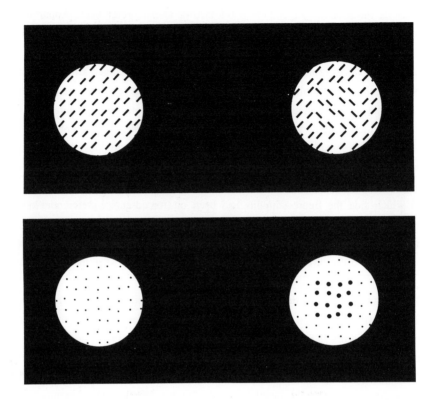

FIG. 2. Stimuli used in the main experiment. The apertures and stimuli are drawn to scale. The parameters were chosen such that the stimuli contain enough elements in both figure and background, and still be within the resolution limits of the youngest infants (Teller et al., 1974; Sireteanu et al., 1984; Sireteanu & Fronius, 1987). Upper panel: the background stimulus contains short lines (1,2° x 0,35°), all having the same oblique orientation. The figure stimulus contains the same elements, with the difference that a group of 16 neighbouring elements have an orthogonal orientation. For adult observers, this group clearly pops-out from the background. To avoid spurious "holes" at the boundaries of the figure, the line elements show a positional jitter. The fundamental spatial frequency of the stimuli was 0,8c/deg. Oblique orientations were used to avoid artifacts due to the transient astigmatism of infancy, which usually affects the horizontal and vertical meridians (Atkinson & French, 1979). Lower panel: the background stimulus is made of randomly arranged small black blobs of the same size (0,3° diameter). The figure stimulus contains a group of 16 neighbouring blobs of larger size (0,8° diameter). This group stands out from the background of small blobs as a dark square. The fundamental spatial frequency of this stimulus was 0,5c/deg.

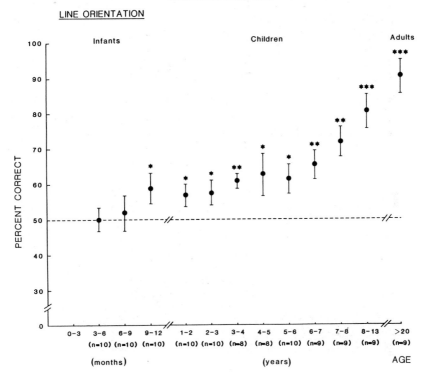

FIG. 3. Mean and standard error of the percentage correct responses for texture segregation based on line orientation, as a function of age. The results based on 112 tests (30 infants, 73 children and 9 adult observers). *: mean is significantly different from chance (p<0,05); **: p<0,005; ***: p< 0,0005.

In an additional experiment, children were tested with textures presented on cardboard cards, similar to the Teller Acuity Cards (Teller, McDonald, Preston, Sebris & Dobson, 1986). Each card contained a uniformly textured field (small dots or oriented lines) with the "figure" (large dots or orthogonal lines) presented on one side. The procedure was adapted from Teller et al., 1986. An observer naive to the side and the identity of the stimulus looked at the child's face through a small peephole at the centre of the card. Based on the reaction of the child, she made a judgement on the side of presentation of the figure. Each card was presented 2-3 times. The judgements were made independently by 2 observers. Inter-observer agreement was 95%. A response was scored positive only when both observers agreed. Infants aged 0-3 months already showed a consistent preference for the side containing the large blobs; in contrast, preference for the figure defined by orientation contrast was first seen at 6-9 months of age (Fig. 5).

The relatively late emergence of the preference for a figure based on orientation contrast was surprising, in view of the fact that orientation per se starts to be discriminable before 2 months of age (Atkinson, 1984), possibly at birth (Atkinson, Hood, Wattam-Bell, Anker, Tricklebank, 1988; Slater, Morison & Somers, 1988). However, inhibition between orientation-selective mechanisms is not functional at birth, but develops only after 6-8 months (Morrone & Burr, 1986).

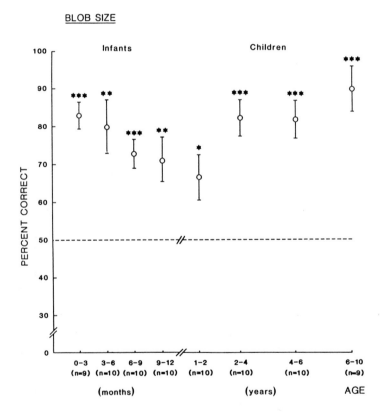

FIG. 4. Mean, standard error and statistical significance for texture segregation based on blob size, as a function of age. Apparatus and procedure as in Figure 2. The results are based on 78 tests (39 infants and 39 children). Symbols as in Figure 3.

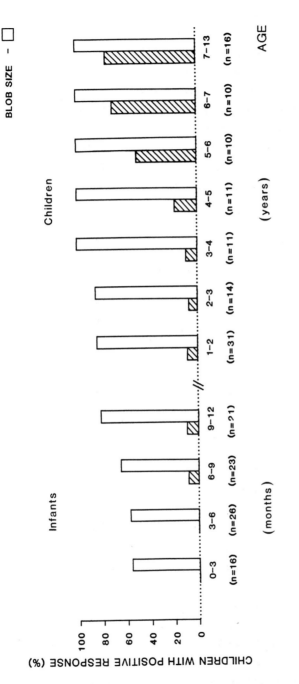

FIG. 5. Percent of children with positive responses for texture segregation based on line orientation and blob size, as a function of age. The stimuli were presented on cardboard cards, 25 cm x 56 cm in size. The orientation stimulus was made of short oriented lines (2, 1° x 0,6°), all having the same oblique orientation. On one side, a field of 9 neighbouring elements had the orthogonal orientation. The fundamental spatial frequency of the stimulus was 0,5 c/deg. The blob stimulus was made of small black blobs of identical size (0,45°), with a group of 9 larger blobs (1, 1°) on ONE side. The fundamental spatial frequency of this stimulus was 0,3 c/deg. The elements of both cards showed a positional jitter. The results were based on 189 tests (86 infants, 103 children).

The lack of preference for a figure defined by orientation in children younger than 6-9 months does not necessarily imply that these children are not able to discriminate the stimuli. Indeed, control experiments show that children aged 3-6 months show a highly significant preference for the figure defined by orientation contrast, after having been habituated to the homogeneously oriented pattern (x=57, 6%; n=27; p<0,0025; one-sided t-test against chance). Rather, it is only at 6-9 months that the pattern containing an orientation contrast acquires the quality of a "figure", which is segregated from the surround and becomes able to elicit the preferential directing of attention, on which the overt, observable eye movements are based. In the absence of global luminance differences, this "figure" has to be generated either by coherent activation (grouping) of individual detectors with similar orientation preferences (Beck, 1966; Grossberg & Mingolla, 1985), or by the orientation contrast along the borders of the figure (Nothdurft, 1991).

The preference for the figure defined by differences in blob size, on the other hand, can be explained by global luminance differences, and does not require the cooperative activation of individual feature detectors.

Segmentation based on differences in blob size is probably based on early-maturing, peripheral neural processes (cells in the lateral geniculate nucleus of the cat respond to borders defined by differences in blob size (Nothdurft, 1990b). Segmentation based on differences in line orientation, on the other hand, was only occasionally seen in LGN cells (Nothdurft, 1990b); only at the cortical level, where most cells are orientation selective, neurons responding more strongly to a texture field containing orientation contrasts than to a uniform texture field have been described (Nothdurft & Li, 1985; van Essen, DeYoe, Olavarria, Knierim, Fox, Sagi & Julesz, 1989).

In the cortex, segmentation based on orientation differences might depend on horizontal connections between columns of neurons with similar orientation preferences (Singer, 1990). The late development of this segregation in infants could be based on the late maturation of far-reaching horizontal connections. Support for this hypothesis is provided by the finding that horizontal connections of neurons in layers II-III in the human primary visual cortex develop after birth and are still immature at 2 years of age (Burkhalter, Bernardo & Charles, 1990). The authors suggest that "circuits that process local features of a visual scene develop before circuits necessary for integrating these features into a continuous and coherent representation of an image" (Burkhalter, personal communication). Such circuits presumably support the correlated cortical oscillations, thought to produce binding and segmentation (Engel, König, Kreiter & Singer, 1991).

Interestingly, preference for a figure embedded in a surround of a different orientation becomes adult-like at an age when children start to acquire reading and writing skills. Further experimentation should clarify whether learning to read is related to the maturation of cortical mechanisms necessary for the segmentation of figures in a cluttered visual environment.

ACKNOWLEDGEMENTS

We are grateful to Christoph Nothdurft for the development of the stimuli used in the study and for fruitful discussions, Jerry Nelson for help with editing and Adriana Fiorentini for helpful comments on an earlier version of the manuscript. We thank Silke Wiemerslage for drawing the figures, Margitt Ehms-Sommer for taking the photographs, and Irmi Pipacs, Helga Reitz and Svenja Wartha for typing the manuscript. Special thanks are due to the parents and the children who took part in the study. This work was supported by the Deutsche Forschungsgemeinschaft.

REFERENCES

Atkinson, J. (1984). Human visual development over the first 6 months of life. A review and a hypothesis. *Human Neurobiology*, 3, 61-74.

Atkinson, J. & French, J. (1979). Astigmatism and Orientation Preference in human infants. *Vision Research*, 1315-1317.

Atkinson, J., Hood, B., Wattam-Bell, J., Anker, S., & Tricklebank, J. (1988). Development of orientation discrimination in infancy. *Perception*, 17, 587-595.

Beck, J. (1966). Perceptual grouping produced by changes in orientation and shape. *Science*, 154, 538-540.

Burkhalter, A., Bernardo, K.L., & Charles, V.C. (1990). Postnatal development of intracortical connections in human visual cortex. *Society for Neuroscience Abstracts*, 16, 1129.

Engel, A.K., König, P., Kreiter, A.K., & Singer, W. (1991). Interhemispheric synchronization of oscillatory neuronal responses in cat visual cortex. *Science*, 252, 1177-1179.

Fantz, R.L. (1964). Visual experience in infants: decreased attention to familiar patterns relative to novel ones. *Science*, 146, 668-670.

Grossberg, S., & Mingolla, E. (1985). Neural dynamics of perceptual grouping: textures, boundaries, and emergent segmentations. *Perception & Psychophysics*, 38(2), 141-171.

Julesz, B., & Bergen, J.R. (1983). Textures, the fundamental elements in preattentive vision and perception of textures. *Bell System Technical Journal*, 62, 1619-1645.

Mayer, D.L., & Dobson, V. (1982). Visual acuity development in young infants and young children, as assessed by operant preferential looking. *Vision Research*, 22, 1141-1151.

Morrone, M.C., & Burr, D. (1987). Evidence for the existence and development of visual inhibition in humans. *Nature*, 321, 235-237.

Nothdurft, H.C. (1990a). Texture segregation by associated differences in global and local luminance differences. *Proceedings of the Royal Society in London*, B Series, 239, 295-320.

Nothdurft, H.C. (1990b). Texture discrimination by cells in the cat lateral geniculate nucleus. *Experimental Brain Research*, 82, 48-66.

Nothdurft, H.C. (1991). Texture segmentation and pop-out from orientation contrast. *Vision Research*, 31, 1073-1078.

Nothdurft, H.C., & Li, C.Y. (1985). Texture discrimination: Representation of orientation and luminance differences in cells of the cat striate cortex. *Vision Research*, 25, 1, 99-113.

Singer, W. (1990). Search for coherence: a basic principle of cortical self- organization. *Concepts Neuroscience*, 1, 1-26.

Sireteanu, R., & Fronius, M. (1987). The development of peripheral visual acuity in human infants. In H. Kaufmann *Transactions-16*. Meeting of the European Strabismological Association. pp. 221-229., Giessen.

Sireteanu, R., Kellerer, R., & Boergen, K.P. (1984). The development of peripheral visual acuity in human infants. A preliminary study. *Human Neurobiology*, 3, 81-85.

Slater, A., Morison, V., & Somers, M. (1988). Orientation discrimination and cortical function with human newborn. *Perception*, 17, 597-602.

Teller, D.Y., McDonald, M., Preston, K., Sebris, S.L., & Dobson, V. (1986). Assessment of visual acuity in infants and children: The acuity card procedure. *Developmental Medicine & Child Neurology*, 28, 779-789.

Teller, D.Y., Morse, R., & Borton, R. (1974). Visual acuity for vertical and diagonal gratings in human infants. Vision Research, 14, 1433-1439.

Treisman, A.M., & Gelade, G. (1980). A feature-integration theory of attention. *Cognitive Psychology* (1), 97-136.

Van Essen, D.C., DeYoe, E.A., Olavarria, J.F., Knierim, J.J., Fox, J.M., Sagi, D., & Julesz, B. (1989). Neural responses to static and moving texture patterns in visual cortex of the macaque monkey. In D.M.K Lam and C.D. Gilbert, *Neuronal Mechanisms of Visual Perception*. Woodlands, TX: Portfolio.

Facets of Dyslexia and its Remediation
S.F. Wright and R. Groner (Editors)

THE ROLES OF SUSTAINED (P) AND TRANSIENT (M) CHANNELS IN READING AND READING DISABILITY

B. G. Breitmeyer

Department of Psychology

University of Houston, USA.

INTRODUCTION

Discourse in written language depends on reading, a skill that draws from a very limited and select source of visual information. Although language is a relatively late evolutionary acquisition, many of the visual functions supporting reading evolved independently of, and prior to, human language. In fact, many higher mammalian species with frontal eyes and a limited area of sharpest vision, such as the area centralis of cat and the fovea of monkey, exhibit properties of visual behaviour which are akin to those found in reading. Due to the spatially limited area of sharpest vision, all of these organisms, if they wish to pick up detailed pattern information from an extensive visual scene, must scan it via saccade-directed sequences of fixations.

The information we pick up depends on the existence of surfaces, objects, and events in the visual world. Although our experience tells us that this information is nearly endless in variety and descriptive detail, our intuition is that all such information is encoded in a relatively small set of perceptual dimensions. This intuition provides the basis for a program evident in much of recent vision research concerned with the study of parallel streams of processing. A main assumption of this program is that each stream of processing is specialized for analyzing one or only a few of the perceptual dimensions. As examples, increasing psychophysical evidence has pointed to the existence in humans of separate pathways specialized for chromatic and achromatic (luminance) vision (de Lange, 1958; Kelly & van Norren, 1977), of mechanisms for the analysis of direction, orientation, and spatial frequency (Blakemore & Campbell, 1969; Campbell & Kulikowski, 1966; Sekuler & Ganz, 1963), and of pathways separately processing light-on-dark and dark-on-light stimuli (Jung, 1973). Moreover, neural analogues of these pathways have been found in recent physiological and anatomical studies of the monkey visual system (Schiller, 1982, 1984; Schiller, Sandell & Maunsell, 1986; Tootell, Silverman, Hamilton, DeValois & Switkes, 1988a; Tootell, Silverman, Hamilton, Switkes & DeValois, 1988c).

Although some of these channels such as the contrast-, orientation-, and spatial frequency-specific ones are clearly important to pattern recognition and, thus, to reading,

I will not detail their properties further except when they are pertinent to the following discussion of sustained and transient channels and their role in reading. Some two and one-half decades ago, Enroth-Cugell and Robson's (1966) discovery of separate classes of X and Y cells in the cat retina provided the impetus to concerted efforts not only to work out the physiological and anatomical details of the separate pathways arising from these retinal cells (reviewed by Lennie [1980a] and Sherman [1985]) but also to search for psychophysical analogues of these pathways in humans (for reviews, see Breitmeyer [1984, in press]). Enroth-Cugell and Robson's (1966) findings were corroborated and elaborated a few years later by Cleland, Dubin, and Levick (1971). Besides using Enroth-Cugell and Robson's (1966) classification scheme based on linear spatial luminance summation across receptive fields, Cleland et al.'s (1971) classification criteria included responses to standing contrast, spatial frequency, size, and speed of stimuli. The class of cells termed "sustained" neurons by Cleland et al. (1971) was characterized by presence of linear summation, sustained response to standing contrast, small and slowly conducting axons, and a preference for high-spatial frequency, slowly moving stimuli. The other class, called "transient" neurons, was characterized by lack of linear spatial summation, lack of a sustained response to standing contrast with transient responses only at stimulus on- and offsets, larger and faster conducting axons, and preference for low-spatial frequency, rapidly moving stimuli. Cleland et al. (1971) took sustained and transient neurons to correspond to Enroth-Cugell and Robson's X and Y cells, respectively. The sustained/transient terminology was initially also adopted by psycho-physicists investigating perceptual signs of these pathways in humans. Tolhurst (1973) and Kulikowski and Tolhurst (1973) were among the first to adopt the sustained/transient distinction to describe pattern- or form-sensitive and flicker- or motion-sensitive channels in human vision. Based on their measurements of separate flicker/motion and form/pattern detection thresholds for moving sinusoidal gratings, these investigators concluded that sustained channels were characterized by preference for higher spatial frequencies but did not prefer temporally modulated stimuli over stationary ones. Transient channels, in contrast, preferred not only low- spatial frequency stimuli but also temporally modulated or moving ones. Other investigators (for review, see Breitmeyer [1984], Green [1984], and Kelly and Burbeck [1984]) adopted related distinctions between sustained and transient channels based on the common feature among their varied findings of consistent differences between the temporal response of the visual system to low and high spatial frequencies.

Although the parallels drawn between the flicker/motion and form/pattern distinction on the one hand and the transient and sustained channel distinction on the other continue to play an important role in vision research, they have not gone uncriticized (Burbeck, 1981; Derrington & Lennie, 1981; Green, 1981, 1984; Kelly & Burbeck, 1984; Lennie,

1980b). I have dealt with these criticisms at length and in detail elsewhere (Breitmeyer, 1992). Here I will only note that most of the criticisms require "cosmetic" rather than substantive changes in the sustained/transient channel distinction in that, rather than being abandoned, it merely needs to be updated by more recent psychophysical and neurophysiological developments, some of which are discussed below.

As noted, early psychophysical models of human sustained and transient channels (e.g. Breitmeyer & Ganz, 1976; Kulikowski & Tolhurst, 1973; Tolhurst, 1973) relied for neural analogues on the X and Y pathways of the cat; little was known of X- and Y-like pathways in primates. However, we know that anatomically, physiologically, and psychophysically the human visual system resembles that of monkey more than that of cat. Particularly pertinent are several investigations (Harwerth, Boltz, & Smith, 1980; Merigan & Eskin, 1986; Merigan & Maunsell, 1990) in which the psychophysical performance of monkeys provides evidence for the existence of sustained and transient channels that parallels similar evidence found in humans. Accordingly, it may be more appropriate to look to the monkey visual system for neural analogues of sustained and transient channels.

However, here a problem arises in that recent studies of parallel pathways in monkey that may qualify as neural analogues of sustained and transient channels have been based on two classification schemes -- the X/Y and the parvocellular (P)/magnocellular (M) distinctions -- which are not equivalent.

Despite these problems (see Breitmeyer, 1992), within a multi-criteria scheme relying on a number of anatomical and physiological distinctions, it is generally agreed among a number of investigators that the M and P cells are characterized by response properties attributed to transient and sustained channels, respectively (Livingstone & Hubel, 1987, 1988; Maunsell, 1987; Maunsell & Schiller, 1984; Schiller & Malpelli, 1978).

As illustrated in Fig. 1, the transient M and sustained P pathways originate in the anatomically and physiologically distinct classes of A and B ganglion cells of the retina (Leventhal, Rodieck, & Dreher, 1981). These two classes of cells project to the separate LGN M and P layers which, in turn, project to area V1 of the cortex from where they branch into three identifiable pathways. In cortical are V1, the P pathway divides into separate P-blob and P-interblob streams of processing which in turn project via the thin and pale stripes of area V2 to V4 and from there to inferotemporal cortex. The cortical M pathway also originates in V1 and projects dorsally via V3 and the thick stripes of V2 to area MT and subsequently to parietal cortex. Interactions exist between the cortical M and P pathways; besides projecting to MT, V3 also projects to V4; and, besides projecting to inferotemporal cortex, V4 is anatomically linked to MT and parietal

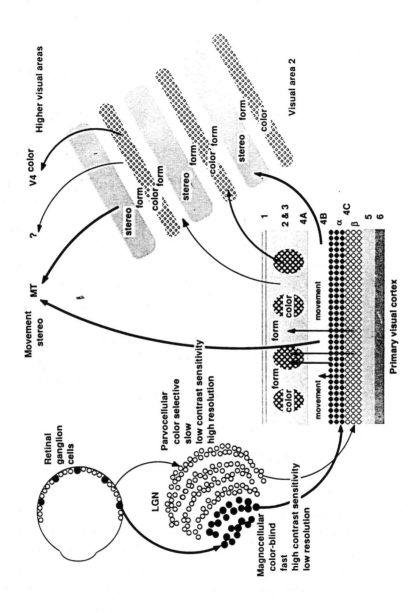

FIG. 1. Schematic representation of the separate magnocelluar (M) & parvocelluar (P) pathways from the retina to higher visual centres in the cerebral cortex. See text for details. (From "Segregation of Form, Colour, Movement, and Depth: Anatomy, Physiology, and Perception" by M. Livingstone and D. Hubel (1988). Science 240, 740-749. Copyright 1988 by the American Association for the Advancement of Science. Reprinted by permission).

areas (Desimone & Ungerleider, 1989; DeYoe & Van Essen, 1988). These cortical processing streams and their cross-linkages play a leading role in current models of visual processing of form, colour, motion, and depth information (Breitmeyer,1992; Cavanagh, 1987, 1989 a,b; Desimone & Ungerleider, 1989; DeYoe & Van Essen, 1988; Grossberg, 1991; Livingstone & Hubel, 1987, 1988; Ramachandran, 1990; Tyler, 1990).

There is general agreement among the various models that the P and M pathways are closely tied to the analysis of colour and motion, respectively; however, there is disagreement as to the specifics of the colour/motion distinction and the roles of these pathways in the processing of form and depth. However, based on Schiller and co-workers' investigations of the effects of selectively lesioning the P- or else the M-cell layers of the LGN on disruption of visual functions in monkeys (Schiller & Logothetis, 1990; Schiller et al., 1990a,b), the following conclusions can be drawn. The P pathway is essential not only for the perception of colour, texture, and high-spatial frequency pattern detail but also fine stereopsis; the M pathway was found to be critical for the perception of fast flicker and motion. Coarse stereopsis and shape discrimination could be supported by either pathway. The P pathway additionally was found to support perception of slow flicker and motion. These results show that the M pathway, besides not dominating fine stereopsis, also is not essential for perception of slow motion and flicker. The spatiotemporal response properties of monkey vision supported by the P and M pathways agree remarkably well with the properties of human sustained and transient channels initially proposed by Kulikowski and Tolhurst (1973), Tolhurst (1973), and Breitmeyer and Ganz (1976). A revision of these original models would require inclusion of additional distinctions between sustained and transient channels on the basis of the differential responses of P and M neurons to colour, texture, and stereopsis.

THE ROLE OF SUSTAINED AND TRANSIENT CHANNELS IN VISUAL MASKING, READING AND READING DISABILITY

Our understanding of the importance of sustained and transient channels for reading and reading disability will become apparent via a slight detour into the field of visual pattern masking, particularly metacontrast masking, and its relation to visual exploratory behaviour characterized by the previously mentioned fixation-saccade sequences. Metacontrast is a type of backward masking in which the contrast and contour visibility of a briefly flashed target is suppressed by a temporally following, spatially flanking, briefly flashed mask stimulus. As shown in Fig. 2, strongest masking occurs when the mask onset is delayed by about 50 msec relative to the target onset, with progressively less masking at progressively shorter or longer stimulus onset asynchronies (SOAs). One class of masking models (Breitmeyer & Ganz, 1976; Matin, 1975; Weisstein, Ozog, &

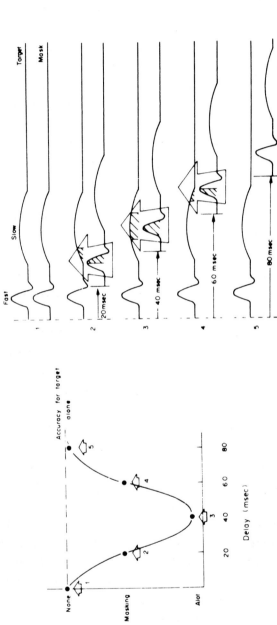

FIG. 2. A schematic U-shaped metacontrast function (left panel) and the corresponding hypothetical responses of transient and sustained channels to the target and the mask stimuli (right panel) at varying target-mask onset delays. See text for details and elaboration. From "The Effect of Perceived Depth and Connectedness on Metacontrast Functions" by M. C. Williams and N. Weisstein (1984). Vision Research, 24, 1279-1288. Copyright 1984 by Pergamon Press.

Szoc, 1975) argues that the U-shaped masking function relating the strength of metacontrast to the SOA results from post-retinal inhibitory interactions between the slower sustained channels activated by the target and the faster transient channels activated by the mask (see Fig. 2). Psychophysical evidence for both the transient-on-sustained and sustained-on-transient channel inhibition has been reported by Breitmeyer, Rudd, and Dunn (1981). A neural analogue of such interactions is the mutual inhibition reported to exist between X and Y cells in the LGN and cortical area 17 of the cat (Singer, 1976; Singer & Bedworth, 1973; Tsumoto & Suzuki, 1976). Up to now analogues of these inhibitory interactions have not been investigated in monkey. Because the P and M cells are anatomically segregated in the LGN, it is likely, as suggested by Lennie (1980a), that in monkey they occur no earlier than the visual cortex.

Singer and Bedworth (1973) proposed that the inhibition of sustained by transient cells provides a mechanism of saccadic suppression when visual scanning of a spatially extensive scene requires fixation-saccade sequences. They propose, as schematized in Fig. 3, that the slowly decaying, trailing activity in retinotopically organized sustained channels activated during one fixation interval is suppressed by the transient activity generated by abrupt and rapid image displacements accompanying a saccade. Hence, this prior sustained activity is prevented from persisting across the saccade interval (20-70 msec) as a form of noise to the sustained activity generated in the following fixation interval. In this way the afferent sustained channels are cleared of activity between fixations, resulting in a series of temporally segregated, retinotopic frames of sustained activity, with each frame corresponding to the pattern information in a given fixation period.

This interpretation was adopted by Matin (1974) and by me (Breitmeyer, 1980, 1984; Breitmeyer & Ganz, 1976) to relate metacontrast to saccadic suppression. However, as I have also noted (Breitmeyer, 1980, 1984), metacontrast provides a spatially local mechanism of saccadic suppression which is weakest in the fovea (Alpern, 1953; Bridgemen & Leff, 1979; Kolers & Rosner, 1960; Saunders, 1977). This would pose a problem since sustained activity, required for analysis of fine patterns, is most heavily concentrated in the fovea (Breitmeyer, 1984), and thus one would actually want to have strong saccadic suppression there. The problem is solved by the jerk effect (Breitmeyer & Valberg, 1979) in which transient activity generated globally by sudden saccadic image shifts in extrafoveal regions of the retinal concenters on the fovea where it powerfully adds to the relatively weak local suppression of sustained activity.

Reading, like other visual scanning of the environment, relies on fixation-saccade sequences. Hence, based on the above considerations, it is evident that efficient pick-up of visual information during reading depends, for one, on a properly functioning sustained pathway which can process the patterns present during a fixation and,

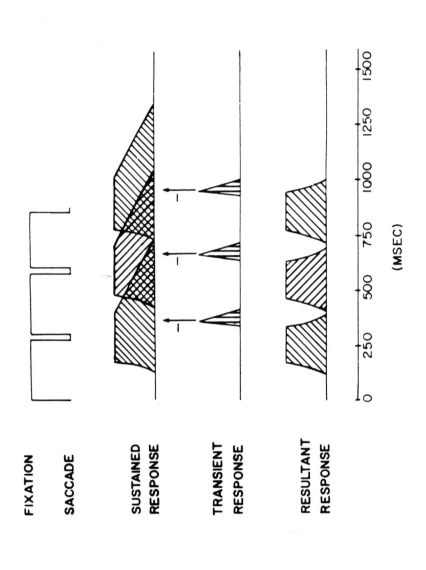

FIG. 3. A hypothetical response sequence of sustained & transient channels and their interactions during three 250-msec fixations separated by two 25-msec saccades. See text for details and elaboration. (From "Unmasking Visual Masking: A Look at the 'Why' Behind the Veil of the 'How'" by B. G. Breitmeyer (1980). Psychological Review 87, 52–69. Copyright 1980 by American Psychological Association. Reprinted by permission).

additionally, on a properly functioning transient pathway whose saccade-produced activity provides a basis for saccadic suppression. Abnormalities in either pathway could therefore contribute to visual deficits in reading. In this regard, the sustained/transient channel approach has contributed to our understanding of dyslexia and specific reading disability (SRD). Psychophysical studies of visual persistence, flicker sensitivity, temporal order judgments, and metacontrast indicate that a transient-channel deficit is found in about 70% of SRD children. These studies, reviewed recently by Lovegrove, Martin, and Slaghuis (1986) and Williams and LeCluyse (1990), are important for their empirical contributions as well as for their challenge to past and prevailing views, e. g., those of Benton (1975) and Vellutino (1979, 1987), claiming that visual deficits do not exist in SRD.

Transient-channel deficits found in SRD are important for several reasons. One consequence of a transient-channel deficit would be a weakened saccadic suppression. The result, according to the scheme sketched above, is at least a partial temporal overlap, rather than clear temporal segregation, of successive frames of retinotopic sustained activity from successive fixations. This could comprise an obvious visually based impediment to the reading process itself. However, besides leading to reading difficulties, it could also play a role in other types of visual dysfunction. In addition to clearing the sustained pattern-analyzing channels between fixations saccadic suppression, as noted by Matin (1974), also prevents the perception of retinal image smear during saccades and, furthermore, is important in maintaining constancy of visual direction and a stable visual world despite continual shifts of the images projected on the retina when scanning the environment. Thus a transient deficit in SRD could contribute to retinal image smear, impairment of visual direction constancy, and an instability of the visual world. These symptoms have been found in about 60-70% of the dyslexics investigated by Stein, Riddell, and Fowler (1989). It may be more than coincidental that this is also the proportion of SRD children noted by Lovegrove et al. (1986) as suffering from a transient-channel deficit.

EXTENSIONS AND IMPLICATIONS OF RECENT DEVELOPMENTS

Recent electrophysiological and anatomical findings corroborate the above theories of a transient-channel deficit in SRD. Livingstone, Rosen, Drislane, and Galaburda (1991) report that visually evoked scalp potentials to rapid, low contrast stimuli, which preferentially activate the M pathway, were diminished in dyslexics relative to normal control subjects whereas the evoked potentials to slow, high contrast stimuli, which preferentially activate the P pathway, were normal. In line with this finding, Livingstone et al. (1991) report that in the LGN of dyslexic brains the M layers are anatomically more disorganized and and their cell sizes are smaller and more variable than in non-dyslexic brains. On the other hand, the P layers of the LGN of dyslexic brains did not differ from the P layers of non-dyslexic brains. These electrophysiological and anatomical results accord well with the psychophysical results reviewed by Lovegrove et al. (1986) which indicate a deficit in the transient but not the sustained channels of SRD children.

At first glance, the combined psychophysical, electrophysiological, and anatomical findings suggest that the visual deficits in SRD are due to neural abnormalities that may not be directly amenable to treatment. Although this may be true, there is reason to be optimistic about devising treatment techniques which will alleviate these deficits. One important property of the transient M pathway not mentioned so far is that its activity is suppressed by diffuse red light (DeMonasterio, 1978; Dreher et al., 1976; Livingstone & Hubel, 1984; Wiesel & Hubel, 1966). Along with several collaborators I recently have looked at psychophysical correlates of this suppression. For example, as would be expected, metacontrast, as shown in Fig. 4, is weaker for stimuli presented on red as compared to white or green backgrounds (Breitmeyer, May, & Heller, 1991; Breitmeyer & Williams, 1990; Williams, Breitmeyer, Lovegrove, & Gutierrez, 1991). An interesting and unexpected finding was that metacontrast was enhanced when stimuli were flashed on blue as compared to white or green backgrounds (Williams et al., 1991). Thus, relative to white or medium-wavelength green backgrounds, long-wavelength red backgrounds suppress and short-wavelength blue backgrounds enhance transient-channel activity.

These findings have clear implications for the use of coloured lenses or overlays in treating visual problems in SRD (Whiting, 1988). In fact, Mary Williams and co-workers (Williams, LeCluyse, & Faucheux, in press) recently showed that relative to a white background, a red one decreases the average reading rate and comprehension score in SRD children whereas a blue background increases both measures of reading performance. These are clearly intriguing and important findings which need to be replicated. Once they are firmly established, they could provide a basis for developing theoretically motivated and empirically validated methods for diagnosing and treating this class of visually based reading deficits in SRD children. Moreover, these treatment procedures may also alleviate some of the other symptoms mentioned above, e. g., the retinal smear, loss of direction constancy, and visual instability reported to plague dyslexics (Stein et al., 1989).

Perceptual phenomena like flicker, visual persistence, metacontrast, and temporal order judgments very likely rely on mechanisms located at relatively early stages of visual processing (Breitmeyer, 1984, Chapter 10). According to Lovegrove et al. (1986), abnormalities in these phenomena point to a low-level transient-channel deficit in SRD. However, as I have noted previously (Breitmeyer, 1989), there is good reason to believe

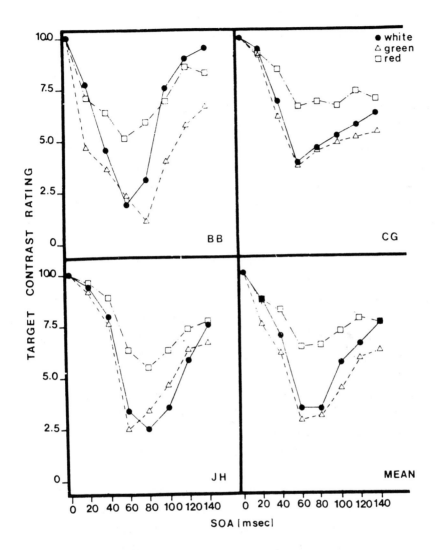

FIG. 4. Metacontrast functions obtained with black stimuli are presented on equiluminant white, green and red backgrounds. (From "Effects of isoluminant-background colour on metacontrast and stroboscopic motion: interactions between sustained (S) and transient (T) channels" by B.G. Breitmeyer & M.C. Williams (1990). Vision Research, 30, 1069-75. Copyright 1990 by Pergamon Press.

that deficits in the transient M pathway should also be found at higher levels. A major recipient site of the dorsal M stream of cortical processing is the middle temporal area (MT) located in the posterior superior temporal sulcus (STS). In humans, analogues of these areas appear to lie near the posterior speech and language areas involved in the processing of written or printed material (Masland, 1981; Ojemann, 1989; Rasmussen & Milner, 1975).

MT responds well to motion, its neurons are highly direction-selective (Desimone & Ungerleider, 1989; Newsome & Wurtz, 1988; Van Essen, 1985) and it projects to frontal eye fields and, via the middle superior temporal area (MST), to the inferior parietal cortex (Seltzer & Pandya, 1989; Ungerleider & Desimone, 1986). MT and the areas to which it projects are involved in the control of spatial attention and eye movements (Mountcastle, 1978; Newsome & Wurtz, 1988; Schiller, True, & Conway, 1979). Hence, if there are abnormalities in these cortical M-pathway areas, one might expect SRDs to show deficits in visuoperceptual functions supported by these areas. Recently, Brannan and Williams (1987) and Fischer & Weber (1990) showed that SRDs are abnormal in their deployment of spatial attention, and Stein et al. (1989) report that dyslexics exhibit a left visual field neglect or inattention akin to that produced by lesions of the right parietal lobe. Moreover, Adler-Grinberg & Stark (1978) report that dyslexics have abnormal pursuit eye movements. Similar abnormalities have also been found in monkey after lesioning the motion-sensitive areas MT and MST in the cortical dorsal M pathway (Newsome & Wurtz, 1988).

Although further arguments can be made for the existence of deficits in higher level visual functioning based on abnormalities in the cortical M pathways (Breitmeyer, 1989), I would like to turn to other intriguing possibilities. In their anatomical studies of the LGN in dyslexic brains, Livingstone et al. (1991) suggest that the abnormalities found in the M pathway of the visual system may also be found in similar fast responding neurons of the auditory and somatosensory systems. Deficits in the fast responding pathways of the auditory system could, as suggested by Livingstone et al. (1991), contribute to the deficiencies in processing rapidly presented acoustic information and acoustic transitions characterizing the spectral patterns of consonants (Tallal, 1976). Since analogues of metacontrast have been found in U-shaped backward masking functions with auditory (Porter, 1975; Studdert-Kennedy, Shankweiler, & Schulman, 1970) as well as tactile (Weisenberger & Craig, 1982) stimuli, such psychophysical phenomena provide ready tools for a closer look at possible deficits in the analogues of transient channels in the auditory and somatosensory systems of SRDs.

REFERENCES

Adler-Grinberg, D., & Stark, L. (1978). Eye movements, scanpaths, and dyslexia. *American Journal of Optometry and Physiological Optics*, 55, 557-570.

Alpern, M. (1953). Metacontrast. *Journal of the Optical Society of America*, 43, 648-57.

Benton, A. (1975). Developmental dyslexia: Neurological aspects. In W. J. Friedlander (Ed.), *Advances in Neurology*, Vol 7: Current Views of Higher Nervous System Dysfunction (pp. 1-47). New York: Raven Press.

Blakemore, C., & Campbell, F. W. (1969). On the existence of neurones in the human visual system selectively sensitive to orientation and size of retinal image. *Journal of Physiology*, 203, 237-260.

Brannan, J. R. & Williams, M. C. (1987). Allocation of visual attention in good and poor readers. *Perception & Psychophysics*, 41, 23-28.

Breitmeyer, B. G. (1980). Unmasking visual masking: A look at the 'why' behind the veil of the 'how'. *Psychological Review*, 87, 52-69.

Breitmeyer, B. G. (1984). *Visual Masking: An Integrative Approach.* New York: Oxford University Press.

Breitmeyer, B. G. (1989). A visually based deficit in specific reading disability. Irish *Journal of Psychology*, 10, 534-541.

Breitmeyer, B. G. (1992). Parallel processing in human vision: History, review, and critique. In J. R. Brannan (Ed.), Applications of Parallel Processing in Vision (pp. 37-78). Amsterdam: Elsevier.

Breitmeyer, B. G., & Ganz, L. (1976). Implications of sustained and transient channels for theories of visual pattern masking, saccadic suppression, and information processing. *Psychological Review*, 83, 1-36.

Breitmeyer, B.G., May. J. G., & Heller, S. S. (1991). Metacontrast reveals asymmetries at red/green isoluminance. *Journal of the Optical Society of America* A, 8, 1324-1329.

Breitmeyer, B. G., Rudd, M., & Dunn, K. (1981). Spatial and temporal parameters of metacontrast disinhibition. *Journal of Experimental Psychology: Human Perception and Performance*, 7, 770-779.

Breitmeyer, B. G., & Valberg, A. (1979). Local, foveal inhibitory effects of global, peripheral excitation. *Science*, 203, 463-465.

Breitmeyer, B. G., & Williams, M. C. (1990). Effects of isoluminant-background colour on metacontrast and stroboscopic motion: Interactions between sustained (P) and transient (M) channels. *Vision Research*, 30, 1069-1075.

Bridgeman, B., & Leff, S. (1979). Interaction of stimulus size and retinal eccentricity in metacontrast masking. *Journal of Experimental Psychology: Human Perception and Performance*, 5, 101-109.

Burbeck, C. A. (1981). Criterion-free pattern and flicker thresholds. *Journal of the Optical Society of America*, 71, 1343-1350.

Campbell, F. W., & Kulikowki, J. J. (1966). Orientation selectivity of the human visual system. *Journal of Physiology,* 187, 437-445.

Cavanagh, P. (1987). Reconstructing the third dimension: Interactions between colour, texture, motion, binocular disparity and shape. *Computer Vision, Graphics and Image Processing*, 37, 171-195.

Cavanagh, P. (1989). Multiple analyses of orientation in the visual system. In D. M.K. Lam & C. D. Gilbert (Eds.), *Neural Mechanisms of Visual Perception: From single Cells to Perception* (pp. 261-279). Houston: Gulf Publishers.

Cleland, B.G., Dubin, M. W., & Levick, W. R. (1971). Sustained and transient neurones in the cat's retina and lateral geniculate nucleus. *Journal of Physiology*, 217, 473-496.

de Lange, H. (1958). Research into the dynamic nature of the human fovea-cortex systems with intermittent and modulated light. II. Phase shifts in brightness and delay in colour perception. *Journal of the Optical Society of America*, 48, 784-789.

DeMonasterio, F. M. (1978). Centre and surround mechanisms of opponent-colour X & Y ganglion cells of retina of macaques. *Journal of Neurophysiology*, 41, 1418-1434.

Derrington, A. M., & Lennie, P. (1984) Spatial and temporal contrast sensitivities of neurones in lateral geniculate nucleus of macaque. *Journal of Physiology*, 357, 219-240.

Desimone, R., & Ungerleider, L. G. (1989). Neural mechanisms of visual processing in monkeys. In F. Boller & J. Grafman (Eds.), *Handbook of Neuropsychology*, Vol. 2 (pp. 267-299). Amsterdam: Elsevier.

DeYoe, E. A., & Van Essen, D. C. (1988). Concurrent processing streams in monkey visual cortex. *Trends in Neuroscience*, 11, 219-226.

Enroth-Cugell, C., & Robson, J. G. (1966). The contrast sensitivity of retinal ganglion cells of the cat. *Journal of Physiology*, 187, 517-552.

Fischer, B., & Weber, H. (1990). Saccadic reaction times of dyslexic and age-matched normal subjects. *Perception*, 19, 805-818.

Green, M. (1981). Spatial frequency effects in masking by light. *Vision Research*, 21, 861-866.

Green, M. (1983). Contrast detection and direction discrimination of drifting gratings. *Vision Research*, 23, 281-289.

Green, M. (1984). Masking by light and the sustained-transient dichotomy. *Perception & Psychophysics*, 35, 519-535.

Grossberg, S. (1991). Why do parallel cortical systems exist for the perception of static form and moving form? *Perception & Psychophysics*, 49, 117-141.

Harwerth, R. S., Boltz, R. L., & Smith, E. L. (1980). Psychophysical evidence for sustained and transient channels in the monkey visual system. *Vision Research*, 20, 15-22.

Jung, R. (1973). Visual perception and neurophysiology. In R. Jung (Ed.), *Handbook of Sensory Physiology*, Vol. VII/3A, Central Processing in the Visual System (pp. 1-152). Berlin: Springer.

Kelly, D. H., & Burbeck, C. A. (1984). Critical problems in spatial vision. *CRC Critical Reviews in Biomedical Engineering*, 10, 125-177.

Kelly, D. H., & van Norren, D. (1977). Two-band model of heterochromatic flicker. *Journal of the Optical society of America*, 67, 1081-1091.

Kolers, P., & Rosner, B. S. (1960). On visual masking (metacontrast):Dichoptic observations. *American Journal of Psychology*, 73, 2-21.

Kulikowski, J. J., & Tolhurst, D. (1973). Psychophysical evidence for sustained and transient detectors in human vision. *Journal of Physiology*, 232, 149-162.

Lennie, P. (1980a). Parallel visual pathways: A review. *Vision Research*, 20, 561-594.

Lennie, P. (1980b). Perceptual signs of parallel pathways. *Philosophical Transactions of the Royal Society* (Series B) 290, 23-37.

Leventhal, A., Rodieck, R. W., & Dreher, B. (1981). Retinal ganglion cell classes in the Old World monkey: Morphology and central projections. *Science*, 213, 1139-1142.

Livingstone, M. S., & Hubel. D. H. (1984). Anatomy and physiology of a colour system in the primate visual cortex. *Journal of Neuroscience*, 4, 309-356.

Livingstone, M. S., & Hubel, D. H. (1987). Psychophysical evidence for separate channels for the perception of form, colour, movement, and depth. *Journal of Neuroscience*, 7, 3416-3468.

Livingstone, M., & Hubel, D. (1988). Segregation of form, colour, movement, and depth: Anatomy, physiology, and perception. *Science*, 240, 740-749.

Livingstone, M. S., Rosen, G. D., Drislane, F. W., & Galaburda, A. M. (1991). Physiological and anatomical evidence for a magnocellular defect in developmental dyslexia. *Proceedings of the National Academy of Science*, 88, 7943-7947.

Lovegrove, W., Martin, F., & Slaghuis, W. (1986). A theoretical and experimental case for a visual deficit in specific reading disability. *Cognitive Neuropsychology*, 3, 225-267.

Masland, R. L. (1981). Neurological aspects of dyslexia. In G. Th. Pavlidis & T. R. Miles (Eds.), *Dyslexia Research and Its Applications to Education* (pp. 35-66). New York: John Wiley & Sons.

Matin, E. (1974). Saccadic suppression: A review and analysis. *Psychological Bulletin*, 81, 899-917.

Matin, E. (1975). Two transient (masking) paradigm. *Psychological Review*, 82, 451-561.

Maunsell, J. H. R. (1987). Physiological evidence for two visual subsystems. In L. Vaina (Ed.), *Matters of Intelligence: Conceptual Structures in Cognitive Neuroscience* (pp. 59-87). Amsterdam: Reidel.

Maunsell, J. H. R., & Schiller, P. H. (1984). Evidence for the segregation of parvo- and magnocellular channels in the visual cortex of the macaque monkey. *Neuroscience Abstracts*, 10, 520.

Merigan, W. H., & Eskin, T. A. (1986). Spatio-temporal vision of macaques with severe loss of PB retinal ganglion cells. *Vision Research*, 26, 1751-1761.

Merigan, W. H., & Maunsell, J. H. R. (1990). Macaque vision after magnocellular lateral geniculate lesions. *Visual Neuroscience*, 5, 347-352.

Mountcastle, V. B. (1978). Brain mechanisms for directed attention. *Journal of the Royal Society of Medicine*, 71, 14-28.

Newsome, W. T., & Wurtz, R. H. (1988). Probing visual cortical function with discrete chemical lesions. *Trends in Neuroscience*, 11, 394-400.

Ojemann, G. A. (1989). Some brain mechanisms of reading. In C. von Euler, I. Lundberg & G. Lennerstrand (Eds.), *Brain and Reading* (pp. 47-59). New York: Stockton Press.

Porter, R. J. Jr. (1975). Effect of delayed channel on the perception of dichotically presented speech and nonspeech sounds. *Journal of the Acoustical society of America*, 58, 884-892.

Ramachandran, V. S. (1990). Visual perception in people and machines. In A. Blake & T. Troscianko (Eds.), *AI and the Eye* (pp. 21-77). New York: Wiley.

Rasmussen, T. & Milner, B. (1975). Clinical and surgical studies of the cerebral speech areas in man. In K. J. Zulch, O. Creutzfeldt & G. C. Galbraith (Eds.), Cerebral Localization (pp. 238-257). Berlin: Springer.

Saunders, J. (1977). Foveal and spatial properties of brightness metacontrast. *Vision Research*, 17, 375-378.

Schiller, P. H. (1982). Central connections of ON and OFF pathways. *Nature*, 297, 580-583.

Schiller, P. H. (1984). The connections of the retinal on and off pathways to the lateral geniculate nucleus of the monkey. *Vision Research*, 24, 923-932.

Schiller, P. H., & Logothetis, N. K. (1990). The colour-opponent and broad-band channels of the primate visual system. *Trends in Neuroscience*, 10, 392-398.

Schiller, P. H., Logothetis, N. K., & Charles, E. R. (1990a). Functions of the colour-opponent and broad-band channels of the visual system. *Nature*, 343, 68-70.

Schiller, P. H., Logothetis, N. K., & Charles, E. R. (1990b). Role of the colour-opponent and broad-band channels in vision. *Visual Neuroscience*, 5, 321-346.

Schiller, P. H., & Malpelli, J. (1978). Functional specificity of lateral geniculate nucleus laminae of the rhesus monkey. *Journal of Neurophysiology*, 41, 788-797.

Schiller, P. H., Sandell, J. H., & Maunsell, J. H. R. (1986). Functions of the ON and OFF channels of the visual system. *Nature*, 322, 824-825.

Schiller, P. H., True, S. D., & Conway, J. L. (1979). Effects of frontal eye field and superior colliculus ablations on eye movements. *Science*, 206, 590-592.

Sekuler, R. W., & Ganz, L. (1963). Aftereffect of seen motion with a stabilized retinal image. *Science*, 139, 419-420.

Seltzer, B. & Pandya, D. N. (1989). Frontal lobe connections of the superior temporal sulcus in the rhesus monkey. *Journal of Comparative Neurology*, 281, 97-113.

Sherman, M. S. (1985). Functional organization of the W-, X-, and Y-cell pathways in the cat: A review and hypothesis. In J. M. Sprague & A. N. Epstein (Eds.), *Progress in Psychobiology and Physiological Psychology*, Vol. II (pp. 233-324). New York: Academic Press.

Singer, W. (1976). Temporal aspects of subcortical contrast processing. *Neuroscience Research Program Bulletin*, 15, 358-369.

Singer, W., & Bedworth, N. (1973). Inhibitory interactions between X and Y units in cat lateral geniculate nucleus. *Brain Research*, 49, 291-307.

Stein, J., Riddell, P., & Fowler, S. (1989). Disordered right hemisphere function in developmental dyslexia. In C. von Euler, I. Lundberg & G. Lennerstrand (Eds.), *Brain and Reading* (pp. 139-157). New York: Stockton Press.

Studdert-Kennedy, M., Shankweiler, D., & Schulman, S. (1970). Opposed effects of a delayed channel on perception of dichotically and monotically presented CV syllables. *Journal of the Acoustical Society of America*, 48, 599-602.

Tallal, P. (1976). Auditory perceptual factors in language and learning disabilities. In R. M. Knights & D. J. Bakker (Eds.), *The Neuropsychology of Learning Disorders* (pp.315-323). Baltimore: University Park Press.

Tolhurst, D. J. (1973). Separate channels for the analysis of the shape and movement of a moving stimulus. *Journal of Physiology*, 231, 385-402.

Tootell, R. B. H., Hamilton, S. L., & Switkes, E. (1988b). Functional anatomy of macaque striate cortex. IV. Contrast and magno-parvo streams. *Journal of Neuroscience*, 8, 1594-1609.

Tootell, R. B. H., Silverman, M. S., Hamilton, S. L., De Valois, R. L., & Switkes, E. (1988a). Functional anatomy of macaque striate cortex. III. Colour. *Journal of Neuroscience*, 8, 1569-1593.

Tootell, R. B. H., Silverman, M. S., Hamilton, S. L., Switkes, E., & De Valois, R. L. (1988c). Functional anatomy of macaque striate cortex. V. Spatial frequency. *Journal of Neuroscience*, 8, 1610-1624.

Tsumoto, T., & Suzuki, D. A. (1976). Effects of frontal eye field stimulation upon activities of the lateral geniculate body of the cat. *Experimental Brain Research*, 25, 291-306.

Tyler, C. W. (1990). A stereoscopic view of visual processing streams. *Vision Research*, 30, 1877-1895.

Ungerleider, L. G. & Desimone, R. (1986). Cortical connections of visual area MT in the macaque. *Journal of Comparative Neurology*, 248, 190-222.

Van Essen, D. C. (1985). Functional organization of primate visual cortex. In E. G. Jones & A. A. Peters (Eds.), *Cerebral Cortex*, Vol. 3 (pp. 259-329). New York: Plenum Press.

Vellutino, F. R. (1979). *Dyslexia: Theory and Research*. Cambridge, Ma.: MIT Press.

Vellutino, F. R. (1987). Dyslexia. *Scientific American*, 256, 34-41.

Weisenberger, J. M. & Craig, J. C. (1982). A tactile metacontrast effect. *Perception & Psychophysics*, 31, 530-536.

Weisstein, N., Ozog, G., & Szoc, R. (1975). A comparison and elaboration of two models of metacontrast. *Psychological Review*, 82, 325-343.

Whiting, P. R. (1988). Improvements in reading and other skills using Irlen coloured lenses. *Australian Journal of Remedial Education*, 20, 13-15.

Wiesel, T. N., & Hubel, D.H. (1966). Spatial & chromatic interactions in the lateral geniculate body of the rhesus monkey. *Journal of Neurophysiology*, 29, 1115-1156.

Williams, M., Breitmeyer, B., Lovegrove, W., & Gutierrez, C. (1991). Metacontrast with masks varying in spatial frequency and wavelength. *Vision Research*, 31, 2017-2023.

Williams, M. C., & LeCluyse, K. (1990). The perceptual consequences of a temporal processing deficit in reading disabled children. *Journal of the American Optometric Association*, 61, 111-121.

Williams, M.C., LeCluyse, K., & Faucheux, A. R. (in press). Effective intervention for reading disability. *Journal of the American Optometric Association.*

Facets of Dyslexia and its Remediation
S.F. Wright and R. Groner (Editors)

DO DYSLEXICS HAVE A VISUAL DEFICIT?

W. Lovegrove,

Department of Psychology,

University of Wollongong, Austrailia.

The research to be described has concerned the question of whether a particular group of children who have trouble learning to read have a visual problem. Children in this group are commonly referred to as specifically reading disabled (SRDs) and are defined as children of normal intelligence, with normal educational opportunities, no brain damage, no gross behavioural problems who nevertheless read at least two years behind the level expected for their age and intelligence.

For some years the commonly accepted view within the reading disability literature has been that reading disability is not attributable to visual deficits and that normal and specifically-disabled readers (SRDs) do not differ systematically in terms of visual processing (Benton, 1962; Vellutino, 1979a, 1979b). Extensive work over the last ten years in a number of laboratories, however, has clearly demonstrated that the two groups do differ in terms of visual processing.

This has been brought about partially by developments in theoretical vision which have been applied to reading thus providing a more meaningful theoretical context in which to consider reading disability and vision. The following section outlines one approach to vision which has been usefully applied.

SPATIAL FREQUENCY PROCESSING

One approach to vision research (Campbell, 1974; Graham, 1980) indicates that information is transmitted from the eye to the brain via a number of separate parallel pathways. The separate pathways are frequently referred to as channels. Each channel is specialised to process information about particular features of visual stimuli.

The properties of channels often have been investigated using patterns like those shown in Fig. 1. These patterns are usually called sine-wave gratings. Two properties of these patterns are of interest in this chapter:

1. Spatial frequency, which refers to the number of cycles (one dark plus one light bar) per degree of visual angle (c/deg). Spatial frequency is higher on the right than the left in Fig. 1. Spatial frequency can be thought of in terms of stimulus size.

2. Contrast, which refers to the difference between the maximum and minimum luminances of the grating. It is a measure of the ratio of the brightest to the darkest section of the pattern.

Research (Campbell, 1974; Graham, 1980) has identified a number of channels each sensitive to a narrow range of spatial frequencies (or stimulus widths) and orientations in cats, monkeys and humans.

low (general)   medium   high (details)

SPATIAL FREQUENCY

FIG. 1 Sine-wave gratings commonly used in vision research concerned with spatial frequency channels. Low spatial frequencies are shown on the left and high spatial frequencies on the right.

Spatial frequency or size-sensitive channels are relevant to reading because when we read we process both general (low spatial frequency) and detailed (high spatial frequency) information in each fixation. We extract detailed information from an area approximately 5-6 letter spaces to the right of fixation. Beyond this we also extract visual information but only of a general nature such as word shape (Rayner, 1975). These two types of size information must in some way be combined.

It has also been shown that the different channels transmit their information at different rates and respond differently to different rates of temporal change. Some channels are sensitive to very rapidly changing stimuli and others to stationary or slowly moving stimuli. Such results have led to the proposal of two subsystems within the visual system. This division is believed to be important in combining the two types of size information involved in reading.

THE SUSTAINED AND TRANSIENT SUBSYSTEMS

It has been shown that spatial frequency channels differ in their temporal properties. In a typical experiment subjects are shown patterns like those in Fig. 1 flickering at various rates. Subjects are required to set contrast levels so that they just can see either flicker or pattern. When low spatial frequency gratings are moving quickly, we see flicker at lower contrasts than we see pattern but we experience the reverse at high spatial frequencies. Separate measures can be taken of our sensitivity to flicker and pattern with a range of different sized stimuli (spatial frequencies) flickering at different speeds. Thus we can plot sensitivity functions for pattern and flicker thresholds at a range of spatial frequencies. With large stimuli (low spatial frequencies) we are more sensitive to rapidly changing stimuli but with small stimuli (high spatial frequencies) we are more sensitive to stationary or slowly moving stimuli. The two functions obtained from such experiments are believed to measure two subsystems in the visual system, the transient and sustained subsystems. An extensive discussion of the properties of these systems and how they are identified can be found in Breitmeyer (1988). Breitmeyer also discusses the evidence indicating the physiological basis of these two systems.

The properties of these two subsystems have been identified and are shown in Table 1.

TABLE 1
General Properties of the Sustained and Transient Subsystems

Sustained System	Transient System
Less sensitive to contrast	Highly sensitive to contrast
Most sensitive to high spatial frequencies	Most sensitive to low spatial frequencies
Most sensitive to low temporal frequencies	Most sensitive to high temporal frequencies
Slow transmission times	Fast transmission times
Responds throughout stimulus presentation	Responds at stimulus onset and offset
Predominates in central vision	Predominates in peripheral vision
The sustained system may inhibit the transient system	The transient system may inhibit the sustained system

It has been demonstrated physiologically (Singer & Bedworth, 1973) and psychophysically that the two systems may inhibit each other (Breitmeyer & Ganz, 1976). In particular if the sustained system is responding when the transient system is stimulated, the transient will terminate the sustained activity. An example of how this may occur is as follows. If we are fixating on the detail of an object and a stimulus moves into the periphery of our vision, the transient system is likely to inhibit or over ride the sustained system until we know what is in our peripheral vision. How this may have evolved is easier to imagine if we consider not a human reading but a rabbit eating and a predator appearing to the side. There would be survival value for the rabbit in having the transient system inhibit the sustained system until the level of threat could be determined. These two subsystems and the interactions between them may serve a number of functions essential to the reading process.

The transient system is predominantly a flicker or motion detecting system transmitting information about stimulus change and general shape. The spatial information it transmits is coarse and thus well suited for transmitting peripheral information in reading. The sustained system is predominantly a detailed pattern detecting system transmitting information about stationary stimuli. In reading the sustained system should be most important in extracting detailed information during fixations and the transient system in extracting general information from the periphery. Below we shall see that the two systems also interact in important ways.

SUSTAINED AND TRANSIENT SUBSYSTEMS AND READING

When reading, the eyes move through a series of rapid eye movements called saccades. These are separated by fixation intervals when the eyes are stationary. Saccadic eye movements are generally in the direction of reading, that is, from left to right when reading English. Sometimes the eyes also move from right to left in what are called regressive eye movements or regressions. The average fixation duration is approximately 200-250 msec for normal readers and it is during these stationary periods that information from the printed page is seen. The average saccade length is 6-8 characters or about 2 degrees of visual angle (Rayner & McConkie, 1976). Saccadic eye movements function to bring unidentified regions of text into foveal vision for detailed analysis during fixations. Foveal vision is the area of high acuity in the centre of vision extending approximately 2 degrees (6-8 letters) around the fixation point on a line of text. Beyond the fovea acuity drops off rather dramatically.

The role of transient and sustained subsystems in reading has been considered by Breitmeyer (Breitmeyer, 1980, 1983, 1988; Breitmeyer & Ganz, 1976). Fig. 2 represents the hypothesised activity in the transient and sustained channels over a sequence of 3 fixations of 250 msec duration separated by 2 saccades of 25 msec duration each.

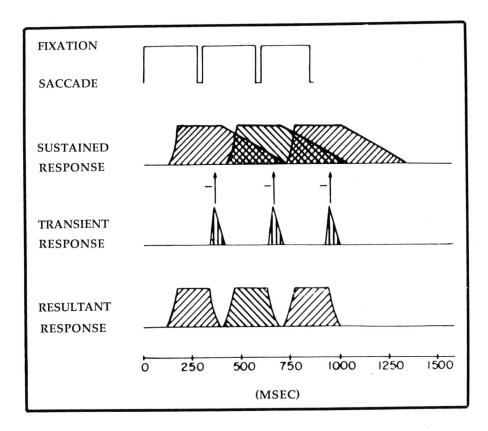

FIG. 2 A hypothetical response sequence of sustained and transient channels during 3 250 msec fixation intervals separated by 25 msec saccades (panel 1). Panel 2 illustrates persistence of sustained channels acting as a forward mask from preceding to succeeding fixation intervals. Panel 3 shows the activation of transient channels shortly after each saccade which exerts inhibition (arrows with minus signs) on the trailing, persisting sustained activity generated in prior fixation intervals. Panel 4 shows the resultant sustained channel response after the effects of the transient-on-sustained inhibition have been taken into account (From "Unmasking Visual Masking : A look at the 'Why' behind the veil of the 'How'" by B.G. Breitmeyer, Psychological Review, 1980, 82, 52-69. Copyright 1980 by the American Psychological Association. Permission to reprint granted.)

The sustained channel response occurs during fixations and may last for several hundred milliseconds. This response provides details of what the eye is seeing. The transient channel response is initiated by eye movements and lasts for much shorter durations. Consequently both systems are involved in reading. The duration of the sustained response may outlast the physical duration of the stimulus. This is a form of visible persistence produced by activating sustained channels. Its duration increases with increasing spatial frequency (Bowling, Lovegrove & Mapperson, 1979; Meyer & Maguire, 1977) and may last longer than a saccade.

If sustained activity (as shown in Fig. 2, panel 2) generated in a preceding fixation persists into the succeeding one, it may interfere with processing there. In this case, what may happen when reading a line of print requiring one, two or three fixations is illustrated in Fig. 3 (adapted from Hochberg, 1978).

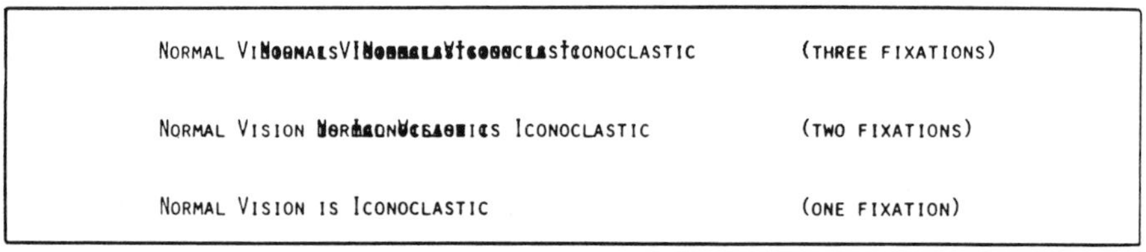

NORMAL VINORMALSVINORMALAVISOODCLASTICONOCLASTIC   (THREE FIXATIONS)

NORMAL VISION NORMONCLAONICS ICONOCLASTIC   (TWO FIXATIONS)

NORMAL VISION IS ICONOCLASTIC   (ONE FIXATION)

FIG. 3 The perceptual masking effects of temporal integration of persisting sustained activity from preceding fixation intervals with sustained activity generated in succeeding ones when the reading of a printed sentence requires one, two or three fixations. Here, as in panel 2 of Fig. 2, the effects of transient-on-sustained inhibition have not been taken into account. (From "Unmasking Visual Masking: A Look at the 'Why' behind the Veil of the 'How'" by B.G. Breitmeyer, Psychological Review, 1980, 82, 52-69. Copyright 1980 by the American Psychological Association.) Permission to reprint granted.

Consequently it is evident that for tasks such as reading, persistence across saccades presents a problem as it may lead to superimposition of successive inputs. Breitmeyer proposes that the problem posed by visible persistence is solved by rapid saccades as shown in the bottom two panels of Fig. 2. Saccades not only change visual fixations, they also activate short latency transient channels (panel 3) which are very sensitive to stimulus movement. This, in turn, inhibits the sustained activity which would persist from a preceding fixation and interfere with the succeeding one (Breitmeyer & Ganz, 1976; Matin, 1974). The result is a series of clear, unmasked, and temporally segregated frames of sustained activity, each one representing the pattern information contained in a single fixation as shown in panel 4 of Fig. 2.

In these terms, clear vision on each fixation results from interactions between the sustained and transient channels. Consequently the nature of transient-sustained channel interaction seems to be important in facilitating normal reading. Any problem in either the transient or the sustained system or in the way they interact may have harmful consequences for reading.

TRANSIENT AND SUSTAINED PROCESSING IN SRD'S AND CONTROLS

The possibility of a visual deficit in SRDs has recently been investigated within a spatial frequency analysis framework. The following is not a complete review of all recent research but a summary of the research carried out in a few laboratories including ours. Much of this research has been directed at the functioning of the transient and sustained systems in normal and specifically-disabled readers.

LOW-LEVEL VISUAL PROCESSING IN CONTROLS AND SRDS

Visible persistence is one measure of temporal processing in spatial frequency channels and refers to the continued perception of a stimulus after it has been physically removed. The top panel in Fig. 2 demonstrates such persistence. This is assumed to reflect ongoing neural activity initiated by the stimulus presentation. In adults, duration of visible persistence increases with increasing spatial frequency (Bowling, Lovegrove & Mapperson, 1979; Meyer & Maguire, 1977). Several studies have compared SRDs and controls on measures of visible persistence. It has been shown that SRDs aged from 8 to 15 years have a significantly smaller increase in persistence duration with increasing spatial frequency than do controls (Badcock & Lovegrove, 1981; Lovegrove, Heddle & Slaghuis, 1980; Slaghuis & Lovegrove, 1985).

When visible persistence is measured in both groups under conditions which reduce transient system activity (using a uniform field flicker mask), persistence differences between the groups essentially disappear (Slaghuis & Lovegrove, 1984). This finding suggests that SRDS may differ from controls mainly in the functioning of their transient systems.

The two groups have also been compared on a task which measures the minimum contrast (refer to Fig. 1) required to see a pattern. Contrast sensitivity (the reciprocal of the minimum contrast required for detection), plotted as a function of spatial frequency, is referred to as the contrast sensitivity function (CSF). Pattern CSFs have been measured in at least 5 separate samples of SRDs and control readers with ages ranging from 8 years to 14 years. It has generally been shown that SRDs are less sensitive than controls at low (1.0 to 4 c/deg) spatial frequencies (Lovegrove, Bowling, Badcock & Blackwood, 1980; Lovegrove et al., 1982; Martin & Lovegrove, 1984). In some studies the two groups do not differ in contrast sensitivity at higher (12 to 16 c/deg) spatial frequencies (Lovegrove et al., 1980) and in others SRDs are slightly more sensitive than controls in that range (Lovegrove et al., 1982; Martin & Lovegrove, 1984). At high luminances SRDs have been found to be less sensitive at high spatial frequencies than controls. Once again inhibition of the transient system by uniform-field masking influenced the SRDs less than controls (Martin & Lovegrove, 1988) thus supporting the notion of a difference between the groups in the transient system.

A third line of research has measured transient system functioning more directly than did the previous two measures. It has been argued that flicker thresholds are primarily mediated by the transient system. Consequently flicker thresholds under a range of conditions have been measured in SRDs and controls. In these experiments subjects are shown a sine-wave grating counterphasing, i.e., moving from right to left and back the distance of one cycle at whatever speed the experimenter chooses. Subjects are required to detect the presence of the flicker. In a number of experiments SRDs have been shown to be less sensitive than controls to counterphase flicker (Brannan & Williams, 1988; Martin & Lovegrove, 1987, 1988). The differences between the groups sometimes become larger as the temporal frequency increases (Martin & Lovegrove, 1987, 1988) and sometimes does not (Brannan & Williams, 1988). What happens depends on the spatial make up of the stimuli. This is a direct measure of transient system processing and distinguishes very well between individuals in the two groups (Martin & Lovegrove, 1987).

Additional support for differences between the groups in terms of spatial frequency processing comes from recent visual evoked potential studies (Galaburda, Drislane & Livingstone, 1991; May, Lovegrove, Martin & Nelson, 1991; May, Dunlap & Lovegrove, 1991). In the study by May et al. subjects were presented with sine-wave gratings ranging in spatial frequency from 0.5 to 8.0 cycles per degree flickering at a rate of two Hertz (Hz). Stimulus duration was 200 msec. This allowed analysis of two components of the VEP elicited by both stimulus onset and by stimulus offset. The major findings indicated that poor readers had significantly lower amplitudes and significantly shorter latencies for components produced by stimulus offsets when low spatial frequency stimuli were used.

Further analyses of these data revealed two factors for both the low and high spatial frequency stimuli (May, Dunlap & Lovegrove, 1991). With the low spatial frequency stimulus, Factor II was associated with the latencies on the first onset component and Factor I with the latencies of all components. These scores were subject to a discriminant analysis which showed that good and poor readers were well differentiated by the factor scores on the low spatial frequency but not the high spatial frequency factor. This is consistent with a problem in the transient system. Further visual-evoked potential data supporting this conclusion but using different conditions have been reported (Livingstone, Drislane & Galaburda, 1991).

Lovegrove and associates have also conducted a series of experiments comparing sustained system processing in controls and SRDs (see Lovegrove et al., 1986). Using similar procedures, equipment and subjects as the experiments outlined above, this series of experiments has failed to show any significant differences between the two groups. This implies that either there are no differences between the groups in the functioning of their sustained systems or that such differences are smaller than the transient differences demonstrated.

In summary, four converging lines of evidence suggest a transient deficit in SRDs. The differences between the groups are quite large on some measures and discriminate well between individuals in the different groups with approximately 75 percent of SRDs showing reduced transient system sensitivity (Slaghuis & Lovegrove, 1985). At the same time evidence to date suggests that the two groups do not differ in sustained system functioning.

HIGHER LEVEL PERCEPTUAL PROCESSES AND SRD

It is known that the transient and sustained systems may be involved in higher-level perceptual processes than those discussed above (Breitmeyer & Ganz, 1976; Weisstein, Ozog & Szoc, 1975). Williams and colleagues have recently investigated the question of how a transient deficit may manifest itself in a range of higher-level perceptual processes. Their general conclusion is that SRDs manifest difficulties on a large number of perceptual tasks, most of which are believed to involve the transient system (Williams & LeCluyse, 1990).

In an important study Williams, Molinet and LeCluyse (1989) plotted the time course of transient-sustained interactions. A standard way of measuring the temporal properties of transient-sustained interactions is to use a metacontrast masking paradigm (Breitmeyer & Ganz, 1976). In metacontrast masking a target is briefly presented followed at various delays by a spatially adjacent masking stimulus. The experiment measures the effect of the mask on the visibility of the target. The target is affected by both the temporal and spatial relationship between the mask and the target. It is normally found that the visibility of the target first decreases and then increases as the pattern mask follows it by longer and longer delays. Breitmeyer & Weisstein have argued that metacontrast masking is due to the inhibition of the sustained response to the target by the transient response to the mask. This happens in much the same way sustained persistence is terminated by transient activity during reading as is shown in Fig. 2.

Maximal masking occurs when the transient response to the mask and the sustained response to the target overlap most in time in the visual system. This occurs in metacontrast when the mask follows the target by a certain interval. The point of maximal masking, then, provides an index of the relative processing rates of the target sustained response and the mask transient response. If the difference in rate of transmission is small, the dip in the masking function occurs after a short delay and vice versa. The magnitude of masking provides an index of the strength of transient-on-sustained inhibition. Additionally metacontrast is normally stronger in peripheral than in central viewing presumably because of the preponderance of transient pathways in peripheral vision. It should be noted that this is the same mechanism proposed to be involved in saccadic suppression (the suppression of sustained activity during eye movements) as discussed earlier in relation to Figs. 2 and 3.

In an experiment using line targets Williams, Molinet & LeCluyse (1989) showed that maximal masking occurred at a shorter delay in SRDs than in controls. This result is direct evidence that SRDs have a slower transient system or at least a smaller difference between the rates of processing for their transient and sustained systems than controls. They also found that in peripheral vision SRDs experienced almost no metacontrast masking which further supports this position. The magnitude of masking was also less in central vision showing that the transient inhibition was also weaker. Further evidence supporting timing differences between the transient and sustained systems in controls and SRDs has been provided in a subsequent metacontrast experiment where subjects had to identify a target letter. The mask was also letters but could combine with the target to form a word or a nonword (Williams, Brannan & Bologna, 1988). Both of these studies provide clear evidence of temporal differences between the two groups contributing to high-level perceptual problems.

In summary there are now a large number of studies which have investigated higher-level perceptual processing in good and poor readers. The results from this wide range of measures confirm the finding of a transient deficit in SRDs. They also suggest that there may be other deficits (visual and higher level) but the precise nature of these is not yet clear.

WHAT ABOUT THE CONFUSION IN THE LITERATURE?

Over the years a substantial number of studies have reported differences in visual processing between good and poor readers. Several researchers (Blackwell, McIntyre, & Murray, 1983; Di Lollo, Hanson, & McIntyre, 1983; Hoien, 1980; Lovegrove & Brown, 1978; Stanley & Hall, 1973) have shown that masking occurs over longer durations in SRDs than in controls. Mason, Pilkington & Brandau (1981) have shown

SRDs to have difficulties with order rather than item information. Hyvarinen & Laurinen (1980) have measured spatial and temporal processing across spatial frequencies. They generally found that disabled readers were less sensitive than controls without specifying whether this difference was greater at certain spatial frequencies.

The difficulty in making sense of this literature, however, is that for almost every study showing differences between the two groups another study may be cited failing to show differences. For example Arnett & Di Lollo (1979), Fisher & Frankfurter (1977), Morrison, Giordani & Nagy (1977) and Manis & Morrison (1982) have all conducted studies with short-duration stimuli without finding any significant differences between groups. Howell, Smith & Stanley (1981) and Smith, Early & Grogan (1986) failed to show spatial-frequency specific differences in visible persistence between the two groups.

An obvious question is whether it is possible to reconcile these different sets of results in terms of the argument presented here. In the context of this chapter it may be suggested that many of the studies which have failed to show differences between dyslexics and normal readers in visual processing may have measured sustained processing and those which have shown differences have measured either transient system processing or transient-sustained interactions. Support for this position has recently been provided by Meca (1985) who has conducted a meta-analysis on a large number of studies investigating vision and reading. He plotted effect size as a function of spatial frequency. As would be expected if SRDs had a transient system problem but not a sustained system problem, effect size was greatest at low spatial frequencies and decreased with increasing spatial frequency (Meca, 1985).

While this is almost certainly over simplistic, it does allow us to make predictions about what should be found on a range of different tasks depending on whether or not transient or sustained processing is being measured. A recent study (Solman & May, 1990) has investigated spatial localisation in dyslexic and normal readers within this context. They predicted not only conditions where dyslexics should be worse than controls but also where dyslexics would perform at least as well or even better than controls. They found that when the targets were close to the fixation points SRDs performed slightly better than controls. This pattern reversed as the targets moved more into peripheral vision (and were presumably processed more by the transient system). It may be stated, therefore, that while there is a large amount of data consistent with the argument presented here there are also substantial data inconsistent with it. The consistent data are, generally, more recent and have formed part of one of a small number of systematic programs of research. If the argument presented here is valid, it is possible to make clear predictions about the types of visual tasks on which SRDs should do worse and/or better than controls.

It is important to note in this context that recent research has not simply demonstrated SRDs performing more poorly than controls on all measures of visual processing. SRDs have been shown to perform at least as well as or even better than controls on some tasks, e.g., high spatial frequency sensitivity, visual acuity and the oblique effect (Lovegrove et al., 1986). Generally this is thought to be the case on tasks measuring sustained system functioning. Further experimentation will determine whether or not this is so.

POSSIBLE RELATIONSHIP TO OTHER PROBLEMS MANIFEST IN SRDS?

There are two issues to be addressed by any theory of specific reading disability which attributes a role to visual factors. The first concerns the relationship between visual deficits and other known processing deficits manifested by SRDs especially those in phonological awareness (Bradley & Bryant , 1983) and working memory (Jorm, 1983). The second is to do with whether a visual approach is able to predict conditions under which normal readers and SRDs may perform differently. Both issues will be briefly discussed.

There is extensive evidence that SRDs perform worse than controls in a number of other areas, especially in aspects of phonological awareness and working memory. It becomes important to ask what, if any, is the relation between the transient system deficits and these other processing areas.

This issue was considered in a recent study of approximately sixty SRDs and sixty controls (Lovegrove, McNicol, Martin, Mackenzie & Pepper, 1988). They took measures of transient system processing, phonological recoding and working memory in each child. These measures were subjected to a factor analysis which showed that some of the phonological recoding measures loaded on the same factor as the measures of transient processing used. This shows some relation between the two processing areas but, of course, does not reveal the precise nature of that relationship. Until this relationship is further clarified it is premature to reject the possibility of a link between visual and phonological processes in reading. The measures of working memory used did not load on the same factor as did the transient system measures and the phonological recoding measures. This study thus provides some preliminary evidence of a link between phonological recoding and visual processing in SRDs but the exact nature of this relation is still to be determined.

It is possible that some of these different deficits are related by virtue of the fact that some SRDs have a problem in processing rapidly presented stimuli in all sensory modalities. Tallal (1980) , for example, has shown similar problems in audition as we have shown in vision. Livingstone, Drislane & Galaburda (1991) have recently noted that the auditory and somatosensory systems may also be subdivided into fast and slow

:omponents like the visual system. They then speculated that problems in each fast ystem may occur in SRDs. Even though this possibility has not yet been directly nvestigated, it is an exciting prospect which may help to integrate a large amount of apparently discrepant data.

The second issue raised above concerns the possibility of predicting conditions which would lead to different levels of performance in the two groups. In terms of Breitmeyer's theory outlined earlier a transient deficit should lead to more errors for SRDs when reading continuous text than when reading isolated words. This is because this task requires integration of peripheral information from one fixation with central information on the next. This has recently been tested by varying the mode of visual presentation. Three conditions of visual presentation on a computer monitor were used while holding the semantic context constant. This was done by presenting stories in three different ways. In the first condition one word at a time was presented in the middle of the screen. Thus the subjects never had to move their eyes and never had information presented to the right of fixation. In the second condition one word was presented at a time but its position was moved across the screen. Here the subjects were required to move their eyes across the screen but still were never presented with information to the right of fixation. The final condition was a whole line presentation which most closely approximated normal reading. Rate of word presentation was held constant across the three conditions. The results (Lovegrove & MacFarlane, 1990) showed normal readers were most accurate in the whole line condition and made more errors in the two one-word at a time conditions. The reverse was true for the SRDs. They read significantly more accurately in both one-word conditions than the whole-line condition. The mode of presentation of written material which maximised reading accuracy in controls , therefore, produced the most errors in SRDs. These findings have both theoretical and practical implications.

CONCLUSIONS

The data reported in the last ten years show that many SRDs have a particular visual deficit. It has also been shown that it is unlikely that the transient deficit results from being unable to read (Lovegrove et al., 1986) although it is not yet known how it may contribute to reading difficulties. This problem appears to be present in a large percentage of disabled readers and not just for a sub-group frequently referred to as visuo-spatial dyslexics. There is still a lot of work to be done before knowing how this processing difficulty relates to other difficulties. The results with different modes of visual presentation will be further investigated in our laboratory and are encouraging.

REFERENCES

Arnett, J. L., & Di Lollo, V. (1979). Visual information processing in relation to age and reading ability. *Journal of Experimental Child Psychology*, 27, 143-152.

Badcock, D.R., & Lovegrove, W. (1981). The effect of contrast, stimulus duration and spatial frequency on visible persistence in normal and specifically disabled readers. *Journal of Experimental Psychology: Human Perception & Performance*, 7, (3), 495-505.

Benton, A.L. (1962). Dyslexia in relation to form perception and directional sense. In J. Money (Ed.), *Reading disability: Progress and Research Needs in Dyslexia.* (pp 81-102). Baltimore: Johns Hopkins Press.

Blackwell, S., McIntyre, D., & Murray, M. (1983). Information processing from brief visual displays by learning disabled boys. *Child Development*, 54, 927-940.

Bowling, A., Lovegrove W., & Mapperson, B. (1979). The effect of spatial frequency and contrast on visible persistence. *Perception*, 8, 529-539.

Bradley, L. & Bryant, P. (1983). Categorising sounds and learning to read-a causal connection. *Nature*, 301, 419-421.

Brannan, J., & Williams, M. (1988). The effects of age and reading ability on flicker threshold. *Clinical Visual Sciences*, 3, 137-142.

Breitmeyer, B. G. (1988). Reality and relevance of sustained and transient channels in reading and reading disability. *Paper presented to the 24th International congress of Psychology*, Sydney.

Breitmeyer, B. G. (1980). Unmasking visual masking: A look at the "why" behind the veil of "how". *Psychological Review*, 87,(1), 52-69.

Breitmeyer, B.G. (1983). Sensory masking, persistence and enhancement in visual exploration and reading. In K. Rayner (Ed.). *Eye Movements in Reading: Perceptual and Language Processes.* New York: Academic Press.

Breitmeyer, B.G., & Ganz, L. (1976). Implications of sustained and transient channels for theories of visual pattern making, saccadic suppression and information processing. *Psychological Review*, 83,1-36.

Campbell, F.W. (1974). The transmission of spatial information through the visual system. In F.O. Schmidt & F.S. Worden (Eds.), *The Neurosciences Third Study Program*, (pp 95-103). Cambridge, Massachusetts: The M.I.T. Press.

Di Lollo, V., Hanson D., & McIntyre, J. (1983). Initial stages of visual information processing in dyslexia. *Journal of Experimental Psychology: Human Perception & Performance*, 9, 923-935.

Fisher, D.F., & Frankfurter, A. (1977). Normal and disabled readers can locate and identify letters: Where's the perceptual deficit? *Journal of Reading Behaviour*, 10, 31-43.

Galaburda, A., Drislane, F. & Livingstone, M. (1991). Anatomical evidence for a magnocellular defect in developmental dyslexia. *Proceedings of the New York Academy of Science*, (in press).

Graham, N. (1980). Spatial frequency channels in human vision. Detecting edges without edges detectors. In C.S. Harris (Ed.), *Visual coding and Adaptability*. (pp 215-262). Hillsdale: Lawrence Erlbaum Associates Inc.

Hochberg, J. E. (1978). *Perception*. Englewood Cliffs, New Jersey: Prentice-Hall.

Hoien, T. (1980). The relationship between iconic persistence and reading disabilities. In Y. Zotterman (Ed.). *Dyslexia: Neuronal, Cognitive and Linguistic Aspects*. (pp 93-107). Oxford: Pergamon Press.

Howell, E. R., Smith, G.A., & Stanley, G. (1981). Reading disability and visual spatial frequency specific effects. *Australian Journal of Psychology*, 33, (1), 97-102.

Hyvarinen, L., & Laurinen, P. (1980). Ophthalmological findings and contrast sensitivity in children with reading difficulties. In Y. Zotternman (Ed.), *Dyslexia: Neural, Cognitive and Linguistic Aspects*. (pp 117-119). Oxford: Pergamon Press.

Jorm, A. (1983). Specific reading retardation and working memory: a review. *British Journal of Psychology*, 74, 311-342.

Livingstone, M., Drislane, F. & Galaburda, A. (1991). Physiological evidence for a magnocellular defect in developmental dyslexia. *Proceedings of the New York Academy of Science,* (in press).

Lovegrove, W. J., Bowling, A., Badcock, D., & Blackwood, M. (1980). Specific reading disability: Differences in contrast sensitivity as a function of spatial frequency. *Science*, 210, 439-440.

Lovegrove, W. J., & Brown, C. (1978). Development of information processing in normal and disabled readers. *Perceptual & Motor Skills*, 46, 1047-1054.

Lovegrove, W., Heddle, M., & Slaghuis, W. (1980). Reading disability: Spatial frequency specific deficits in visual information store. *Neuropsychologia*, 18, 111-115.

Lovegrove W. & Macfarlane, T. (1990). *How can we help SRDs in learning to read?* Unpublished honours thesis. University of Wollongong.

Lovegrove, W., Martin, F., Bowling, A., Badcock, D., & Paxton, S. (1982). Contrast sensitivity functions and specific reading disability. *Neuropsychologia,* 20, 309-315.

Lovegrove, W., Martin, F., & Slaghuis, W. (1986). A theoretical and experimental case for a residual deficit in specific reading disability. *Cognitive Neuropsychology*, 3, 225-267.

Lovegrove, W., McNicol, D., Martin, F., Mackenzie, B., & Pepper, K. (1988). Phonological recoding, memory processing and memory deficits in specific reading disability. In D. Vickers and P. Smith (Eds), *Human Information Processing: Measures , Mechanisms and Models* . North-Holland: Amsterdam, 65-82.

Manis, F. R., & Morrison, F. J. (1982). Processing of identity and position information in normal and disabled readers.*Journal of Experimental Child Psychology*, 33, (1), 74-86.

Martin, F., & Lovegrove, W. (1984). The effects of field size and luminance on contrast sensitivity differences between specifically reading disabled and normal children. *Neuropsychologia*, 22, 73-77.

Martin, F., & Lovegrove, W. (1988). Uniform & field flicker in control and specifically-disabled readers. *Perception*, 17, 203-214.

Martin, F., & Lovegrove, W. (1987). Flicker contrast sensitivity in normal and specifically-disabled readers. *Perception*, 16, 215-221.

Mason, M., Pilkington, C., & Brandau, R. (1981). From print to sound: Reading ability and order information. *Journal of Experimental Psychology: Human Perception & Performance*, 7, 580-591.

Matin, E. (1974). Saccadic suppression: A review and an analysis. *Psychological Bulletin*, 81, 899-915.

May, J., Dunlap, W. & Lovegrove, W. (1991). Visual evoked potentials latency factor scores differentiate good and poor readers. *Clinical Vision Sciences*. (in press).

May, J., Lovegrove, W., Martin, F., & Nelson, W. (1991). Pattern-elicited visual evoked potentials in good and poor readers. *Clinical Vision Sciences*, 2, 131-136.

Meca, J. (1985). La Hipotesis del deficit perceptivo del retraso especifico en lectura : un estudio meta-analitico. *Anales de Psicologia*, 2, 75-91.

Meyer, G. E., & Maguire, W. M. (1977). Spatial frequency and the mediation of short-term visual storage. *Science*, 198, 524-525.

Morrison, F., Giordani, B., & Nagy, J. (1977). Reading disability: An information processing analysis. *Science*, 196, 77-79.

Rayner, K. (1975). The perceptual span and peripheral cues in reading. *Cognitive Psychology*, 7, 65-81.

Rayner, K., & McConkie, G. W. (1976). What guides a reader's eye movements? *Vision Research*, 16, 829-837.

Singer, W., & Bedworth, N. (1973). Inhibitory interaction between X and Y units in the cat lateral geniculate nucleus. *Brain Research*, 49, 291-307.

Slaghuis, W., & Lovegrove, W. J. (1984). Flicker masking of spatial frequency dependent visible persistence and specific reading disability. *Perception*, 13, 527-534.

Slaghuis, W., & Lovegrove, W. J. (1985). Spatial-frequency mediated visible persistence and specific reading disability. *Brain and Cognition*, 4, 219-240.

Solman, R. & May, J. (1990). Spatial localisation discrepancies: a visual deficit in reading. *American Journal of Psychology*, 103, 243-263.

Smith, A., Early, F., & Grogan, S. (1986). Flicker masking and developmental dyslexia. *Perception*, 15, 473-482.

Stanley, G., & Hall, R. (1973). Short-term visual information processing in dyslexics. *Child Development*, 44, 841-844.

Tallal, P. (1985). Auditory temporal perception, phonics and reading disabilities in children. *Brain and Language,* 9, 182-198.

Vellutino, F. R. (1979a). The validity of perceptual deficit explanations of reading disability: a reply to Fletcher and Satz. *Journal of Learning Disabilities, 12*, 160-167.

Vellutino, F. R. (1979b). *Dyslexia: Theory and Research.* London: M.I.T. Press.

Weisstein, N., Ozog, G., & Szoc, R. (1975). A comparison and elaboration of two models of metacontrast. *Psychological Review*, 2, 325-342.

Williams, M., Brannan, J., & Bologna, N. (1988). Perceptual consequences of a transient subsystem visual deficit in the reading disabled. *Paper presented to the 24th International congress of Psychology, Sydney.*

Williams, M & LeCluyse, K. (1990). Perceptual consequences of a temporal processing deficit in reading disabled children. *Journal of the American Optometry Association*, 61, 111-121.

Williams, M., Molinet, K., & LeCluyse, K. (1989). Visual masking as a measure of temporal processing in normal and disabled readers. *Clinical Vision Sciences, 4*, 137-144.

Facets of Dyslexia and its Remediation
S.F. Wright and R. Groner (Editors)

MANIFESTO ON DYSLEXIA

G. Geiger and J. Lettvin
Department of Biomedical Engineering
Rutgers University, Piscataway, N.J. USA

APPROACH

Whatever we speculate to be the cause of dyslexia we still must try to explain the process. Our study concerns only the testing of a specific performance and the inferences to be drawn about process; it does not address structural cause.

Dyslexics can foveally identify isolated letters presented tachistoscopically against a blank background, performing as well as ordinary readers. But when short letter-strings or words are presented foveally dyslexics do considerably worse than readers (see also Bouma and Legein, 1977). They complain that letters are "crowded" so that neither letter-identity nor letter-order can be had in the interior of the string, whether the string is a word or a random series of letters. Thus because they speak language and understand the language they hear, it cannot be language that is compromised but something more primitive since it is not judgment, but perception itself, that seems confused. Since each letter can be identified in isolation the acuity is sufficient. By standard optometric tests, the optics of dyslexics (corrected for any refraction error) and the sensory apparatus do not differ from the normal.

The problem in dyslexia is then somewhere between early visual process (i.e., perceptual) and late process (i.e., interpretive). However, it is difficult to decide what is the case by a priori or conventional argument, and there is no reason for assuming one or the other without evidence. Disorder of eye movement as causal is ruled out by the problem surviving under tachistoscopic presentation. Disorder of visual word comprehension as primary and causal is ruled out by the difficulty with letter strings that are not words; for. how can one visually interpret what is not comprehended? Disorders of form recognition of symbols or letters is ruled out because in isolated presentation they are as easily identified by dyslexics as by ordinary readers. Thus there must be some visual degrading property in the seen aggregate which is not inferable from the aggregate as a collection of individuals.

There is an ancient principle that all the information in perception is provided by the senses and no contribution is had from expectation, will or belief of the perceiver. Sharpened over two millenia, that principle in one or another phrasing keeps us from the philosophic vacuity of pure solipsism. Under this rule no information can be added by the perceiver, but there is nothing that says that all sensory information must be preserved in process. Indeed, the concept of information processing involves the sacrifice of information to gain intelligibility; it is always degenerative of information. (e.g., the data which are fitted by a function cannot be recovered from the function). Accordingly there is no prohibition to keep the perceiver from actively reducing the information supplied by perception so long as none is added. And so perception itself can be systematically degraded by the perceiver without violating the principle stated above.

This is all the fiat needed to investigate perception directly for signs of a shaping influence from higher functions, not contributory to the content of perception but degradingly modulatory of it, diminishing the content to facilitate visually guided action. We suspected that such shaping influences were different between ordinary readers and dyslexics. We proposed that there are task-directed strategies of vision that shape, but do not augment the content of visual perception, that these strategies are learned in the course of achieving expert task-performance, and that dyslexia represents such a strategy misused.

EMPIRICS

The empirics sketched here are given in detail by our recent paper (Geiger, Lettvin & Zegarra-Moran, 1992) which is too long to be rewritten for inclusion in this chapter. We direct the reader to it as a reference paper.

The method is a simple perceptual test, easily replicated. Pairs of letters are projected tachistoscopically on a screen (we do not use a CRT display in working with the peripheral visual field). One letter of the pair is always presented at the centre of gaze and the other letter in the peripheral field along the horizontal axis. In successive presentations the peripheral letters are presented at different eccentricities (angular distances to either side from the centre of gaze). Each test consists of 200 such presentations. The response to each pair is recorded and the plot of the average of correct peripheral letter identification against eccentricity is called the "form-resolving field" (FRF).

We have redrawn one of the figures from Geiger et al. (1992) to show how the FRF of 10 ordinary readers differs from that of 10 severe dyslexics (Fig. 1). The FRF of ordinary readers is monotonic, narrow and symmetric to the left and right of the centre of gaze. That is, eccentric letter identification for ordinary readers is best in and near the centre of gaze and it falls off rapidly with eccentricity. The FRF of the English-native severe dyslexics (Fig. 1a) is similar to that of ordinary readers on the left side but is markedly different on the right side, i.e. it is distinctly asymmetric. Their FRF on the right side is not monotonic and it is wider. For these dyslexics, best recognition of the letter pairs occurs when the peripheral letter on the right side is at 5º-7.5º eccentricity. Recognition extends farther in the peripheral field than it does for ordinary readers. And with 2.5º eccentricity on the right, the recognition of the letter pairs is significantly lower than with 5º eccentricity (Geiger & Lettvin, 1989).

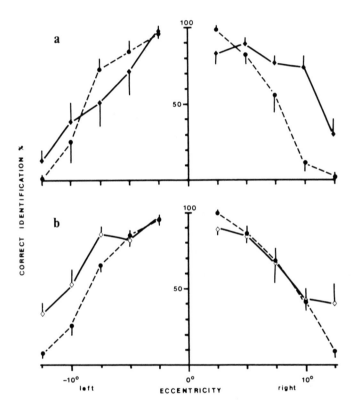

FIG. 1. Depicts the % correct identification of the peripheral letter as a function of eccentricity. The solid lines are for adult severe dyslexics, the dashed lines for adult ordinary readers. **a**. English-natives. On the right, at all eccentricities except for 5º, the differences between ordinary readers and dyslexics are statistically significant. On the left side the plots are not significantly different. **b**. Hebrew-natives. The plots are similar on the right side except at 2.5º where they are significantly different. On the left side, the plots are significantly different except at 2.5º and 5º eccentricities. The recognition of the central letter of the pairs in all cases has been above 90%. The vertical bars denote the standard deviation.

On the other hand, the FRF of 5 adult Hebrew-native severe dyslexics (Fig. 1 b) is also asymmetric but inversely directed with respect to that of English-native dyslexics. It is wide to the left, but narrow and similar to that of ordinary readers on the right. There is a dip of recognition near the centre on the right as with English-native severe dyslexics. The FRF of 5 adult Hebrew-native ordinary readers is symmetric and narrow and is similar to that of English-native ordinary readers.

During the last few years we have measured 47 adult ordinary readers and 11 ordinary reading children, altogether 58 ordinary readers. These subjects were reading according to their developmental levels and did not show any other learning problems. The FRFs of some of these subjects were tested a few times during the years and showed no changes. All the ordinary readers had FRFs similar to that of the ordinary readers in Fig. 1. Over the same time we have measured 10 adult residual dyslexics (Geiger & Lettvin, 1987), 10 adult severe dyslexics, and 34 other dyslexics of whom 20 were children, i.e. 54 dyslexics in all. The FRFs of the dyslexics were similar to what is depicted in Fig. 1 for severe dyslexics and were distinct from the FRF of ordinary readers individually and as a group.

The dyslexic subjects had all been diagnosed by independent psychologists and neurologists prior to their arrival in our laboratory. Except for the 10 residual dyslexics who were assigned to us by their teachers, all the dyslexics arrived in our laboratories (in the USA, Italy, Israel and Germany) upon hearing about us. We did not select or choose them and, as far as we know, we were not recommended to them according to any specific sub-type of dyslexia. All these dyslexics had normal or above normal intelligence, they comprehended language and spoke without any difficulties. None of them had any evident neurological symptoms but all had marked handicap in reading. The adult severe dyslexics had grave reading difficulties and had not been in any remedial training for at least three years prior to our testing. In order to read at their low level of competence they had adopted many kinds of reading methods (letter-by-letter, skimming, mirror image reading etc.). The adult residual dyslexics who had been in long term remedial training for reading were in college and had serious reading problems although not as extreme as those of the severe dyslexics.

The dyslexic children had reading levels at least 2 grades below their age and the level of their other academic achievements. Each dyslexic had a personal way of reading and these ways or styles were most diverse. However, the FRFs of all the dyslexic individuals were similar, asymmetric with wide eccentric-letter recognition in the direction of reading (to the right for English and Italian natives, and to the left for Hebrew natives). There was lessened letter pair recognition in and near the centre of gaze where the other letter always appeared.

The differences in the FRFs between adult ordinary readers and dyslexics were independently verified by Perry, Dember, Warm & Sacks (1989) who have tested the subjects by using an optical projection display. On the other hand, when we performed the same tests using a CRT display the differences between dyslexics and ordinary readers were smaller and not significant (see also Klein, Berry, Briand, D' Entremont & Farmer, 1990). The reasons for the differences between results with the CRT and the image projection methods are interesting technically (Zegarra-Moran & Geiger, in press) but outside the realm of this chapter.

We attribute the differences in shape of the FRF between ordinary readers and dyslexics to different distributions of a feed-back influence on perception. Later discussion will enlarge on this point. This influence is best known by one of its manifestations, lateral masking (as described, e.g., by Bouma, 1970), sometimes called "crowding". It is the loss of identifiability of a form due to the presence of neighbouring forms in the visual field, and was first studied in peripheral vision. The effect is demonstrated by Figure 2. If instead of single eccentric letters in the FRF test we use strings of three letters, ordinary readers show a steep increase of lateral masking with eccentricity, strongest for the middle letter. Dyslexics, however, show a more shallow increase of lateral masking with eccentricity and beyond 5° eccentricity, the middle letter is markedly less masked than for readers (Geiger & Lettvin, 1987, 1989, Geiger et al., 1992).

As a diagnostic measure the FRF is an easier test to perform than analysis of lateral masking within short strings of letters at different eccentricities. It distinguishes the same populations although it does not give the same detail of interaction between neighbouring symbols. In the peripheral field isolated letters against a blank background have a "self-masking" property that is less in foveal presentation. The parts of a complex letter seem to interact to ambiguate the letter shape (Fig. 2). But in the fovea self-masking is not obvious. Bouma & Legein (1977) have shown that dyslexics recognize isolated letters in or near the centre as well as ordinary readers do, but dyslexics recognize strings of letters or words in and near the centre worse than do ordinary readers.

The different shapes of the FRFs between ordinary readers and dyslexics and their associated distribution of lateral masking indicate how these two groups differ in visual perception. When ordinary readers gaze at a word, the letters comprising the word mask each other very little but the surrounding text away from the direction of gaze is strongly masked. On the other hand when dyslexics gaze at a word the letters in and near the centre of gaze mask each other. However, dyslexics perceive the letters best in the near periphery in the direction of reading but at the same time they perceive a large portion of the surrounding text, since it is poorly masked. Hence they do not perceive words in isolation as ordinary readers do. Consequently dyslexics do not recognize well the form of words, cannot isolate words adequately; the text seems to be "seen all at once" as they say.

As appears from these measurements, the FRF has one shape for ordinary readers and another for dyslexics. Another shape of the FRF can be shown in a group of 5 Italian-native "speed readers" whose FRF is wider to the right than that of ordinary readers; but it does not resemble that of dyslexics (Geiger & Lettvin, 1991). Other measurements have shown differences in the FRFs related to different stimulus types (letters vs. line drawings of objects) and for different age groups as related to different tasks (Zegarra-Moran & Geiger, in press).

We have determined that the differences between ordinary readers and dyslexics correlate with the differences in their FRFs and it occurred to us that the FRF measures a visual strategy that is learned. To show causal relations between a learned visual strategy and the skill of reading we had to show that it is possible for a dyslexic to learn a new visual strategy, that this learning also improves the reading skills as a consequence, and that the change is reflected in the FRF.

Four severe dyslexics volunteered to participate in a practise regimen which is described in detail in the reference paper. The regimen was not devised as a remedy but as a test for our hypothesis of visual strategy. This regimen had two complementary activities. One was to devote two hours a day to novel small scale hand-eye coordination activities like painting, drawing, clay modeling, etc. The other was to practise recognizing words at that eccentricity where letter strings were best recognized. We asked the subjects to use a specially designed mask which they laid on the text to be read. The mask was a blank sheet with a rectangular window, cut to be somewhat larger than a long word in the text. Left of the window and at the optimal eccentricity determined by the individual FRF of the participant we drew a fixation mark. The subjects laid this mask on the text to be read, fixed gaze on the mark, and read the word which appeared in the window on the right They shifted the mask along the text and read it word by word.

After 2 1/2 to 4 months (different for each participant) of this combined practise which we purposely did not monitor or supervise, we measured their FRFs and observed how well they read and with what effort. The shape of the FRF for each individual and for the group had shifted to resemble that of ordinary readers (Figs. 6, 7, 8 in the reference paper). At the same time their reading had improved dramatically and took much less effort.

This combined practise achieved two complementary objectives: exercise in construction of new task-related visual spaces and, in proximate time, exercise in form recognition for words in the part of the visual field where letter strings were most distinct. The practise of the novel small scale hand-eye coordination tasks provided new maps of operational space through active coordination of sensory-motor with visual control (Held & Gottlieb, 1958; Held & Hein, 1958; Kohler, 1962). At the same time the peripheral reading through the window provided blanking of the text surrounding the words and so enabled the perception of the isolated words as distinct forms in that region of the visual field where lateral masking for them was least. This also ensured that the new practise did not clash with an entrenched existing strategy of masking in the centre.

The subjects used the text mask in the beginning of the practise. But soon they claimed that at last they saw "the forms of the words" and not long after they did not use the mask any more except when they were very tired. According to their report they gazed at the fixation mark as long as they used the mask while reading. We had no way of verifying that. What is clear however, is that when they began the regimen they gazed at the fixation mark while they read, and a few months later they seemed to gaze directly at the words they read.

All but one of the group stopped the practise after we saw and tested them at the end of the few months of training. They felt that in learning this new visual strategy they had gained reading at the expense of other skills involving multi-attentional performances and artistic competences which they valued more. A few months later we measured again the ones who stopped practice and found that the FRFs had reverted to the shapes they had when the subjects came to us first. Their reading ability deteriorated markedly at the same time that the FRFs reverted. Perhaps by psychotherapy and other reinforcements we might have kept them reading. But it was not clear to us that it was our province to make decisions for them. The one case that chose not to revert was a college student who had every incentive to continue.

The last point to cover is a small set of subjects (3) that we call "conditional dyslexics". They fluctuate between being ordinary readers some of the time and being obviously dyslexic some of the time. One case serves to illustrate the character of the change, but the details of timing differ considerably from one case to another. This subject is an adult man whose profession is in the graphic arts. When he wakes in the morning he reads easily and well but his artistic skill is not at its best. By afternoon he cannot read easily at all and feels "tired". At the same time his graphic skill improves and he has no trouble at all in pursuing his work. By the FRF test he is an ordinary reader in the morning but a definite dyslexic in the afternoon. Under a simple regimen he found it easy to switch into the dyslexic mode in the morning so that he could work all day. But he could not switch back to the reading mode in the afternoon. The FRF changes appropriately with his voluntary switching (Geiger & Lettvin, 1989, Geiger et al., 1992).

Three main empirical results emerge from our study:

1. Dyslexics and ordinary readers are systematically different under a specific visual perception test (the FRF) and the difference is in the hemifield corresponding to the direction of reading. Dyslexics who were first taught to read from left to right (English, Italian) are symmetrically different by this test from dyslexics who were first taught to read from right to left (Hebrew).

2. Dyslexics can learn a new visual (perceptual) strategy and as a result their test records come to resemble that of ordinary readers at the same time that their reading skills markedly improve.

3. There exist "conditional dyslexics". They are ordinary readers part of the time and dyslexics at other times. The switch of phases takes time, an hour to a few hours, and the FRF test shows the correlated change.

Now we turn to the concept of task-determined visual strategy and the tactics of lateral masking.

ARGUMENT

From the results we infer that there are two grossly described visual strategies one for ordinary readers and one for dyslexics. Both strategies demonstrably exist for the conditional dyslexics and for the English-native severe dyslexics who learned to read under the regimen we devised. In the dyslexic strategy lateral masking is strong in the fovea and immediate parafoveal region, and it is least at 5°-7.5^o to the right of the centre of gaze. Further away from the centre, 7.5^o and beyond, the eccentric letter recognition stays significantly better and decays slower with eccentricity than for ordinary readers. In the acquired reading strategy the FRF comes to match that of the ordinary reader; lateral masking in the near foveal region is much lessened. and instead appears mainly in the peripheral field, steepening the decay of the FRF with eccentricity.

That lateral masking can be relieved in peripheral vision was shown by us earlier before the work on dyslexia (Geiger & Lettvin, 1986). That establishes the variable nature of lateral masking action. The present work establishes that the distribution of lateral masking can be controlled.

It would be useful here to give a clear concise verbal description of lateral masking, but in the language of visual psychology, indeed even in common language, the effect is not easily described and is better given by illustration. Fig. 2 will serve to demonstrate what is meant by lateral masking. It is evident that under lateral masking the spatial order of parts in an individual thing and the spatial order between things have become innominate in the interior of an arrangement. There is no blurring of boundary nor decay of contrast, but simply a degradation from a collection of forms, individually describable, to an aggregate that has a texture, i.e. some sort of collection of things that cannot be well told apart or well related spatially. Much less information is provided by the collection as a spatial distribution than as a set of distinct forms related by distinct distances between them.

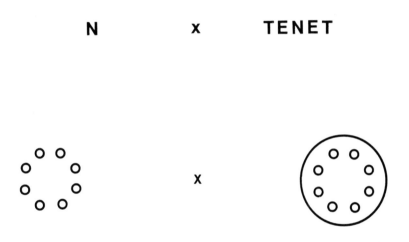

FIG. 2. Two different demonstrations of lateral masking. Fix your gaze on the X in the top line. Without moving your gaze you will recognize the N on the left, however, the N on the right will not be recognizable. The letters surrounding the N on the right "laterally mask" each other. If you fix your gaze on the lower X you will see the small circles on both sides. However the one on the left will suggest the spatial arrangement of a large ring or diamond. The circles on the right surrounded by the large circle will appear as a distinct heap of small circles without a particular spatial order.

How the image of a local arrangement of distinct separate things in perception can be reduced to the impression of a texture in the same part of the visual field is a technical process that would be idle to pursue here. More to the point is that the regional process of form-to-texture degradation is demonstrable and can be shown by experiment to be regionally reversible under training. Lateral masking can be controlled in its distribution; it is the tactical arm of visual strategy; different distributions define different strategies. Dyslexia now becomes crudely understandable in terms of the how dyslexics describe what they see. Consider first a vague argument. If indeed there is lateral masking between letters in central vision and little masking peripherally, how can a dyslexic ever see individual words in textual context ? And if they can't see them how can they reinforce learning to read in central vision? Rather, the failure reinforces itself as is known to happen elsewhere, and the condition seems intractable since the problem is in perception not in action. If they but knew what to perceive, the words as forms had from specific serial arrangements of letters, the task would be defined and they could improve. It is a strange concept. But the four severe dyslexics whom we brought to read called us independently after a few weeks to report one way or another that finally they saw "the shape of a word" in the eccentric window of the text masker. Only then, when they knew what to look for, did the task of reading become experientially defined.

There is no way of dealing with task-driven form-texture transitions (as were described in the previous paragraph) by conventional theories of visual perception and certainly no way of excluding intention in its various forms. But an elementary approach through task performance is useful. We each have an internal model of the world by which we decide what actions to take on the information received from perception. Those actions are taken to change what we currently perceive to what we want to perceive. To be able to operate at all in a complex world we have to relegate to background what is not relevant to attaining that new perceptual state as a goal, otherwise the amount of informational processing needed is enormous. It is not that we attend what is relevant by enhancing it in perception --that is regenerative, unstable, and violates the canon of perception by adding something to it. Rather we choose cognitively what is relevant and degrade everything else, reducing information about everything else and relegating it to background. This is degenerative, stable and consistent with the canon.

Lateral masking is the way of degrading form to background. For this to work as a system there must be a representation of what we intend by which to degrade everything else. We practise acting not only to refine what we do but to limit the content given by perception so as to speed our action. It is that practise that shapes the particular distribution of lateral masking. To act towards a goal there must be some representation of it, some model of what we want to perceive, and if there is none we flounder with inappropriate action. This can lock us into a mode of degrading precisely what is needed for conceiving the goal. To break this block it is necessary to by-pass it so as to evolve a new strategy. Battering against the block only reinforces it. That guided what we did with the four severe dyslexics.

Emerging from this loose formulation is a specific point. The strategy of degrading a region to background by the distribution of lateral masking is not shaped by cognition unaidedly acting back on perception. As in any action it is necessary to predict what is expected in the immediate future where in the visual field it will occur and how it is to be processed. Degrading to background is not specific but merely applies to everywhere in the field except where intended or expected change is to happen. This involves two factors in our internal model of the world, a notion of where the predicted perception is to occur and what the nature of the content will be. In turn these two factors are developed only by practise in the real world for which the model is to be developed in cognition. Suppose in over-simplification we consider the visual field divided into two regions, the central or foveal field, and the surround or peripheral field. And imagine that form-to-texture degradation can be switched between the two fields depending on whether we are interested in predicting detail of the stationary scene (foveal vision) or detecting what moves in the scene (peripheral vision). It is easy to see how, without prior knowledge of what to look for, a wrong strategy can be acquired, an inappropriate state of the switch. And with the practise of wrong strategy, the "what to look for" cannot be conceived and a vicious cycle freezes the practise against certain specific tasks. The important point is that the switch of distribution of degrading-to-background is a switch of visual perceptual strategy that is task-determined to be sharpened and shaped by active practice, not by gedanken-experiment. It is acquired by training, not by an arbitrary effort of the mind to conceive what has never been experienced so as to experience it.

By this crudely put argument we justify the concept of task-determined visual strategy. By itself, the strategy does not determine competence which is gained only by practise under the strategy. Expertise is another matter. For example, young ordinary reading children who have first learned to read have the same FRF as adult ordinary readers (Zegarra-Moran & Geiger, in press). Thus, there is a wide distribution of reading competence among ordinary readers (Shaywitz, Escobar, Shaywitz, Fletcher & Makurch, 1992).

Visual strategies can be delicately designed but our evidence for this will take much too long to develop here. The idea of learnable visual strategies shaped by task-performance and practise is the central point of our analysis of dyslexia as a learned perceptual dysfunction rather than a fault in cognition. By "learning" we do not mean the acquisition of some abstract explanation but the competence to perform a task --the ability to use what is had in concept. Without use there is no knowledge.

We believe, in line with what has been discussed above, that all of us possess many discrete task-determined visual strategies between which we switch easily and rapidly as task changes. Checking for specific strategy is hard to reduce to experiment unless there is a detectable absence of a required strategy as occur in dyslexics, or a "sticky switch" as in the conditional dyslexics or simply an inappropriate working strategy.

We do not deny that there can be traits, organic predispositions, hereditary, neurological etc., that can be involved with dyslexia, but the expression of the trait is not a necessary consequence of having it, and having the expression does not attest the existence of the trait in the individual. Our study was designed to show that developmental dyslexia is a learned visual strategy, that it can be acquired without any organic cause, and that it need not prevent acquiring a reading strategy.

POSITION

Developmental dyslexia is generally a learned visual strategy, rather than the product of neurological or genetic predestination. There are ways of establishing a separate reading strategy for dyslexics. Both strategies can coexist in the armamentarium of vision but only one can be used at a time. For the task of ordinary reading there seems to be a basic strategy and for dyslexia several aberrant strategies: but the crude classification into two general types of strategy serves our purpose in this paper.

REFERENCES

Bouma, H. (1970). Interaction effects in parafoveal letter recognition. *Nature* 226, 177-178.

Bouma, H., & Legein, Ch. P. (1977). Foveal and parafoveal recognition of letters and words by dyslexics and by average readers. *Neuropsychologia*, 15, 69-80.

Geiger, G., & Lettvin, J Y. (1986). Enhancing the perception of form in peripheral vision. *Perception* , 15, 119-130.

Geiger, G., & Lettvin, J.Y. (1987). Peripheral vision in persons with dyslexia. *New England Journal of Medicine* , 316, 1238-1243.

Geiger, G., & Lettvin, J. Y. (1989). Dyslexia and reading as examples of alternative visual strategies. In C. von Euler, I. Lundberg, & G. Lennerstrand (Eds.), *Brain and Reading,* London: Macmillan Press Ltd. pp 331-343.

Geiger, G., & Lettvin, J. Y. (1991). The form-resolving field (FRF) as a measure of specific visual strategies. *Invest. Ophthal. Vis. Sci*, 32(4), 1238.

Geiger, G., Lettvin, J. Y. & Zegarra-Moran, O. (1992). Task-determined strategies of visual processing in readers and dyslexics". *Cognitive Brain Research* , 1(1), 39-52.

Held, R. & Gottlieb, N. (1958). Technique for studying adaptation to disarranged hand-Eye Coordination. *Perceptual and Motor Skills,* 8, 83-86.

Held, R. & Hein, A.V. (1958). Adaptation of disarranged hand-eye coordination contingent upon re-afferent stimulation.. *Perceptual and Motor Skills*, 8, 87-90.

Klein, R., Berry, G., Briand, K., D'Entremont, B. & Farmer, M. (1990). Letter identification declines with increasing retinal eccentricity at the same rate for normal and dyslexic readers. *Perception and Psychophysics*, 47(6), 601-606.

Kohler, I. (1962). Experiments with goggles. *Scientific American*, 206, 62-72.

Perry, A. R., Dember, W. N., Warm, J. S. & Sacks, J. G. (1989). Letter identification in normal and dyslexic readers: a verification. *Bulletin of the Psychonomic Society.*, 27, 445-448.

Shaywitz, S. E., Escobar, M. D., Shaywitz, B. A., Fletcher, J. M. & Makuch, R. (1992). Evidence that dyslexia may represent the lower tail of a normal distribution of reading ability. *New England Journal of Medicine*, 326, 145-150.

Zegarra-Moran, O. & Geiger, G. Visual recognition in the peripheral field: letters vs. symbols and adults vs. children. *Perception* , (in press).

Facets of Dyslexia and its Remediation
S.F. Wright and R. Groner (Editors)

ONE WORD AT A TIME: A SOLUTION TO THE VISUAL DEFICIT IN SRDs?

R. Hill & W. J. Lovegrove

Department of Psychology, University of Wollongong,

Australia

INTRODUCTION

Many children have difficulty reading, despite having average or above average intelligence, normal educational opportunities and no serious behavioural problems. Because many researchers have failed to find a visual deficit in those with Specific Reading Disability (SRD), it has been widely accepted that visual problems do not play an important role in specific reading disability. However, recent research suggests that the processing of temporal and pattern information is performed by two parallel visual pathways: the sustained visual system and the transient visual system (Campbell, 1974; Graham, 1980). The characteristics of each system are shown in Table 1.

TABLE 1

General Properties of the Sustained and Transient Subsystems

From Lovegrove, Martin & Slaghuis (1986).

SUSTAINED SYSTEM	TRANSIENT SYSTEM
* Most sensitive to fine detail	* Most sensitive to large objects and overall shape
* Most sensitive to stationary or slow moving objects	* Most sensitive to fast moving or flickering objects
* Transmits messages to the brain slowly	* Transmits messages to the brain quickly
* Responds continously while the stimulus is present	* Responds only at the appearance and disappearance of a stimulus
* Predominates in central vision	Predominates in peripheral vision
* The sustained system may suppress messages from the transient system	* The transient system may suppress messages from the sustained system

Reading is initially a visual task requiring the processing and integration of visual information across fixation-saccade sequences (Lovegrove, Martin & Slaghuis, 1986). The role of the transient and sustained subsystems in reading has been considered by

Breitmeyer (Breitmeyer & Ganz, 1976). Figure 1 represents a response sequence of sustained and transient channels during three 250 msec fixation intervals separated by 25 msec saccades (Panel 1 of Fig. 1).

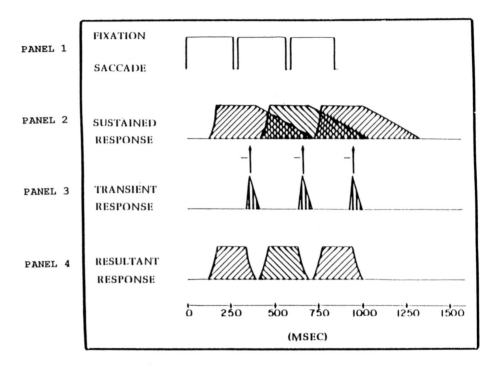

FIG. 1. A hypothetical response sequence of sustained and transient channels during three 250 msec fixation intervals (From "Unmasking Visual Masking: A look at the `Why' Behind the Veil of the 'How'" by B.G. Breitmeyer, *Psychological Review*, 1980, 82, 52-69).

The sustained system transmits continuous detailed information about the appearance of the word to the brain (see panel 2). The transient system is stimulated when the eyes move to fixate the next word (see panel 3). The sustained response, which may outlast the physical duration of the stimulus, is terminated by the transient response (see panel 4). The interaction between the sustained and transient systems prevents information from the previous fixation interfering with the processing of the fixated word. Consequently, a problem in one of these visual channels may have an adverse effect on reading.

There is evidence that approximately 75% of SRDs have a visual deficit which lies in the transient subsystem (Lovegrove et al., 1986). If the sustained messages from a previous fixation are not suppressed by the transient system, superimposition of the words from successive fixations may occur. Since the transient system is involved with processing the textual information surrounding the fixated word, removing peripheral information from text presentation should result in improved reading in those SRDs who have a transient deficit. Three text conditions were used to test this hypothesis: in the no eye-movement condition text was presented one word at a time in the centre of the screen; in the forced eye-movement condition text was presented one word at a time from left to right across the screen; and in the eye-movement condition text was presented a line at a time across the screen.

METHOD

Subjects

SRDs and normal readers attending primary schools in the Wollongong area participated in the two experiments. Initial assessment was made using the tests listed below and the results are shown in the Table 2.

TABLE 2
Mean Reading Age, IQ and Nonsense Word Scores (SD in brackets) for SRDs and Controls for Experiments 1 & 2.

	Reading Age	IQ	Nonsense Word
Experiment 1			
SRDs (N=10)	7.9 (1.1)	95 (8.6) **	19 (11.0) *
Normal Readers (N=5)	8.7 (0.8)	118 (8.8)	35 (7.3)
Experiment 2			
SRDs (N=12)	8.2 (1.1)*	99 (9.6)	20 (10.4) *
Normal Readers (N=12)	9.7 (1.3)	102 (14.9)	36 (9.6)

$*p<.05$ $**p<.01$

WISC-R (IQ) Performance Sub-Scale: a Test of General Intelligence.
Neale Analysis of Reading Ability (1966): a reading test which estimates a child's reading age for accuracy, comprehension and reading rate. Accuracy scores were used to determine reading age.

Nonsense Word Test (Woodcock, 1987): a test of phonetic ability, in which the child reads aloud a list of meaningless pronounceable letter strings.

Transient System Functioning: Sensitivity to flicker, which is mediated by the transient system (Kulikowski & Tolhurst, 1973), was used to measure transient system functioning. Two spatial frequencies (.3 and 6 cpd) at a temporal frequency of 20 Hz were used. Space averaged luminance was 11 cd/m^2 for Experiment 1, and 36 cd/m^2 for Experiment 2. For details of procedure and apparatus refer to Martin & Lovegrove (1987).

Text Presentation: Each child read aloud three stories presented on a computer screen. Rate of presentation and level of difficulty of the stories was determined on the basis of performance on the Neale reading test. Each story was presented in one of the three conditions: one word per line; one word at a time from left to right across the screen; and one full line of text at a time. Number of errors were recorded and comprehension was assessed for each of the three conditions.

RESULTS

Transient System Functioning: Experiment 1

Figure 2 shows mean log contrast sensitivity scores when average luminance was 11 cd/m^2 at spatial frequencies of .3 and 6 c/degree. A two-way univariate repeated measures analysis of variance (using SPSS-X) showed the main effect of spatial frequency to be significant [$F(1,13)=199.15$, $p<.001$], with sensitivity being lower in the case of high spatial frequency. The groups x spatial frequency interaction was also significant [$F(1,13)=5.20$, $p<.05$], indicating that the differences in sensitivity between the 2 groups was greatest when spatial frequency was 6 c/d. SRDs were as sensitive as normals to flicker at the low spatial frequency (means=2.23, SD=0.31 and 2.25, SD=0.251 respectively), but were less sensitive than normals at the higher spatial frequency (means=1.17, SD=0.3 for SRDs and 1.52, SD=0.1 for Normals). These results support the findings of Martin & Lovegrove (1987).

Transient System Functioning: Experiment 2

Figure 3 shows mean log contrast sensitivity scores when average luminance was 36 cd/m^2 at spatial frequencies of .3 and 6 c/degree. A two-way univariate repeated measures analysis of variance (using SPSS-X) showed the main effect of spatial frequency to be significant [$F(1, 22)=66.34$, $p<.001$], with sensitivity decreasing with increasing spatial frequency. However, the groups x spatial frequency interaction was not significant [F(1,22)=1.05, p>.05], indicating no differences in sensitivity between the two groups at either spatial frequency. SRDs were as sensitive as normals to flicker at both the low spatial frequency (means=2.39, SD=0.10 and 2.34, SD=0.09 respectively), and the high spatial frequency (means=1.97, SD=0.27 for SRDs and 1.80, SD=0.33 for Normals).

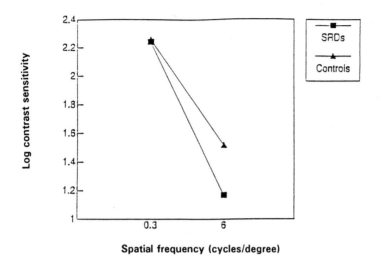

FIG. 2. Mean log contrast sensitivity scores for SRDs and Normal controls for Experiment 1 for two spatial frequencies (.3 and 6 c/degree) when average luminance was 11 cd/m².

Accuracy Scores

The number of reading errors were recorded for the three text conditions. The results for the two groups are shown in the figures below.

Experiment 1

A two-way univariate repeated measures analysis of variance (using SPSS-X) showed the groups by conditions interaction of error scores was not significant (p>.05), indicating that the number of errors across the three text conditions did not significantly differ for the two groups. However, there was a main effect for text [F(2,26)=4.24, p<.05] indicating that more errors were made when text was presented a full line at a time than when text was presented from left-to-right or one word per line. Planned comparisons revealed a significant difference between the combined one word-per-line and left-to-right conditions versus the full-line presentation [F=4.85(1,13),p<.05], with

SRDs making significantly more errors on the full-line condition. Figure 4 shows that SRDs made most errors on the full-line condition (mean=17.4, SD=7.2), and least errors on the left-to-right condition (mean=10.7, SD=2.9), whereas normal controls made most errors on the one word-per-line condition (mean=14, SD=4.7) and fewest errors on the left-to-right condition (mean=9.8, SD=2.4).

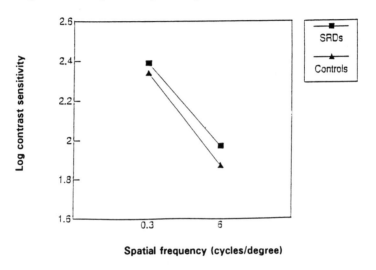

FIG. 3. Mean log contrast sensitivity scores for SRDs & Normal controls for Experiment 2 for two spatial frequencies (.3 and 6 c/degree) when average luminance was 36 cd/m^2.

Experiment 2

A two-way univariate repeated measures analysis of variance (using SPSS-X) showed the groups by conditions interaction of error scores was significant [F(2,44)=4.83, p<.05), indicating that the number of errors across the three text conditions differed significantly for the two groups. This difference between the two groups was not significant in Experiment 1.

Planned comparisons revealed that SRDs made significantly more errors than normal controls on the full-line condition [F(1,22)=4.81, p<.05]. Figure 5 shows that SRDs made most errors on the full-line condition (mean=18.8, SD=9.3), and fewer errors on the left-to-right condition (mean=10.5,SD=2.7), and one word-per-line condition (mean=10.6, SD=4.4). Normal controls, made fewest errors on the left-to-right condition (mean=10.7, SD=2.8) and more errors on the one word-per-line condition (mean=11.9, SD=4.7), and the full-line condition (mean=12.1, SD=5).

The hypothesis that removing peripheral information from text presentation would improve reading performance in SRDs was supported.

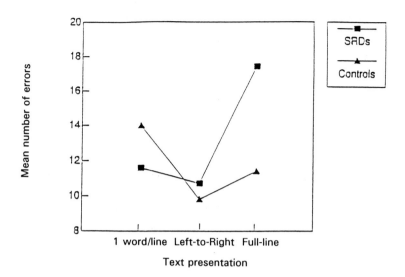

FIG. 4. Mean number of errors for SRDs and Normal controls as a function of type of text presentation (one word-per-line, left-to-right & full-line) for Experiment 1.

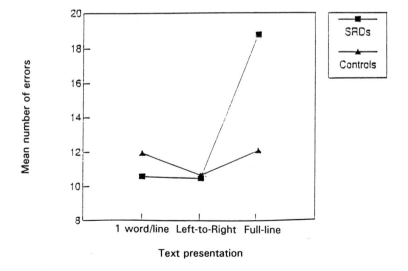

FIG. 5. Mean number of errors for SRDs and Normal controls as a function of type of text presentation (one word-per-line, left-to-right & full-line) for Experiment 2.

One Word Advantage and Flicker Sensitivity: Experiment 1
One word advantage was calculated by subtracting the mean number of errors made on the one word per line condition and the left to right condition from the number of errors made on the full line condition. The relationship between flicker sensitivity and one word advantage, which is shown in Fig. 6, indicates that one word advantage increased as flicker sensitivity decreased.

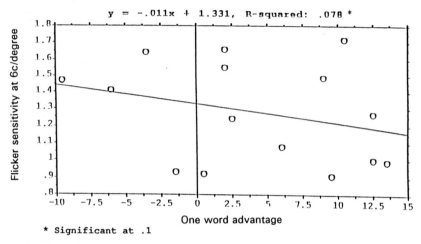

FIG. 6. Regression plot of one word advantage against log flicker sensitivity at 6 c/degree for Experiment 1.

One Word Advantage and Flicker Sensitivity: Experiment 2
There was no significant correlation between one word advantage and flicker sensitivity in Experiment 2 (p> .1).

CONCLUSIONS

Transient System Functioning
SRDs demonstrated a transient system deficit in Experiment 1 at the higher spatial frequency (6 c/degree) when luminance level was low (11 cd/m^2), but not in Experiment 2 when luminance level was high (36 cd/m^2).

One Word Advantage
In both Experiments 1 and 2, SRDs performed better when potentially disruptive visual context was removed and least transient activity was involved. Their performance was worst in the full line condition which involved the most transient activity. These results support the "transient system deficit" theory, and indicate that the problem posed by a

transient system deficit is related to the integration of peripheral information within a saccade, rather than persistence of information across saccades.

Inter-relationship between Flicker Sensitivity and One Word Advantage.
There was a trend for one word advantage to increase with decreasing sensitivity to flicker at 6 c/degree, in Experiment 1 when luminance level was low (11 cd/m^2), but not in Experiment 2 when luminance level was similar to that of a normal page of print (36 cd/m^2).

The Solution?
While presenting text one word at a time was found to improve reading performance of children with specific reading disability, it cannot be concluded from the results of this study that presenting text in this manner is the solution to their visual deficit. The one word advantage for SRDs could alternatively be explained in terms of eye-movements or attentional differences.

FUTURE RESEARCH

A more sensitive measure of sustained-transient functioning at luminance levels similar to those of a normal page of print could be used to assess the role of vision in relation to one word at a time presentation. In addition, while it has been established that SRDs have a transient deficit, whether this deficit affects the reading process has yet to be determined. A transient deficit may play a direct causal role in the perception, discrimination and analysis of the visual features of letters and words, or it may be that a transient deficit and reading disorders are unrelated symptoms of a common neurological disorder (Smith, Early & Grogan, 1990).

REFERENCES

Breitmeyer, B. G. & Ganz, L. (1976). Implications of sustained and transient channels for theories of visual pattern masking, saccadic suppression and information processing. *Psychological Review,* 83 (1) 1-37.

Kulikowski, J. J. & Tolhurst, D. J. (1973). Psychophysical evidence for sustained and transient detectors in human vision. *Journal of Physiology*, 232, 149-162.

Lovegrove, W.J. & Macfarlane, T.L. (1989). *How can we help SRDs in learning to read?* Unpublished thesis submitted to Wollongong University.

Lovegrove, W.J., Martin, F. & Slaghuis, W.   (1986). A theoretical and experimental case for a visual deficit in specific reading disability.*Cognitive Neuropsychology*, 3, 225-267.

Martin, F. & Lovegrove, W. (1987). Flicker contrast sensitivity in normal   and specifically-disabled readers. *Perception*, 16, 215-221.

Neale, M.D. (1966). *The Neale Analysis of Reading Ability*. London: Macmillan.

Wechsler, D. (1974). *Wechsler Intelligence Scale for Children - Revised*. New York.: The Psychological Corporation.

Williams, M. & LeCluyse, K. (1990). Perceptual consequences of a temporal processing deficit in reading disabled children. *Journal of the American Optometric Association,* 61 (2), 111-121.

Woodcock, R.W. (1987). *Reading Mastery Tests-Revised*. Minnesota: American Guidance Service, Inc.

APPENDICES

COPIES OF ORIGINAL FIGURES

Experiment 1

Individual Data on subjects

Subject	Age	Performance I.Q.	Neale R.A.	Reading Lag	Rate	Woodcock Score
SRD						
SR	12.50	93	8.67	3.83	93	19
PH	10.25	92	8.75	1.5	88	32
SE	8.92	117	6.5	2.42	32	9
DR	12.33	92	7.08	5.25	60	21
IR	12.42	88	8.25	4.17	50	16
TR	11.92	93	8.92	3.00	51	29
MR	9.58	90	6.42	3.16	25	1
TI	9.67	96	6.58	3.09	34	7
MA	12.33	88	8.42	3.91	42	30
KG	11.33	100	8.92	0.34	86	30
Mean		94.9	7.85			19.4
(SD)		(8.56)	(1.07)			(11.0)
CONTROLS						
PJ	8.33	111	9.08	-0.75	43	40
RE	9.58	124	9.75	-0.17	47	43
ST	7.17	115	7.58	-0.41	49	24
LR	7.17	111	8.6	-1.43	79	33
RC	7.75	131	8.42	-0.67	55	36
Mean		118.4	8.68			35.2
(SD)		(8.82)	(0.80)			(7.33)

Experiment 2

Individual Data on Subjects

Subject	Age	Performance I.Q.	Neale R.A.	Reading Lag	Rate	Woodcock Score
SRD						
PH	10.25	92	8.75	1.5	88	32
SE	8.92	117	6.5	2.42	32	9
IR	12.42	88	8.25	4.17	50	16
TR	11.92	93	8.92	3	51	29
TI	9.67	96	6.58	3.09	34	7
JM	11.67	93	9.5	2.17	45	37
BH	11.08	106	9.58	1.5	68	34
DB	11.33	109	7.92	3.41	51	20
CN	10.92	102	9.08	1.84	31	24
OW	11.17	93	8.08	3.09	40	17
SC	11.5	109	8.50	3	60	12
CD	12.8	87	6.75	6.05	26	10
Mean		98.75	8.2			20.58
(SD)		(9.6)	(1.09)			(10.5)
CONTROLS						
RE	9.58	124	9.75	-0.17	47	43
LR	7.17	111	8.6	-1.43	79	33
RC	7.75	131	8.42	-0.67	55	36
KG	11.33	100	10.83	0.34	86	30
LL	8.08	104	9.0	-0.92	80	26
AB	8.33	101	8.58	-0.25	90	26
MM	8.0	81	10.5	-2.5	67	48
AC	8.58	95	9.5	-0.92	94	31
GH	7.92	101	11.92	-4.0	56	41
MH	8.83	91	9.83	-1.0	47	38
TT	8.33	104	11.50	-3.17	78	47
KGi	8.08	88	8.08	0	28	17
Mean		101.8	9.73			35.75
(SD)		(14.87)	(1.26)			(9.6)

Experiment 1

Log Contrast Sensitivity Scores for SRDs and Normal Controls for .3 and 6 cycles per degree.

	SRDs			Normal Controls	
ID	.5 c/d	8 c/d	ID	.5 c/d	8 c/d
SR	2.62	1.27	PJ	2.48	1.55
PH	2.35	1.00	RE	1.92	1.66
SE	2.00	0.92	ST	2.41	1.49
DR	2.11	1.24	LR	2.41	1.41
IR	1.93	0.99	RC	2.05	1.47
TR	2.15	0.91			
KG	2.57	1.72			
MR	1.74	1.08			
TI	2.21	0.93			
MA	2.68	1.64			
Mean	2.24	1.17		2.26	1.52
(SD)	(0.31)	(0.3)		(0.25)	(0.09)

Experiment 2

Log Contrast Sensitivity Scores for SRDs and Normal Controls for .3 and 6 cycles per degree.

	SRDs			Normal Controls	
ID	.3 c/d	6 c/d	ID	.3 c/d	6 c/d
PH	2.29	1.63	RE	2.28	1.80
SE	2.29	1.93	LR	2.42	1.82
IR	2.26	1.81	RC	2.37	1.92
TR	2.24	1.63	KG	2.48	2.13
TI	2.41	1.71	LL	2.37	1.93
JM	2.52	2.07	AB	2.42	1.84
BH	2.44	2.06	MM	2.36	2.00
DB	2.44	1.18	AC	2.38	1.90
CN	2.50	2.05	GH	2.29	2.08
OW	2.46	2.52	MH	2.21	1.76
SC	2.30	2.28	TT	2.20	2.01
CD	2.48	2.11	KGi	2.28	1.54
Mean	2.39	1.97		2.34	1.89
(SD)	(.102)	(.269)		(.086)	(.159)

Facets of Dyslexia and its Remediation
S.F. Wright and R. Groner (Editors)

THE EFFECTS OF WAVELENGTH ON VISUAL PROCESSING AND READING PERFORMANCE IN NORMAL AND DISABLED READERS

A. Rock-Faucheux, K. LeCluyse & Mary Williams
Department Of Psychology, University of New Orleans,
New Orleans, USA

VISUAL FACTORS IN SPECIFIC READING DISABILITY

Specific reading disability is a broad term which encompasses reading disabilities arising from a number of sources. A specific reading disabled child (SRD) is defined here as one of normal or better intelligence with no known organic or behavioural disorders who, despite normal schooling and average progress in other subjects, has a reading disability of at least 2.5 years (Badcock & Lovegrove, 1981; Critchley, 1964; Lovegrove, Billings & Slaghuis, 1978, Lovegrove, Martin & Slaghuis, 1986; Slaghuis & Lovegrove, 1984; Stanley, 1975). Since reading involves a dynamic visual processing task that requires the analysis and integration of visual pattern information across fixation-saccade sequences, studies in the area of reading disability have explored the possibility that visual processing abnormalities contribute to reading difficulties. A number of studies have provided evidence for basic visual processing differences between normal and disabled readers, especially at early stages of visual processing. Differences have been reported in visual information store duration (Lovegrove & Brown, 1978; Stanley, 1975; Stanley & Hall, 1973a), in the rate of transfer of information from visual information store to short term memory (Lovegrove & Brown, 1978; Stanley & Hall, 1973a), and in the characteristics of visual short term memory itself (Stanley & Hall, 1973b). These results indicate that some disabled readers process information more slowly and have a more limited processing capacity than normal readers.

Studies that have used tasks relying less on dynamic visual processing and temporal resolution, and more on pattern-formation processes and long-term memory, however, have failed to show visual processing differences between normal and disabled readers (Benton, 1962, 1975; Vellutino, 1977, 1979a, 1979b, 1987; Vellutino, Steger, DeSetto & Phillips, 1975a, Vellutino, Steger, Kaman & DeSetto, 1975b), although the validity of these studies has been called to question (Fletcher & Satz, 1979a, 1979b). Thus the long-standing debate as to whether visual factors play a significant role in reading difficulties has been complicated by the differences in methodological factors and the failure to distinguish between the measurement of temporal versus pattern-formation processes.

TRANSIENT-SUSTAINED THEORY OF VISUAL PROCESSING

It has been suggested that the processing of temporal and pattern information is accomplished by two separate but interactive subsystems in vision with different spatiotemporal response characteristics (see Breitmeyer, 1980, 1983, 1985; Breitmeyer & Ganz, 1976; Weisstein, Ozog & Szoc, 1975 for reviews). A short-latency transient system is most sensitive to low spatial frequencies, has a high temporal resolution and short response persistence, and is thought to be involved in the perception of motion, the control of eye movements, and the localization of targets in space. A longer latency sustained system is most sensitive to high spatial frequencies and stationary or slowly moving targets, has a long response persistence or integration time and low temporal resolution, and is thought to be involved in the identification of patterns and resolution of fine detail. According to this theory, the transient system is a fast-operating early warning system that extracts large amounts of global information.

The sustained system, on the other hand, responds more slowly and subsequent to the transient response, and is dependent to an extent on the output of the transient system. This transient-sustained processing distinction has recently been reconceptualized in light of evidence for separate parvocellular (P) and magnocellular (M) pathways in primates, which have been found to have largely sustained and transient response properties, respectively (Cavanagh, 1991; Livingstone & Hubel, 1987, 1988; Maunsell, 1987). The sustained P pathway appears to be additionally involved in the perception of colour, texture, and fine stereopsis, while the transient M pathway is involved in the perception of flicker and motion.

TRANSIENT-SUSTAINED PROCESSING AND READING DISABILITY

There is evidence that this transient-sustained relationship is different in normal and disabled readers. Lovegrove and coworkers have shown that visual processing differences between normal and disabled readers are evident when transient system processing is involved, but fail to surface under sustained processing conditions. For example, disabled readers are less sensitive than normal readers to low spatial frequencies, but equally or more sensitive to high spatial frequencies (Lovegrove et al., 1980, 1982; Martin & Lovegrove, 1987). Additionally, disabled readers show lower overall temporal sensitivity (Martin & Lovegrove, 1987) and a different pattern of temporal processing across spatial frequencies (Badcock & Lovegrove, 1981; Lovegrove et al., 1980b; Slaghuis & Lovegrove, 1984), but these temporal processing differences between normal and disabled readers are eliminated when transient system activity is reduced (Slaghuis & Lovegrove, 1984). These findings indicate that disabled readers have a deficient transient system. Measures of sustained channel processing, such as orientation bandwidth, spatial frequency bandwidth, and the oblique effect, did not provide evidence of differences between normal and disabled readers (Lovegrove et al., 1978, 1986), suggesting that the integrity of the sustained system is intact.

Williams and coworkers (Brannan & Williams, 1987, 1988a, 1988b; Williams & Bologna, 1985; Williams & LeCluyse, 1990; Williams et al., 1987, 1989, 1990, 1992) have expanded on Lovegrove's work by studying the perceptual consequences of a transient deficit in disabled readers. These studies have demonstrated the existence of a number of perceptual deficits in disabled readers that would be predicted by a transient deficit, i.e., the perceptual skills affected are those that are most likely to be mediated by the transient system. The evidence suggests that the nature of the transient deficit is a slowed or sluggish response from the transient system, such that it does not clearly precede and provide output to the sustained system.

VISUAL MASKING STUDIES

More direct measures of the temporal aspects of visual processing in normal and disabled readers have been obtained in recent visual masking studies (Williams et. al., 1989, 1990). Williams et. al. (1990) utilized a metacontrast masking paradigm to index the processing rate in both foveal and peripheral vision. In metacontrast, a target is briefly presented, and is followed at various delays by a spatially adjacent masking stimulus. Accuracy of detecting the target is measured as a function of the delay between the target and the mask. The time course of the accuracy function is thought to represent the time course of the processing of the target and mask. The accuracy functions typically obtained in metacontrast experiments are U-shaped, much like the schematic one shown in Fig. 1a. Accuracy first decreases, reaches a low point at an intermediate delay, and then increases again to baseline level.

Two-component metacontrast theories (Breitmeyer & Ganz, 1976; Matin, 1975; Weisstein, 1968, 1972; Weisstein et. al., 1975) attribute U-shaped metacontrast functions to the interaction of transient and sustained components of visual response. These models posit metacontrast masking as the result of the faster transient response to the later occurring mask catching up with, and inhibiting, the slower sustained response to the target. For this to occur, the mask must be delayed in time relative to the target. Figure 1b illustrates these timing assumptions. The dip, or lowest accuracy point in the function, is the point of maximum inhibition. As the dip shifts rightward toward longer delays, it could be assumed that some aspect of the transient (inhibitory) response to the mask is travelling faster. This is simply because a response that is activated later has to travel faster to catch up. Thus, according to these models, dips at long delays between the target

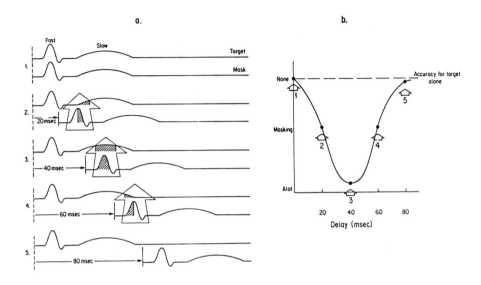

FIG. 1. A schematic U-shaped metacontrast function together with hypothetical visual response to a target and to a mask. (a) Schematic U-shaped metacontrast function: accuracy is plotted against delay. (b) Hypothetical visual responses. (1) simultaneous onset of target and mask. Transient responses do not overlap with sustained responses, and there is no masking, as shown by the arrow labelled 1 on the left. (2) Targets leads the mask by 20 msec. The transient response to the target and some interference occurs. There is a small amount of masking, as shown by arrow 2 on the left. (3) Difference in onsets of target and mask produce maximum overlap of transient and sustained components, and thus, the greatest amount of interference. (4) Target leads mask by 60 msec. The transient response to the mask again only slighlty overlaps the sustained response to the target, and the amount of the interference is again small. Masking begins to decrease, as shown by arrow 4 on the left. (5) Target leads mask by a long delay. No interference occurs, and from this point on, no masking occurs either. (Williams M., Weisstein, N., "The effect of perceived depth and connectedness on metacontrast functions". Vision Research, 1984; 24 (10), 1279-88.) Pergamon Press. Reprinted with permission.

and mask imply faster transient processing, and dips at short delays imply slower transient processing.

Williams et al. (1989), using diagonal line segments as targets and a surrounding outlined square as a masking stimulus, obtained the metacontrast functions shown in Fig. 2. The difference in dip location in the functions obtained from adults, normal readers, and disabled readers indicated that the foveal transient response was fastest in normal adults, slowest in reading disabled children, and intermediate in normal reading children. These findings are consistent with previous reports of increased temporal resolution with age (Brannan & Williams, 1988a, 1988b), and additionally suggest that temporal processing in disabled readers is sluggish compared to that of normal readers.

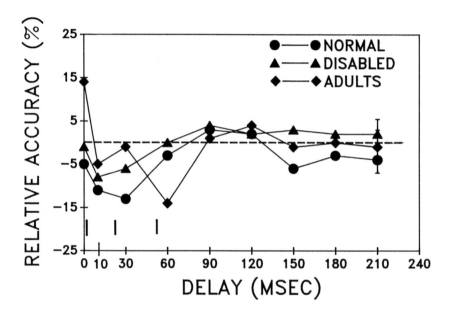

FIG. 2. Metacontrast functions obtained from adults, normal and disabled readers. Accuracy (measured as percent correct) for detecting the target lines when followed by the masking stimulus after various delays is plotted relative to target lines-alone accuracy level (horizontal line), which was set at a level between 70 and 80% before each session. (Williams et al. Visual masking as a measure of temporal processing in normal & disabled readers. *Clinical Vision Sciences*, 1989; 4 (2), 137-144.) Permission to reprint granted.

The magnitude of metacontrast masking increased in the peripheral retina in adults and normal readers (Figs. 3a, b), which is consistent with previous reports of increased masking effects in the periphery (Kolers & Rosner, 1960; Stewart & Purcell, 1970;

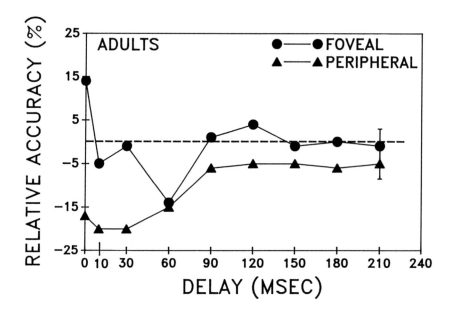

FIG. 3a. Metacontrast functions obtained from adults with foveal (circles) and peripheral (triangles) viewing. Accuracy (measured as percent correct) for detecting the target lines when followed by a masking stimulus at various delays is plotted relative to target lines-alone accuracy level (horizontal line), which was set at a level between 70 and 80% before each session. (Williams et al. Visual masking as a measure of temporal processing in normal and disabled readers." *Clinical Vision Sciences*, 1989; 4 (2), 137-144. Permission to reprint granted.

Williams & Weisstein, 1981). There was, however, an absence of metacontrast masking in disabled readers with peripheral presentations (Fig. 3c), a finding which is compatible with Geiger & Lettvin's (1987) finding that dyslexic subjects show a smaller magnitude of simultaneous lateral masking in the periphery. Geiger & Lettvin attribute the reduced masking effect to an attentional strategy of dyslexic subjects to allocate more processing capacity to peripheral as compared to foveal areas of the visual field. An alternate explanation can be derived from the two-component masking theories described above, which attribute metacontrast masking to the inhibition of slow sustained channels by fast transient channels. These theories would predict that a transient channel deficit would lead to an attenuation or elimination of metacontrast masking.

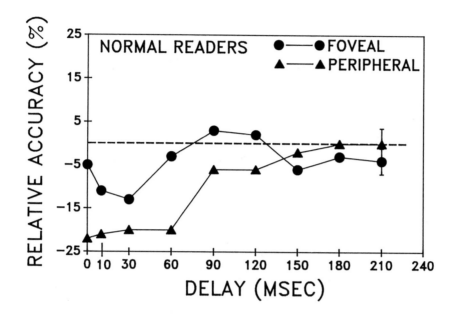

FIG. 3b. Metacontrast functions obtained from normal readers (b; above) and disabled readers (c: below) with foveal (circles) and peripheral (triangles) viewing.

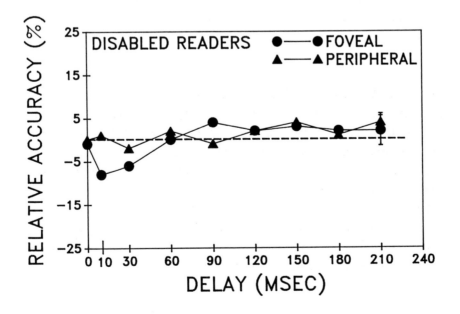

The masking study described above suggests that the nature of the transient system deficit in disabled readers is a slowed or sluggish response from the transient system. Additional evidence for temporal processing differences between normal and disabled readers is provided by studies of the effect of wavelength on temporal aspects of visual processing. Recent psychophysical and physiological data indicate that the wavelength of a stimulus differentially affects the response characteristics of transient and sustained processing channels. Physiological investigations of the primate visual system indicate that there are differences in the colour selectivity of these systems (Livingstone & Hubel, 1988), and that a steady red background light attenuates the response of transient channels (Dreher, Fukuda, & Rodieck, 1976; Kruger, 1977; Schiller & Malpeli, 1978). A recent investigation by Breitmeyer & Williams (1990) provides evidence that variations in wavelength produce similar effects in the human visual system. They found that the magnitude of both metacontrast and stroboscopic motion was decreased when red as compared with equiluminant green or white backgrounds were used. According to transient-sustained theories of metacontrast and stroboscopic motion, these results indicate that the activity of transient channels is attenuated by red backgrounds. Williams, Breitmeyer, Lovegrove, & Gutierrez (1991), using a metacontrast paradigm, additionally found that the rate of processing in transient channels increases as wavelength decreases, and that red light enhances the activity of sustained channels.

In the context of the above findings, Williams, Rock-Faucheux & LeCluyse (1992b) utilized a metacontrast paradigm to obtain direct measures of the effects of wavelength on temporal visual processing in normal and disabled readers. Using white diagonal lines as targets, and a white, red, or blue 12 c/deg flanking grating as a mask, Williams et. al., obtained the metacontrast functions shown in Figs. 4 and 5. Normal readers showed differences in both enhancement and dip location with the different coloured masks (Fig. 4). In line with Williams et al.,'s (1991) results, the fact that the delay of maximum masking occurred at a shorter delay for the red as compared with the other masks suggests that the processing rate in transient channels is slowest for the red masks. This finding may be related to previous findings that red light inhibits the activity of transient channels (Dreher, Fukuda & Rodieck, 1976; Breitmeyer & Williams, 1990). Along the same lines, the fact that the delay of maximum masking occurred at a longer delay for the blue as compared with the other masks suggests that blue light may enhance the processing rate in transient channels.

Next, consider the differences found in the magnitude of masking at the dips in the functions. Again, this is the point in the metacontrast function where the transient response to the mask maximally overlaps with, and inhibits, the sustained response to the target. Since the target was always the same, differences in the magnitude of masking at the dip can be attributed to differences in the response magnitude of transient channel activity generated by the mask. The fact that there was a smaller magnitude of masking with the red as compared with the shorter wavelength masks suggests that transient channels respond less vigorously to long wavelength stimuli.

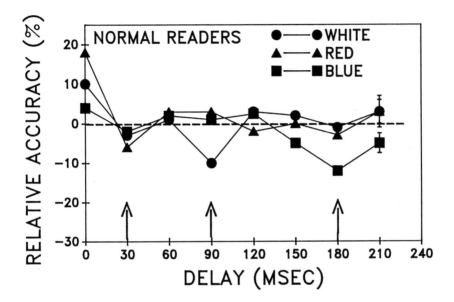

FIG. 4. Metacontrast functions collected on normal readers with masks varying in wavelength. Accuracy for the target lines when followed by a flanking grating mask is plotted relative to accuracy for the target lines-alone (horizontal line). Positive accuracy indicated that the mask enhanced the visability of the target, and negative accuracy indicated that the mask impaired the visability of the target.

At simultaneous presentation of the target and mask, accuracy for detecting the target was enhanced over the accuracy level for the targets when they were presented alone. This finding is consistent with previous reports of contextual information enhancing the detectability of briefly presented targets (Weisstein & Harris, 1974; Williams & Weisstein, 1981, 1984). According to masking models based on transient-sustained theory, this is the part of the function where figural information carried by the sustained components of response to the target and mask can interact. Since the enhancement effect varied with the wavelength of the mask, it appears that the sustained component of visual response is sensitive to variations in wavelength. The results indicate that the sustained channels respond with greater sensitivity to red light as compared with blue and white light. Disabled readers also showed differences in dip location and magnitude of masking with the wavelength of the mask (Fig. 5). Overall, dip locations occurred at shorter delays for disabled as compared with normal readers, suggesting that the processing rate in transient channels is slower in disabled readers. As with the normal readers, however, the processing rate in transient channels appears to be slowest with the red mask and fastest with the blue mask. Disabled readers generally showed a smaller magnitude of masking than normal readers, again suggesting that, overall, transient channels respond less vigorously in disabled readers. As with the normal readers, there was a smaller magnitude of masking with the red as compared with the shorter wavelength masks, suggesting that transient channels respond less vigorously to long wavelength stimuli. Finally, it is interesting to note that the function produced by the blue mask in disabled readers is similar in time course to the function produced by the white mask in normal readers. This finding suggests that blue light may produce a normal time course of processing in disabled readers, and is consistent with the contention that blue light may enhance the processing rate in transient channels.

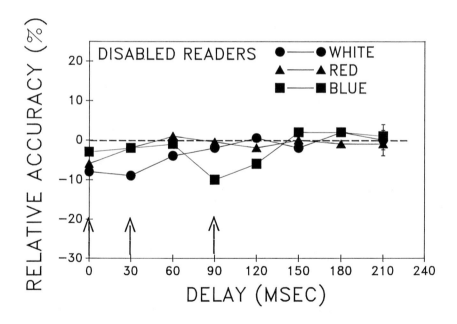

FIG. 5. Metacontrast functions collected on disabled readers with masks varying in wavelength. Accuracy for the target lines when followed by a flanking grating mask is plotted relative to accuracy for target lines-alone (horizontal line). Positive accuracy indicated that the mask enhanced the visability of the target, and negative accuracy indicated that the mask impaired the visability of the target.

THE USE OF COLOUR AS AN INTERVENTION FOR READING DISABILITY

Given the systematic effects of wavelength on the visual processing of the reading disabled, and the fact that blue light can render their performance comparable to that of normal readers, Williams, LeCluyse & Rock-Faucheux (1992a) investigated the effects of wavelength on actual reading performance. In this study, passages from graded reading books were covered with blue, green, red, grey, and clear plastic overlays and presented to normal and reading disabled children. The grade level of the passages was determined by each subject's performance on a standardized reading test. Each subject read passages at their current measured reading level, whether normal or delayed. Subjects were instructed to read each passage at a rate that was comfortable for them, and to pay close attention to what they were reading, since they would be asked to answer questions at the end of each passage. Multiple choice questions of literal comprehension were presented after the reading passages to check reading comprehension. An experimenter read each question to the subject to ensure that the subjects understood the questions.

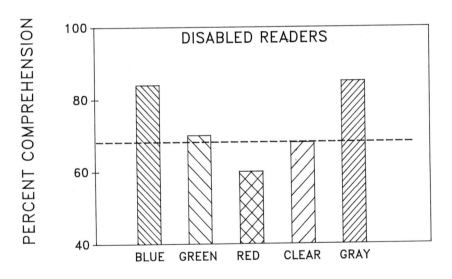

FIG. 6a. Reading comprehension (as measured by questions of literal recall) for passages covered with various coloured overlays for disabled readers.

Thirty-six subjects (18 normal readers and 18 disabled readers) participated in the study. Subjects read one story in each colour condition, with the presentation order of the colour conditions being counterbalanced across subjects. Percent correct on the reading comprehension questions and reading rate were the dependent measures. Figures

6a & 6b show the percent correct on reading comprehension questions obtained with the different coloured overlays and the different subject groups.

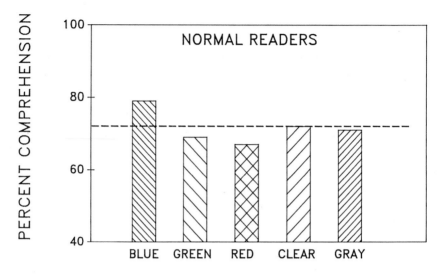

FIG. 6b. Reading comprehension (as measured by questions of literal recall) for passages covered with various coloured overlays for normal readers.

As can be seen, the blue and grey overlays produced significant improvement in the reading comprehension of disabled readers compared to the clear overlay, while the red overlay produced a significant decrement in performance. Performance with the green overlay did not significantly differ from that with the clear overlay. Seventy five percent of the disabled readers tested showed this pattern of results. Normal readers showed significant improvement with the blue overlay only, and a significant decrement in performance with the red overlay. Reading rate increased with blue overlays as compared with the clear overlay in both normal and disabled readers. Disabled readers also showed an increase in reading rate with the grey overlay, and a significant decrease in reading rate with the red overlay. Normal readers showed a decrease in reading rate with the red overlay as compared with the clear overlay, but this effect did not reach significance. In disabled readers, blue and grey overlays had equivalent effects on reading performance. It is likely, however, that this is due to effects on different visual processing mechanisms produced by these two overlays. While the use of short wavelength backgrounds has been found to increase response magnitude and speed in transient channels (Williams et. al., 1991), the use of grey filters, by reducing stimulus luminance or contrast, may instead slow the response of sustained channels more than that of transient ones (Breitmeyer, Clark, Hogben & DiLollo, 1991; Harwerth & Levi, 1978).

Thus, although the mechanisms may be different, the effect of using blue and grey overlays on reading performance in reading disabled subjects are the same: a temporal separation between transient and sustained responses may be created, which apparently compensates for the visual deficit of disabled readers. Normal readers, on the other hand, show reading gains only with the blue overlay. This may be due mainly to the increased magnitude of the transient response observed in the visual processing of short wavelength stimuli in normal readers (Williams et. al., 1991), resulting in greater efficiency of visual processing. Reducing contrast, on the other hand, may not be expected to benefit normal readers, since the creation of a temporal separation is not necessary in this subject group, and delaying sustained processing is not likely to result in greater efficiency in visual processing.

The use of a red overlay produced a significant decrease in the reading performance of both normal and disabled readers. This finding may be related to the previous finding that the use of red stimuli weakens and slows transient processing (Breitmeyer & Williams, 1990; Williams et. al., 1991). This would serve to exacerbate the visual deficit of disabled readers, and to simulate a transient processing deficit in normal readers.It should be noted that there are some individual differences which are obscured by the averaged data. For example, approximately 80% of disabled readers demonstrated an effect of colour, the remaining 20% performing best with clear overlays. Further, not all disabled readers showed best performance with the blue overlay; a small subgroup of approximately 5% shows beneficial effects of red overlays. It is believed that this subgroup is representative of another type of visual syndrome called "visual discomfort" (Wilkins, in press; Wilkins, Nimmo-Smith, Tait, McManus, Sala, Tilley, Arnold, Barrie, & Scott, 1984; see LeCluyse & Williams, 1992 for discussion). Impaired reading, due to the perception of visual illusions and distortions, is only one of the consequences associated with this syndrome. Additional symptoms include headaches, eye strain, and hypersensitivity to glare and high contrast stimuli. This occurs in individuals with otherwise normal or corrected-to-normal visual acuity. Preliminary research with subjects experiencing a high degree of visual discomfort indicate that their transient systems may be hypersensitive rather than deficient (Lovegrove, 1990). Thus, the use of red overlays, which would function to suppress transient function, would be expected to reduce visual discomfort symptoms and improve reading performance.

SUMMARY

SRDs are a relatively large population of individuals who suffer subtle visual deficits. The families of SRDs usually suffer years of frustration and inordinate financial burden as they attempt to remedy the disorder. SRD children often endure repeated evaluations, undertake expensive treatment programs, and continue to experience frustration and disappointment as each new attempt fails. Since the underlying deficit of the reading disability has not been well understood, past attempts at intervention have been ill-focused and often expensive. We have found that simply reducing the contrast of text materials, or using colour overlays in books, at practically no expense to the client, produces definite and measurable results. We caution that, since there are individual differences in the effects of different colours, an objective measure of the reading performance of each individual must be measured with various coloured and with grey overlays before an assignment is made. Additionally, a thorough evaluation and appropriate diagnosis of reading disability must be established before any intervention is employed.

REFERENCES

Badcock, D. & Lovegrove, W. (1981). The effects of contrast, stimulus duration, and spatial frequency on visible persistence in normal and specifically disabled readers. *Journal of Experimental Psychology, Human Perception and Performance*, 7, 495-505.

Benton, A. (1962). Dyslexia in relation to form perception and directional sense. In J. Money (Ed.) *Reading Disability: Progress and Research Needs in Dyslexia*. Baltimore: John Hopkins Press.

Benton, A. (1975). Developmental dyslexia: Neurological aspects. In W. J. Freidman *Advances in Neurology*, Vol. 1, Current Reviews of Higher Order Nervous System Dysfunction. New York: Raven Press.

Brannan, J. & Williams, M. (1987). Allocation of visual attention in good and poor readers. *Perceptual Psychophysics*, 41, 23-8.

Brannan, J. & Williams, M. (1988a). The effects of age and reading ability on flicker thresholds. *Clinical Vision Sciences*, 3,(2), 137-142.

Brannan, J. & Williams, M. (1988b). Developmental versus sensory deficit effects on perceptual processing in the reading disabled. *Perception and Psychophysics*, 44 (5), 437-444.

Breitmeyer, B. (1980). Unmasking visual masking: A look at the "why" behind the veil of the "how". *Psychological Review*, 87, 52-69.

Breitmeyer, B. (1983). Sensory masking, persistence, and enhancement in visual exploration and reading. In Raynor K. (Ed.) *Eye Movements in Reading: Perceptual and language Processes*. New York: Academic Press.

Breitmeyer, B. (1985). *Visual Masking: An Integrated Approach*. Oxford University Press, Oxford.

Breitmeyer, B. & Ganz, L. (1976). Implications of sustained and transient channels for theories of visual pattern masking, saccadic suppression, and information processing. *Psychological Review*, 83, 1-36.

Breitmeyer, B., Clark, C.D., Hogben, J.H. & DiLollo, V. (1990). The metacontrast masking in relation to stimulus size and intensity. *Revue Suisse de Psychologie*, Vol. 50, 87-96.

Breitmeyer, B. & Williams, M. (1990). Effects of isoluminant-background colour on metacontrast and stroboscopic motion: Interactions between sustained (P) and transient (M) channels. *Vision Research*, in press.

Cavanagh, P. (1991). The contribution of colour to motion. In A. Valberg and B.B. Lee (Eds.) *From Pigment to perception.*. New York: Plenum Press.

Critchley, M. (1964). *Developmental Dyslexia*. Heinemann, London.

Dreher, B., Fukuda, Y. & Rodieck, R. (1976). Identification, classification and anatomical segregation of cells with X-like and Y-like properties in the lateral geniculate nucleus of old-world primates. *Journal of Physiology*, 258, 433-452.

Fletcher, J. & Satz, P. (1979a). Unitary deficits hypothesis of reading disability: Has Vellutino led us astray? *Journal of Learning Disabilities*, 12 (3), 155-159.

Fletcher, J. & Satz, P. (1979b). Has Vellutino led us astray? A rejoinder to a reply. *Journal of Learning Disabilities*, 12, 168-171.

Geiger, G. & Lettvin, J. (1987). Peripheral vision in persons with dyslexia. *New England Journal of Medicine*, 316, 1238-1243.

Harwerth, R. & Levi, D. (1978). Reaction time as a measure of suprathreshold grating detection. *Vision Research*, 18, 1579-1586.

Kolers, P. & Rosner, B. (1960). On visual masking (metacontrast): dichoptic observations. *American Journal of Psychology*, 73, 2-21.

Kruger, J. (1977). Stimulus dependent colour specificity of monkey lateral geniculate neurones. *Experimental Brain Research*, 30, 297-311.

LeCluyse, K., Williams, M. & Rock-Faucheux, A. (1992). Towards an effective intervention for specific reading disabilities. *Journal of Learning Disabilities*. Submitted.

Livingstone, M.S. & Hubel, D.H. (1987). Psychophysical evidence for separate channels for the perception of form, colour, movement and depth. *Journal of the Neurosciences*, 7, 3416-3468.

Livingstone, M.S. & Hubel, D.H. (1988). Segregation of form, colour, movement, and depth: Anatomy, physiology and perception. *Science*, 240, 740-749.

Lovegrove, W. (1990). Personal communication.

Lovegrove, W., Billings, G. & Slaghuis, W. (1978). Processing of visual contour orientation information in normal and disabled reading children. *Cortex*, 14, 268-278.

Lovegrove, W. & Brown, C. (1978). Development of information processing in normal and disabled readers. *Perceptual and Motor Skills,* 4 6, 1047-1054.

Lovegrove, W., Heddle, M. & Slaghuis, W. (1980). Reading disability: spatial frequency specific deficits in visual information store. *Neuropsychologica*, 18, 111-115.

Lovegrove, W., Martin, F., Bowling, A., Blackwood, M., Badcock, D. & Paxton, S. (1982). Contrast sensitivity functions and specific reading disability. *Neuropsychologia*, 20, 309-315.

Lovegrove, W., Martin, F. & Slaghuis, W. (1986). A theoretical and experimental case for a visual deficit in specific reading disability. *Cognitive Neuropsychology*, 3, 225-267.

Martin, F. & Lovegrove, W. (1987). Flicker contrast sensitivity in normal and specifically disabled readers. *Perception*, 16, 215-221.

Matin, E. (1975). The two-transient (masking) paradigm. *Psychological Review*, 82, 451-461.

Maunsell, J.H.R. (1987). Physiological evidence for two visual subsystems. In L. Vaina (Ed.) *Matters of Intelligence*. Amsterdam: Reidel.

Schiller, P. & Malpeli, J. (1978). Functional specificity of lateral geniculate nucleus laminae of the rhesus monkey. *Journal of Neurophysiology*, 41, 788-797.

Slaghuis, W. & Lovegrove, W. (1984). Flicker masking of spatial frequency dependent visible persistence and reading disability. *Perception*, 13, 527-534.

Stanley, G. (1975). Visual memory process in dyslexia. In D. Deutsch & J. Deutsch (Eds.) *Short-term memory*. New York: Academic Press.

Stanley, G. & Hall, R. (1973a). A comparison of dyslexics and normals in recalling letter arrays after brief presentations. *British Journal of Educational Psychology*, 43, 301-304.

Stanley, G. & Hall, R. (1973b). Short-term visual processing information in dyslexia. *Child Development*, 44, 841-844.

Stewart, A. & Purcell, D. (1970). U-shaped masking functions in visual backward masking: effects of target configuration and retinal position. *Perceptual Psychophysics*, 7, 253-256.

Vellutino, F. (1977). Alternative conceptualizations of dyslexia: Evidence in support of verbal-deficit hypothesis. *Harvard Educational Review,* 47, 334-354.

Vellutino, F. (1979a). The validity of perceptual deficit explanations of reading disability: A reply to Fletcher and Satz. *Journal of Learning Disabilities*, 1, (3), 160-167.

Vellutino, F. (1979b). *Dyslexia: Theory and Research.*. M.I.T.Press, London.

Vellutino, F. (1987). Dyslexia. *Scientific American*, 256 (3), 34-41.

Vellutino, F., Steger, J., DeSetto, L. & Phillips, F. (1975a). Immediate and delayed recognition of visual stimuli in poor and normal readers. *Journal of Experimental Child Psychology*, 19, 223-232.

Vellutino, F., Steger, J., Kaman, M. & DeSetto, L. (1975b). Visual form perception in deficient and normal readers as a function of age and orthographic-linguistic familiarity. *Cortex*, 11, 22-30

Weisstein, N. (1968). A Rashevsky-Landahl neural net: Simulation of metacontrast. *Psychological Review*, 75, 494-521.

Weisstein, N. (1972). Metacontrast. In Handbook of sensory physiology, Vol, 7/4, *Visual psychophysics* (Ed. D. Jameson & L.M. Hurvich) pp. 233-72.

Weisstein, N. & Harris, C. (1974). Visual detection of line segments: and object-superiority effect. *Science*, 186, 752-755.

Weisstein, N., Ozog, G. & Szoc, R. (1975). A comparison and elaboration of two models of metacontrast. *Psychological Review*, 82, 325-342.

Wilkins, A. (1991). Visual discomfort and reading. In J. Stein (Ed.) *Reading and Reading Disabilities*. In press.

Wilkins, A., Nimmo-Smith, I., Tait, A., McManus, C., Sala, S.D., Tilley, A., Arnold, K., Barrie, M. & Scott, S. (1984). A neurological basis for visual discomfort. *Brain*, 107, 989-1017.

Williams, M. & Weisstein, N. (1981). Spatial frequency response and perceived depth in the time-course of object-superiority. *Vision Research*, 21, 631-645.

Williams, M. & Weisstein, N. (1984). The effects of perceived depth on metacontrast functions. *Vision Research*, 24, (10), 1279-1288.

Williams, M. & Bologna, N. (1985). Perceptual grouping in good and poor readers. *Perceptions and Psychophysics*, 38, 367-374.

Williams, M., Brannan, J. & Lartigue, E. (1987). Visual search in good and poor readers. *Clinical Visual Science*, 1, 367-71.

Williams, M., Molinet, K. & LeCluyse, K. (1989). Visual masking as a measure of temporal processing in normal and disabled readers. *Clinical Visual Science*, 4, 137-44.

Williams, M. & LeCluyse, K. (1990). Perceptual consequences of a temporal deficit in disabled reading children. *Journal of the American Optometric Association*, 61, 111-121.

Williams, M., Breitmeyer, B., Lovegrove, W. & Gutierrez, C. (1991). Metacontrast with masks varying in spatial frequency and wavelength. *Vision Research*, 31, 2017-2033.

Williams, M., LeCluyse, K. & Rock-Faucheux, A. (1992a). Effects of image blurring and colour on reading performance in normal and disabled readers. *Journal of the American Optometric Association*, 63, 411-417.

Williams, M., Rock-Faucheux, A. & LeCluyse, K. (1992b). The time-course of processing of stimuli with different wavelengths in normal and disabled readers. In preparation.

Facets of Dyslexia and its Remediation
S.F. Wright and R. Groner (Editors)

THE ROLE OF VISUAL PROCESSING IN GOOD AND POOR READERS'
UTILIZATION OF ORTHOGRAPHIC INFORMATION IN LETTER STRINGS.

E. Corcos

Glendon College, York University

and

D. M. Willows

The Ontario Institute for Studies in Education

Toronto, Canada

INTRODUCTION

The literature examining possible causes of the differences in the word recognition processes between normal and disabled readers has focused primarily on phonological and linguistic factors (Stanovich, 1988; Vellutino, 1979, 1987) and convincingly argues that these types of information are central to efficient word recognition. Recently, evidence has been accumulating to suggest that disabled readers may also have difficulty in the early stages of visual information processing (Geiger & Lettvin, 1987; Lovegrove, Martin & Slaghuis, 1986; Willows, 1991).

Visual information is available in words at the level of individual letters and groups of letters. Certain letter sequences occur more frequently than others, and certain letters are more frequently located in specific spatial positions within words. These frequency and spatial position patterns in words provide a source of orthographic information (OI). Knowledge of OI, or orthographic knowledge, generates expectancies of letter and letter groupings which contribute to the speed and accuracy of word recognition.

Beginning readers appear to have a fully developed "sensitivity" to the orthographic structure contained in words by the third-grade (Allington, 1978; Leslie & Shannon, 1981; Niles & Taylor, 1978). There is evidence, however, to indicate further development of orthographic knowledge up to the sixth-grade (Corcos & Willows, 1989; Samuels, Bremer & Laberge, 1978; Samuels, Miller & Eisenberg, 1979) when performance is not measured by a lexical decision task but by more demanding methods which encourage intraword processing.

Knowledge of spelling patterns assists in identifying words both in and out of context (Stanovich, 1980), and in spelling (Schwartz, 1983). Poor readers are less proficient in their use of orthographic information (Corcos & Willows, 1989) which may explain why they typically have more difficulty with context-free word recognition when they are required to rely exclusively on OI without the benefit of semantic and syntactic information.

It is not clear how readers acquire orthographic knowledge and how they apply it to word recognition. Samuels, Miller and Eisenberg (1979) have suggested that OI usage alters the unit of perception by allowing readers to move from processing words letter-by-letter to longer letter groupings and Samuels, Bremer and Laberge (1978) demonstrated that readers in Grade 2 were likely to process words using a component analysis while Grade 6 readers relied on holistic analyses. Grade 4 readers were found to use both types of analyses depending on word familiarity.

The underlying processes which allow some readers to make greater use of orthographic information and which differentiate readers at different stages of reading acquisition have yet to be examined. This study attempted to explore the possible role of various cognitive processes in the development and use of orthographic knowledge. Among adult skilled readers, the utilization of orthographic information is strongly associated with visual processes (Bruder, 1978; Krueger, 1975; Mason, 1975; Pollatsek, Well & Schindler, 1975). Doctor & Coltheart (1980), however, suggested that novice-readers may use a phonemic code while expert-readers use a visual code in word recognition. So, there was an expectation that phonemic processes may be prominent in Grade 2 performance while visual processes are more pronounced in the performance of older readers. Considering the processes underlying the performance of good and poor readers, it was expected that if a phonemic code is typical of early readers, then reader group differences may relate to phonemic ability. In contrast, reader group differences with older readers may be based on the variance accounted for by visual processing abilities.

An alternative view would consider that in the early stages of reading acquisition, the child may be very dependent on an analysis of the individual letters and interletter information in words (OI), to facilitate their word recognition; while at a later stage, they may process words more holistically with less attention to parts of words. In this case, it would be expected that the importance of visual processes in word recognition would decrease across stages of reading development (Chall, 1983) as the reader's word recognition skills became more automatic. A larger proportion of the variance associated with the processing of OI by younger readers would be accounted for by visual processes, and would decrease as readers became more proficient.

A sample of good and poor readers in Grades 2, 4 and 6 who had previously completed an experimental task measuring the speed and accuracy of processing pseudowords containing different levels of orthographic information were subsequent-ly given an extensive battery of standardized tests tapping a variety of cognitive processes including visual processing and memory, receptive and expressive language, memory span and speed of retrieval, and reading and spelling achievement. Performance on a these tests was used to account for the variance in the readers' OI performance.

METHOD

Subjects

From an original sample of approximately 400 subjects, a final sample of 90 children (39 boys and 51 girls), with 15 good and 15 poor readers in each of Grades 2, 4 and 6 were matched on the basis of several screening measures (see Table 1). The two groups represented skilled and less skilled readers found in the context of the regular class. They were matched on measures of non-verbal intelligence and receptive vocabulary while their reading ability differed by one standard deviation on the Gates-MacGinitie Reading Test. T-tests confirmed that the two reader groups did not differ significantly on age, PPVT-R or Raven scores, but a highly significant difference at each grade level on the reading comprehension measure (p<.001) was obtained.

TABLE 1

Group Performance on the Screening Measures

	Good Readers		Poor Readers	
GRADE	MEAN	SD	MEAN	SD
Grade 2				
Age in months	93.6	3.5	93.9	3.2
Peabody Picture Vocabulary	103.5	9.2	103.3	9.2
Raven Progressive Matrices	21.6	6.1	20.1	6.4
Gates MacGinitie Reading Test	59.0	5.8	41.1	4.1
Grade 4				
Age in months	117.7	5.0	116.5	4.4
Peabody Picture Vocabulary	102.9	8.7	101.9	8.5
Raven Progressive Matrices	27.3	5.8	26.5	5.2
Gates MacGinitie Reading Test	57.1	6.0	47.0	6.4
Grade 6				
Age in months	141.5	3.2	143.0	3.4
Peabody Picture Vocabulary	98.4	10.5	97.0	10.0
Raven Progessive Matrices	28.0	3.6	28.6	3.6
Gates MacGinitie Reading Test	56.8	5.5	45.3	3.0

Stimuli

Experimental Task. The stimuli used in the experimental task consisted of 3-letter and 6-letter strings constructed to control the frequency and spatial position information in each item. Frequency information is related to the redundancy of a letter sequence in a language, in this case English. Some letter sequences occur with high frequency (e.g., whe, ine) while other letter sequences occur rarely (e.g. cht , rhy). The assumption is the more frequently a letter sequence is contained in words, the more likely it is to be encountered in print and thus the more familiar it is likely to be. Spatial information deals with expectations of where a letter sequence is located within words. Some letter sequences have very high expectancies at the beginning of words (e.g. sta, reg) while other letter sequences rarely if ever occur at the beginning of words but are expected at the end of words (e.g. ion, ure).

Mayzner and Tresselt trigram frequency tables (Mayzner, Tresselt & Wolin, 1965) were used to construct stimuli varying in frequency and spatial information. Letter strings of 3-letter and 6-letters containing five levels of OI were generated:

1. High frequency/high-spatial (HFHS)
2. High frequency/low spatial (HFLS)
3. Low frequency/high spatial (LFHS)
4. Low frequency/low spatial (LFLS)
5. Consonant strings (CS)

In the 6-letter condition, pseudowords were constructed by combining two 3-letter strings to create each of the five combinations. For example, to construct HFHS pseudo-words, two 3-letter strings of high frequency were combined with the first normally expected at the beginning of a word and the second, at the end of a word (e.g., regure). In the HFLS, trigrams of high frequency were combined so that the first is normally expected at the end of words and the second, at the beginning (e.g., ionrhy).

Test Measures. Standardized tests of visual processing and memory, linguistic processing and memory, and reading and spelling achievement were administered (see Table 2) to all subjects.

Procedure

Subjects completed 12 blocks of experimental trials, each containing 20 trials, during two 30-minute sessions. Three-letter strings were presented in half of the trials while 6-letter strings were displayed in the other half. All subjects completed a block of 3-letter strings followed by a block of 6-letter strings. Subjects were required to judge whether a target-

and test-item were the "same" or "different" (see Fig. 1). Subjects were alerted to an upcoming target-item by a tone followed by a fixation point. A target-item (a pseudo-word) was then displayed for 60 msec. After a delay of 1 sec or 5 sec, a test-item was shown on the screen until one of two keys was pressed. The subjects were required to judge whether the target-item was the "same" or "different" from the test-item. Accuracy and latency measures were collected.

TABLE 2

Tests Administered to all Subjects

General Cognitive Ability	Tests	Specific Ability
Visual Processing and Memory		
Visual Processing	Tests of Visual Perception Skills (TVPS) Ravens Coloured Progressive Matrices	visual discrimination visual spatial relations visual perception and reasoning
Visual Memory	Memory for Designs TVPS	visual memory visual memory
Visual-Motor Skill	Bender Coding, -R	visual-motor perception visual-motor mapping
Language Processes		
Linguistic Processing	PPVT-R Vocabulary, -R CELF DTLA-2	receptive vocabulary expressive vocabulary sound discrimination word opposites
Verbal Memory		
Verbal Memory	CELF Digit Span, WISC-R Rapid Automatised naming	sentence repetition auditory sequencing speed of word retrieval
Achievement Measures		
Reading and Spelling	WRAT-R Durrell Analysis of Reading Gates MacGinitie Reading	word identification spelling dictation oral reading silent reading comp.

Subjects were also administered a comprehensive test battery involving tests of language, visual processing, visual-motor perception, memory and reading and spelling ability (see Table 2). Each subject was tested in two forty-five minute sessions. Testing was conducted under formal conditions as outlined in test instructions. Children were awarded a sticker following each test session.

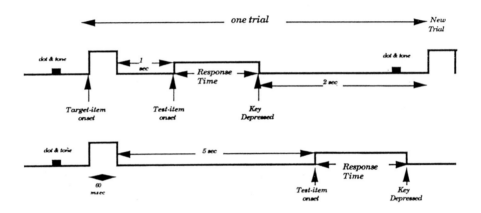

FIG. 1. The Structure of a Trial

RESULTS AND DISCUSSION

As reported in Corcos and Willows (1989) Grade 4 and 6 did not differ in their accurate use of OI to process letter strings. Grade 2 students, however, were less accurate than the two older grades on both 3-letter and 6-letter strings. The performance of good readers was more accurate than poor readers' on both the longer and shorter letter strings and poor readers across all three grades demonstrated less efficient OI usage even on the most word-like letter strings (HFHS). Reader group differences at the Grade 2 level were much larger than those found at the Grade 4 and 6 levels which suggested a greater disparity between good and poor readers at the early stages of reading acquisition. Latency data indicated differences at all three grade levels with Grade 6 subjects responding fastest and Grade 2, slowest. Despite anticipated developmental differences in children's reaction time, treatment effects were still evident. For example, Grade 6 readers processed 3-letter and 6-letter strings as quickly, and their performance was unchanged when letter-strings were held in memory for 1 or 5 seconds.

In contrast, the latency data of Grade 2 subjects indicated both string-length and response-delay effects while Grade 4 performance was somewhat affected by these variables. These findings led Corcos and Willows (1989) to conclude that Grade 2 readers still lacked the accuracy and speed benefits of OI while Grade 6 readers demonstrated these advantages of OI in their performance. Grade 4 readers were at an intermediate level in their ability to use OI in word recognition since the accuracy but not the speed benefits were acquired.

These findings led to a stepwise multiple regression being performed on the measures of visual processing and memory, linguistic processing and verbal memory, and reading and spelling achievement as predictors of OI accuracy, in the present study. This analysis was undertaken on 3-letter and 6-letter string performance separately for each grade (see Table 3) because Corcos and Willows (1989) found the 3-letter condition to represent an "easier" condition especially for the Grade 4 and 6 students whose performance was near ceiling.

Contribution to Grade 2 Performance. On 3-letter strings, 39% of the variance associated with the processing of frequency and spatial information in pseudowords was accounted for by the cognitive measures. The high performance (85%) on the OI task (Corcos & Willows, 1989) left only a limited amount of variance to be accounted for. Verbal memory explained 18% of the overall variance as it would appear that the ability to repeat sentences of increasing length contributed 4% while the remaining 14% related to the speed of retrieving the names of objects and numbers as measured by the Rapid Automatized Naming test (Denckla & Rudel, 1976).

Visual processing and memory explained 16% of the variance associated with OI. Most of this variance (Memory for Designs, TVPS-Visual Memory) was related to the ability to look at a geometric pattern, to hold it in memory for a brief period of time and then to reproduce it. Visual-motor integration as measured by the Bender-Gestalt explained the remaining variance. On the harder 6-letter strings, 63% of the variance of Grade 2 OI performance was accounted for by the cognitive measures. Verbal memory measures explained 48% of this variance (Digit Span-33%; Sentence Repetition-7%; accuracy in retrieving object name-8%). The large contribution of the Digit Span subtest suggested that Grade 2 readers were likely to use letter-by-letter strategies as they attempted to process the 6-letter strings.

Visual memory accounted for 15% of the variance (Memory for Designs-7%; TVPS-Visual Memory-8%) and this finding reinforced the notion that beginning readers initially process the visual information contained at the letter level (Adams, 1990).

TABLE 3

Amount of Variance Explained (R^2) by Each Measure

	3-LETTER			6-LETTER		
	Grade 2	Grade 4	Grade 6	Grade 2	Grade 4	Grade 6
Visual Measures						
Visual Discimination (TVPS)	-	-	-	-	.06	-
Visual Spatial (TVPS)	-	-	-	-	-	-
Raven	-	-	-	-	-	-
Memory for Designs	.05	-	-	.07	.04	-
Visual Memory (TVPS)	.08	.24	-	.08	-	-
Bender	.03	.03	-	-	.06	-
Coding (WISC-R)	-	.09	-	-	.39	-
Language Measures						
PPVT-R	-	-	-	-	-	-
Vocabulary (WISC-R)	-	-	-	-	-	.10
Sound Discrimination (CELF)	-	-	-	-	-	-
Word Opposites (DTLA-2)	-	.03	-	-	-	-
Verbal Memory						
Sentence Repetition (CELF)	.04	.05	-	.07	-	-
Digit Span (WISC-R)	-	-	.22	.33	-	-
Word Retrieval (R.A.N.)						
Objects:time	.07	-	-	-	-	-
Objects: error	-	-	-	.08	-	.08
Letters: time	-	.05	-	-	-	-
Numbers: time	.07	-	-	-	-	.19
Numbers: error	-	-	-	-	.09	-
Colours	-	-	-	-	-	-
Achievement Measures						
Word Identification (WRAT-R)	-	-	.09	-	.05	.05
Spelling (WRAT-R)	-	.11	-	-	.03	.09
Reading Comprehension (GMRT)						

Contribution to Grade 4 Performance. The cognitive measures explained 76% of the variance associated with the processing of OI in 3-letter strings. Visual processing and memory measures explained 36% of this variance (TVPS-Visual Memory-24%; Coding-9%; Bender-3%). It was apparent that these subjects made serious attempts to process these 3-letter strings holistically. This was borne out by the 27% contribution of achievement measures (Reading Comprehension-16%; Spelling-11%). Verbal memory explained 10% of the variance (Sentence Repetition-5%; Speed of retrieving letter names-5%) and the remaining 3% were associated with the ability to provide antonyms.

The cognitive measures explained 72% of the variance associated with processing OI in 6-letter strings. OI performance was largely the result of visual processing and memory (55% of the explained variance). These processes included the ability to associate visual symbols (Coding-39%), visual-motor integration (Bender-6%), and visual memory (TVPS-Visual Memory-6%; Memory for Designs-6%). So, the full complement of visual processes was applied to processing the frequency and spatial information contained in the letter strings. Verbal memory explained 9% of the variance on the basis of accurate retrieval of number names while the contribution of reading experience was evident as achievement measures accounted for 8% (Reading-5%; Spelling-3%).

Contribution to Grade 6 Performance. The processing of 3-letter strings was found to be an extremely easy task for these older readers who reached ceiling levels. As a result, only 31% of the variance was explained by the cognitive measures. Verbal memory, more specifically Digit Span, accounted for 22% of the explained variance and it is likely that these subjects opted for encoding the words by letter names because the task was so simple. Reading ability, however, accounted for an additional 9%.

On 6-letter strings, 51% of the variance in the OI task was explained by the cognitive measures. Verbal memory factors explained 27% of the variance: the speed of retrieving number name accounted for 19% and the accuracy of retrieving object names contributed a further 8%. Achievement measures explained 14% of the variance as spelling ability contributed 9% and reading ability 5%. Finally, 10% of the variance was explained by an expressive language factor. The processing of OI by the older Grade 6 readers was somewhat dependent on their reading ability and their speed and accuracy in retrieving words from memory.

It is evident that the easier 3-letter OI task did not generate as much variability in the performance of the older readers as did the 6-letter task. However, regardless of the amount of variance to be explained by the cognitive measures as a result of the performance variability in the OI task, the significance of the findings relates primarily to the differential distribution of variance as shown in Table 3. The distinctive pattern found for each age group implies different processes applied to using OI.

CONCLUSION

The results of the multiple regression analysis shed important new light on the types of cognitive processing factors that contribute to the variance in the processing of OI. For the younger, less experienced reader in Grade 2, both verbal and visual processing factors made significant contributions to OI variance. The larger contribution of verbal memory to the Grade 2 performance may be associated with letter-by-letter (rather than holistic) processing of letter-strings of increasing length while visual memory abilities facilitate the retention of letters as visual forms. These findings relating to Grade 2 performance were inconsistent with the expectation that younger more novice readers rely heavily on phonemic processes (Doctor & Coltheart, 1980). Instead, beginning readers may be very dependent on an analysis of the individual letters and interletter information in words (OI) to facilitate their word recognition in the early stages of reading acquisition as suggested by Adams (1990) and Chall (1983).

For the Grade 4 students who had more experience with the orthographic patterns of language through their reading and spelling experience, individual differences in visual processing and memory were the most powerful factors accounting for the differences between children in the accuracy of their OI usage. These findings were consonant with the expectation that more capable readers would rely more heavily on visual processes. The performance of the oldest Grade 6 readers, however, did not support this increased dependence on visual processes.

For the older highly experienced sixth grade readers, much less overall variance was accounted for by individual differences factors. On the easy 3-letter strings, there was very little OI variance to be accounted for, and on the 6-letter strings, most of the variance was accounted by the speed and accurate retrieval of word names. A more difficult task placing greater demands on the information processing system may have yielded more meaningful findings.

Grade 2 and 4 students displayed an increasing reliance on visual processes when they processed pseudowords containing different levels of orthographic information which implies that the visual code is relatively important in reading acquisition when a new word is first encountered. This would indicate, therefore, that reader group differences also pertain to visual processing abilities. In the case of older readers (Grade 6), it is likely that this reliance decreases as words gain more visual familiarity and can be directly retrieved from the reader's lexicon.

REFERENCES

Adams, J. M. (1990). *Beginning to Read: Thinking and Learning about Print.* Cambridge, MA: The MIT Press.

Allington, R. L. (1978). Sensitivity to orthographic structure as a function of grade and reading ability. *Journal of Reading Behaviour*, 10, 437-439.

Bruder, G. A. (1978). Role of Visual Familiarity in the Word Superiority Effects Obtained with the Simultaneous-matching Task. *Journal of Experimental Psychology: Human Perception and Performance*, 4, 88-100.

Chall, J. (1983). *Stages of Reading Development*. New York: McGraw-Hill.

Corcos, E. & Willows, D. M. (1989, November). A Developmental Study of the Processing of Orthographic Information in Children with Varying Reading Ability. Paper presented at the Annual Meeting of the National Reading Conference. Austin, Texas.

Denckla, M. B. & Rudel, R. (1976). Rapid "automatized" naming (RAN): Dyslexia differentiated from other learning disabilities. *Neuropsychologia*, 14, 471-479.

Doctor, E. A., & Coltheart, M. (1980). Children's use of phonological encoding when reading for meaning. *Memory and Cognition*, 8, 195-209.

Dunn L. M. (1979). *Peabody Picture Vocabulary Test.*. Minnesota: American Guidance Services Inc.

Gardner, M. (1982). *Test of Visual Perceptual Skills (Non-motor)*. Seattle: Special Child Publications.

Geiger, G., & Lettvin, J. Y. (1987). Peripheral Vision in Persons with Dyslexia. *The New England Journal of Medicine*, 316, 1238-1243.

Krueger, L. E. (1975). Familiarity Effects in Visual Information Processing. *Psychological Bulletin*, 82, 949-974.

Leslie, L., & Shannon, A. J. (1981). Recognition of orthographic structure during beginning reading. *Journal of Reading Behaviour*, 13(4), 313-324.

Lovegrove, W., Martin, F., & Slaghuis, W. (1986). A Theoretical and Experimental Case for a Visual Deficit in Specific Reading Disability. *Cognitive Neuropsychology*, 3, 225-267.

MacGinitie, W. H., Kamous, J., Kowalski, R. L., MacGinitie, R. K., & McKay, T. (1978). *Gates-MacGinitie Reading Test (second edition)*. Chicago, Illinois: The Riverside Publishing Company.

Mason, M. (1975). Reading Ability and Letter Search time: Effects of Orthographic Structure Defined by Single-letter Positional Frequency. *Journal of Experimental Psychology: General*, 104, 146-166.

Mayzner, M. S., Tresselt, M. E., & Wolin, B. R. (1965). Tables of trigram frequency counts for various word-length and letter position combinations.*Psychonomic Monograph Supplement,* 1(3).

Niles, J. A., & Taylor, B. M. (1978). The development of orthographic sensitivity during the school years by primary grade children. In P. D. Pearson and J.

Hansen (Eds.). *Reading: Disciplined Inquiry in Process and Practice* (pp. 41-44). Celmson, S.C: National Reading Conference.

Pollatsek, A., Well, A. D., & Schindler, R. M. (1975). Familiarity Affects Visual Processing of Words. *Journal of Experimental Psychology: Human Perception and Performance*, 1, 328-338.

Raven, J. C. (1947). *Progressive Matrices*. London: Lewis

Samuels, S. J., Bremer, C. D., & Laberge, D. (1978). Units of word recognition: Evidence for developmental changes. *Journal of Verbal Learning and Verbal Behaviour*, 17, 715-720.

Samuels, S. J., Miller, N., & Eisenberg, P. (1979). Practice effects on the unit of word recognition. *Journal of Educational Psychology*, 71, 524-530.

Semel-Mintz, E., & Wiig, E. (1982). *Clinical Evaluation of Language Functioning*. Columbus, Ohio: Charles E. Merrill.

Stanovich, K. E. (1980). Toward an interactive-compensatory model of individual differences in the development of reading fluency. *Reading Research Quarterly*, 16, 32-71.

Stanovich, K. E (1988). The Right and Wrong Places to Look for the Cognitive Locus of Reading Disability. *Annals of Dyslexia*, 38, 154-177.

Schwartz, S. (1983). Spelling disability: A developmental linguistic analysis of pattern abstraction. *Applied Psycholinguistics*, 4, 303-316.

Vellutino, F. R. (1979). *Dyslexia: Theory and Research*. Cambridge, Mass: The MIT Press.

Vellutino, F. R. (1987). Dyslexia. *Scientific American*, 256(3), 34-41.

Wechsler, D. (1974). *Wechsler Intelligence Scale for Children-Revised*. New York: Psychological Corporation.

Willows, D. M. (1991). Visual Processes in Learning Disabilities. In B. Y. L. Wong (Ed.), *Learning about Learning Disabilities*. New York: Academic Press.

Facets of Dyslexia and its Remediation
S.F. Wright and R. Groner (Editors)
© 1993 Elsevier Science Publishers B.V. All rights reserved.

THE OPTIMAL VIEWING POSITION FOR CHILDREN
WITH NORMAL AND WITH POOR READING ABILITIES

M. Brysbaert and C. Meyers
University of Leuven, Leuven, Belgium

INTRODUCTION

Eye movements during reading are characterized by a succession of saccades and fixations. Saccades bring new information into the centre of the visual field; fixations serve to take up the information. Erdmann & Dodge (1898) already noted that fixations are not uniformly distributed over a text line, but fall almost exclusively on the words rather than on the gaps between the words, and "probably almost always on the middle of the word". These findings have recently received considerable attention (McConkie, Kerr, Reddix, & Zola, 1988; Rayner, 1978; Vitu, O'Regan, & Mittau, 1990) and it is now well established that there is a preferred landing position between the beginning and the middle of a word.

Part of the reason why the eye prefers to land between the beginning and the middle of a word probably is because this is the best place to recognize a word (Brysbaert & d'Ydewalle, 1988, 1991; Nazir, 1991b; Nazir, O'Regan, & Jacobs, 1991; O'Regan & Lévy-Schoen, 1987; O'Regan, Lévy-Schoen, Pynte, & Brugaillère, 1984). Nazir, O'Regan, & Jacobs (1991), for instance, found that the probability of recognizing a tachistoscopically presented nine-letter word amounted to 77% if the subjects were forced to look at the first letter, 90% for fixations on the third and the fifth letter, 77% for fixations on the seventh letter, and only 47% for fixations on the last letter. The same result is obtained when response latencies are used as the dependent variable, as can be seen in Fig. 1 (see also Brysbaert, 1991c). This figure displays response latencies for tachistoscopically presented five-letter words as a function of the letter fixated. The left part shows data of a word naming experiment; the right part shows data of a lexical decision experiment. Subjects were asked to fixate a gap between two vertically aligned lines and to process a verbal stimulus displayed between those lines. The stimulus was presented horizontally for 160 msec in such a way that a different letter fell between the fixation lines (for more information on the method, see below). In the first part of the experiment subjects had to read words aloud; in the second part they had to decide whether the presented stimulus was a word or a non-word. If we confine the discussion to the subjects with left cerebral hemisphere dominance (the most frequent case; see solid lines), we find that the optimal viewing position (OVP) for both tasks was situated on the second letter.

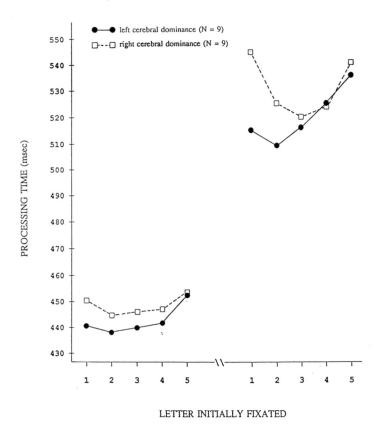

LETTER INITIALLY FIXATED

FIG. 1. Processing time (msec) of isolated five-letter words as a function of cerebral dominance of the subjects, and of the initially fixated letter. Left part: data of a naming task; right part: data of a lexical decision task.

The pattern of data displayed in Fig. 1 is explained by the joint contribution of four factors. The first factor has to do with the drop of visual acuity outside the centre of fixation. It has been shown (Alpern, 1962; Anstis, 1974; Jacobs, 1979; Wertheim, 1894) that the resolution of the visual system decreases rapidly for stimuli presented outside the fixation location, and certainly when these stimuli are flanked by other stimuli (i.e. the phenomenon of lateral masking or inhibition; Bouma, 1970). This is even true for distances of less than one degree, that is, for stimuli well within the foveal area. If the drop in visual acuity were the only significant factor, the OVP would lie in the middle of a word and processing time would be a perfect U-shaped curve of the letter initially fixated (McConkie, Kerr, Reddix, Zola, & Jacobs, 1989).

Two additional factors have been invoked to explain why the OVP lies between the beginning and the middle of a word, and why processing time usually looks more like a J-shaped than a U-shaped curve. The first of these additional factors involves the fact that there is more information in the beginning of a word than at the end, and that in most languages studied processing happens in a left-to-right order (O'Regan & Lévy-Schoen, 1987). This favours fixations towards the beginning. The second additional factor has to do with cerebral asymmetry and interhemispheric transfer, as is shown in Fig. 1 (see also Brysbaert & d'Ydewalle, 1988, 1991). Because most people have better language capacities in the left cerebral hemisphere (Bradshaw & Nettleton, 1983), fixations towards the beginning of a word demand less interhemispheric transfer than fixations towards the end of a word. This is because fixations at the beginning make the whole word fall into the right visual half field which has direct communication to the left hemisphere. On the other hand, persons with language capacities mainly situated in the right cerebral hemisphere (about 5% of the population) need less interhemispheric transfer after fixations at the end of a word. A final factor determining the OVP has to do with task requirements (Brysbaert & d'Ydewalle, 1991; Nazir, 1991a, 1991b), as can also be seen in Fig. 1. The curves are flatter for naming tasks than for lexical decision tasks. Similarly, the OVP lies more to the middle when refixation probability is taken as dependent variable than when response accuracy or latency are considered (Nazir, 1991a).

The above analysis suggests that the OVP paradigm may be used to investigate the importance of low-order processes (such as visual acuity and interhemispheric transfer) in reading. The experiment below presents the first of a series of studies in which the OVP effect for different subject samples is examined. More specifically, data of children with poor reading abilities are compared with data of age-controls and adult readers. Because the study is the first of a series, it was decided to use a free-vision task (see below) similar to normal reading.

Three outcomes were plausible on different grounds. First, it could be expected that the OVP pattern of primary school children would be very similar to that of adults, certainly for the children with normal reading capacities. This is because the basic visual functions of 8- to 12 year old children are equivalent to those of adults (Gwiazda, Bauer, & Held, 1989). Second, it could be expected that the OVP effect would be smaller for children. Children need more time to read a word (see below), so that more than one fixation is needed to process the word. This might weaken the effect of the first fixation location. Finally, it could be hypothesized that the OVP effect would be stronger for children, because they need more visual input before they are able to recognize a word.

With respect to the children with reading difficulties, an additional hypothesis of large interindividual differences might be stated. Agreement is rising that reading difficulties can be due to deficiencies in different stages of the reading process, either in the visual or in the linguistic system; and even if the same stage of the process is affected, the inadequacy need not be the same. For example, with respect to the decrease of visual acuity outside the centre of fixation, two different bases of reading difficulties have been proposed: Bouma & Legein (1977) argued that for their sample reading problems were caused by too sharp a decrease of acuity, so that the subjects experienced a kind of tunnel vision. Geiger & Lettvin (1987; see also Perry, Dember, Warm, & Sacks, 1989; Rayner, Murphy, Henderson, & Pollatsek, 1989; but see Klein, Berry, Briand, d'Entremont, & Farmer, 1990), on the other hand, maintained that for their subjects reading difficulties were due to an excessively shallow decrease of acuity, so that information over an overly large region of the visual field was sampled and peripheral stimuli interfered with the processing of the foveal stimulus.

METHOD

Subjects

There were three samples of subjects. The first sample consisted of 15 children with reading difficulties, between 8 and 12 years old (mean age 9.9 years). They were of normal intelligence and, at the time of the experiment, received treatment for their reading problems. The centre with which we collaborated had the tendency to give us their children with the mildest difficulties in order to make a good impression. This means that the subject sample was more in line with a group of poor readers than with true dyslexics. All subjects were boys. The second sample of subjects consisted of 21 male age-matched controls (mean age 10.1) who were considered to be normal readers by their school teachers. Twenty-one male undergraduate students constituted the third group of subjects.

Stimuli

The stimuli consisted of 700 five-letter words randomly divided over seven groups. Care was taken to incorporate words with which young children were likely to be familiar. Subjects were distributed over seven latin-square groups with each group seeing the words at a different fixation location (see below). All subjects processed the whole sample of stimuli, but in a different order (for the randomization procedure used, see Brysbaert, 1991a). Stimulus presentation was controlled by an IBM XT microcomputer and displayed on a Philips monochrome CRT screen.

Procedure

Subjects were placed in front of the CRT display, so that they were sitting at a normal viewing distance between 50 and 70 cm (there were no head restraints). Subjects were told to fixate a gap between two vertically aligned lines. In order to ensure adequate fixation, at a random time interval a small figure (ASCII codes 1 or 15) was presented between the fixation lines for 50 msec. Subjects had to indicate whether the figure was a laughing face or a star. They were warned by a tone if they made a mistake on these fixation control stimuli. The same procedure using digits had been applied successfully before (Brysbaert & d'Ydewalle, 1988, 1991) and is explained in greater detail in Brysbaert (1991a). As soon as the stimulus word appeared, subjects had to name it. Voice onset time was detected by a voice trigger. Unlike in the experiments leading to Figure 1, the stimuli were not presented tachistoscopically but remained on the screen until a reaction was made. This was done to assess the importance of the OVP effect in a normal reading task (see above). Incorrect responses or reaction times smaller than 200 msec and larger than 3500 msec led to a second presentation of the stimulus at a random time later in the series. A mistake on the second occasion was considered as a failure.

Words were presented in such a way that subjects initially fixated at one of seven different locations. They either looked at the blank space in front of the word, the first letter of the word, the second, the third, the fourth, the fifth letter, or the blank space behind the word. The place where the subject fixated varied randomly from trial to trial. After the response, the experimenter typed in whether the subject was correct or incorrect. The subjects did not get immediate feedback about the correctness of their answers to the words. A new trial started automatically half a second after the experimenter typed in the response. The series could be interrupted any time, in order to give the subjects a rest. Especially for the children with reading difficulties this was necessary. The experiment was divided in two sessions. After each session, the subjects were given a short rest and an overall appreciation of their performance. This consisted of feedback about the number of incorrect responses, the mean reaction time, and the number of fixation stimuli (see above) missed. The duration of the experiment varied between 40 minutes for the adult subjects to three hours for one of the poor readers (who had to be studied in two separate periods with one week in between).

RESULTS

Group Analyses

ANOVAs. Percentages of incorrect responses amounted to 0.1%, 1.1%, and 3.5% for the undergraduate students, the normal children and the children with reading difficulties respectively, which confirms the easiness of the stimuli and the rather mild deficiency of the poor readers. There were no differences between the fixation locations or the sessions, except for the poor readers who made significantly less errors in the second session (2.8%) than in the first session (4.1%).

Percentages of stimuli that had to be administered twice before they were correctly responded amounted to 2.6%, 9.5%, and 8.4% for the undergraduate students, the normal children and the children with reading difficulties, respectively. There were again no differences between fixation locations and sessions.

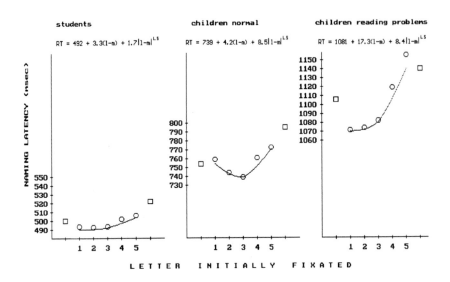

FIG. 2. Naming latency (msec) of isolated five-letter words as a function of the letter initially fixated, and the reading level of the subjects. Left part: graduate students; Middle part: normally reading children; right part: children with reading difficulties.

Reaction times of the three groups as a function of the letter location first fixated are shown in Fig. 2. These data are corrected for outliers. That is, data which after logarithmic transformation were larger or smaller than the mean plus or minus three times the standard deviation, were omitted. Percentage of outliers amounted to 1.6%, 1.7%, and 1.1% for the students, the normal children and the children with reading difficulties respectively. Visual inspection of Fig. 2 reveals that the U/J- shaped function of Fig. 1 is present for all three samples. To some extent this is surprising as naming latencies exceed 730 msec for the normal children and 1060 msec for the poor readers, which makes it quite unlikely that the words were processed within a single fixation. Still the effects remain and as we will see below become even stronger.

Separate analyses of variance for each subject sample with one between-subjects variable (latin-square Group) and two repeated measures (Session and Letter location fixated) yielded a significant main effect of Letter location for all three groups [students: $F(6,84) = 26.23$, $p < .001$; normal children: $F(6,84) = 5.48$, $p < .001$; poor readers: $F(6,48) = 4.57$, $p < .01$], and a significant Session effect for the students [session 1: RT = 518, session 2: RT = 481; $F(1,14) = 10.22$, $p < .01$]. No other effects were significant.

Analysis with a Mathematical Model.
Brysbaert and d'Ydewalle (1991) argued that ANOVAs of raw data are not well suited to investigate the effects of the different factors underlying an OVP pattern. Instead, they proposed the following model:

$$RT = a + b(l - m) + c \, |l - m|^{1.5}$$

in which   RT   = reaction time
          l   = letter location fixated
         m   = middle position of the word (i.e. the 3rd letter for 5-letter words)
       $|l-m|$ = the absolute value of l-m,
       a,b,c   = weights to be estimated via regression analysis.

In this model, the last component with the c-weight represents the importance of visual acuity. It starts from a minimum in the centre of the word, where acuity constraints are smallest, and raises symmetrically towards the beginning and the end of the word. The exponent 1.5 was preferred to the more familiar exponent 2 (i.e., the square value), because empirical data showed that an exponent of 1.5 resulted in c weights of relatively constant size for words ranging from 3 to 9 letters while an exponent of 2 would have resulted in unequal c weights decreasing with word length The second component of the model, with the b value, stands for the effect due to left-to-right processing and cerebral asymmetry (see above). If this component equals zero, the OVP pattern is perfectly U-shaped; a negative b-value results in a mirrored J-shaped curve with a minimum at the right side of the word; a positive b-value leads to a normal J-shaped curve with minimum between beginning and centre of the word. As indicated in Figure 1, the b-value usually is positive for adults. Finally, the constant a roughly coincides with reaction times after initial fixation in the middle of the word; that is, reaction times independent of the OVP manipulation.

Figure 2 contains the results of the analysis with the mathematical model for the three groups of subjects. In the analysis the two extreme positions, namely the blank space before and after the word, were omitted. Analyses of variance with two between-subject variables (reading level and latin-square group) gave a significant main effect of reading level for the constant a [students a= 492, normal children a= 739, poor readers a= 1080; $F(2,36) = 20.59$, $p < .001$], the linear component b [students b = 3.27, normal children b= 4.15, poor readers b= 17.34; $F(2,36) = 3.61$, $p < .05$], but not for the quadratic component c [students c= 1.73, normal children c= 8.50, poor readers c= 8.41; $F(2,36) = 1.68$, $p > .10$]. A posteriori tests (due to Spjøtvoll & Stoline, see Kirk, 1982, pp. 118-119) indicated that all groups differed from one another at the .05 level for the constant, but that the students and the normal children did not differ significantly from each other on the linear component. Mean percentages of variance explained by the model for individual subjects amounted to 65% for the students, 56% for the normal children, and 72% for the children with reading difficulties. The difference between fixations towards the blank space behind the words and fixations towards the blank space in front of the words amounted to 22 msec for the students, 41 msec for the normal children, and 31 msec for the children with reading problems. The differences between the three groups were not significant [$F(2,54) = 0.534$, $p > .50$].

Individual Analyses

Group analyses like the ones just described are claimed to be of limited value in neuropsychology because they are likely to obscure the large individual differences that are present in the samples (Caramazza, 1986; Caramazza & McCloskey, 1988; Shallice, 1988). Therefore Figs. 3-6 present histograms of individual data.

The constant: Fig. 3 shows the distribution of reaction times when subjects fixated in the middle of the words (i.e., the constant a). The distribution of the undergraduate students is one that should be expected from a homogeneous group: unimodal and with a reasonable scatter. The distribution of the normal children gives rise to more hetero-geneity, which can be expected from the fact that children of different ages (8 till 12) participated in the experiment. Fifty percent of the variance is explained by age differ-ences (correlation between school grade and reaction time equals -.69, n = 21, p < .01). Finally, the distribution of the children with reading difficulties shows a very large scatter. Although as a group these children are worse than the normal children (see group analyses), some of them are well within the range of children without reading problems. Two children are even in the range of the students. The fact that some children with read-ing problems do have difficulties with the processing of isolated words, and others do not, strengthens the argument that it is unwarranted to treat reading disability as a uniform syndrome. Only a non-significant 9% of the variance was explained by age differences between the poor readers (correlation between school grade and reaction time is -.29, n = 15, p > .10).

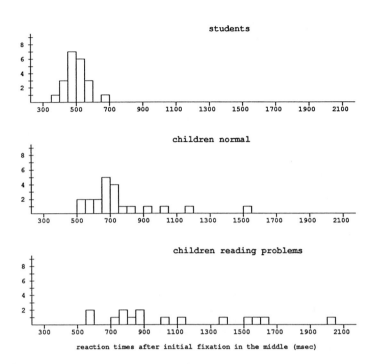

FIG 3. Distribution of the constant a for the different samples of subjects.

The linear component: Fig. 4 shows the distribution of linear components, the b values. The distribution of the students is again unimodal and with a reasonable standard deviation. The average value 3.3 is in line with the values obtained in comparable experiments (Brysbaert & d'Ydewalle, 1991). The linear component has a significant positive correlation with the constant (r = 0.65, n = 21, p < .01), so that the difference found between the linear components of the groups is also present within the sample of the students: The more time it takes for a person to name a word the larger the difference becomes between fixations at the beginning and fixations at the end of a word.

In contrast with the students, the distribution of linear components for the normal children appears to be bimodal, with one peak around -7.5 msec and another around +7.5 msec. That is, some children name words faster after initial fixations on the beginning of a word, whereas other children show better performance after initial fixations on the end of a word. We have indicated above that the linear component is due to left-to-right processing and/or to cerebral asymmetry. The linear component is not correlated with school grade (r = -.23, n = 21, p > .10), but just like for the student sample, is related to the constant a (r = 0.48, n= 21, p< .05). The latter correlation is higher when the absolute value of the linear component is taken into account rather than the raw value (r = 0.73, n= 21, p < .01).

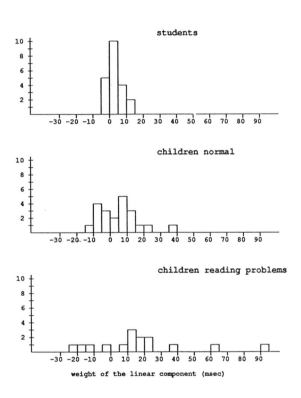

FIG. 4. Distribution of the linear slope b for the different samples.

The scatter of the data of the children with reading problems is in clear contrast with the results based on the group analysis (larger linear component for the poor readers than for the normal children and the students, see above). Data range from - 24.7 msec to +87.4 msec. As for the normal children there seem to be two groups, one with positive linear components and one with negative linear components. Also analogous to the normal children is the larger positive correlation between the absolute value of the linear slope and the constant (r = .59, n = 15, p < .05) than between the raw value of the linear slope and the constant (r = .29, n = 15, p > .10).

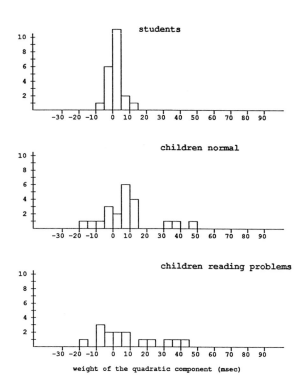

FIG. 5. Distribution of the quadratic component c for the different samples.

The quadratic component: Fig. 5 displays the distribution of the quadratic components (i.e., the c-values). For the students it is unimodal and quite narrow. Though not as clear as for the linear component, the distributions of the two children samples again appear to be bimodal. For some children (normal and poor readers) the quadratic component is clearly negative, whereas for others it is positive. A negative quadratic component means that words are processed more rapidly after a fixation at the very beginning or the very end of a word, and probably indicates that those children unlike adults prefer to process words by means of two separate fixations at the extremes. It would be interesting to test the above statement with an eye-tracking device, but because of practical limitations this has not yet been done. A further indication of the validity of the statement, however, is to be found in the correlation between the quadratic component and the school grade of the normal children. This correlation is negative and larger for the absolute value of the quadratic component than for the raw value (absolute value: r = -.50, n = 21, p < .05; raw value: r = -.37, n = 21, p < .05), which means that the importance of visual acuity decreases with increasing age. This correlation is not significant for the reading disabled children (absolute value: r = 0.13, n = 15, p > .10; raw value: r = .05, n = 15, p > .10). There is no covariation between negative scores on the linear component and negative scores on the quadratic component, either for the normal children (chi-square = 1.64, df= 1, p > .10), or for the poor readers (chi-square = 2.78, df = 1, p > .05).

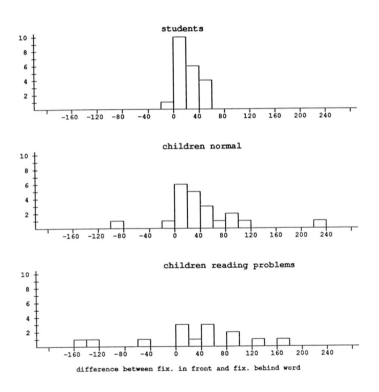

FIG. 6. Distribution of differences in reaction time after initial fixation behind and in front of the stimulus words.

The difference between fixations behind and in front of the words: Fig. 6 displays histograms of the differences in reaction times for fixations behind minus fixations in front of the words for all three groups. It was our expectation that this difference would correlate with the linear component of the model, as both components are the result of the same processes. However, only for the student population there is a moderate positive correlation (r = .34, n = 21, p < .08); For the children, the correlation is negative, and even significantly so for the children with reading problems (normals: r = -.26, n = 21, p > .10; poor readers: r = -.51, n = 15, p < .05). On the contrary, the difference between fixations behind and fixations in front of the words is positively correlated with the quadratic component for all three samples (students: r = .36, n = 21, p < .06; normal children: r = .36, n = 21, p < .06; poor readers: r = .68, n = 15, p < .01). This seems to indicate that the difference between fixations to the blank spaces in front of and behind the words is more influenced by acuity constraints, and therefore by the tendency to make an eye-movement (certainly for the child populations). Again it would be interesting to have some eye-movement data on this (but see above).

DISCUSSION

This study compared undergraduates, normal 8 to 12 year olds, and 8-12 year old poor readers on a task where subjects fixated in different conditions on each letter position in five letter words. Naming latencies for each letter position were recorded. Adults have been shown to recognize words faster when allowed to fixate between the first and middle letters of a word (Fig. 1). This U/J shaped curve is assumed to be the result of a decrease of visual acuity outside the fixation location, left-to-right processing of the words, and cerebral dominance of the subjects.

The U/J shaped curve was also found in this study with free vision for adults, poor readers and their chronological age controls (Fig. 2). Data of the adults were very similar to those obtained in other studies with tachistoscopic (Fig.1) and free vision (Brysbaert & d'Ydewalle, 1991). Manipulation of the children's initial fixation location seemed to have a larger effect than manipulation of the adults' initial fixation location, despite the fact that the children's naming latencies were much longer, even after fixation on the optimal location. This was true both for the good and the poor readers. Although multiple fixations on the words were possible, children's performance depended to a large extent on the letter first fixated. This implies that more constraints are put on a child's oculomotor behaviour during reading, if a relationship between the optimal viewing position and the optimal landing position is assumed (Nazir, 1991b).

A complicating factor, however, was, that there were large interindividual differences between the children (both good and poor readers): Some children were faster after initial fixations on the beginning of the stimulus words, others after initial fixations on the end. Similarly, some children performed better after initial fixations on the middle of the stimulus, whereas others were faster after initial fixations on the extremes. This gave rise to bimodal distributions respectively on the linear (Fig. 4) and the quadratic (Fig. 5) component. Because these differences were not expected, at the moment we can only speculate about their origin. The linear component is thought to be the result of (a) left-to-right processing of words, and (b) the cerebral dominance of the subjects (see the Introduction). Both factors lead to expect a positive component in most of the cases, as was indeed true for the students. However, a considerable percentage of the children (about 33%) showed pronounced negative components, which indicate better performance after fixations on the end of the words. So, for these children either the last half of a word was the most informative part, or they took advantage of the word being presented left of fixation (i.e., in the left visual half field). It is not clear what the basis of the former statement could be, but the latter explanation is in agreement with Bakker's theory (e.g. Bakker, 1991) that hemispheric functioning alters during reading acquisition. According to this theory, initial reading skills are controlled primarily by the right cerebral hemisphere, but as readers become more advanced, the left hemisphere becomes dominant. More controlled studies are, however, needed before firm statements can be made about this matter, also because there was no significant relation between chronological age and magnitude of the linear component for either group of children; a relationship that might have been expected on the basis of Bakker's theory.

The bimodal distribution of the quadratic component, on the other hand, seems to be less difficult to account for, because both phenomena can be explained by the same principle: If the decrease of visual acuity outside the fixation location is substantial, then a considerable penalty time will be needed for fixations on the extremes of a word. However, from a certain point on it will become more advantageous to use a dual-fixation strategy to process the word; that is, processing time will be smaller after a fixation on the beginning and the end of a word than after a single fixation on the middle. Eye-movement registrations could accredit the validity of this statement.

The interindividual differences among the children do not invalidate the statement that the OVP effect is stronger for children than for adults. They only indicate that there is more variability with respect to the optimal viewing position itself in young subjects. This can also be demonstrated by considering the absolute values of the weights rather than the raw values: For the student sample, values remain approximately the same (the linear component increases from 3.3 to 3.7, the quadratic component from 1.7 to 3.3). For the samples of children, however, weights increase dramatically (normal children: linear component 9.4 instead of 4.2, quadratic component 12.8 instead of 8.5; poor readers: linear component 24.6 instead of 17.3, quadratic component 13.7 instead of 8.4). The differences between the student and the children samples in these analyses are all reliable ($p < .01$).

In the introduction it was mentioned that three outcomes could be expected with respect to the OVP effect in children on different grounds: either a similar effect, a smaller effect, or a larger effect. The last alternative was supported by the data. This indicates that the children's performance is dominated to a larger extent by rather low level processes such as the drop of visual acuity outside the fovea and interhemispheric communication, despite the fact that their low order visual functioning is equivalent to that of adults (Gwiazda et al., 1989). Two mechanisms can account for this: Either there are less top-down influences in primary school children than in adults, or threshold values of the word recognition units are higher so that the system needs a better input before it can figure out what has been presented (the difference between the two mechanisms is that the latter can be implemented in a purely bottom-up system). Whatever the mechanism, it follows that the input must be of a better quality for a learning system than for a mature system, and more so for poor readers than for good readers. This has consequences for the interpretation of experimental results. For instance, the finding that (dyslexic) children recognize fewer letters and words after parafoveal presentation (Bouma & Legein, 1977) needs not indicate that something is wrong in the visual system. This study suggests that the same outcome may be found for deficiencies in the linguistic system. Similarly, the finding that the perceptual span during reading is smaller for children than for adults (Rayner, 1986) needs not be interpreted as a mere focusing of attention on foveal processing; It could also be due to the immature system requiring a higher quality of visual input. Other stimuli than letters and words must be utilized, if an experimenter wants to make firm statements about the importance of an adequate visual system or the focusing of attention in beginning readers. Another conclusion the results point to is that the same deficiency of the visual system will be more harmful for young readers than for skilled readers. Or to put it more positively, minor defects of the visual system that hamper reading in the beginning stages, may become tractable as the child's reading skills grow. This idea may have implications for remedial teaching, because it suggests that strengthening the linguistic capacities of a child may help overcome otherwise incurable weaknesses in the visual system.

ACKNOWLEDGMENTS

The authors would like to thank 'Het Centrum voor Neurologie en Leerstoornissen' from Neerpelt and 'Het Sint-Pieterscollege' from Leuven for their kind cooperation. They also wish to thank Ingrid Gielen for helpful comments on an earlier draft of the manuscript.

REFERENCES

Alpern, M. (1962). Muscular Mechanisms. In H. Dawson (Ed.), *The Eye*, Vol. 3. New York: Academic Press.

Anstis, S. M. (1974). A chart demonstrating variations in acuity with retina position. *Vision Research*, 14, 589-592.

Bakker, D. J. (1991). *Neuropsychological Treatment of Dyslexia.* Corby, Northants: Oxford University Press.

Bouma, H. (1970). Interaction effects in parafoveal letter recognition. *Nature,* 226, 177-178.

Bouma, H. & Legein, C. P. (1977). Foveal and parafoveal recognition of letters and words by dyslexics and by average readers. *Neuropsychologia,* 15, 69-80.

Bradshaw, J. L. & Nettleton, N. C. (1983). *Human Cerebral Asymmetry.* Englewood Cliffs, N. J.: Prentice Hall.

Brysbaert, M. (1991a). Algorithms for randomness in the behavioural sciences: A tutorial. *Behaviour Research Methods, Instruments, & Computers,* 23, 45-60.

Brysbaert, M. (1991b). In search of normal subjects with right cerebral hemisphere dominance: A case study. Manuscript submitted for publication.

Brysbaert, M. (1991c). Cerebral dominance and the processing of foveally presented words. Manuscript in preparation.

Brysbaert, M. & d'Ydewalle, G. (1988). Callosal transmission in reading. In G. Lüer, U. Lass, & J. Shallo-Hoffmann (Eds.), *Eye Movement Research: Physiological and Psychological Aspects.* Göttingen: Hogrefe.

Brysbaert, M. & d'Ydewalle, G. (1991). A mathematical analysis of the convenient viewing position hypothesis and its components. In R. Schmid & D. Zambarbieri (Eds.), *Oculomotor Control and Cognitive Processes: Normal and Pathological Aspects.* Amsterdam: North-Holland.

Caramazza, A. (1986). On drawing inferences about the structure of normal cognitive systems from the analysis of patterns of impaired performance: The case for single-patient studies. *Brain and Cognition,* 5, 41-66.

Caramazza, A. & McCloskey, M. (1988). The case for single-patient studies. *Cognitive Neuropsychology,* 5, 517-528.

Erdmann, B. & Dodge, R. (1898). *Psychologische Untersuchungen über das Lesen.* Halle: Niemeyer.

Geiger, G. & Lettvin, J. Y. (1987). Peripheral vision in persons with dyslexia. *New England Journal of Medicine,* 316, 1238-1243.

Gwiazda, J., Bauer, J., & Held, R. (1989). From visual acuity to hyperacuity: A 10-year update. *Canadian Journal of Psychology,* 43, 109-120.

Jacobs, R. J. (1979). Visual resolution and contour interaction in the fovea and periphery. *Vision Research,* 19, 1187-1196.

Kirk, R. E. (1982). *Experimental Design: Procedures for the Behavioural Sciences.* Monterey, CA: Brooks/Cole.

Klein, R., Berry, G., Briand, K., d'Entremont, B., & Farmer, M. (1990). Letter identification declines with increasing retinal eccentricity at the same rate for normal
and dyslexic readers. *Perception & Psychophysics* ,46, 601-606.

McConkie, G. W., Kerr, P. W., Reddix, M. D., & Zola, D. (1988). Eye movement
control during reading: I. The location of initial fixations on words.*Vision Research*, 28, 1107-1118.

McConkie, G. W., Kerr, P. W., Reddix, M. D., Zola, D., & Jacobs, A. M. (1989). Eye
movement control during reading: II. Frequency of refixating a word. *Perception
& Psychophysics*, 46, 245-253.

Nazir, T. A. (1991a). On the role of refixations in letter strings: The influence of
oculomotor factors. *Perception & Psychophysics*, 49, 373-389.

Nazir, T. A. (1991b). On the relation between the optimal and the preferred viewing
position in words during reading. Paper presented at the Sixth European Conference on Eye Movements, Leuven.

Nazir, T. A., O'Regan, J. K., & Jacobs, A. M. (1991). On words and their letters.
Bulletin of the Psychonomic Society, 29, 171-174.

O'Regan, J. K. & Lévy-Schoen, A. (1987). Eye-movement strategy and tactics in word
recognition and reading. In M. Coltheart (Ed.), *Attention and Performance XII:
The Psychology of Reading*. London: Erlbaum.

O'Regan, J. K., Lévy-Schoen, A., Pynte, J., & Brugaillère, B. (1984). Convenient
fixation location within isolated words of different length and structure. *Journal of
Experimental Psychology: Human Perception and Performance*, 10, 250-257.

Perry, A. R., Dember, W. N., Warm, J. S. & Sacks, J. G. (1989). Letter identification
in normal and dyslexic readers: A verification.*Bulletin of the Psychonomic
Society*, 27, 445-448.

Rayner, K. (1978). Eye movements in reading: A tutorial review. *Psychological Bulletin*,
85, 618-660.

Rayner, K. (1986). Eye movements and the perceptual span in beginning and skilled
readers. *Journal of Experimental Child Psychology*, 41, 211-236.

Rayner, K., Murphy, L. A., Henderson, J. M., & Pollatsek, A. (1989). Selective
attentional dyslexia. *Cognitive Neuropsychology*, 6, 357-378.

Shallice, T. (1988). *From Neuropsychology to Mental Structure*. Cambridge:Cambridge
University Press.

Vitu, F., O'Regan, J. K., & Mittau, M. (1990). Optimal landing position in reading
isolated words and continuous text. *Perception & Psychophysics*, 47, 583-600.

Wertheim, T. (1894). Ueber die indirekte Sehschärfe. *Zeitschrift für Psychologie*, 7,172.

Facets of Dyslexia and its Remediation
S.F. Wright and R. Groner (Editors)
© 1993 Elsevier Science Publishers B.V. All rights reserved.   125

VISION IN DYSLEXICS: LETTER RECOGNITION ACUITY, VISUAL CROWDING, CONTRAST SENSITIVITY, ACCOMMODATION, CONVERGENCE AND SIGHT READING MUSIC

J. Atkinson

Visual Development Unit,

University of Cambridge, UK

INTRODUCTION

Both in sight reading music and in reading text the same general strategy is used: significant structural units must be visually identified in successive fixations and translated into a different code (discussed in Sloboda, 1989). In reading text the phonological code is used to form words, in music pitch changes form the melody. Fluent readers preview whole words, while competent music sight readers preview whole phrases made up of a number of bars. An eye-voice span can be calculated for reading both text and music, with text comprehension being analogous to tune recognition. This span is claimed to be between 4-6 words for competent readers (Levin & Addis, 1980) and 4-6 notes for proficient sight readers (Sloboda, 1974). Of course both visual parameters (e.g. contrast) and complexity of material will alter the ability of the reader to 'chunk' the text into performance units. Efficiency on both tasks can be reduced by poor acuity, contrast sensitivity, and eye movement control. Given these underlying similarities between sight reading and reading text, it seems likely that the type of problems encountered by dyslexics in reading text might carry over to sight reading music. There have been alternative suggestions that because of hemisphere specialization, difficulties in reading text aloud and sight reading music could be dissociated in individuals. Gordon & Bogen (1974) found that left hemisphere intervention caused speech disturbances whereas right hemisphere intervention caused disruption in singing. However, Wertheim & Botez (1961) reported a vascular left brain injury causing right hemiplegia in a right handed professional violinist who claimed prior to the injury to have perfect pitch. After the injury he lost perfect pitch, together with the ability to name musical intervals. He seemed to be able to read musical notes separately but could not integrate them into a pattern or melody. Others have claimed that differential hemisphere control in music is dependent in the level of musical ability. Bever and Chiarello (1974) presented melodies to the right and left ear and found a superiority in non-musicians in the left ear and the reverse in trained musicians.

It seems clear that both hemispheres are involved in the very complex process of text or music recognition and it is quite plausible that deficits in different stages of the process might be found in different dyslexics and for that matter in different controls. Nevertheless, the possibility exists that dyslexics might be able to use a strategy involving musical coding to help them to read and to remember text. There are anecdotal accounts to this end - one folk tale tells of a dyslexic medical student who could only read his anatomy text if he sang it out loud to Gilbert and Sullivan melodies!

In this chapter two preliminary studies of dyslexics' ability on visual tasks are reported. In the first study their ability to recognize melodies from both sight reading music and from hearing the tunes is compared with that of musically-matched and chronological age-matched controls. In the second study a number of basic visual capacities are compared in a group of dyslexics and in a group of controls matched for non-verbal intelligence. In both studies the dyslexic children studied had been initially diagnosed 'dyslexic' in that they showed two years' minimum difference between their mental age according to their non-verbal intelligence scores (WISC and Ravens Progressive Matrices) and their reading age. At the time of testing these children still showed at least 12 months' difference between reading age (assessed by using Test A of the British Abilities Scale) and the performance age norm attained from the Standard Progressive Matrices (Ravens), having had intensive remedial teaching. Many of these children showed Ravens scores above their chronological age, suggesting that the general level of intelligence in these groups was high. The children in these studies attended a school in which all children were given musical instruction (King's College Choir School, Cambridge).

STUDY 1: MUSICAL TEXT

Twenty-one dyslexic children were tested along with 21 musically-matched controls (age range 7-13 years, 19 male, 2 female). All children had started musical tuition on at least one instrument. All children had some experience of attempting to read music. Age matching was within 18 months of chronological age. The mean age of the dyslexic group was 10.2 years and that of the controls 9.8 years, the difference between the groups being insignificant. Musical match was on the basis of the same number of years' teaching on the same musical instrument (or in some cases the same 'type' of instrument e.g. both string or both wind instruments). For example a nine year old dyslexic child who had had piano instruction for one year could be matched with an eight-year old control with the same training on the piano. Whenever possible (in over 70% of the group) dyslexics were matched with children from the same academic year and same class (classes being academically 'streamed'). Although children were not formally matched in this study for nonverbal IQ, the fact that the majority of children were matched for the same mathematics stream and academic stream (excluding English) means that their general level of intelligence was comparable across groups. It was not possible to match IQ strictly and at the same time maintain a good music-match across instruments in a school of this sort. Many were matched for the same music teacher, although 23 different teachers were involved. The experimenters were 'blind' as to whether the child they were testing was from the dyslexic or control group. Each child carried out three tasks :

Auditory recognition task. Melodies of 10 familiar tunes (e.g. The National Anthem, Happy Birthday to You) were presented on a BBC Microcomputer in random order for auditory recognition. The children were instructed to press a key when they were ready to listen to the tune and to press the space bar as soon as they recognised the tune. Several practice trials were given. All children commented that this task was 'easy' but some commented that they could not remember the words. They were encouraged to press as soon as the tune was 'recognised' even if they could not remember the words. A number of different responses (saying some of the words in the appropriate order and rhythm, continuing to sing the tune) were taken as evidence of recognition. The reaction times for the tunes were recorded in seconds from the computer record. This score was normalized across different tune lengths by calculating the percentage of notes of the complete melody heard before the space bar was depressed. The same scoring system was used for the sight reading tasks (see below).

Sight reading tasks. Relatively large, high contrast musical manuscript notation was used. The stave was 1 cm in height, with each note 0.8 cm high and 0.2 cm wide. The inter-note space was 1 cm. From pilot work the stave used was that supposed to be optimal for visibility. The music was written in different clefs appropriate to different instruments. The children were required to sight read both by singing and by playing their instruments. They were asked to choose which task they attempted first. Nearly all the children chose to sight play first rather than sight sing, but all the children completed both tasks. They were asked to stop playing or singing the tune as soon as they recognized it. The 10 tunes were given in random order after 3 practice trials.

From comparisons across tasks and groups, analysis of variance showed a significant difference between dyslexics and controls (p< 0.05). The group means are shown in Fig. 1. Post hoc (Newman-Keuls) comparison of means revealed no significant difference between the two groups on the auditory recognition test but a significant difference on both sight reading tests (at the 0.05 level). Interestingly, both groups found the auditory task easier than either sight reading task, and sight playing was found to be easier than sight singing. This was in spite of the fact that all but 2 subjects chose sight playing before sight singing and in spite of the fact that the auditory recognition task was completed before either sight reading task.

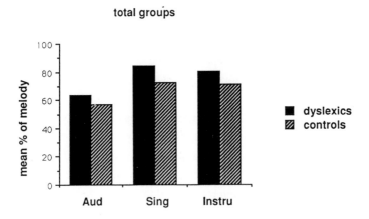

FIG. 1. Mean percentage of melody required for recognition in auditory task and two sight reading tasks for all subjects.

It was noted that six dyslexic children did not recognize any of the tunes on either sight reading task in spite of having a minimum of 3 school terms (approximately 9 months) musical tuition - in fact these children could not sight read at all. A reanalysis comparing the remaining groups, with the non-sight readers and their partners removed from the analysis, revealed no significant difference between the two groups (dyslexics and controls) on any of the tasks (Fig. 2), although the mean scores were still considerably lower for the dyslexics on both sight reading tasks.

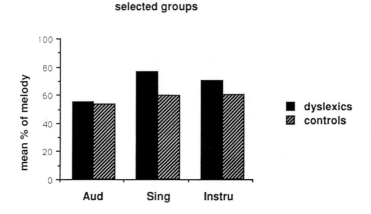

FIG. 2. Mean percentage of melody required for recognition in auditory task and sight reading tasks for subset of dyslexics (and their controls) who attempted sight reading.

In these remaining groups (children with some ability to sight read) the mean scores for auditory recognition were very close for dyslexics and controls and much closer than in the first analysis of the whole group.

Comparing the results from Figs. 1 and 2, suggests that many dyslexics show marked impairment on sight reading tasks and that some subset may also show poor auditory recognition compared to music-matched controls (causing the auditory score mean to be lower in the whole group analysis than in the subgroup analysis of the dyslexics). The remaining dyslexics do not show impairments in the auditory recognition task but are still poorer on average than controls on sight reading tasks. Presumably these individuals may attempt to compensate for their difficulties in sight reading by learning music in the auditory domain and remembering it. Although no further formal analysis of sight reading problems was carried out in this study, many spontaneous comments made by the children were revealing and require further, more detailed studies. Dyslexics commented approximately five times as often as controls on 'notes running into each other', 'jumping around on the stave', and 'becoming blurry', and it was also noticed that dyslexics frequently played two notes in reverse order.

These comments suggest that there is a close link between the visual parameters in the sight reading task and those in other text reading tasks. Consequently a second study was run (described below) in order to look at more basic visual capacities in dyslexics, including single letter recognition.

STUDY 2: VISUAL CAPACITIES IN DYSLEXICS

The Cambridge Crowding Cards Test is standardized for testing acuity at distance at 3 metres viewing and gives a value in adults equivalent to Snellen acuity (Atkinson, Anker, Evans, Hall & Pimm-Smith, 1988; Anker, Atkinson & MacIntyre, 1989). It has also been standardized as a near acuity test at 30cm viewing. In both tests the cards contain one of five letters, varied in size, and surrounded by four adjacent letters of the same size. The spacing of the adjacent letters is a constant ratio of the letter size. A psychophysical staircase procedure is used to measure the acuity threshold. The smallest central letter correctly identified in this procedure is taken as the crowded acuity measure and can be compared with the size of letter identified without the surrounding letters (single letter test). In general, crowded letter acuity is somewhat poorer than single letter acuity. The difference is small in adults and children over 6 years and large in amblyopes (e.g. Atkinson, Pimm-Smith, Evans, Harding & Braddick, 1986; Mayer & Gross, 1990; Stager, Everett & Birch, 1990).

Flom (1991) has argued that the term 'crowding' includes contour interaction, attentional factors and problems of fixational eye movements. He states that the main difference between contour interaction and crowding is that in the former the eye is stationary, fixating a single target, whereas in a crowding task eye movements are required to fixate particular elements in the array. This distinction is a little hard to apply to the Cambridge Crowding Cards. Indeed, a single target must be fixated in the centre of the circular array, but that target has to be selected and the other surrounding letters ignored. We have already demonstrated that for adults, surrounding letters and surrounding line segments produce similar degrees of crowding, whereas for preschool children more marked degradation is produced by letters (Anker et al., 1989). This result would seem to suggest that an additional attentional selective process is immature in preschool children, the magnitude of the effect being dependent on the nature of the unattended parts of the display and the comparison in terms of perceptual category between the target and distractors. For these reasons it would seem preferable to maintain the term 'crowding' when referring to the Cambridge Crowding Cards task rather than using the more limited term 'contour interactions'.

As well as possible differences in attentional factors between adults and children, it is also likely that eye movement control is poorer in children than adults (e.g. Kowler & Martins, 1982). It seems likely that all these effects included under crowding would be increased in dyslexics and play a role in poor reading (see discussions in Wolford & Chambers, 1983; Geiger & Lettvin, 1987; Atkinson, 1991a). In a previous study (Atkinson, 1991a) a subgroup of dyslexics was shown to have high levels of crowding compared to both age-matched controls and to a group of much younger children (6-7 years) with a lower reading age than any of the dyslexics. Interestingly, the increased crowding was much more marked in monocular testing than in binocular. In monocular viewing convergence cues cannot be used to aid the accurate accommodation which is necessary for this task.

Preschool children who have had uncorrected marked hyperopic refractive errors throughout infancy also show abnormal levels of crowding, similar to this subset of dyslexics (Atkinson, in press; Atkinson, Braddick, Wattam-Bell, Durden, Bobier, Pointer & Atkinson, 1987; Atkinson & Braddick, 1988). In addition many of these children who had been infant hyperopes showed poor control of accommodation and a high incidence of strabismus. Because of this clustering of visual defects in preschool children, we thought it was plausible that similar results might be found in a variety of measures in a subset of dyslexics. However, it is not possible to infer clear causal links in either of the studies of the dyslexics, as their pre-school vision history is not known in detail. Consequently the study described below was regarded as a preliminary one to investigate a small group of dyslexics across a relatively wide range of vision tests. The following tests were carried out on 13 dyslexic children and 13 reading age-matched controls. Two dyslexic children did not complete two of the tests, while two control children were absent for one test.

Letter recognition acuity for displays with both single letters and crowded letters (Cambridge Crowding Cards) at near and far distances. The tests were conducted with monocular and binocular viewing. For the crowded and single letters a value equivalent to 6/6 or better was taken as the norm for this age group. Most children had better single letter acuity than this value.

Pelli-Robson Contrast sensitivity letter chart. This test (Pelli et al., 1988) can be used to measure monocular and binocular contrast sensitivity in any individual who is able to name the letters of the alphabet. The chart shows a random array of letters with systematic reduction in contrast from top to bottom of the chart. For binocular viewing, a Weber contrast of 1.5% was taken as the norm; for monocular viewing this value was 2.2%. These values are slightly below the norms for student adults (Pelli et al., 1988).

Changes of accommodation and convergence were assessed in two ways: (a) videorefraction and (b) RAF rule. We have previously used photorefraction and videorefraction to look at the consistency and accuracy of accommodation in a number of different populations (Atkinson, 1985, 1989; Braddick et al., 1979). In the Cambridge Infant screening programme accommodative behaviour was gauged from photorefraction without cycloplegia, which was carried out at 9 months of age. The infant looked at a target close to the camera at 75 cm distance (Braddick et al., 1988). While the control infants (who did not have marked refractive errors under cycloplegia) generally focused close to the camera and sometimes slightly in front of it, a significant proportion of the infants with marked hyperopic refractive errors maintained a more distant position of focus than the target at 75 cm even without cycloplegia. This implies that, at least in some cases, infant hyperopes are not accommodating to the full extent necessary to achieve maximally clear vision.

This increase in accommodative lag in hyperopic infants has been confirmed by Bobier, Atkinson & Braddick, (1988). Bobier also found that while the accommodative lag increases for emmetropic infants when they view near targets (30-40 cm), some hyperopic infants show a decreased lag and may even overaccommodate at these distances. It appears, then, that the accommodative behaviour of hyperopes may be more complicated than would be expected from a simple account of failing to meet excessive accommodative demands. Recent studies in our laboratory of the reliability of accommodative responses in infants with marked hyperopic errors reveal fluctuations in accommodation so that on some occasions sufficient accommodative effort is made, whereas on other occasions hyperopic focus persists. Studies on the reliability of accurate accommodative responses in children with severe neurological deficits show two characteristic modes of behaviour in different sub-groups (Atkinson, 1985). In one group the accommodative responses resemble newborn behaviour. Here the child tends to focus at a near distance and make some changes of accommodation in the correct direction for changes in target distance in the near range but little change of accommodation for targets beyond 75 cm. The second group tend to show marked fixed hyperopic accommodation resembling their cycloplegic refractions. Here we looked at the reliability of responses to targets at distances between 20 and 150 cm. Three tests at each target distance were given, with inconsistent accommodation being recorded if on any of these occasions accommodation did not change in the appropriate direction for target distance compared with accommodation at 75 cm.

The near point of convergence and near point of accommodation were tested using the RAF Near Point Rule. The convergence was tested binocularly, both objectively and subjectively, the subject being asked to state when he experienced diplopia, and the observer noting when convergence failed. The near point of accommodation was measured monocularly and binocularly using N5 print. The print was moved closer to the eyes and the subject was asked to indicate when the print became too blurred to read. The test was repeated three times for each condition (ie binocular, right eye only, left eye only), and the blur point noted. If accommodation is below normal both monocularly and binocularly then accommodative insufficiency is indicated, and progressive recession of the near point of accommodation would indicate the presence of accommodative fatigue. We took as the normal for the age group near point of convergence of 6 cms, and near point of accommodation of 8 cms or better. None of the children tested was known to have marked uncorrected refractive errors as ascertained from recent optometric examinations carried out outside the study.

Orthoptic examination. A standard orthoptic examination was carried out including a Hirschberg test (looking at the corneal reflexes to gauge manifest strabismus), a cover test, ability to overcome 20 dioptre base out prisms with each eye, ability to make full eye movements, and the TNO Test for measuring stereoscopic acuity.

RESULTS

An overall comparison of the results for the dyslexics and controls is shown in the histogram in Fig. 3. For each part of the examination a criterion pass score has been applied, as stated above, and the percentage of children in each group reaching this criterion has been calculated. As some children did not complete all tests, in some cases the total number of children compared is 11 or 12, rather than 13.

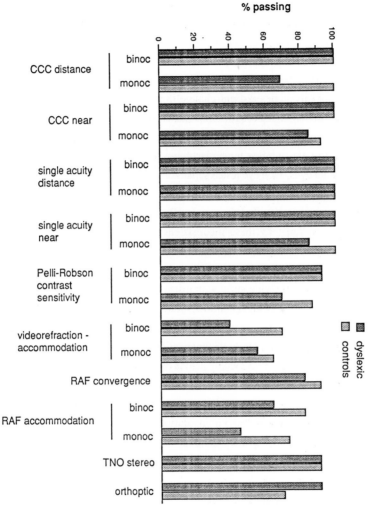

FIG. 3. Percentage of children in each group who reached passing criterion on each visual test.

Single and crowded letter acuities. None of the children showed reduced binocular acuity (worse than 6/6 equivalent) at distance or at near on the Cambridge Crowding Cards. However, four children from the dyslexic group showed reduced monocular acuity on the crowded test at distance and two of these showed the reduction in both eyes. At monocular near testing one of the controls showed reduced acuity in one eye, and two of the dyslexics failed the test in both eyes. This result is similar to that found in a previous study of a different group of dyslexics in which a subset of children showed reduced monocular acuity (Atkinson, 1991a).

Contrast sensitivity . On binocular testing one child out of each group failed to reach the pass criterion; on monocular testing four dyslexics and two controls failed the test.

Changes of accommodation. On the videorefractive tests of changes of accommodation 8 out of the 13 dyslexics did not change their focus appropriately and consistently for targets at different distances for binocular viewing, and 5 out of 11 did not accommodate consistently monocularly (three children failed in both eyes). In comparison, 4 out of 13 in the control group failed binocularly and 4 out of 11 monocularly. One control child failed in both eyes, each eye tested separately.

It may be argued that failure on monocular acuity tasks at distance and failure to make consistent accommodative shifts may be all part of the same deficiencies in eye movement control. However, why these inconsistencies should be so much worse for accurate accommodation in the absence of a convergence cue (i.e. in monocular viewing) for one test (acuity) but also be poor when accommodation can be aided by convergence for nearby targets (in the videorefraction tests carried out binocularly) remains a puzzle. In general the failures to shift accommodation were for a target at 150 cm rather than for the target at 30-40 cms. No attempt was made in this study to measure the accuracy of accommodation at each target distance, although such measures can be made using the videorefractive technique. Two dyslexics and one control child failed to reach the norm on the convergence test using the RAF rule. Four dyslexics compared to 2 controls showed very variable accommodation on the 3 binocular measures made in this test. On monocular testing, variable accommodation was shown by 6 dyslexics and 3 controls. Some showed poorer responses the first time the test was given whereas others seemed to decline in performance as testing continued, showing marked accommodative fatigue.

Orthoptic examination. The dyslexic group showed no failures on the orthoptic tests. Two of the control children showed exophoric responses on the Cover Test and one control child failed to make a corrective eye movement with one eye in the prism test. One control and one dyslexic child showed stereo acuity poorer than 60 seconds on the TNO Stereo Test.

SUMMARY OF RESULTS

So far it appears on a number of visual tasks that some dyslexics show mild visual deficits (e.g. only one line down on monocular acuity testing from the age norms).

However, on most tests where there were failures among the dyslexics there were also one or two failures in the control group. The most sensitive tests would appear to be those making highest demands on consistent accommodation. However, it is possible that problems of controlling accommodation are indicative of more fundamental underlying problems of general poor visuo-motor control, possibly linked to control of visual attention. This has been suggested from a number of different studies on eye movement control in dyslexics (e.g. Stein, 1991). The causal links for these associated faults have yet to be fully determined.

CONCLUSIONS

As both these studies can be regarded only as starting points, our conclusions to date are very limited. In the melody recognition study we have shown a clear decrement in a subset of dyslexics in sight reading compared to controls, matched in terms of musical instrument and training. However, although there is a small subset of dyslexics who show very poor recognition in any of the three domains tested (auditory recognition, sight singing, sight reading), on total group comparisons we did not find significantly poorer performance in the dyslexics on the auditory recognition task. It seems quite plausible that many dyslexics avoid sight reading tasks and prefer instead to learn melodies 'by ear' so that they can easily reproduce them. We have yet to explore the details of the deficits shown on the sight reading tasks in both controls and dyslexics, to determine whether the dyslexics are merely slower to learn and to use appropriate sight reading strategies or whether they show aberrations in performing some part of the task. It is quite possible that severe visual discomfort accompanies sight reading for some dyslexics and that this in itself deters them from practising sight reading skills (e.g. Wilkins, 1991).

We have also shown mild deficits on sensory and perceptual vision tasks in a subgroup of dyslexics. These deficits seem to appear in acuity tasks with monocular viewing. Even more marked are deficits in some dyslexics in simple changes of focus for targets at different distances, and accommodative fatigue was seen in a number of this group in the RAF test. Whether the fault is in the accommodative mechanisms *per se* or in links between accommodative and convergence mechanisms or between accommodation and attentional stability, or in combinations of these problems, remains to be discovered.

ACKNOWLEDGEMENTS

I would like to thank all members of the Visual Development Unit who have discussed, carried out and assisted in this research - in particular Mrs Shirley Anker (for vision testing including orthoptic tests), Ms Kate Watkins and Ms Susie Fowler Watts (for programming and testing in the music study) and Dr Oliver Braddick (for helpful discussions), I would like to thank the Headmaster, Mr Gerald Peacocke, and Staff of King's College School for assisting in the running of the dyslexic studies and in particular Mrs Beryl Wattles (Dyslexia Centre) and Ms Charmian Farmer (Music Director) for their help and assistance. I also thank Professor Alan Baddeley (Director, MRC Applied Psychology Unit, Cambridge) and Ms Claudia de Val for their collaboration concerning intelligence and reading tests and the selection of appropriate control groups, and Dr Arnold Wilkins for helpful discussions concerning visual parameters in the musical text.

This research was supported by the Medical Research Council, UK.

REFERENCES

Anker, S., Atkinson, J. & MacIntyre, A. M. (1989). The use of the Cambridge Crowding Cards in preschool vision screening programmes, ophthalmic clinics and assessment of children with multiple disabilities. *Ophthalmic and Physiological Optics*, 9.

Atkinson, J. (1985). Assessment of vision in infants and young children. In S. Harel & N. J. Anastasiow (Eds.) *The At-Risk Infant: Psycho/Socio/Medical Aspects*. Baltimore: Paul H Brookes Publishing Co.

Atkinson, J. (1989). New tests of vision screening and assessment in infants and young children. In J.H. French, S. Harel and P. Casaer (Eds.) *Child Neurology and Development Disabilities*: 219-227. Baltimore: Paul H. Brookes Publishing Co.

Atkinson, J. (1991a). Review of human visual development: crowding and dyslexia. In J. F. Stein (Ed.)*Vision and Visual Dyslexia*: 44-57. Vol 13 of *Vision and Visual Dysfunction* edited by J. R. Cronly-Dillon. Macmillan Press.

Atkinson, J. (in press) Infant vision screening: prediction and prevention of strabismus and amblyopia from refractive screening in the Cambridge photorefraction programme. In K. Simons and D. L. Guyton (Eds.) *Infant Vision: Basic and Clinical Research*. Oxford University Press.

Atkinson, J. & Braddick, O.J. (1988). Infant precursors of later visual disorders: correlation or causality ? In A.Yonas (Ed.) *20th Minnesota Symposium on Child Psychology*. Hillsdale, N.J.: Lawrence Erlbaum Associates.

Atkinson, J., Pimm-Smith, E., Evans, C., Harding, G. & Braddick, O. J. (1986). Visual crowding in young children. *Documenta Ophthalmologica Proceedings Series 45*: 201-213.

Atkinson, J., Braddick, O. J., Wattam-Bell, J., Durden, K., Bobier, W., Pointer, J. & Atkinson, S. (1987). Photorefractive screening of infants and effects of refractive correction. *Investigative Ophthalmology and Visual Science (supplement)* 28: 399.

Atkinson, J., Anker, S., Evans, C., Hall, R. & Pimm-Smith, E. (1988). Visual acuity testing of young children with the Cambridge Crowding Cards at 3 and 6 months. Acta *Ophthalmologica* 66: 505-508.

Bever, T. G. & Chiarello, R. J. (1974). Cerebral dominance in musicians and non-musicians. *Science* 185: 537-539.

Bobier, W. R., Atkinson, J. & Braddick, O. J. (1988). Eccentric photorefraction: a method to measure accommodation in highly hypermetropic infants. In *Technical Digest of the 4th Topical Meeting on Non-Invasive Assessment of Visual Function*. Nevada, USA: Optical Society of America.

Braddick, O. J., Atkinson, J., French, J. & Howland, H. C. (1979). A photo-refractive study of infant accommodation. *Vision Research* 19: 319-330.

Braddick, O.J., Atkinson, J., Wattam-Bell, J., Anker, S. & Norris, V. (1988). Video-refractive screening of accommodative performance in infants. *Investigative Ophthalmology and Visual Science* (supplement) 29: 60.

Flom, M.C. (1991). Contour Interaction and the Crowding Effect. *Problems in Optometry* 3(2): 237-257.

Geiger, G. & Lettvin, J. Y. (1987). Peripheral vision in persons with dyslexia. *New England Journal of Medicine* 316: 1238-1243.

Gordon, H.W. & Bogen, J. E. (1974). Hemispheric lateralisation of singing after intracarotid sodium amylobarbitone. *Journal of Neurology, Neurosurgery & Psychiatry,* 37: 727-738.

Kowler, E. & Martins, A. J. (1982). Eye movements of preschool children. *Science ,* 215: 997.

Levin, H. & Addis, A. B. (1980). *The Eye-Voice Span.* Cambridge, Mass: MIT Press.

Mayer, D. L. & Gross, R.D. (1990). Modified Allen pictures to assess amblyopia in young children. *Ophthalmology ,* 97: 827.

Pelli, D.G., Robson, J. G. & Wilkins, A.J. (1988). Designing a new letter chart for measuring contrast sensitivity. *Clinical Vision Sciences ,* 2: 187-199.

Sloboda, J. A. (1974). The eye-hand span: an approach to the study of sight-reading. *Psychological Music,* 2: 4-10.

Sloboda, J. A. (1989). *The Musical Mind.* Oxford: Clarendon Press

Stager, D.R., Everett, M. E. & Birch, E.E. (1990). Comparison of crowding bar and linear optotype acuity in amblyopia. *American Orthoptic Journal,* 40: 51.

Stein, J. F. (1991). Visuospatial sense, hemispheric asymmetry and dyslexia. In J. F. Stein (Ed.),*Vision and Visual Dyslexia*. Vol 13 in the series *Vision and Visual Dysfunction*. Basingstoke: Macmillan.

Wertheim, N. & Botez, M. I. (1961). Receptive amusia: a clinical analysis. *Brain* , 84: 19-30.

Wilkins, A. (1991). Visual discomfort and reading. In J. F. Stein (Ed.), *Vision and Visual Dyslexia*. Vol 13 in the series*Vision and Visual Dysfunction*. Basingstoke: MacMillan Press Ltd.

Wolford, G. & Chambers, L. (1983). Lateral masking as a function of spacing. *Perception & Psychophysics* , 33: 189-226.

Facets of Dyslexia and its Remediation
S.F. Wright and R. Groner (Editors)
139

FIXATION, CONTRAST SENSITIVITY AND CHILDREN'S READING

P. Cornelissen,

Physiology Department, Oxford University

INTRODUCTION

Fluent reading demands that visual formation, gleaned from a sequence of binocular fixations and saccades, is rapidly integrated with linguistic information. Though some children acquire this skill before they begin school, as many as 3-10% of children in the U.K. fail to learn to read despite several years schooling (Rutter & Yule, 1975). Given that reading is both a visual and a linguistic task, it is plausible to ask whether both visual and language processing problems may contribute to these childrens' difficulties.

It is now clear that deficient language processing, in particular poor phonemic awareness, does contribute to childrens' reading problems (Bradley & Bryant, 1978; Lundberg, Olofsson & Wall, 1980; Wagner & Torgeson 1987; Liberman, Shankweiler, Fischer & Carter, 1974). Furthermore, the elegant single case study of J.M. by Snowling, Hulme, Wells & Goulandris (1991) revealed an intimate link between abnormal speech development and this subjects' poor phonological skills, hence his poor spelling and reading. But, the question whether visual problems might also contribute to childrens' reading difficulties remains unanswered.

On the one hand, many investigators have failed to discriminate normal readers from disabled readers on the basis of their performance on visual tasks (Benton, 1962, 1975; Vellutino, 1973, 1975a, 1975b). On the other hand, when either the dynamic processing of the retinal image (Brannan & Williams, 1988; Lovegrove, Martin & Slaghuis, 1986; Williams, Molinet & LeCluyse, 1989; Breitmeyer, 1980; Geiger & Lettvin, 1987; Ruddock, 1990), or eye-movement control, (Pirozzolo & Rayner, 1988; Fischer & Weber, 1990; Stein & Fowler, 1982; Bigelow & Mckenzie, 1985) have been studied, strong correlations between poor reading and poor visual performance emerge.

These opposing conclusions are probably not the result of genuine conflict, because the kinds of experiments from which the conclusions were drawn were methodologically very different. Broadly, in the first group of experiments where no differences were found, poor and normal readers were compared on tasks of static pattern processing. By contrast, differences between normal and poor readers do emerge once they are compared on visual tasks which depend on events changing rapidly in time (Williams et al., 1989; and see Lovegrove et al., 1986 for review). Similarly, it is the temporal aspects of auditory perception which have also been shown to differ between normal and poor readers (Tallal, 1976). However, as Hulme (1988) and Bishop (1989) emphasize, even if good and poor readers can be distinguished on the basis of their visual performance, we are still left asking whether there is any causal relationship between poor visual processing and poor reading.

In this chapter we try to address this issue of causality directly. In doing so we have adopted a somewhat unusual approach in that we have not attempted to compare the performance of dyslexics with normal readers on a visual task. Instead we have asked a much simpler question: 'Do childrens' visual problems affect how they read?'. For, if visual problems really do interfere with children's reading, then this influence should be independent of their overall reading ability. Any child should find reading easier if s/he can reliably see what is printed on the page.

THE DUNLOP TEST AND READING

The visual impairment dealt with here is unstable binocular fixation (Stein, Riddell & Fowler, 1987). Children with unstable binocular fixation are said to find it more difficult to maintain steady gaze than children with normal vision. Stein and Fowler have used the Dunlop Test (DT) as an indirect way of grouping children into those who do and those who do not have unstable binocular fixation. (For details of test administration see Stein & Fowler, 1982). Used in this way, there are two main findings to date which relate DT performance to childrens' reading ability. They are:

1) 753 unselected primary school children were divided into four chronological age groups (Stein & Fowler, 1986). They performed the DT and had their reading abilities assessed. As can be seen from the histogram in Fig. 1, within each chronological age group, children who failed the DT were significantly worse readers than children who passed it

2) Occlusion of one eye for close work, for an average period of 3 to 6 months, improves performance in the DT. This improvement is correlated with a subsequent, marked increase in reading ability. (Stein & Fowler (1985), but see Bishop (1989) for an alternative interpretation).

Clearly, poor performance in the DT correlates with worse reading performance in school children. But these data do not allow us to distinguish between a number of possible explanations for these correlations. For example: superior performance in the DT may be a result of the wider reading experience of better readers; there may be no relationship between inefficient low level visual processing and reading, or visual problems may indeed cause reading difficulties.

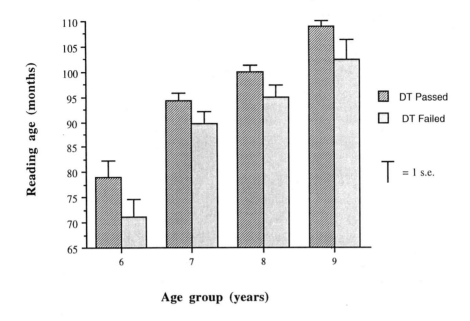

FIG. 1. Histogram of the mean reading ages of 753 primary school children grouped into 4 chronological age bands. Within each age band, children were further divided into those who passed and those who failed the DT. DT failure was always significantly correlated with worse reading performance. Error bars represent 1 standard error.

Furthermore, the DT has only been used to assess binocular fixation indirectly. Therefore, before addressing the main issue of causality, we must first show that children who fail the DT really do have more unstable fixation when they are reading.

THE DUNLOP TEST AND FIXATION STABILITY DURING READING

To do this Neil Munro and I have recorded the binocular eye movements of children and adults while they read single words. The data presented here are part of an on-going study. So far we have recorded from 8 children who have passed the DT and 8 who have failed it. These data were also compared with recordings from 4 fluent adult readers.

Subjects
The subjects in all the experiments reported in this chapter were recruited from children referred for orthoptic assessment at the Royal Berkshire Hospital, Reading U.K.,

because of their reading difficulties. In drawing subjects from such a clinical population, our samples were likely to be biased in favour of children with visual problems.

TABLE 1

Subject characteristics from eye movement recording experiment, comparing children who passed the Dunlop Test with children who failed the Dunlop Test.

		PERFORMANCE IQ (B.A.S.)		READING AGE (B.A.S.) (yrs:mnth)		CHRONOLOGICAL AGE (yrs:mnth)	
		MEAN (SD)	RANGE	MEAN	RANGE	MEAN	RANGE
DUNLOP TEST	PASSED n=8	106.0 (11.5)	91 – 124	9:3 (2:8)	7:4 – 12:4	10:10 (1:0)	9:0 – 12:3
	FAILED n=8	103.5 (10.7)	89 – 119	8:10 (1:9)	7:3 – 11:7	10:9 (1:1)	8:11– 12:4

N.B.   No significant differences between children who passed the DT and children who failed it were found for any of these measures.

Table 1 shows the characteristics of children from the eye movement study. It can be seen that children who passed and failed the DT were matched as closely as possible for chronological age, reading age and IQ (t-test comparisons showed no significant differences on all three measures). Therefore any difference in the fixation stability of the two groups of children could not be attributed to differences in age, intelligence or reading experience. Furthermore, the wide ranges of both the chronological and reading ages indicates that the children we see at the Orthoptic clinic in Reading are of mixed abilities, emphasizing that we deliberately did not set out to compare dyslexic and normal readers. Lastly, the two groups of children in this study did not differ significantly on standard clinical measures of visual acuity, near accommodation, local stereopsis or global stereopsis.

METHODS

Subjects wore a helmet on which both the text screen and the infra-red eye movement recorder (Skalar IRIS EM6500) were mounted. The text comprised 8 lines of single real words mounted on a screen which subtended 30° of visual angle. The screen was curved around an isovergent circle (radius 15.7 cm, assuming an average interpupillary distance of 58 mm) in the horizontal plane. This curvature of the screen ensured that the vergence demands of the reading task were held constant for all eccentricities. Thus we could compare the vergence drift during fixation at all eccentricities.

Following separate calibration of left and right eyes, binocular eye movements were recorded as subjects read two lines of words to themselves. Each subject read a total of 4 pairs of lines (approximately 90 words in total). After each pair of lines was read, post-calibrations of the left and right eyes were performed. Such frequent calibrations were required to reliably estimate any DC drift in the system.

The single words that children read were appropriate for their reading ability as measured by the British Ability Scales (B.A.S.) reading test. Adults were asked to read from a set of low frequency words selected to be difficult. All words were printed in lower case 12 point Helvetica font with single line spacing.

We used the same velocity threshold algorithm to search all the eye movement records for saccade/fixation/saccade sequences. For each fixation we measured the size of the preceding saccade,the fixation duration and the largest change in the vergence angle (vergence error) of the two eyes. Return sweeps from the first to the second line of words in each run were not included in this analysis. Our measure of interest, peak vergence error, comprised mostly vergence drift but included other phenomena like microsaccades and square wave jerks.

RESULTS

Collewijn, Erkelens & Steinman (1988) have shown that the vergence error at the beginning of fixation increases as a function of preceding saccade size. We therefore plotted peak fixational vergence error as a function of preceding saccade size (Fig. 2). As expected the 4 adults showed the most stable binocular fixation, i.e. the least change in vergence position during fixation. On average their vergence position shifted by only approximately 0.2 degrees which is equivalent to almost one character space in this experiment.

The children who failed the DT showed significantly worse fixational stability than children who passed the DT. Not surprisingly all children had significantly more unstable fixation than adults. To ensure that the difference between the 2 groups of children could not be attributed to systematic differences in reading age, chronological age, fixation duration, preceding saccade size or calibration errors, we performed multiple regression analyses. After controlling for all these variables we found that children who failed the DT had significantly more unstable fixation during reading than children who passed the DT ($p < 0.001$).

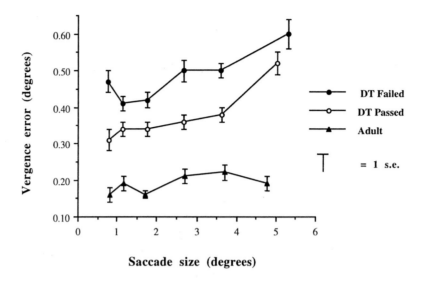

FIG. 2. Plot of mean fixational vergence error as a function of preceding saccade size. Each data point from the adult data represents the mean vergence error from about 40 fixations. Each point from the childrens' data represents the mean vergence error from about 180 fixations. Error bars represent 1 standard error.

Interestingly, as might be predicted by extrapolating the superior performance of adults, we also found that fixational stability in all children significantly improved as a function of increasing reading age. This is shown in Fig. 3 in which peak fixational vergence error is plotted as a function of reading age. Note this effect was independent of the DT effect.

THE DUNLOP TEST AND CONTRAST SENSITIVITY

At the start of this chapter we described how reading requires both eyes to be moved rapidly from one fixation to the next. Once steady fixation is achieved the retina can then sample the detailed pattern information of the text. So, visual processing during reading can be separated into the control of eye movements and image analysis. As far as eye movement control is concerned, we have shown that children who fail the DT have more unstable fixation than children with normal vision. But do they analyse the retinal image differently? Certainly, many disabled readers are less able to detect flicker (Martin & Lovegrove, 1987). They also find it harder to see static, low spatial frequency sinusoidal grating patterns (Lovegrove, Martin, Bowling, Blackwood, Badcock & Paxton 1982).

This leads us to ask how children who fail the DT perform on tests of low level image analysis? To answer this question Alexandra Mason and I have measured both the static and flicker contrast sensitivity functions of children who failed the DT, and compared them with children who had normal vision.

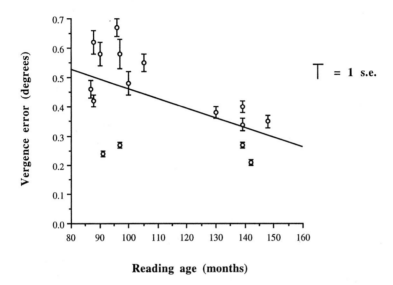

FIG. 3. Plot of fixational vergence error as a function of reading age, from the childrens' data. Each point represents the mean vergence error of all fixations made by one child. Error bars represent 1 standard error.

Subjects

We recruited mixed ability children from the same clinical population as in the eye movement recording experiment described above. In the first experiment we measured the static contrast ·sensitivity functions of 20 children who failed the DT (mean (sd) chronological age 116.6 (23.7) months; mean (sd) reading age 104.0 (27.7) months; mean (sd) IQ 108.0 (11.1)). We compared them with 20 children who passed the DT (mean (sd) chronological age 110.9 (21.5) months; mean (sd) reading age 84.3 (25.5) months; mean (sd) IQ 108.2 (15.2)).

In the second experiment we measured the flicker contrast sensitivity functions of 15 children who failed the DT (mean (sd) chronological age 107.5 (14.7) months; mean (sd) reading age 97.5 (27.1) months; mean IQ 114.8 (15.1)). We compared them with 15 children who passed the DT (mean (sd) chronological age 109.2 (20.5) months; mean (sd) reading age 92.2 (17.0) months; mean (sd) IQ 108.4 (16.6)).

Thus, in both experiments, children who failed the DT were well matched for age, reading age and IQ with children who passed the DT. In addition, we checked that the two groups of children in both experiments did not differ on standard clinical measures of Snellen visual acuity, near accommodation or stereopsis.

Methods

From a distance of 1 1/2 metres, subjects viewed a CRT display (Joyce screen). The edges of the screen were masked with dark card, so that only a central, circular portion of the screen, subtending 8 degrees was visible. Children viewed the screen with natural pupils and without a fixation target. Screen luminance averaged 3 cd/m^2.

We used a signal generator to present sinusoidal grating patterns of 0.5, 1.5, 3 & 6 cycles/degree. In both experiments, the tester pressed a button to make the grating patterns appear for exactly 1000 msec.. Therefore, children saw a gray field, then a grating pattern with the same mean luminance, and then a gray field again. Stimulus onset and offset had a square wave profile.

In the first experiment, the gratings were static, except for some temporal contamination introduced by their sudden onset and offset. In the second experiment, the grating patterns were sinusoidally modulated in counterphase at 20 Hz. To measure contrast threshold we used a four alternative forced choice protocol. Stimulus contrast was reduced according to a stair case method. At each point on the stair case, children had to decide in which of four randomly selected orientations the grating had appeared. The smallest contrast decrement that children could reliably detect was 3 dB. The order in which we measured the threshold for each spatial frequency was randomised across subjects.

RESULTS

Fig. 4 shows that, on average, the 20 children who failed the DT found it more difficult to see the static 0.5 cycle/degree grating than the 20 children with normal vision. Conversely, children who failed the DT found it easier to see the static 6 cycle/degree grating. Analysis of variance confirmed that though this effect was small, there was a significant interaction between spatial frequency and DT performance ($F=3.51$, $p<0.05$).

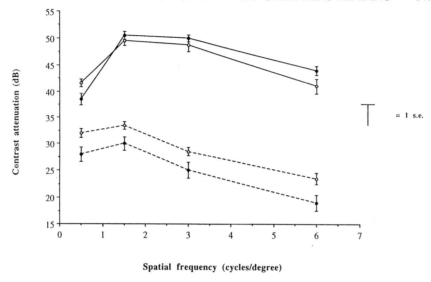

FIG. 4. Contrast sensitivity (plotted as mean contrast attenuation in decibels (dB)) as a function of spatial frequency. Open circles are children who passed the DT. Filled circles are children who failed the DT. The solid lines are the contrast sensitivity functions for static gratings presented for 1000 msec.. Dashed lines are the contrast sensitivity functions for gratings whose counterphase modulated at 20Hz. The latter were also presented for 1000 msec..

Figure 4 also shows that when contrast sensitivity was measured using gratings flickering in counterphase, the 15 children who failed the DT found all 4 grating patterns more difficult to see than the 15 children with normal vision. Analysis of variance revealed a very significant effect of DT (F=18.1, p<0.001). Both of these results are qualitatively very similar to those of Lovegrove et al. (1982) and Martin et al. (1987). Therefore, one possible interpretation of these findings might be that children who fail the DT have less efficient transient visual subsystems than children with normal vision. This point will be considered in the discussion.

We included two control experiments in this study. First, we tested the static contrast sensitivity of childrens' left and right eyes separately. In a subsample of 24 children, we did not find any systematic left/right eye differences when comparing children who passed and failed the DT. Therefore we felt that the results reported above could not be attributed to systematic, subtle differences between the visual acuity of the two eyes, for example. Secondly, we checked that our method of measurement was able to reproduce the findings of Lovegrove et al., who compared dyslexic readers with age matched normal readers. Therefore we measured the static contrast sensitivities of 34 normal readers (mean (sd) reading age 117.4 (28.6) months; mean (sd) chronological age 108.1

(19.8) months; mean (sd) IQ 109.9 (15.9)). We compared them with with 56 chronologically age matched dyslexic children (mean (sd) reading age 89.6 (22.9); mean (sd) chronological age 111.8 (19.6) months; mean (sd) IQ 110.4 (14.2)).

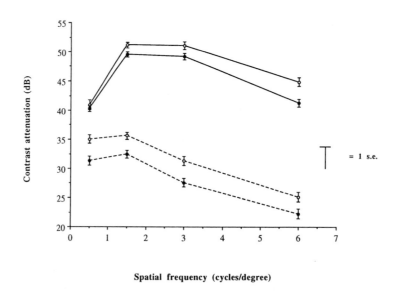

Spatial frequency (cycles/degree)

FIG. 5. Contrast sensitivity (plotted as a mean contrast attenuation in decibels (dB)) as a function of spatial frequency. Open circles are normal readers. Filled circles are age matched dyslexics. The solid lines are the contrast sensitivity functions for static gratings presented for 1000msec.. Dashed lines are the contrast sensitivity functions for gratings whose counterphase modulated at 20Hz. The latter were also presented for 1000 msec..

In a second control experiment we measured the flicker contrast sensitivities of 22 normal readers (mean (sd) reading age 119.8 (28.6) months; mean (sd) chronological age 110.2 (17.4) months; mean (sd) IQ 107.2 (16.5)). We compared them with with 28 chronologically age matched dyslexic children (mean (sd) reading age 93.1 (21.9); mean (sd) chronological age 113.3 (14.6) months; mean (sd) IQ 108.0 (16.6)). Children were defined as dyslexic if their B.A.S. reading ages fell 2 or more s.d.'s behind that expected on the basis of their performance I.Q.. Figure 5 shows that both dyslexic childrens' flicker and static contrast sensitivities were reduced at all four spatial frequencies. Therefore we concluded that our method of measurement was at least as sensitive as that used by Lovegrove et al.

Finally, Kulikowski (1971) and Dickinson & Abadi (1985) have shown that drifting eye movements (produced by tension tremor or nystagmoid drift) tend to reduce contrast sensitivity for the higher spatial frequency components of static grating patterns (above 3.0 cycles/degree). But we found that children who fail the DT show increased sensitivity to high (6.0 cycles/degree) frequency gratings, and reduced sensitivity to low (0.5 cycles/degree) frequency gratings. Therefore it is unlikely that our contrast sensitivity results could be accounted for by differences in the amount of fixational vergence drift between the two groups of children.

UNSTABLE BINOCULAR CONTROL AND VISUAL SYMPTOMS

So far, we have shown that failing the DT predicts poorer fixational stability during reading. Furthermore, poor DT performance is correlated with both reduced sensitivity to low spatial frequency static gratings, and to flickering gratings. But does unstable binocular fixation affect what children see? Anecdotal answers to this question come from the kinds of symptoms described by children who fail the DT. For example, one ten year old boy, V.C., reported that print often appeared "blurry" or "zig-zaggy". When asked to explain precisely what he meant he said that letters from the same line of text, or from the lines above or below, sometimes "moved over each other". Often he would see a string of letters which were very different from those actually printed. He said that this "blurring" might happen two or three times in any line of text.

These symptoms need not be restricted to reading. Another 9 year old child, B.D., described visual confusion when he tried to colour in small objects. Most of the time he was able to apply colour up to the object's margin very accurately. But sometimes, he saw the margin splitting into two. When this happened he was so unsure about which of the two 'apparent' margins to use, that he allowed colour to spill across the object's borders. Compelling reports like these provide anecdotal support for the idea that unstable binocular fixation can affect what children see. But we still have to address the fundamental question of whether it affects how children read.

LANGUAGE CLAMPING REVEALS THAT UNSTABLE FIXATION DOES AFFECT READING

The experimental principle. One way to demonstrate whether unstable binocular fixation affects the way children read is to create a situation in which the visual system is stressed when they are reading. If we can show that children with unstable fixation read differently when their vision is stressed, but that the same stress has no effect on children with normal vision, we could suggest a direct link between the efficiency of visual processing and reading.

To achieve this we used an experimental design that we call 'language clamping'. The principle was to give children real word lists to read, appropriate for their reading ability, and change only the visual appearance of the lists. Across the lists the average word frequency, spelling regularity, the average number of syllables and letters per word, as well as the average word concreteness were kept at a constant level. By 'clamping' these parameters at the same level, the demands made of childrens' language processing were kept constant with each new list. In contrast, while linguistic factors were held constant, the demands made on childrens' vision were varied with each new list. We did this in one of two ways. First, we changed the print size and line spacing of the lists. Second, we asked children to read with both eyes open and with one eye covered. Using this method, we could ensure that any change in childrens' reading performance during an experiment could only have been due to something visual, because the visual appearance of the lists was the only thing that changed.

Specifically, we aimed to show that children who fail the DT read differently when print size or viewing conditions were changed, but that these manipulations had no effect on children with normal vision. Then we would infer that unstable binocular fixation certainly does affect reading. In short, we aimed to show that only children who fail the DT made different kinds of reading errors when print size or viewing conditions were changed.

Nonword errors. There are many different ways of classifying reading errors, so what kinds of errors might we look for? Let us consider a child, who has failed the DT, viewing the word 'DOG'. This is schematized in Fig. 6 (A). The left eye is foveating the centre of the letter 'O'. However, because of vergence drift during fixation, the right eye has turned inward, so that it is foveating the letter 'D', instead of the letter 'O'. The two images from the left and right eyes are then momentarily combined in binocular vision. By comparing corresponding retinal points on the two retinae (Fig. 6 (B)), we can see that the resulting binocular image would be a very confused version of 'DOG'. This situation is comparable with physiological diplopia, and could be further complicated by retinal rivalry between components from the non-corresponding images (Davson, 1980). Therefore it is clear that binocular instability of this sort could lead to children experiencing visual confusion and perceiving incorrect, nonsense letter sequences.

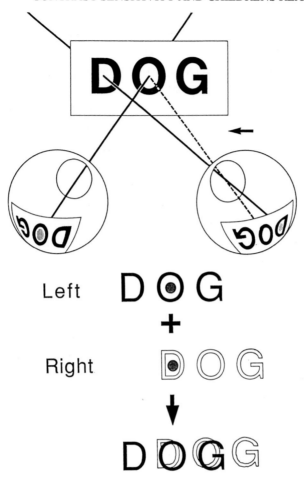

FIG. 6. (A) Diagram of the two eyes viewing the word DOG. See text for explanation. (B) Diagram of the combination of the images from the left and right eyes when corresponding retinal points are compared, producing a visually confused perception of DOG. The filled grey circles represent the anatomical centre of the fovea.

How might children who fail the DT deal with such a visually confused version of 'DOG'? Frith (1985) suggests that during the earliest stages of learning to read, children adopt a logographic strategy, in which 'letter order is largely ignored and phonological factors are entirely secondary'. They may recognize words from salient graphic features alone. So when a child who fails the DT sees a scrambled version of a real word, s/he might refuse to attempt it. Or s/he might make a wild guess which could be nonsense. Later on in their development, during Frith's 'alphabetic' stage, children try to work words out using a grapheme to phoneme conversion strategy. They probably fixate each component in a word they do not know. Therefore at this stage, a child who fails the DT could be at a particular disadvantage. S/he might well be expected to translate a sequence of letters they perceive incorrectly as a nonword. Put simply, if children see visual 'nonsense' then they might make 'nonword' reading errors.

At this point it should be noted that many children make nonword errors for phonological reasons alone (Gough, Juel & Roper-Schneider, 1983). This is easily predicted if we consider a child in Frith's alphabetic stage of reading development, who can apply grapheme-phoneme conversion rules. If asked to read the word 'SPHERE', s/he may accurately break the letter string into two graphemic units, i.e. 'SPH-' and 'ERE-'. But s/he might incorrectly translate the first unit as 'SPUH-' instead of 'SF-'. Thus s/he might utter the nonword error 'SPUH-HERE'. For any given level of reading ability, effects of this kind would depend on the linguistic complexity of words. So this is why we introduced language clamping. By keeping linguistic factors constant across our reading lists, the proportion of nonword errors made for phonological reasons should also remain constant across the lists. Therefore, any change in the the proportion of nonword errors across the reading lists must have a visual basis, because this is the only parameter that has been varied.

In summary, we set up the following hypothesis. If unstable binocular fixation affects reading, then the proportion of nonword errors made by children who fail the DT should change as print size and viewing conditions were manipulated. By contrast there should be no effect of print size or viewing condition on the nonword errors made by the children who had normal vision.

GENERAL DESCRIPTION OF SUBJECTS AND METHODS

Subjects
Subjects were recruited from the same pool of children as in the previous experiments. Details of the children who took part in the two language clamping experiments are given in Table 2.

Methods
Details concerning clinical testing, word list design and administration are described in Cornelissen, Bradley, Fowler & Stein (1991) and Cornelissen, Bradley, Fowler & Stein (1992). However there are three points worth emphasizing. First, the overall difficulty of the experimental lists was chosen so that each child would make approximately 45 errors out of the 90 words they read. This ensured that we would harvest sufficient numbers of errors to make our analyses reliable. Second, both experiments were balanced reading age match designs, so that children who failed the DT read exactly the same words, and under

exactly the same conditions, as children who had normal vision. Third, in our statistical analyses of the effects of print size and viewing conditions on nonword errors, we controlled for reading age, chronological age, IQ and phonological skill. IQ was calculated from the means of the similarities and matrices subtest scores of the B.A.S.. Phonological skill was assessed using a version of Bradley & Bryant's (1983) rhyme oddity task.

TABLE 2

Subject Characteristics from the 2 Language Clamping Experiments, Comparing Children Who Failed the DT with Children who Passed the DT.

		EXPERIMENT 1: CHANGING PRINT SIZE		EXPERIMENT 2: READING WITH ONE AND TWO EYES	
		DUNLOP TEST		DUNLOP TEST	
		PASSED N=45	FAILED N=45	PASSED N=32	FAILED N=32
IQ (B.A.S.)	MEAN (SD)	107 (13.8)	109 (11.7)	107 (10.6)	106 (15.3)
	RANGE	87–133	89–131	88–133	80–130
READING AGE	MEAN (SD)	7:9 (1:6)	7:8 (1:5)	8:2 (1:7)	7:11 (1:1)
yrs:mth	RANGE	5:6–14:0	5:7–12:9	6:1–11:7	6:6–10:10
CHRNLGCL AGE	MEAN (SD)	9:3 (1:5)	9:5 (1:1)	9:9 (1:4)	9:8 (1:8)
yrs:mth	RANGE	6:7–11:10	7:2–11:10	7:7–12:5	7:2–14:5
RHYME SCORE	MEAN (SD)	10.4 (3.4)	10.7 (3.1)	10.9 (3.0)	10.7 (3.2)
	RANGE	3–16	4–15	5–16	3–16

N.B.   1) There were no significant differences between children who passed and failed the DT on any of these measures, in either experiment.

2) Rhyme score = number of correct responses out of 16 items

RESULTS

Experiment 1: Changing Print Size and Line Spacing.

In the first experiment children were given 3 lists of words to read. While the linguistic difficulty of the words was clamped across each list, print size and line spacing was changed. The largest print condition used 24 point lower case Helvetica font with quadruple line spacing and a minimum of 3 character spaces between each word. The medium print condition used 12 point lower case Helvetica font with single line spacing and 1 character space between each word. The small print condition used 9 point lower case Helvetica font with half line spacing and 1 character space between each word.

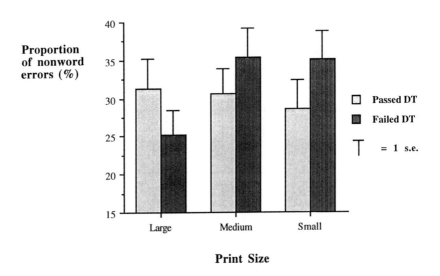

FIG. 7. Histogram showing the proportion of nonword errors (nonword/total errors) that children made at each of 3 print sizes. Light bars represent children who passed the DT. Dark bars represent children who failed the DT. When print size was changed from 'medium' to 'small', the proportion of non-word errors made by children who failed the DT fell from 35.4% to 25.2%. Error bars represent 1 standard error.

As can be seen from Fig. 7, children who failed the DT made significantly fewer nonword errors in the large print condition, whereas the proportion of nonword errors made by children with normal vision was unaffected by changes in print size. Thus, even after controlling for chronological age, reading age, IQ and phonological skill, we found a significant interaction between DT performance and print size (p<0.01).

Experiment 2: Reading with one Eye as Opposed to Two.

In the second experiment, children were given two lists of words to read, using the same technique of language clamping across lists. Both lists were printed in 12 point Helvetica font with single line spacing. Each child read one of the lists with both eyes open, and the other list with one eye occluded.

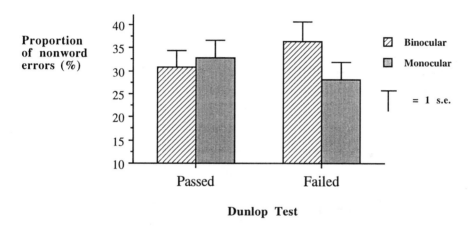

FIG. 8. Histogram showing the proportion of nonword errors (nonword errors/total errors) that children made when they read with one eye closed (dark bars) and with both eyes open (hatched bars). When children who failed the DT read with one eye only the proportion of non-word errors fell from 36.4% to 28.2%.

Figure 8 shows that only children who failed the DT made significantly fewer nonword errors when they read with one eye instead of two. Again after controlling for chronological age, reading age, IQ and phonological skill, we found a significant interaction between DT performance and viewing conditions (p<0.01).

DISCUSSION

Previous work has shown that poor DT performance is correlated with poor reading in school children. The new results from our eye movement recordings show that children who fail the DT do have more unstable binocular fixation than children with normal vision. But, most importantly, our 'language clamping' experiments show that it is only these children with unstable fixation who read differently when print size or viewing conditions are changed. Because the only feature that was varied in these experiments was the visual appearance of the reading lists, this difference must have a visual basis. Therefore we conclude that unstable binocular fixation does affect how children read.

TABLE 3

Relationship between Nonword Error Score, Print Size, Phonological Ability and DT

| | | MEAN NONWORD ERROR SCORE | |
| | | DUNLOP TEST | |
		PASS	FAIL
PRINT SIZE	PHNOLOGICAL ABILITY		
LARGE	GOOD	35.3	34.7
	POOR	26.4	16.1
MEDIUM	GOOD	35.6	41.4
	POOR	24.8	29.7
SMALL	GOOD	34.7	44.7
	POOR	21.4	25.9

N.B. 1) Nonword error score=(Nonword Errors/ Total Errors)x100
      2) Phonological Ability is 'GOOD' if a child made ≤ 5 errors out of 16 items in the rhyme task. Phonological Ability is 'POOR' if a child made > 5 errors out of 16 items in the rhyme task.

The data in Table 3 is taken from the 'print size' experiment. It shows that at each print size, children who had better phonological skills always made more nonword errors than children with worse phonological skills. This result confirms the findings of Gough, Juel and Roper-Schneider (1983). These workers used a pseudoword reading task to measure children's phonological skill. They found that children's ability to read pseudowords was positively correlated with the proportion of nonword errors they made in an oral reading test. But, it is also clear in our data that the nonword error discrepancy between children with good and poor phonological skills was significantly more marked for children who failed the DT. This finding suggests that there may be important dynamic interactions between poor visual processing and phonological processing during reading.

We would like to propose the following scheme to explain the results in Table 3. Children with good phonological skills are more likely to work out words that they do not recognize at first sight by using a grapheme-phoneme strategy. To do this successfully they must look at, or fixate, each component in a word separately. As Gough et al. have shown, they may sometimes misapply grapheme- phoneme conversion and make a nonword error. In this case children see what is on the page but decode it wrongly.

However, children who fail the DT and who use a phonological reading strategy may be further disadvantaged. Like other children they may misapply phonological rules and make nonword errors. But, in addition, because they have to fixate each grapheme separately, they are at risk of suffering visual confusion. In this case they may perceive a scrambled version of what is actually on the page, which they also decode as a nonword. In short, children who fail the DT may have both a phonological and a visual/phonological route for making nonword errors. Further multiple regression analysis supported this explanation, because we found a powerful and significant three way interaction between DT, print size and phonological skill, as measured by the rhyme test (p<0.0001). Although we repeated this finding in our 'viewing condition' experiment, the effect was weaker for two reasons. Firstly there were only 128 observations in that experiment as opposed to 270 in the 'print size' experiment. Secondly the children in the 'viewing condition' experiment were older, and their reading ages were higher (see Table 2). Therefore more of them would have been in Frith's orthographic stage of reading development, where children make less reliance on explicit grapheme-phoneme conversion rules.

If unstable binocular fixation affects reading then we would expect children who fail the Dunlop Test to make more reading errors overall than normal children of the same chronological age. Clearly, this effect was masked in our 'language clamping' experiments because we used a reading age match design. However, in a survey of 753 primary school children aged between 6 & 11 years, Stein et al. (1986) did find that children who failed the DT had lower reading ages than normal children matched for chronological age (see Fig. 1). In other words children with unstable binocular fixation are more prone to making reading errors.

In the 'language clamping' experiments, increasing print size or monocular viewing conditions did not reduce the overall number of errors made by children with unstable binocular fixation; it only altered the proportion of nonword errors that they made. Though this may at first seem surprising, it does suggest that visual confusion of words did not prevent children from recognizing familiar words and therefore reading them correctly. Certainly, there is evidence from adult studies that word recognition is more robust for familiar words, even when words are presented in a visually mutilated form. Connine, Mullennix, Shernoff & Yelen (1990) showed a reaction time advantage for high- familiarity, high-frequency words presented visually to adults. Furthermore, Pollatsek, Well & Schindler (1975) asked adults to judge whether pairs of words, which were printed in mixed upper and lower case letters, were the same or different. These workers found that the accuracy of subject's same-different judgements was greater for visually mutilated real words than for pseudowords (which by definition must be less familiar than real words). Lastly, Paap, Newsome & Noel (1984) investigated how efficiently adults could detect letters embedded in real words and pseudowords of both common and rare word shapes. Again subjects displayed an advantage for real, familiar words.

So, assuming we can liken visual mutilation in these adult studies to the visual confusion experienced by the children in the present studies, these results suggest that unstable binocular fixation is more likely to prevent children from working out unfamiliar words than from recognizing familiar ones. In turn, this means that, during their development, unstable binocular fixation would reduce the rate at which children add new words to their lexicon of familiar words. Hence we would expect these children to have lower reading ages than their chronologically age matched peers who have normal binocular vision, as reported by Stein et al. (1986).

Finally, what conclusions can we draw from our contrast sensitivity results? We suggest that the same mechanism may underlie both our results and those of Lovegrove and others, namely relatively inefficient processing of low spatial, high temporal frequency information. As Livingstone and Galaburda point out (Livingstone, Rosen, Drislane & Galaburda, 1991), such information is preferentially carried by the magnocellular subdivision of the visual pathway of primates, whose visual systems are very similar to humans'. Based on electrophysiological and anatomical data, Livingstone et al. have recently suggested an explicit link between abnormal magnocellular functioning in the lateral geniculate nucleus and dyslexia. This finding raises the possibility that the contrast sensitivity differences described by Lovegrove and others, which they have ascribed to deficiencies in the psychophysically defined 'transient' visual subsystem, may indeed reside anatomically in the magnocellular pathway.

To conclude, we would tentatively like to suggest that there may be an important interaction between magnocellular functioning, low spatial, high temporal frequency contrast sensitivity and the control of binocular fixation. There are three lines of evidence which support this possibility. First, it is the magnocellular population of retinal ganglion cells which projects almost exclusively to the three pathways which are primarily involved in the control of eye movements. These are the retinotectal pathway (Schiller & Malpelli, 1977), the geniculo-collicular pathway (Schiller, Malpelli & Schein, 1979) and the major geniculo-striate pathway, which ascends to the posterior parietal cortex (Van Essen & Maunsell, 1983; DeYoe & Van Essen, 1988; Livingstone & Hubel, 1988).

Second, the magnocellular pathway is considerably more sensitive than the parvocellular pathway to high frequency flicker and motion (Livingstone & Hubel, 1988; Schiller, Logothetis & Charles, 1990; Derrington & Lennie, 1984). Furthermore, the magnocellular pathway is said to be selectively masked by uniform field flicker (Badcock & Smith, 1989). Third, there is some evidence that uniform field flicker renders both fixation (West & Boyce, 1967; Woodruff & Neill, 1990) and saccadic eye movements during reading inaccurate (Kennedy & Murray, 1991). Therefore it is tempting to suggest that developmental or pathological abnormality in the magnocellular system may underlie the poor visual performance we find in children who fail the DT.

REFERENCES

Badcock, D.R. & Smith, D. (1989). Uniform field flicker: masking and facilitation. *Vision Research*, 29, 803-808.

Benton, A. (1962). Dyslexia in relation to form perception and directional sense.In Money, J. (Ed.) *Reading Disability: Progress and Research Needs in Dyslexia*. Johns Hopkins: Baltimore.

Benton, A. (1975). Developmental dyslexia: neurological aspects. In Friedman, W.J. (Ed.) *Advances in Neurology, Vol. 1, Current Reviews of Higher Order Nervous System Dysfunction*. New York: Raven Press.

Bigelow, E.R. & McKenzie, B.E. (1985). Unstable ocular dominance and reading ability. *Perception*, 14, 329-335.

Bishop, D.V.M. (1989). Unstable vergence control and dyslexia - a critique. *British Journal of Ophthalmology*, 72, 77-79.

Bradley, L. & Bryant, P.E. (1978). Difficulties in auditory organization as a possible cause of reading backwardness.*Nature*, 271, 746-747.

Bradley, L. & Bryant, P.E. (1983). Categorizing sounds and learning to read - a causal connection. *Nature*, 301, 419-421.

Brannan, J. & Williams, M.C. (1988). The effects of age and reading ability on flicker threshold.*Clinical Vision Science*, 3 (2), 137-142.

Breitmeyer, B.G. (1980). Unmasking visual masking: A look at the veil of "why" behind the veil of "how". *Psychological Review*, 87(1), 52-69.

Collewijn, H., Erkelens, C.J. & Steinman, R.M. (1988). Binocular co-ordination of human horizontal saccadic eye movements. *Journal of Physiology*, 404, 157-182.

Connine, C.M., Mullennix, J., Shernoff, E., Yelen, J. (1990). Word familiarity and frequency in visual and auditory word recognition. *Journal of Experimental Psychology: Learning, Memory and Cognition*, 16 (6), 1084-1096.

Cornelissen, P.L., Bradley, L., Fowler, M.S. & Stein, J.F. (1992). Covering one eye affects how children read. *Developmental Medicine and Child Neurology*, 34, 296-304.

Cornelissen, P.L., Bradley, L., Fowler, M.S., Stein J.F. (1991). What children see affects how they read. *Developmental Medicine and Child Neurology*, 33, 755-762.

Davson, H. (1980). *Physiology of the eye*. London: Churchill Livingstone.

Derrington, A.M. & Lennie, P. (1984). Spatial and temporal contrast sensitivities of neurones in lateral geniculate nucleus of macaque.*Journal of Physiology*, 357, 219-240.

DeYoe, E.A. & Van Essen, C.C., (1988). Concurrent processing streams in monkey visual cortex. *Trends in Neuroscience*, 11 (5), 219-226.

Dickinson, C.M. & Abadi, R.V. (1985). The influence of nystagmoid oscillation on contrast sensitivity in normal observers.*Vision Research*, 25, 1089-1096.

Fischer, B. & Weber, H. (1990). Saccadic reaction times of dyslexic and age-matched normal subjects. *Perception*, 19, 805-818.

Frith, U. (1985). Beneath the surface of developmental dyslexia. In Patterson, K.E., Marshall, J.C., Coltheart, M. (Eds.) *Surface Dyslexia*. London: Routledge and Kegan Paul.

Geiger, G., Lettvin, J.Y. (1987). Peripheral vision in persons with dyslexia. *New England Journal of Medicine*, 316, 1238-1243.

Gough, P.B., Juel, C. & Roper-Schneider, D. (1983). A two stage model of initial reading acquisition. In Niles, J. A., Harris L.A. (Eds.) *Searches for Meaning in Reading /Language Processing and Instruction*. Rochester, New York: National Reading Conference.

Hulme, C. (1988). The implausibility of low-level visual deficits as a cause of children's reading difficulties. *Cognitive Neuropsychology*, 5(3), 369-374.

Kennedy, A. & Murray, W.S. (1991). The effect of flicker on eye movement control. *The Quarterly Journal of Experimental Psychology*, 43A (1), 79-99.

Kulikowski, J.J. (1971). Effect of eye movements on the contrast sensitivity of spatio-temporal patterns.*Vision Research*, 11, 261-273.

Liberman, I. Y., Shankweiler, D., Fischer, F.W. & Carter, B. (1974). Explicit syllable and phoneme segmentation in the young child. *Journal of Experimental Child Psychology,* 18, 201- 212.

Livingstone, M. & Hubel, D., (1988). Segregation of form, colour, movement, and depth: anatomy, physiology, and perception. *Science*, 240, 740-749.

Livingstone, M.S., Rosen, G.D., Drislane, F.W. & Galaburda, A. (1991). Physiological and anatomical evidence for a magnocellular deficit in developmental dyslexia. *Proceedings of the National Academy of Science*, 88, 7943-7947.

Lovegrove, W.J., Martin, F. & Slaghuis, W. (1986). A theoretical and experimental case for a visual deficit in specific reading difficulty. Cognitive Neuropsychology, 3(2), 225-267.

Lovegrove, W., Martin, F., Bowling, A., Blackwood, M., Badcock D. & Paxton, S. (1982). Contrast sensitivity functions and specific reading disability. *Neuropsychologia,* 20 (3), 309-315.

Lundberg, I., Olofsson, A., Wall, S. (1980). Reading and spelling skills in the first school years predicted from phonemic awareness skills in kindergarten. *Scandinavian Journal of Psychology*, 21, 159-173.

Martin, F. & Lovegrove, W. (1987). Flicker contrast sensitivity in normal and specifically disabled readers. *Perception*, 16, 215-221.

Paap, K.R., Newsome, S.L., Noel, R.W. (1984). Word shape's in poor shape for the race to the lexicon. *Journal of Experimental Psychology: Human Perception and Performance*, 10 (3), 413-428.

Pollatsek, A., Well, A.D. & Schindler, R.M. (1975). Effects of segmentation and expectancy on matching time for words and nonwords.*Journal of Experimental Psychology: Human Perception and Performance*, 1, 328-338.

Pirozzolo, F.J. & Rayner, K. (1988). Dyslexia: The Role of Eye Movements in Developmental Reading Disabilities. In Johnston, C.W., Pirozzolo, F.J. (Eds.) *Neuropsychology of Eye Movements*. Hillsdale, NJ: Lawrence Erlbaum.

Ruddock, K.H. (1990). Visual search in dyslexia. In Stein, J.F. (Ed.) *Vision and Visual Dyslexia, Vol. 13, Vision and Visual Dysfunction*. Macmillan Press.

Rutter, M. & Yule, W. (1975). The concept of specific reading retardation. *Journal of Child Psychology and Psychiatry*, 16, 181-197.

Schiller, P.H., Logothetis, N.K. & Charles, E.R. (1990). Role of the colour-opponent and broad-band channels in vision.*Visual Neuroscience*, 5, 321-346.

Schiller, P.H. & Malpelli, J.G. (1977). Properties and tectal projections of monkey retinal ganglion cells. *Journal of Neurophysiology*, 40, 428-445.

Schiller, P.H., Malpelli, J.G. & Schein, S.J. (1979). Composition of geniculostriate input to superior colliculus of the rhesus monkey. *Journal of Neurophysiology*, 42, 1124-1133.

Snowling, M.J., Hulme, C., Wells, B. & Goulandris, N. (1991). Continuities between speech and spelling in a case of developmental dyslexia. *Reading and Writing: an Interdisciplinary Journal*, 4, 19-31.

Stein, J.F. & Fowler, M.S. (1982). Diagnosis of dyslexia by means of a new indicator of eye dominance. *British Journal of Ophthalmology*, 66, 332-336.

Stein, J.F. & Fowler, M.S. (1982). Effect of monocular occlusion on visuomotor perception and reading in dyslexic children. *Lancet*, July, 69-73.

Stein, J.F., Riddell, P.M. & Fowler, M.S. (1987). Fine binocular control in dyslexic children. *Eye*, 1, 433-438.

Stein, J.F., Riddell, P.M. & Fowler, M.S. (1986). The Dunlop Test and reading in primary school children. *British Journal of Ophthalmology*, 70, 317-320.

Tallal, P. (1976). Auditory Perceptual Factors in Language and Learning Disabilities. In R.M. Knights & D.J. Bakker (Eds.) *The Neuropsychology of Learning Disorders :Theoretical Approaches*. London: University Park Press.

Van Essen, C.C. & Maunsell, J.H.R. (1983). Hierarchical organization and functional streams in the visual cortex.*Trends in Neuroscience*, September, 370-375.

Vellutino, F., Pruzek, R., Steger, J. & Meshoulam, U. (1973). Immediate visual recall in poor and normal readers as a function of orthographic-linguistic familiarity. *Cortex*, 9, 370-386.

Vellutino, F., Steger, J., DeSetto, L. & Phillips, F. (1975). Immediate and delayed recall of visual stimuli in poor and normal readers. *Journal of Experimental Child Psychology*, 19, 223- 232.

Vellutino, F., Steger, J., Kaman, M. & DeSetto, L. (1975). Visual form perception in deficient and normal readers as a function of age and orthographic-linguistic familiarity. *Cortex*, 11, 22-30.

West, D.C. & Boyce, P.R. (1967). The effect of flicker on eye movements. *Vision Research*, 8, 171-192.

Wagner, R. & Torgeson, J. (1987). The nature of phonological processing and its causal role in the acquisition of reading skills. *Psychological Bulletin*, 101, 192-212.

Williams, M.C., Molinet, K. & LeCluyse, K. (1989). Visual masking as a measure of temporal processing in normal and disabled readers. *Clinical Vision Science*, 4 (2), 137-144.

Woodruff, C.J. & Neill, R.A. (1990). Detecting target shifts under uniform field flicker. *Vision Research*, 30, 479-487.

Facets of Dyslexia and its Remediation
S.F. Wright and R. Groner (Editors)
© 1993 Elsevier Science Publishers B.V. All rights reserved.

PERCEPTUAL AND COGNITIVE FACTORS IN DISABLED AND NORMAL READERS' PERCEPTION AND MEMORY OF UNFAMILIAR VISUAL SYMBOLS

D. M. Willows, E. Corcos & J. R. Kershner
Department of Instruction and Special Education
Ontario Institute for Studies in Education

INTRODUCTION

Despite arguments by some reading theorists that visual perception and visual memory deficits have no significant role in reading disability (Calfee, 1983; Stanovich, 1985; Vellutino, 1979, 1987), there is a growing body of evidence indicating that, relative to normal readers, disabled readers are deficient in basic visual processes (DiLollo, Hanson & McIntyre, 1983; Lovegrove, Martin & Slaghuis, 1986; Willows, 1991). A number of researchers claim to have demonstrated that visual processing factors are not implicated in reading disability, but the work of Vellutino and his colleagues is widely viewed as the "critical" research on the issue. In particular, three studies investigating disabled and normal readers' processing of unfamiliar visual symbols (Hebrew letters) have been most influential (Vellutino, Pruzek, Steger & Meshoulam, 1973; Vellutino, Steger, Kaman & DeSetto, 1975; Vellutino, Steger, DeSetto & Phillips, 1975). These studies have, however, been subject to serious methodological, statistical and conceptual criticism (Fletcher & Satz, 1979a, 1979b; Gross & Rothenberg, 1979; Singer, 1979).

Willows, Corcos & Kershner (1988) recently undertook an experiment designed to compare the visual perception and visual memory abilities of 6-, 7- and 8-year old disabled and normal readers. This work also used Hebrew letters as unfamiliar stimuli, but the research was designed to avoid the various experimental limitations of the work by Vellutino and his associates. The result of the research by Willows et al. (1988) clearly demonstrated differences between disabled and normal readers on a task designed to assess basic visual processes. Disabled readers were less accurate and slower than normal readers at recognizing single Hebrew letters (Item task) and at remembering where they saw a particular Hebrew letter in a three-letter string (Spatial-Order task). These effects were greatest for the 6-year old disabled readers, but were also apparent to a lesser degree for the older disabled readers. Given this evidence of differences between disabled and normal readers on tasks assumed to be assessing basic visual processes, the purpose of the present research was to examine the contribution of a range of perceptual, cognitive and linguistic factors in accounting for these differences.

Although there is quite a large body of evidence supporting the conclusion that disabled and normal readers perform differently on tasks designed to assess basic visual processing (Lovegrove et al., 1986; Willows, Kruk & Corcos, in press), it is not yet clear what types of processing deficits might account for the observed differences in performance. Despite the use of unfamiliar visual stimuli in these studies attempting to assess visual processing differences between disabled and normal readers, it is not possible to completely exclude the possibility that differences in cognitive and linguistic processes might have had some role in the performance differences. Of the numerous studies examining the visual processes of disabled readers, none have specifically attempted to account for the variance in performance on the basis of visual, linguistic and cognitive factors. In the present research, the 6-, 7-, and 8-year old children who participated in the Willows et al., (1988) study were tested on a large battery of standardized and experimental tasks designed to assess individual differences in perceptual, cognitive and linguistic processing. The purpose was to relate performance on these various measures to accuracy and speed of performance on tasks assessing perception and memory for unfamiliar visual symbols (on the Item and Spatial-Order tasks) and also to word recognition and reading comprehension on standardized tests.

METHOD

The research was conducted in three phases: Phase I (a screening phase to select disabled and normal readers), Phase II (an experimental testing phase), and Phase III (a psychoeducational assessment phase). Phase I involved the administration of tests of basic verbal and nonverbal cognitive ability as well as reading and spelling achievement, Phase II (previously reported by Willows et al, 1988) involved testing each child on a computer-based task assessing visual perception and visual memory for item and spatial-order information in unfamiliar symbols (Hebrew letters), and Phase III involved the administration of a large battery of standardized and experimental tests measuring various aspects of visual, phonological, linguistic and memory processes.

Phase 1: Screening of Subjects

Children experiencing reading difficulties and normal readers from similar backgrounds were screened on measures of general cognitive and linguistic ability, and academic achievement in reading and spelling. Based on these measures, 45 disabled readers (DR) and 45 normally achieving readers (NR) in the age range from 6 to 8 years of age were selected as subjects. All subjects were of average to above average intelligence with no hyperactivity or behavioural disturbances. Children with any previous exposure to the Hebrew alphabet were excluded from the sample.

The reading disabled children were referred to the Psycho-Educational Clinic at the Ontario Institute for Studies in Education by parents and teachers following advertisements requesting participants in a study of the "causes of children's reading and writing difficulties." The control group children were selected from local elementary schools on the basis of average to above average scores on a test of reading comprehension, and similarity to the reading disabled children in chronological age, socio-economic status, nonverbal cognitive abilities and receptive and expressive vocabulary.

Definition of Disabled Readers. The traditional operational definition of reading disability -- average or above intelligence and below age/grade level reading ability, despite normal educational opportunity -- was adopted in this research (Vellutino, 1979). Although for older children, it is conventional to select as "reading disabled" children who are at least 2 years below age/grade expectations, this criterion cannot be applied to children younger than 8 years since formal reading instruction only begins at the age of six. Hence, for children who are below the age of 8, and therefore cannot be more than 2 years below grade level, the practice commonly followed is to designate an individual who is at least one year below age/grade at age 7 or at least six months below age/grade expectation at age 6 as being reading disabled (Lyle, 1969; Willows, 1991).

Measures of Nonverbal and Verbal Cognitive Abilities. Screening measures were administered in order to select disabled and normal reading groups that were as similar as possible in their nonverbal and verbal cognitive abilities. Raven's Coloured Progressive Matrices (Raven, 1957) were used to equate the groups on nonverbal reasoning ability. The Peabody Picture Vocabulary Test-Revised (PPVT-R), a test of receptive vocabulary (Dunn & Dunn, 1981), and the Vocabulary subtest of the Wechsler Intelligence Scale for Children-Revised, WISC-R, (Wechsler, 1974), a measure of expressive vocabulary, were administered to insure that the groups were similar in their language abilities.

Measures Reading and Spelling Achievement. The disabled and normal readers were also screened on measures of reading comprehension, word recognition, and spelling. Reading comprehension was measured by the small-group administration of the Gates-MacGinitie Reading Tests, GMRT, (MacGinitie, 1980). Word recognition abilities of subjects were assessed with the Reading Subtest of the Wide Range Achievement Test - Revised, WRAT-R, (Jastak & Wilkinson, 1984), a test involving reading words in isolation. Spelling ability was evaluated with the Spelling subtest of the WRAT-R, a test which utilizes a spelling dictation format.

Final Sample. Based on the above operational definition, the selected subjects were 45 reading disabled children (27 boys, 18 girls) and 45 normal readers (27 boys, 18 girls) attending English-speaking public elementary schools in the Metropolitan Toronto area. These children ranged in age from 6.0 to 8.11 years (15 disabled and 15 normal readers at each of ages 6.0 to 6.11, 7.0 to 7.11, and 8.0 to 8.11, respectively). Table 1 presents a summary of the screening data for the reading disabled and normal reading children, grouped by age levels. At each age level, the disabled and normal reader groups did not differ (p>.10) on measures of nonverbal intelligence (Raven), receptive vocabulary (PPVT-R) and expressive vocabulary (WISC-R, Vocabulary subtest), but did differ significantly (p<.001) on measures of reading comprehension (GMRT, Comprehension subtest), word recognition (WRAT-R, Reading subtest), and spelling (WRAT-R, Spelling subtest).

TABLE 1

Screening Measures to Select Disabled and Normal Reader Groups

Age Group	6		7		8	
	DR	**NR**	**DR**	**NR**	**DR**	**NR**
	(N=15)	(N=15)	(N=15)	(N=15)	(N=15)	(N=15)
Measures*: Cognitive Linguistic						
Raven	20.3	21.5	25.3	24.9	26.1	24.7
PPVT-R	89.9	89.8	90.1	93.4	93.9	101.8
WISC-R Vocab	11.5	12.7	11.5	12.7	11.4	12.4
Measures*: Reading / Spelling						
GMRT, Comp	3.8	31.0	12.3	31.5	20.9	34.0
WRAT-R Read	30.9	51.9	36.2	62.4	42.5	64.8
WRAT-R Spell	22.1	31.2	24.9	38.4	27.8	40.1

* Note: Raw scores except for WISC-R Vocabulary (scaled scores)

Phase II: Experimental Testing

Each child was also tested on two computer-based experimental tasks designed to assess visual processing of item and spatial-order information. Presented in the context of a "computer game", these tasks involved a same/different paradigm. Accuracy and response time were the dependent measures on both tasks. The details of the design and outcome of this aspect of the research are presented elsewhere (Willows et al., 1988), but the methodology is described briefly here in order to relate the findings of the procedure to the screening and psycho-educational measures.

Item Task. In the Item task (180 trials), the child was shown a single unfamiliar target item (a Hebrew letter, selected from the pool of 18 potential targets shown in Figure 1) followed, after a variable delay (500, 1500, 3000 msec), by either the same item or by a visually-similar different item from the same set.

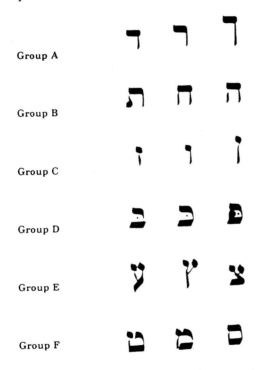

Group A

Group B

Group C

Group D

Group E

Group F

FIG. 1. Six sets of Hebrew letters used in Item and Spatial-Order tasks.

Spatial-Order Task. In the Spatial-Order task (240 trials), a target trigram of 3 visually-distinct unfamiliar symbols (each selected from one of the six sets of visually distinct Hebrew letters shown in Fig. 1) was presented for a brief exposure. At a variable delay after a target trigram had been presented (500, 1500, 3000 msec), one of the letters in the trigram was presented as a test item in the same or in a different spatial position compared to the letter in the target string. On each trial either the first, second or third spatial-position of the trigram was selected for testing. The child's task was to press a key to indicate whether the test item was in the same or a different position compared with its location in the target item.

Apparatus. An IBM XT microcomputer was programmed to display the stimulus and response sets for the experimental tasks, to record response accuracy and latency, and to manage the randomization of the stimuli and the presentation order. The displays were presented on an IBM colour monitor. For both the Item and Spatial-Order tasks, a same-different paradigm was employed. Two keys (one inch apart) on the keypad were used by subjects to indicate their responses. The keys were labelled with a green and a red dot to indicate "same" and "different" response, respectively. The position of the red and green keys was counterbalanced across subjects. A voice-synthesizer, interfaced with the microcomputer, was connected to earphones worn by subjects. The words "right" or "error" followed each response to provide feedback to the subject on the status of his/her response choice.

Phase III: Psycho-educational Testing

In addition to the screening measures (6 tests), each of the 90 selected children was individually tested on a large battery of psycho-educational measures assessing various aspects of visual, phonological, linguistic, and memory processes. The complete set of tests is presented in Table 2.

Measures of Visual Processing and Memory. Five subtests of the Test of Visual Perception Skills, TVPS, (Gardner, 1982) were administered. Each of these subtests consisted of 16 multiple-choice items involving unfamiliar visual stimuli which progressively increased in difficulty across the 16 items. The Visual Discrimination subtest required that subjects look at a target symbol and find an identical match from a pool of five response-items while the target-item remained in view. In the Visual Spatial Relations subtest, five forms were presented simultaneously and subjects were to identify the one form which was "facing the wrong way". The Visual Memory subtest required subjects to look at a target symbol for five seconds and then choose, from memory, its identical match from a pool of five response-items. The Visual Form Constancy subtest involved having the child select a matching geometric figure from a set that differs in size or orientation from the given target form. The Visual Sequential Memory subtest involved viewing a string of geometric figures for a number of seconds and than selecting the identical sequence from among alternatives.

The Block Design subtest of the Wechsler Intelligence Scale for Children - Revised, WISC-R, (Wechsler, 1974) was used to assess the child's ability to reproduce geometric designs from a model using coloured blocks and the Coding subtest of the WISC-R was used to assess their ability to form grapheme-to-grapheme associations, as well as psychomotor speed.

The Developmental test of Visual-Motor Integration (Beery & Buktenica, 1981), involves copying of a series of 24 geometric forms of increasing difficulty, and was included in the battery to assess visual-motor processing.

TABLE 2

Psychoeducational Battery Administered to All Subjects

Tests of Visual Processing and Memory

Wechsler Intelligence Scale for Children - Revised (WISC-R)
    Block Design (BD)
    Coding (COD)
Test of Visual Perceptual Skills (TVPS)
    Visual Discrimination (VD)
    Visual Memory (VM)
    Visual Spatial Relations (VSR)
    Visual Form Constancy (VFC)
    Visual Sequential Memory (VSM)
Test of Visual Motor Integration (TVMI)

Tests of Linguistic Processing

CELF, Producing Model Sentences (Sentence repetition)
WISC-R Digit Span (Auditory sequencing)
Rapid Automatized Naming (Word retrieval)
    Objects, (RAN-O)
    Colours, (RAN-C)
    Letters, (RAN-L)
    Numbers, (RAN-N)

Tests of Letter and Word Knowledge / Learning

    Encoding (Sound-Letter)
    Decoding (Letter-Sound)
    Word Discrimination
Goldman-Fristoe-Woodcock
    G-F-W, Sound-Symbol

Measures of Linguistic Processing and Memory. The Producing Model Sentences subtest the Clinical Evaluation of Language Fundamentals, CELF, (Semel-Mintz &

Wiig, 1982) was administered to assess subjects' productive control of sentence structure and their verbal memory. This sentence-repetition task consisted of 30 sentences of increasing length and complexity which varied semantically and syntactically.

The WISC-R Digit Span (forward and backward) subtest evaluated auditory sequencing and memory. This test requires that subjects repeat back strings of digits of increasing length, in the same order and in reverse order.

Four subtests of the Rapid Automatized Naming Test, RAN, (Denckla & Rudel, 1974) were included in the battery to test speeded word retrieval. Subjects are required to name a set of common objects (RAN-O), colours (RAN-C), letters (RAN-L), and numbers (RAN-N) as quickly as possible. Typically, subjects are familiarized with the set of five items in each task which they will have to retrieve as they scan a random listing of 100 items. The time taken to respond to the entire list, and the retrieval errors are recorded. Although this test is grouped with verbal measures because it involves the naming of words, the test may also reflect a basic speed of retrieval.

Tests of Letter and Word Knowledge/Learning. Experimenter-created tests of Phoneme Encoding, Decoding, and Word Discrimination were also included in the battery to assess the children's beginning reading skills. These tasks were designed to test sound-letter and letter-sound knowledge of the basic orthographic patterns in English, as well as word discrimination involving a basic sight vocabulary of words (each embedded in a list of 3 nonwords, similar to the target word except for minor differences in visual, phonemic, and letter order information).

The Symbol Learning subtest of the G-F-W (Goldman, Fristoe & Woodcock, 1974) Sound-Symbol test was also administered to assess the children's ability to learn nonsense symbols and associate them with pseudowords.

RESULTS AND DISCUSSION

The results were examined by analyzing the relation between the experimental measures of visual processing and visual memory (as measured by the Item and Spatial-order tasks), and the screening and psycho-educational measures of perceptual, cognitive and linguistic functioning. First the findings of the experimental tasks are presented in summary form.

Experimental Tasks. As reported by Willows et al. (1988), there were significant disabled/normal reader differences on both the Item and Spatial-Order tasks, but these differences were greatest among the 6 year olds. As shown in Figures 2 and 3, there were significant Group and Age main effects on the Item-Task on both the accuracy

(Group: F [1,84]=6.05, p<.05 and Age: F [2,84]=5.40, p<.01) and response time
(Group: F [1,84]=3.63, p<.06 and Age: F [2,84]=3.78, p<.05) measures.

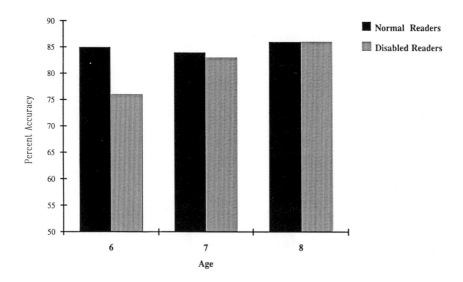

FIG. 2. Accuracy scores for Item task.

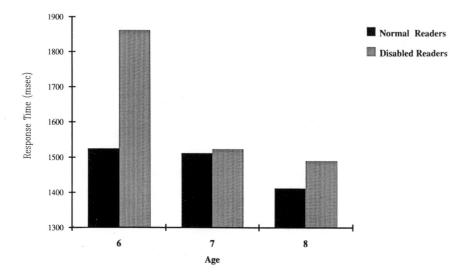

FIG. 3. Response time scores for Item task.

There was also a significant Group X Age interaction on the accuracy measure (F [2, 84]=3.28, p<.05) reflecting the fact that the 6-year old disabled readers performed less accurately than the older children. In other words the younger disabled readers were less accurate than older and normal readers in their responses on a task in which they had to press a key to indicate whether an unfamiliar visual symbol (a Hebrew letter) was the same or different from one they had just seen.

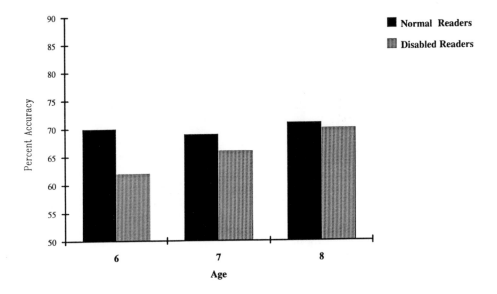

FIG. 4. Accuracy scores for Spatial-Order task.

The results of the Spatial-Order Task on the accuracy and response time measures are shown in Figs. 4 and 5 respectively. These findings demonstrate that overall the disabled readers were not as accurate (F [1,84]=4.70, p<.05) or as fast (F [1,84]= 6.90, p<.01) as the normal readers at remembering where in a three-letter string they had just seen a particular Hebrew letter. The disabled/normal reader differences on the Spatial-Order task were greatest among the 6-year old children. Although all children performed less accurately and slower when there was a longer delay between the target and test displays, there were no differential effects of delays on the disabled readers. Thus, it appears that the disabled readers' visual processing difficulty in these tasks was at the level of perceptual input, rather than in visual memory (similar to the findings in several studies reported by Willows, 1991).

Thus, the evidence from the Willows et al. experiments demonstrated disabled/normal reader differences on tasks that were designed to assess basic visual processes. The purpose of the next phase of the study was to examine the relation between performance on a range of perceptual, cognitive and linguistic measures and performance on the Item and Spatial-Order Tasks. To this end, all children's scores (residual scores derived from raw scores, corrected for chronological age in months) on the psycho-educational measures, on Raven's Progressive Matrices, as well as on both accuracy and response time on the Item and Spatial-Order tasks, were entered into a factor analysis.

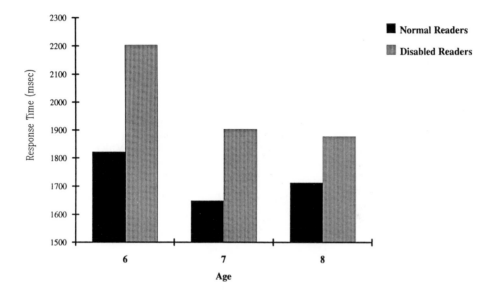

FIG. 5. Response time scores for Spatial-Order task.

Factor Analysis. The results of a principal component factor analysis with varimax rotations are shown in Table 3. The analysis produced four factors, first a visual processing factor, then a retrieval/ hook-up factor, then a verbal/language factor, and finally a speed of processing factor. Performance on both the item and spatial-order tasks had primary loadings on the visual processing factor and response time on both tasks loaded on the speed of processing factor. Thus, accuracy of performance on the Item and Spatial-order tasks appears to be reflecting visual processing rather than retrieval or language processes, as some might argue. Speed of processing appears to represent a separate independent factor.

Multiple Regression Analysis for Overall Sample. Multiple regression analyses were performed to determine the amount of variance in reading ability that was accounted for by the perceptual, cognitive, and linguistic measures in the psycho-educational battery. Measures that loaded significantly on the first factor in the factor analysis (visual processing) were entered into the analysis as a group first and the measures that loaded on the second factor (retrieval/hookup) were entered as a group second. Word recognition (WRAT-R) and reading comprehension (GMRT) scores were used as criterion measures. All scores were residuals with age partialled out.

TABLE 3

Results of Principal Component Factor Analysis (Varimax Rotation)

	SUBTEST	LOADING
Factor 1. Visual/Spatial		
	Item (Accuracy)	.43
	Order (Accuracy)	.45
	WISC-R, Block Design (BD)	.49
	WISC-R, Coding (COD)	.34
	TVPS, Visual Discrimination (VD)	.67
	TVPS, Visual Memory (VM)	.73
	TVPS, Visual Spatial Relations (VSR)	.55
	TVPS, Visual Form Constancy (VFC)	.55
	TVPS, Visual Sequential Mem.(VSM)	.61
	Raven	.68
	Beery, TVMI	.58
Factor 2. Retrieval & Hookup		
	RAN-O, errors	-.44
	RAN-N, errors	-.58
	RAN-C, errors	-.41
	Phoneme Encoding	.75
	Phoneme Decoding	.62
	Word Discrimination	.84
	(CELF), Sentence Imitation	.78
Factor 3. Verbal / Language		
	WISC-R, Vocabulary (VOC)	.57
	WISC-R, Digit Span (DS)	.50
	PPVT-R	.82
	RAN-L, time	.69
	GFW, Sound-Symbol	.38
Factor 4. Speed of Processing		
	Item Task (Response Time)	.71
	Order Task (Responses Time)	.73
	RAN-N, time	.44
	RAN-C, time	.47
	RAN-O, time	.45

The results of the multiple regression analysis for the overall sample showed that: 18% (Adjusted R^2) of the variance in word recognition scores was accounted for by the visual processing factor and the retrieval/hook-up factor added an additional 69% (Adjusted R^2); and that 17% (Adjusted R^2) of the variance in reading comprehension scores was accounted for by the visual processing factor, while the retrieval/hook-up factor accounted for an additional 62% (Adjusted R^2).

Multiple Regression Analysis for Six Year Olds. Because the performance of both the normal and disabled readers in the 7- and 8-year-old groups was relatively high on both the Item and Spatial-order tasks, it was considered of interest to analyze the scores of the 6-year olds separately. However, the multiple regression analysis on the data from the 6 year-olds, who demonstrated the greatest difficulty on the experimental visual processing tasks, produced results quite similar to the overall analysis. The results indicated that a visual processing factor accounted for 18% (Adjusted R^2) of the variance in word recognition, with the retrieval/hook-up factor adding another 69% (Adjusted R^2); and that the visual processing factor accounted for 26% (Adjusted R^2) of the variance in the reading comprehension scores, while the retrieval/hook-up factor added another 42% (Adjusted R^2).

CONCLUSION

Taken together then, the findings indicate that although a range of perceptual, cognitive and linguistic factors are involved in accounting for variance in both word recognition and reading comprehension performance among normal and reading-disabled children, the results of this research demonstrate that visual processing factors are also implicated to some degree, especially among younger children. Consistent with the previous literature, memory and linguistic factors clearly account for the largest portion of variance in this study, but the amount of variance accounted for by visual processing factors is large enough to warrant further study.

REFERENCES

Beery, K. E., & Buktenica, N. A. (1981). *Developmental Test of Visual-Motor Integration.* Cleveland: Modern Curriculum Press.

Calfee, R. C. (1983) *The mind of the dyslexic.* Annals of Dyslexia, 33, 9-28.

Denckla, M. B. & Rudel, R. G. (1974). Rapid "automatized" naming of pictured objects, colours, letters and numbers by normal children. *Cortex,* 10, 186-202.

Di Lollo, V., Hanson, D., & McIntyre, J. S. (1983). Initial stages of visual information processing in dyslexia. *Journal of Experimental Psychology: Human Perception and Performance,* 9, 923-935.

Dunn, L. M. & Dunn, L. M. (1981). *Peabody Picture Vocabulary Test - Revised.* Circle Pines, MN: American Guidance Services.

Fletcher, J. M. & Satz, P. (1979a). Unitary deficit hypotheses of reading disabilities: Has Vellutino led us astray? *Journal of Learning Disabilities,* 12(3), 155-159.

Fletcher, J. M. & Satz, P. (1979b). Has Vellutino led us astray? A rejoinder to a reply. *Journal of Learning Disabilities*, 12(3), 168-171.

Gardner, M. F. (1982). *Test of Visual-Perceptual Skills (Non-motor)*. Seattle: Special Child Publications.

Goldman, R., Fristoe, M. & Woodcock, R. W. (1974). *G-F-W Sound-Symbol Tests*. Minnesota: American Guidance Service, Inc.

Gross, K. & Rothenberg, S. (1979). An examination of methods used to test the visual perceptual deficit hypothesis of dyslexia. *Journal of Learning Disabilities*, 12, 670-677.

Jastak, S. & Wilkinson, G. S. (1984).*Wide Range Achievement Test - Revised*. Wilmington, DE: Jastak, Assoc., Inc.

Lovegrove, W., Martin, F. & Slaghuis, W. (1986). A theoretical and experimental case for a visual deficit in specific reading disability. *Cognitive Neuropsychology*, 3, 225-267.

MacGinitie, W. H. (1980). *Gates-MacGinitie Reading Tests*. Toronto: Nelson, Canada.

Raven, J. C. (1957). *Raven's Coloured Progressive Matrices*. Cambridge, MA: University Press.

Semel-Mintz, E. & Wiig, E. (1982). *Clinical Evaluation of Language Functions*. Columbus, OH: Charles E. Merrill.

Singer, H. (1979). On reading, language and learning. *Harvard Educational Review*, 49, 125-128.

Stanovich, K. E. (1985). Explaining the variance in terms of psychological processes: What have we learned? *Annals of Dyslexia*, 35, 67-96

Vellutino, F. R. (1979). *Dyslexia: Theory and Research*. Cambridge, Mass.: The MIT Press.

Vellutino, F. R. (1987). Dyslexia. *Scientific American*, 256 (3), 34-41.

Vellutino, F. R., Pruzek, R., Steger, J. A. & Meshoulam, U. (1973). Immediate visual recall in poor and normal readers as a function of orthographic-linguistic familiarity. *Cortex*, 9, 368-384.

Vellutino, F. R., Steger, J. A., DeSetto, L. & Phillips, F. (1975). Immediate and delayed recognition of visual stimuli in poor and normal readers. *Journal of Experimental Child Psychology*, 19, 223-232.

Vellutino, F. R., Steger, J. A., Kaman, M. & DeSetto, L. (1975). Visual form perception in deficient and normal readers as function of age and orthographic linguistic familiarity. *Cortex*, 11, 22-30.

Wechsler, D. (1974). *Wechsler Intelligence Scale for Children - Revised*. New York: The Psychological Corporation.

Willows, D. M. (1991). Visual processes in learning disabilities. In B. Y. L. Wong (Ed.) *Learning about Learning Disabilities*. New York: Academic Press.

Willows, D. M., Corcos, E., & Kershner, J. R. (1988, August). Disabled and normal readers' visual processing and visual memory of item and spatial-order information in unfamiliar symbol strings. Paper presented as part of the symposium, *Visual factors in learning disabilities*, at the XXIV International Congress of Psychology, Sydney, Australia.

Willows, D. M., Kruk, R., & Corcos, E. (in press). *Visual Processes in Reading and Reading Disabilities*. Hillsdale, NJ: Lawrence Erlbaum Associates.

Facets of Dyslexia and its Remediation
S.F. Wright and R. Groner (Editors)

LINKING THE SENSORY AND MOTOR VISUAL CORRELATES OF DYSLEXIA

B. J.W. Evans, N. Drasdo, Vision Sciences
& I. L Richards, Applied Psychology
Aston University, Birmingham

INTRODUCTION

The role of visual problems in dyslexia continues to be an extremely controversial subject. The work of Galaburda and colleagues (Geschwind & Galaburda, 1985; reviewed by Hynd & Semrud-Clikeman, 1989) could provide an anatomical framework for the many correlates of dyslexia. These correlates need not all be causally related to the reading disability, and could include the visual correlates discussed below.

The literature relating to traditional measures of visual function (see Evans & Drasdo, 1991) is extensive and equivocal, but suggests that the following optometric problems are common in dyslexia: convergence insufficiency, accommodative dysfunction, and unstable eye movements. Other problems that the literature suggested, although with less certainty, to be correlates of dyslexia are hypermetropia, exophoria at near, low prism vergences, and poor stereopsis. The many inconsistent findings can often be attributed to problems in defining dyslexia, failure to include control groups matched for IQ, and inappropriate optometric techniques.

In addition to these traditional correlates, some more recent studies have re-emphasised the importance of visual factors. Stein and colleagues (see Stein, Riddell & Fowler, 1989 for review) have suggested that poor binocular localisation and visuomotor control result in perceptual difficulties during reading in two-thirds of dyslexic children. There is also convincing evidence for a deficit of the transient visual system in dyslexia (Lovegrove, Martin & Slaghuis, 1986a). There is some feedback through the visual system by which the binocular alignment of the eyes is controlled. It is possible, although rather speculative, that this feedback is achieved through the transient visual system. Hence, binocular dysfunction in dyslexia could be the result of a transient system deficit.

One final theory is Irlen's (1983) claim that about 70% of learning disabled individuals have a visual dysfunction that can be corrected with tinted lenses. A review of this therapy shows a lack of rigorous scientific evidence despite a large body of anecdotal evidence and the recent emergence of several potential theoretical explanations (Evans & Drasdo, 1991).

The literature showed that these individual theories were being studied in isolation, and a small pilot study (Evans, Drasdo & Richards, 1990) suggested that it would be profitable to look for links between the findings of different research teams. The aim of the present research, therefore, was to investigate the visual factors that the literature suggested may be present more often in dyslexia; and to study the relationship between these different factors. Our control group was matched for age, sex, and intelligence; and in view of the likelihood of sub-groups within the dyslexic population, we used fairly large groups.

This paper provides an overview of our research. Further details can be found in Evans, Drasdo & Richards, (1992 a & b) and Evans (1991).

METHOD

Subjects

The subjects were 43 control and 39 dyslexic children, aged between 7 years 6 months and 12 years 3 months. All children had a performance IQ over 85, and the dyslexic group had a WISC-R full-scale IQ over 90. The reading retardation of the dyslexic group was at least 16 months as calculated from Yule, Landsdown & Urbanowitz's 1982 regression equation (cited by Thomson, 1984):

Expected reading age= (0.63* FSIQ) + (0.78 * age) -38.86
>   where expected reading age relates to Neale Accuracy
>   FSIQ= WISC-R Full Scale IQ
>   Age= Chronological age in months

This equation was based on a "primary school population". The difference between the expected reading age and the actual reading age from the Neale Accuracy was calculated as the reading retardation. The age and IQ of the groups were not significantly different (unpaired t-test, two-tailed; $p > 0.7$ and 0.3 respectively), but the Schonell reading and Vernon spelling results were significantly different (unpaired t-test, two-tailed; $p \leq 0.0001$).

Psychometric Assessment

The psychometric testing was carried out by IR and included tests of intelligence, reading, spelling, and auditory and visual sequential memory. One aspect of the study was to investigate the relationship between these psychological correlates of dyslexia and the visual correlates.

Optometric Examination

The optometric examination was carried out by BE and is summarised in Table 1.

TABLE 1

Tests Comprising the Optometric Examination*.

ophthalmoscopy	20
basic refractive tests	
distance vision	2
near vision	3
retinoscopy	15
subjective refraction and visual acuities	16
binocular vision and orthoptic tests	
cover test (distance & near)	4
near point of convergence	5
dissociation test - distance	14
dissociation test - near	7
AC/A ratio	8
near associated heterophoria	9
binocular status	11
stereopsis	12
relative vergences	19
tests of accommodation	
amplitude of accommodation	6
accommodative lag	18
dynamic assessment of accommodation	17
and convergence	
modified Dunlop Test	10/13
pattern glare	23
investigations of parallel visual processing	
spatial contrast sensitivity function	21
temporal contrast threshold	24
visual search task	22
symptoms and history	1

* Numbers represent the order in which the tests were carried out for each subject. To minimise subject boredom, the modified Dunlop Test was carried out in 2 sections.

This commenced with symptomatology, which included conventional questions about refractive correction and history of ocular problems and additional questions about personal and family history of headache, migraine, and epilepsy. The visual acuities were measured at distance and, using a Logmar chart (Bailey & Lovie, 1976) at near. The refractive error was determined objectively and subjectively. Apart from the fact that some studies have shown a correlation between long-sightedness and reading disability, binocular co-ordination is influenced by the refractive error and these two variables should not be studied in isolation.

The binocular vision tests included 3 ways of investigating the heterophoria. This is the tendency for the eyes to become misaligned when one is covered or an optical contrivance is used to make the monocular retinal images incompatible. Most people have a heterophoria at distance, and/or near. In an open-loop system, such as a completely dark room, the eyes automatically adopt a resting position where they point at, and are focussed on, a distance of about 1 metre away. Distance vision may be considered as an active divergence away from this resting point and near vision as an active convergence away from it. There is some evidence that dyslexic children show a reduced ability to move the eyes away from this resting point, and we investigated this hypothesis by measuring the prism vergence amplitude. This is the amplitude between the maximum divergence and convergence that a person can exhibit.

Accommodation is the ability of the eyes to change their focus as an object moves closer or further away. As the eyes change their vergence they normally also change their accommodation, and one of these should not be considered without some assessment of the other. Our pilot study found, like some other workers, that many dyslexic children have relatively reduced accommodation and convergence amplitudes (Evans et al., 1990). We therefore investigated several tests of accommodation, including an objective measure of accommodative lag.

The Dunlop test can be used to provide a sensitive measure of visuomotor ocular dominance, although this test has been shown to have severe limitations (Bishop, 1989; Buckley & Robertson, 1991). Previous research (Evans et al., 1990) suggested that a more meaningful response could be obtained by using a polarised fixation disparity instrument, which was less disruptive of normal binocular vision than the synoptophore. Although we found that even with this instrument the results were still strongly correlated with the child's IQ, we did find an interesting interaction between poor performance on this test, binocular dysfunction, and reading disability. We included, therefore, our modified version of this test in the main study.

Wilkins & Nimmo-Smith (1984, 1987) suggested that lines of print on a page may resemble a striped pattern that can cause visual discomfort through "pattern glare". Pattern glare can result in anomalous visual effects and Wilkins & Neary (1991) suggested that these phenomena may be linked to the use of tinted lenses in dyslexia. We modified Wilkins, Nimmo-Smith, Tait, McManus, Della Sala, Tilley, Arnold, Barrie &

Scott's (1984) pattern glare test to include a control condition and the results were described by Evans, Drasdo, Cook & Richards (1991).

Although there has been a considerable amount of research using tests that are designed to measure the functioning of the transient visual system, it should be noted that the psychophysical properties and cellular basis of parallel sub-systems in man is the subject of current debate (Wilson, Levi, Maffei, Rovamo & DeValois, 1990; Kaplan, Lee & Shapley, 1991). Moreover, none of the studies of reading retardates have ever applied 2 of these tests to the same group of children to assess the reliability of these measures. Recent research (Merigan & Maunsell, 1990) suggests that a very specific test for transient function is to measure the contrast modulation threshold for a homogeneous field of light that is flickering at 10 Hz. This type of method was used with reading retardates by Brannan & Williams (1988) and we replicated their procedure.

It may seem paradoxical that static gratings with exposures of up to 1 second have been used to demonstrate a "transient system deficit" (Lovegrove, Slaghuis, Bowling, Nelson & Geeves, 1986b). Therefore, we investigated this phenomenon further with prolonged viewing of static gratings. We used a commercially available contrast sensitivity system (Vistech VCTS Near Vision Test, Vistech Consultants Inc.) to replicate the experimental conditions of Lovegrove, Martin, Bowling, Blackwood, Badcock & Paxton (1982) within the tolerances established by Martin & Lovegrove (1984). An advantage of this test is that previous studies had only shown a low spatial frequency deficit, and hence the high spatial frequency response could act as a control against inter-group criterion differences. Unlike the flicker test, we thought it unlikely that static low spatial frequency contrast sensitivity would be a good measure of transient function; so we were keen to investigate the relationship between these two tests.

Finally, our subjects carried out a non-verbal simulated reading visual search task under two conditions, monocular and binocular. The task consisted of 4 subtests comprising numbers in a pseudo-random array. The children were asked to search for a given numeral, for example the number 1, and the time taken for them to count how many number 1's there were was recorded. The test was repeated for 4 search numbers in each subtest, and the conditions were interwoven in an ABBA design. If a blurred or confused image was interfering with reading then it should also impair performance at this test. The design of the final subtest may also induce pattern glare and required accurate saccadic programming.

RESULTS

Where an assessment of the frequency distribution of data suggested that they may not be normally distributed they were subjected to a test of normality. The test that was used employed the Kolmogorov-Smirnov test to compare the samples with a theoretical normal distribution. This was applied to the complete subject sample and parametric and non-parametric statistics were then used as appropriate. Where there was a non significant tendency for the data to differ from a normal distribution then both parametric and, wherever possible, non-parametric analyses were used.

Two findings from the binocular vision data were significant. Firstly, when the eyes were dissociated at near, the dyslexic group exhibited significantly more vergence instability (Mann -Whitney U-test, one-tailed; p= 0.023). Secondly, the vergence amplitude was significantly lower in the dyslexic group (unpaired t-test, one-tailed, p= 0.004). This may represent a reduced ability to turn the eyes away from their natural resting point. The fact that other measures of binocular function were normal may suggest that these were in most cases sub-clinical correlates of dyslexia.

The dyslexic group demonstrated a significantly reduced ability to detect flicker at 10 Hz (unpaired t-test, one-tailed, p= 0.0005) and a significantly reduced contrast sensitivity at 1, 2, 4, 6, and 8 cycles per degree (unpaired t-test, two-tailed; p < 0.001). The contrast sensitivity at 12 cycles per degree was not significantly different in the two groups (unpaired t-test, two-tailed; p= 0.16) (Fig. 1).

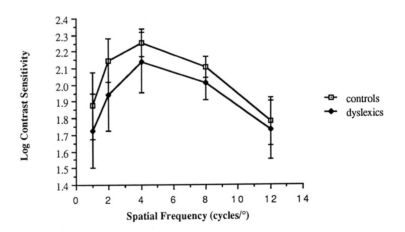

FIG. 1. Graph of Spatial Frequency v. Log Contrast Sensitivity for the control and dyslexic groups. The points represent the mean value for each group, and the error bars are ± 1 standard deviation.

Like Lovegrove's team (Lovegrove et al., 1986a), we found a low spatial frequency deficit in the dyslexic group, but we also found a reduced sensitivity at medium spatial frequencies. The literature suggests that this may be explained by differences in exposure times (Lovegrove et al., 1980, 1982; Martin & Lovegrove, 1984).

Although the dyslexic group performed worse on both of these functions that have been reported as indicating a transient system deficit, the 2 parameters do not correlate well (r= 0.265, goodness of fit (ANOVA), p= 0.027) and, therefore, do not seem to be measuring the same thing. It was noted above that the flicker test is much more likely to be an appropriate measure of transient function.

We next investigated our hypothesis that the results in the tests of parallel processing were related to the binocular function. The prism vergence amplitude was regressed on the log temporal contrast threshold and gave a significant correlation for both groups together (r= 0.329; two-tailed p= 0.006), and a highly significant correlation for the dyslexic group by itself (= 0.627; two-tailed p= 0.0006) (Fig. 2).

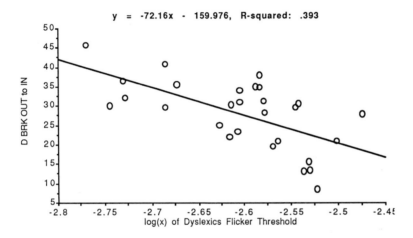

FIG. 2. Regression of vergence amplitude ("D BRK OUT to IN"), in Δ, on log temporal contrast threshold for the dyslexic group. No points are hidden by overlap, and the equation of the regression line is given above the graph.

So, we did find a slightly lower level of vergence reserves in the dyslexic group, which may have been caused by a deficit of the transient visual system. We investigated whether either of these results was causally related to the poor reading skills by using the visual search task. In visual terms this is a simulated reading task; although unlike reading, it does not require high level verbal skills. The average search times for all the subtests and conditions combined was significantly slower for the dyslexic group (Mann-Whitney U-test, one-tailed, corrected for ties; p= 0.012).

Some authorities believe that low-level visual problems are a major cause of dyslexia. The differences between the groups in search task performance were relatively small and further analyses showed that at least a quarter of the variance in the search task performance could be accounted for by psychometric variables. These results suggest, therefore, that visual distortions or eye movements abnormalities were not major factors impairing the scanning of printed matter.

We next calculated how much quicker each subject was with both eyes than with one eye only, and we called this the *binocular factor*. This was not significantly different between the two groups (Mann-Whitney U-test, one-tailed, corrected for ties; $p= 0.12$). This is another important point, since although immediate occlusion is different to long-term occlusion, one might have expected from Stein et al.'s (1989) theory that occlusion would have some beneficial effect on performance at this simple task. When we regressed the binocular advantage on the results from the four most relevant binocular vision tests we also found that they were not significantly related. The only exception was the vergence instability, which was related in the combined and dyslexic groups, but this correlation was in fact in the opposite direction to that expected.

The dyslexic group had a significantly lower amplitude of accommodation (Mann-Whitney U-test, one-tailed, corrected for ties; $p= 0.0003$) that did not originate from an uncorrected refractive error. This was significantly correlated with the vergence amplitude ($p < 0.05$). Unlike the vergence, however, the accommodative amplitude was not strongly linked to a transient system deficit. The poorer accommodative amplitude did not seem to result in slower search task performance, and this may suggest that it did not impair detailed near visual tasks. This was supported by the results of the other tests of accommodation, which were similar in the two groups.

In the Dunlop Test the eyes are diverged until, just before the person sees double or suppresses one image, one eye starts to give up following the target. This slip of fixation is called a fixation disparity and the eye in which this starts to occur is said to be the non-referent eye. It is claimed (Stein & Fowler, 1985; Stein, 1989) that the number of times out of 10 trials that the eye that is most often the referent eye *is* the referent eye, is a measure of binocular control or binocular stability. Although this result is collected using a discrete interval scale, it is usually converted to a binary nominal scale; i.e., stable or unstable. Some information will be lost in this transformation, and we analysed the data both with and without the transformation. Using the interval scale, the groups were not significantly different (Mann Whitney U-test, two-tailed $p = 0.20$). When transformed to the binary nominal scale, 60.5% of the control and 80.6% of the dyslexic group had "unfixed" reference; this approached significance (chi-square, $p= 0.091$).

On some trials, a subject already had a fixation disparity before prism was introduced, and the result was obtained without diverging the eyes. Whilst this should reflect more closely the normal reading situation, Stein (1991b) has suggested that forced divergence is essential to obtain a valid result with the Dunlop Test. We re-analysed our data, therefore, when all subjects who had not required prism to produce the divergence were excluded (the new groups comprised 30 control and 19 dyslexic children). This did not seem to have an overwhelming effect on the significance of our data (interval scale: Mann Whitney U-test, two-tailed p= 0.048; binary nominal scale: chi-square, p= 0.14).

Some subjects on this test seem to give an unreliable or contrived response. The experimenter, therefore, gave a subjective score for the quality of response; and this was much worse in the dyslexic group (Mann-Whitney U-test, two-tailed; p= 0.0096). Whilst one should be careful in interpreting what is, after all, a subjective grading of subjectivity, these results support the now considerable literature criticising the reliability of this test.

Unlike Stein et al. (1988), we found that vergence amplitude did not significantly correlate with the Dunlop Test result, whichever scale of measurement was used. The Dunlop Test results did not correlate with the spatial or temporal contrast sensitivity functions, but there was a tendency for children with a less stable referent eye to have relatively slower binocular search times. This tendency approached significance for both groups combined using the binary nominal scale (Mann-Whitney U-test, two-tailed; = 0.066) but did not approach significance using the interval scale or for either group individually.

The literature on what the Dunlop Test theoretically measures has, over the years, drifted away from ocular dominance towards binocular function. Stein (1991a), stated that the Dunlop Test detects *binocular instability*. This term has been used since the 1950's and '60's to describe a condition that is characterised by low vergence amplitudes and vergence instability (Giles, 1960, p. 464-467; Gibson, 1955, p.330 and 385-394). These abnormalities did occur more often in our dyslexic group, and we agree that binocular instability is a correlate of dyslexia. However, we, like many others, found that the Dunlop Test was too unreliable to detect this condition.

This paper will only include a brief discussion of some of our results on symptomatology. The dyslexic group reported more transient blurring at near than the control group (Mann-Whitney U-test; p < 0.05). Reports of diplopia were also more common in the dyslexic group; although this finding was of border-line significance (chi-square with continuity correction; p= 0.081). Transient blurring of near vision was associated with vergence instability and reduced amplitude of accommodation in the control group (p < 0.05), but not in the dyslexic group. This may suggest that transient blurring in dyslexia is more likely to result from cognitive than visual problems. Conversely, the incidence of episodes of diplopia was significantly related in both groups to a decreased prism vergence amplitude (p < 0.05).

CONCLUSIONS

Some measures of binocular function were reduced in our dyslexic population, and this seemed to be related to the transient system deficit and to unusually low amplitudes of accommodation. There was no evidence to suggest that the mild binocular dysfunction was a major cause of the reading disability, but this type of difficulty can lead to eye strain when reading. Our pilot study, and case studies, have suggested that this problem does affect slightly more dyslexic children than controls.

Our findings have linked the transient deficit with binocular dysfunction in dyslexia. Several workers (e.g., Holland, Tyrell & Wilkins, 1991) have linked this binocular dysfunction with the use of tinted lenses and Solman, Dain, & Keech, (1991) and Williams (1991) have linked the transient deficit with the tinted lens therapy. It seems that the confusing, diverse, visual aspects of reading disability are slowly falling into place to give a cohesive picture of this facet of dyslexia.

An issue that has to be resolved is whether these visual factors are a major cause of dyslexia, or whether they are mainly a non-causal correlate. Recent neuroanatomical models of dyslexia could be used to support either of these possibilities.

One final aspect of our study is the interaction between psychometric and optometric correlates of dyslexia. We have evidence that, even when IQ is controlled for, other psychometric factors still significantly interact with visual factors. Indeed, it is possible, although from our data unlikely, that some optometric tests could be simply measuring attentional differences between the groups. More work is needed to investigate this possibility.

ACKNOWLEDGEMENT

We are grateful to the British College of Optometrists for a research scholarship awarded to Bruce Evans.

REFERENCES

Bailey, I.L. & Lovie, J.E. (1976). New design principles for visual acuity letter charts. *American Journal of Optometry and Physiological Optics*, 53, 740-745.

Bishop, D.V.M. (1989). Unfixed reference, monocular occlusion, and developmental dyslexia - a critique. *British Journal of Ophthalmology* , 73, 209-215.

Brannan, J. & Williams, M. (1988). Allocation of visual attention in good and poor readers. *Perception and Psychophysics* , 41, 23-28.

Buckley, C.Y. & Robertson, J.L. (1991). The Dunlop test - fact or fiction. *British Orthoptic Journal* , 48, 39-40.

Evans, B.J.W. (1991). *Ophthalmic factors in dyslexia.* Unpublished PhD thesis. Aston University, Birmingham.

Evans, B.J.W. & Drasdo, N. (1990). Review of ophthalmic factors in dyslexia. *Ophthalmic Physiol. Opt.* , 10, 123-132.

Evans, B.J.W., Drasdo, N., & Richards, I.L. (1990). Optometric aspects of reading disability. *Frontiers of vision. Transactions of British College of Optometrists 10th Anniversary Conference*, Butterworths, London. pp. 14.

Evans, B.J.W. & Drasdo, N. (1991). Tinted lenses and related therapies for learning disabilities: a review. *Ophthalmic Physiol. Opt.,* 11, 206-217.

Evans, B.J.W., Drasdo, N., Cook, A., & Richards, I.L. (1991). *Pattern glare, tinted lenses, and reading performance.* Presented at Coloured Spectacles and Reading Difficulties, Apothecaries Hall, London.

Evans, B.J.W., Drasdo, N., & Richards I.L. (1992a). Sensory visual correlates of dyslexia *In preparation.*

Evans, B.J.W., Drasdo, N., & Richards I.L. (1992b). Motor visual correlates of dyslexia *In preparation.*

Geschwind, N. & Galaburda, A.M. (1985). Cerebral Lateralization Biological Mechanisms, Associations, and Pathology: A Hypothesis and a Program for Research. *Archives of Neurology*, 42, 429-457.

Gibson, H.W. (1955). *Textbook of Orthoptics.* London: Hatton Press Ltd.

Giles, G.H. (1960). *The Principles and Practice of Refraction.* London: Hammond, Hammond and Company Ltd.

Holland, K.C., Tyrell, R., & Wilkins, A. (1991). *The effect of Irlen coloured overlays on saccadic eye movements and reading - a preliminary report.* Presented at: Coloured Spectacles and Reading Difficulties: a Joint Meeting Between the Applied Vision; Association and the Colour Group, Apothecaries Hall, London.

Hynd, G.W. & Semrud-Clikeman, M. (1989). Dyslexia and brain morphology. *Psychological Bulletin* , 106, 447-482.

Irlen, H. (1983). *Successful treatment of learning difficulties.* Presented at The Annual Convention of the American Psychological Association, Anaheim, California.

Kaplan, E., Lee, B.B. & Shapley, R.M. (1991). New views of primate retinal function. In N. Osborne & J. Chader (Eds.) *Progress in Retinal Research, Volume 9.* Oxford: Pergamon, pp. 275-336.

Lovegrove, W., Martin, F., Bowling, A., Blackwood, M., Badcock, D., & Paxton, S. (1982). Contrast sensitivity functions and specific reading disability. Neuropsychologia 20, 309-315.

Lovegrove, W., Martin, F., & Slaghuis, W. (1986a). A theoretical and experimental case for a visual deficit in specific reading disability. *Cognitive Neuropsychology*, 3, 225-267.

Lovegrove, W., Slaghuis, W., Bowling, A., Nelson, P., & Geeves, E. (1986b). Spatial frequency processing and the prediction of reading ability: a preliminary investigation. *Perception & Psychophysics*, 40, 440-444.

Lovegrove, W.J., Bowling, A., Badcock, D., & Blackwood, M. (1980). Specific reading disability: differences in contrast sensitivity as a function of spatial frequency. *Science* , 210, 439-440.

Martin, F. & Lovegrove, W. (1984). Note: the effects of field size and luminance on contrast sensitivity differences between specifically reading disabled and normal children. *Neuropsychologia* , 22, 73-77.

Merigan, W.H. & Maunsell, J.H.R. (1990). Macaque vision after magnocellular lateral geniculate lesions. *Visual Neuroscience* , 5, 347-352.

Solman, R.T., Dain, S.J., & Keech, S.L. (1991). Color-mediated contrast sensitivity in disabled readers. *Optometry and Visual Science*, 68, 331-337.

Stein, J. (1989). Visuospatial perception and reading problems. *Irish Journal of Psychology,* 10, 521-533.

Stein, J.F. (1991a). Vision and Language. In M. Snowling & M. Thomson (Eds) *Dyslexia: Integrating Theory and Practice* . London: Whurr pp. 31-43.

Stein, J.F. (1991b). Personal communication.

Stein, J. & Fowler, S. (1985). The effect of monocular occlusion on visuomotor perception and reading in dyslexic children. *The Lancet*, 13th July, 69-73.

Stein, J.F., Riddell, P.M., & Fowler, S. (1988). Disordered vergence control in dyslexic children. *British Journal of Ophthalmology*, 72, 162-166.

Stein, J., Riddell, P., & Fowler, S. (1989). Disordered right hemisphere function in developmental dyslexia. In C. Von Euler, I. Lundberg, and G. Lennerstrand (Eds.), *Brain and Reading*. New York: Stockton pp. 139-157.

Thomson, M.E. (1984). *Developmental Dyslexia*, Edward Arnold, London.

Wilkins, A. & Neary, C. (1991). Some visual, optometric and perceptual effects of coloured glasses. *Ophthalmic and Physiological Optics*, 11, 163-171.

Wilkins, A.J. & Nimmo-Smith, M. I. (1984) On the reduction of eye-strain when reading. *Ophthalmic and Physiological Optics*, 4, 53-59.

Wilkins, A.J. & Nimmo-Smith, M.I. (1987). The clarity and comfort of printed text. *Ergonomics*, 30, 1705-1720.

Wilkins, A.J., Nimmo-Smith, I., Tait, A., M[c] Manus, C., Della Sala, S., Tilley, A., Arnold, K., Barrie, M., & Scott, S. (1984). A neurobiological basis for visual discomfort. *Brain*, 107, 989-1017.

Williams, M. (1991). Effective interventions for reading disability. Presented at: *Rodin Remediation 18th Scientific Conference - Reading and Reading Disorders: Interdisciplinary Perspectives*, Berne, Switzerland.

Wilson, H.R., Levi, D., Maffei, L., Rovamo, J., & DeValois, R. (1990). The perception of form: retina to striate cortex. In L. Spillmann & J.S. Werner (Eds.), *Visual Perception: The Neurophysiological Foundations*. San Diego: Academic Press, pp. 231-272.

Yule, W., Landsdown, R. & Urbanowitz, M. (1982). Predicting educational attainment for WISC-R. *British Journal of Clinical Psychology*, 21, 1, 43-47.

Facets of Dyslexia and its Remediation
S.F. Wright and R. Groner (Editors)

TOWARD AN ECOLOGICALLY VALID ANALYSIS OF VISUAL PROCESSES IN DYSLEXIC READERS

R. S. Kruk and D. M. Willows
Ontario Institute for Studies in Education
Department of Instruction and Special Education

INTRODUCTION

Current approaches to reading instruction in the classroom and remedial intervention in the clinic assume that individual differences in reading acquisition and in reading disabilities are based primarily on differences in psycholinguistic development. Thus, little attention is paid to possible underlying visual processing factors. Children whose main area of difficulty lies in visual processing, who demonstrate few apparent phonological or linguistic deficits, may be receiving the same instructional approach as others who legitimately require intervention for difficulties in verbal abilities.

VISUAL PROCESSING DEFICITS IN DISABLED READERS

Reading involves eye movements that are separated by fixations during which new visual information is taken in by the reader. Within this process of eye movement and information intake, the visual processing system plays a significant role. A given piece of information, read during one fixational pause, can still be perceived for a period of time after the eye has moved to read another section of text (Coltheart, 1980). This visible persistence can last for about one-quarter second beyond the fixational pause, and raises a potential problem for reading. If it occurs, how can text be read from one fixation to the next without successive visual input overlapping with previously read material, making the text appear like a jumble of lines? Researchers have argued that, at the onset of each new eye movement, the input from the last fixational pause is erased or inhibited by the activity of the transient system; a process called saccadic suppression (Breitmeyer, 1984).

Although several contemporary theorists have asserted that visual processing deficits are not significantly involved in reading disabilities (Hulme, 1988; Stanovich, 1985; Vellutino, 1987), research on basic visual processes has identified differences in some aspects of visual processing between normal and disabled readers (Lovegrove, Martin, & Slaghuis, 1986; Willows, 1991; Willows, Kruk, & Corcos, in press). The evidence shows that disabled readers have longer visible persistence durations than normal readers

(Badcock & Lovegrove, 1981; DiLollo, Hanson, & McIntyre, 1983; Lovegrove, Heddle, & Slaghuis, 1980; Lovegrove & Slaghuis, 1989; Slaghuis & Lovegrove, 1984; Stanley & Hall, 1973), which could have serious consequences for disabled readers' ability to see textual material with enough clarity to read words correctly. Words read in succession across a line of text could be superimposed visually, resulting in a jumbled perception. The finding of longer visible persistence durations among disabled readers has been interpreted within Breitmeyer's framework of transient and sustained channels for visual processing (Breitmeyer, 1984) .

Using Breitmeyer's terminology, both normal and disabled readers have similar "sustained" visual functioning, but different "transient" functioning. The transient system which responds only at the onset and offset of a stimulus is thought to function at a non-optimal level for disabled readers (Lovegrove et al., 1986). A normally-functioning transient system inhibits the persisting activity of the sustained system, ensuring that visual information from one eye fixation does not interfere with input from the next fixation (Breitmeyer, 1984). In reading, a transient system deficit could result in the superimposition, or masking, of words from one fixation to the next, and in inefficient eye movements across a line of text. Much of the research demonstrating transient system deficits among disabled readers has involved the use of basic psychophysical paradigms and stimuli such as spatial frequency gratings that have very limited similarity to text characteristics.

BACKWARD MASKING EFFECTS IN DISABLED READERS

One frequently used approach for investigating the visual processes of disabled readers has been to use backward masking to assess visual processing of a target stimulus. Masking tasks provide a measure of the rate of input of visual information at the initial stages of visual processing. The typical paradigm involves presenting a stimulus, called the target, for a brief duration followed by another stimulus, called the mask, in close temporal contiguity. Normally the interval between the target and the mask is varied. Experiments utilizing this method have shown that disabled readers take longer than normal readers to process visual information, as indicated by the need for a longer time interval between the target and the mask to perform at a pre-set criterion level. Results like these suggest that disabled readers do not process visual information as quickly as normal readers (Willows, 1991).

In a study of masking of consonants and vowels, Stanley & Hall (1973; Experiment 2) found that disabled readers required longer intervals between the presentation of a target letter and a mask to reach a criterion level of letter naming performance than did normal readers. This showed that disabled readers processed visual information slower than normal readers.

DiLollo et al. (1983) carried out a study with normal and disabled readers in which single letters were masked, and found slower rates of visual information processing in disabled readers. They suggested that their findings reflected differences in the recovery time from the after effects of neural activity that was evoked by the presentation of the targets.

A number of masking studies have explored specific predictions based upon a transient deficit explanation of reading disability. Using a different type of masking procedure, Slaghuis & Lovegrove (1984) tested the hypothesis of a weak transient inhibitory mechanism in disabled readers. The paradigm, developed by Breitmeyer, Levi, & Harwerth (1981), involved flicker masking of sinusoidal gratings. Slaghuis et al. found that under unmasked conditions, disabled readers performed differently from normal readers. But with a mask, which reduced the involvement of the transient system, both normal and disabled readers performed the same. The visible persistence differences between the two groups were essentially minimized.

Williams, LeCluyse & Bologna (1990) masked a shape displayed on a computer screen by a block of light to explore speed of visual information processing differences between normal and disabled readers. They measured detection accuracy as a function of stimulus onset asynchrony (SOA), the time between the initial presentation of the first item and the time of the onset of the mask stimulus. They found prolonged masking in disabled readers, indicating a transient deficit. Moreover, Williams, Molinet & LeCluyse (1989) found disabled readers displayed strongest masking effects at shortest SOA's with foveal presentation of materials, again demonstrating a transient deficit in disabled readers.

The degree to which this masking effect manifests itself with textual material has been relatively little investigated within controlled reading contexts with reading-disabled children. Williams and her associates, however, have conducted a series of experiments designed to explore the possible consequences of a transient system deficit on the reading performance of disabled and normal readers. Although the results of these experiments can only be considered as tentative, since they have been reported without any accompanying statistical analysis (e.g. Williams & LeCluyse, 1990), they warrant consideration here because they represent the only systematic attempt to move from basic psychophysical findings to more externally valid procedures for comparing the basic visual processes of disabled and normal readers.

Williams & LeCluyse (1990) reported a study of masking involving letters, wherein the target letters were either presented alone, or in the context of a three-letter mask, which, when integrated with the mask, formed words. Unlike normal readers, disabled readers failed to show improved performance for letter recognition in the context of a word; that is, the simultaneous presentation of the target and mask. This result showed a different pattern of visual processing between the two groups of readers, because the shapes of the masking functions indicated that the time course for processing words was different for each group. The investigators also performed the same investigation under conditions that blurred the image, which reduced contrast particularly for higher spatial frequencies. They found that the performance of disabled readers, with blurred materials, paralleled the performance of the disabled readers with clear stimuli. Image blurring may have decreased the strength of the sustained response. The authors argued that the data indicated "sluggish transient activity" in disabled readers, resulting in lowered temporal separation between transient and sustained responses. The positive effects of blur might improve reading performance for disabled readers.

Research with backward masking showed that disabled readers may have deficits in their visual processes that could impair reading. However, studies involving continuous text are necessary to demonstrate that the observed visual processing deficits can have a direct effect on reading.

RESEARCH WITH READING-LIKE TASKS

A final study reported by Williams & LeCluyse (1990) involved a direct test of image blurring effects on reading. Reading comprehension was measured for standardized reading passages, under three word presentation conditions representing different levels of eye movement control. The presentation rates of the passages were selected according to subjects' reading abilities as measured by a standardized test. The mode of presentation did not have an effect on the comprehension performance of normal readers, but disabled readers did worse in a free-eye-movement condition than in the other two conditions. Their performance improved in the free-eye-movement condition with blurred text, which effectively established a normal pattern of interactions between transient and sustained channels.

The results reported by Williams and her colleagues indicate that a transient deficit may indeed manifest itself in reading activities. However, because their studies can only be viewed as preliminary due to the lack of reporting of statistical evidence, additional research is needed to examine more explicitly the functional consequences of a transient system deficit on disabled readers. Moreover, the research reported by Williams & LeCluyse (1990) had a number of limitations. For instance, with the use of comprehension measures in studying transient system deficits and reading disabilities, factors other than those involved with visual processing may have played a role in mediating the patterns of results obtained. A greater degree of control over the presentation rates of words, and a more objective, or direct, measure of the operation of the visual processing system may clarify some of the potential sources of additional effects. These issues point to the need to explore other aspects of the reading process, in addition to comprehension performance; it cannot be assumed that comprehension reflects the full range of processes involved in reading, and by implication the complete pattern of effects of visual processing deficits upon reading. Techniques that permit a wide range of control over the manner in which stimulus material are presented may provide a means to explore directly activities that are part of the reading process, like recognition and form discrimination, that may not involve the same degree of complexity as those involved in comprehension.

Research that has examined reading with disabled populations has made use of a means of displaying textual material called Rapid Serial Visual Presentation (RSVP). Text is presented at a single point, and thus it eliminates the need for eye movements in reading. RSVP has been cited by a number of researchers for its potential utility in the study of reading disability (e.g., Potter, 1984; Pollatsek, 1983) and as a way of compensating for potential disabling effects. Indeed, as Pollatsek states:

> "One technique that could be used to distinguish among eye control problems is to assess whether the reading difficulty is much less when the requirement to move the eyes is removed. Normal subjects can read text presented rapidly and sequentially word by word in a single location .... It would be interesting to see whether dyslexics with oculomotor problems could also read well using this rapid serial visual presentation (RSVP) technique. If so, it would suggest that their problem is one of eye movement guidance as opposed to a problem in maintaining fixation. Such a technique might also be of some value as an aid to such dyslexics."
> (p. 518).

However, very little has been done to investigate the utility of RSVP in the study of reading disability in children. Research has shown that normal readers do better with RSVP than full-page presentation on tasks involving visual search (Lawrence, 1971), and sentence recall (Potter, 1984). The elimination of inefficient eye movements, it is argued, would raise the limit of the rate and effectiveness of text processing. In a case study of its use in compensating for reading difficulties brought about by retinitis pigmentosa, so-called "tunnel vision", RSVP was found to be effective in improving adult reading performance (Williamson, Muter, & Kruk, 1986). RSVP has also been used in a comparison study of good and poor adult readers (Chen, 1986). Recognition performance for text in RSVP was better than in full-page presentation for the poor readers. The results indicated that RSVP presentation compensated for the negative effects of inefficient eye movements, and made reading less cognitively demanding for the poor readers. However, in eliminating the need for eye movements in reading, the effects of erratic eye movements, observed in disabled readers (Pavlidis, 1983), may have been removed. Performance might have been improved simply because text was being perceived more clearly with RSVP.

The results of research on the visual processes of disabled readers in reading-like contexts point to possible explanations for the nature of visual deficits and their effects on reading, but there is a clear need to substantiate the claims made within a strictly-controlled methodology, using a wider range of dependent measures. In this way definitive conclusions regarding visual processing deficits in disabled readers can be made.

BACKWARD MASKING RESEARCH WITH READING-LIKE TASKS

There is a great deal of evidence to show that normal and disabled readers differ substantially in their visual processing abilities (Willows, Kruk, & Corcos, in press). Research findings have indicated a possible explanation for a visual processing deficit in disabled readers, based upon inefficient transient system functioning, and possible interventions for this deficit based upon the physical characteristics of text. However, given the limitations of the research, there is a need to investigate these issues further, particularly in the context of more direct measures of visual processes in reading.

Kruk (1991) carried out a study to examine the possible functional consequences of a transient visual system deficit in disabled readers. The inability of a deficient transient system to inhibit sustained system activity might result in visible persistence that interferes with the processing of subsequent visual input. Furthermore, deficient visual processing can result in erratic eye movements that could impair reading (Pavlidis, 1983). However, much of the research studying disabled readers' eye movements is controversial, given the difficulties in replicating some early findings (Olson, in press).

Three experiments, involving gradual approximations toward more reading-like situations, were designed to examine how visual processing differences might be manifested between normal and disabled readers using textual materials, and whether a transient deficit explanation could account for possible differential patterns of performance by the two reader groups. All three experiments involved visual form discrimination in a backward masking paradigm. SOA was varied to provide a measure of masking effects. In the context of a "computer game" the child's task was to press a key to indicate whether a target and test item were the same or different.

The general hypothesis was that with a longer visible persistence duration, disabled readers should experience different rates of release from the effects of masking than normal readers. Normal readers would be released from masking effects at a quicker rate because transient-on-sustained inhibition would minimize the negative consequences of superimposition of successive visual stimuli presented at shorter SOA's. Disabled readers, on the other hand, would not benefit from a normally functioning transient-on-sustained inhibition mechanism, and hence stronger masking effects would be observed at longer SOA's. Disabled readers were expected to require longer SOA's to escape masking effects than normal readers. This result would be indicated by a significant interaction effect between the variables of reader group and SOA, showing faster improvement by normal readers than disabled readers as SOA increased. A deficit in transient activity might also result in impaired integration of visual information across fixations. If low spatial frequency information plays a role in facilitating this process, then a deficient transient system might disrupt the way peripheral and central visual information are integrated. Furthermore, given the role that the transient system has been hypothesized to play in saccadic suppression (Breitmeyer, 1984), this process too might not be operating at an optimal level in disabled readers. Hence, in comparisons with mode of presentation of textual materials, disabled readers should perform less well with a display that requires eye movements within a masking paradigm, than a mode that does not require eye movements.

Kruk's (1991) experiments were carried out with groups of disabled and normal readers (aged eight to eleven years) attending English-language elementary schools. Criteria for involvement in the study for all participants included average to above average scores on measures of verbal ability, no complications due to language disorders or behavioural disturbances, and ability to recognize the stimulus words used in the study. Those participants scoring at least two years below grade level on the reading measures, the comprehension subtest of the Gates-McGinitie Reading Test (McGinitie, 1980) and the reading subtest of the Wide Range Achievement Test -Revised (Jastak & Wilkinson, 1984), were chosen as disabled-reader group subjects. The normal-reader group was selected to match as closely as possible the disabled-reader group on the basis of age, gender, verbal ability, and family background. The normal readers had to perform at or above their expected levels on the measures of reading ability.

In the first experiment, pairs of unfamiliar visual stimuli, Japanese letters with spatial frequency characteristics similar to English letters, were shown in rapid succession on a computer screen, with a mask presented after the first stimulus. Figure 1 illustrates schematically the time course of events of a typical trial. The experiment involved 168 trials for each of the 17 normal and 16 disabled readers who participated. Items and conditions were fully randomized. The mean proportion accuracy was the primary dependent measure.

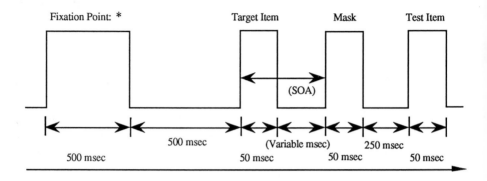

FIG. 1. A Schematic representation of a trial in Experiment 1

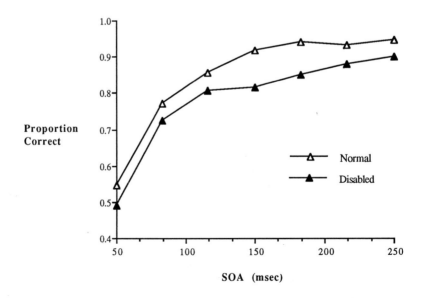

FIG. 2. Experiment 1 accuracy performance of normal and disabled readers at each SOA level

The results, depicted in Fig. 2, showed significant Group differences (F 1, 31= 7.44, p<.01), with normal readers scoring higher than disabled readers, and a progressive increase in overall performance to an asymptote level was found as SOA increased. These data supported the hypothesis of visual processing differences between normal and disabled readers, because of the observed group differences. However, as there was no indication of a Group by SOA interaction, a transient deficit explanation was not supported.

The second experiment carried out by Kruk (1991) involved familiar materials, English letters and words, as well as unfamiliar non-words made up of English letters. Like the first experiment, a mask was inserted, at varying SOA's, after the first stimulus. The accuracy scores for letter data are plotted in Fig. 3. Significant differences were found between the two reader groups (F 1, 35 = 15.87, p < .001), which were made up of 19 normal and 18 disabled readers. Similar Group differences for word and nonword items were also obtained.

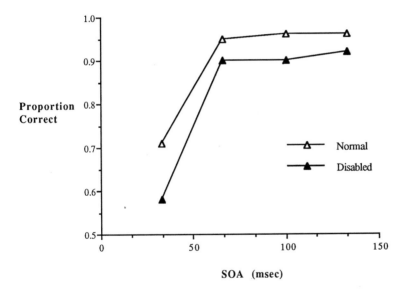

FIG. 3. Experiment 2 accuracy performance for Letter items of normal and disabled readers at each SOA level.

Performance on a masking task involving letters showed a strikingly similar pattern of results to Experiment 1, despite the added degree of familiarity. Again, it was concluded that the data indicated a visual processing difference between normal and disabled readers. However, a transient system deficit explanation for these data could not provide a complete account, not only because of the absence of a significant Group by SOA interaction, but also because of the possibility that additional non-visual factors, like labelling, may have mediated performance.

The third experiment conducted by Kruk (1991) introduced complete sentences, with target words embedded, to provide a closer approximation to "natural" reading. Sentences were presented in one of two modes in order to examine possible differences in eye movement efficiency. RSVP displayed words sequentially at a fixed location, and a Moving Window mode presented single words in temporal sequence across the screen from left to right. SOA between the target word and the next word in the sentence was varied, in order to produce backward masking. The independent variables in the third experiment involved Group (normal and disabled), Trial Type (Same and Different), SOA (100, 166, 233, and 300 msec), Target Word Type (Word and Nonword), and Mode of Presentation (RSVP and Moving Window). A total of 192 trials were administered to the 17 normal and 14 disabled readers. A schematic depiction of a trial in Experiment 3 is given in Fig. 4.

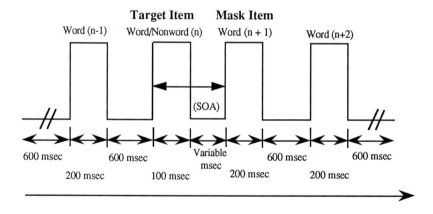

FIG. 4. A schematic representation of a trial in Experiment 3

Results showed significant differences in accuracy between normal and disabled readers (F 1, 29 = 50.03, p < .001). Overall, normal readers performed better than disabled readers in accuracy, but there were no Group by SOA or Group by Mode interactions. The mean proportion correct scores over SOA levels for normal and disabled readers are given in Fig. 5.

The results indicated that visual processing differences between the two groups of readers can manifest themselves within the context of "natural" reading; using materials like those encountered in reading text. However, hypothesized mechanisms that mediate performance, based upon a transient system deficit, may not provide a complete account for the effects obtained.

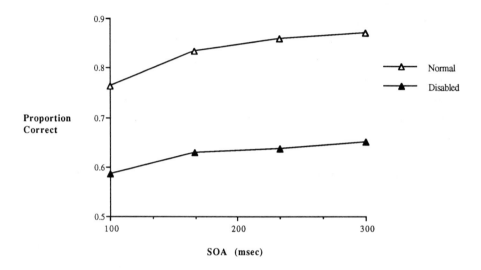

FIG. 5. Experiment 3 accuracy performance of normal and disabled readers at each SOA level.

The results showed that as the quality of the visual input improved as SOA increased, accuracy performance improved. The rate of improvement did not differ between normal and disabled readers. Nevertheless, a general pattern emerged in the outcome of the masking effect. Typically, both groups reached asymptote levels of performance at high SOA's, with disabled readers performing at a lower level than normal readers.

The results of Kruk (1991) give a strong indication of what appears to be a deficit in the visual processes of disabled readers within the context of a "reading-like" environment. Although his study had many limitations, procedures were implemented to make it as "clean" a test of visual processing involvement with reading materials as possible. It represents a step towards an understanding of visual processing differences between normal and disabled readers within a natural reading context, since it explored a basic process involving reading; namely, visual form discrimination. The results, though, do not give clear support for a transient-deficit explanation.

CONCLUSION

There is a great deal of evidence to show that normal and disabled readers differ in their visual processing abilities, in a variety of contexts involving abstract patterns and textual information. Studies of basic visual processes, involving psychophysical methods, have provided compelling evidence for a possible explanation for differences in visual processes, based upon a transient system deficit in disabled readers. Research looking at the speed of visual information processes, using backward masking techniques has shown that disabled readers' visual systems operate at a less efficient rate than that of normal readers; some of the data also supporting a transient system deficit explanation.

Several studies have taken a more ecologically valid perspective, seeking to examine how differences in visual processes between normal and disabled readers might be observed under conditions that approximate "natural" reading. Visual processing differences under these circumstances were found in tasks that involved recognition, comprehension, and visual form discrimination. Although the studies found evidence for differences in the visual processes of normal and disabled readers, the findings were interpreted in different ways. Williams and her colleagues interpreted their evidence, based upon comprehension measures, as consistent with a transient system deficit. However, the measure used gave indirect evidence of the effects of visual processing differences between the two groups. The data from Kruk (1991) indicated that differences existed using a more direct measure of visual processing. However, the results of his study could not be interpreted using a transient deficit explanation of the differences in visual processing between normal and disabled readers.

The research carried out to this point shows that, contrary to previous assumptions, visual processes may play a role in reading disability. However, the manner in which those processes are related to reading difficulties needs to be investigated further. In addition, given that visual processing differences have been identified in comparisons of normal and disabled readers, there is a need to examine how these differences can be explained.

REFERENCES

Badcock, D., & Lovegrove, W. (1981). The effects of contrast, stimulus duration, and spatial frequency on visible persistence in normal and specifically disabled readers. *Journal of Experimental Psychology: Human Perception and Perfor-mance,* 7, 495-505.

Breitmeyer, B.G. (1984). *Visual Masking: An Integrative Approach.* New York: Clarendon Press.

Breitmeyer, B.G., Levi, D.M., & Harwerth, R.S. (1981). Flicker masking in spatial vision. *Vision Research,* 21, 1377-1385.

Coltheart, M. (1980). Iconic memory and visible persistence. *Perception and Psychophysics,* 27, 183-228.

Chen, H. C. (1986). Effects of reading span and textual coherence on rapid-sequential reading. *Memory and Cognition,* 14, 202-208.

DiLollo, V., Hanson, D., & McIntyre, J.S. (1983) Initial stages of visual information processing in dyslexia. *Journal of Experimental Psychology: Human Perception and Performance,* 9, 923-935.

Hulme, C. (1988). The implausibility of low-level visual deficits as a cause of children's reading disabilities. *Cognitive Neuropsychology,* 5, 369-374.

Jastak, S., & Wilkinson, G. (1984). *The Wide Range Achievement Test-Revised.* Wilmington, DE: Jastak Associates.

Kruk, R.S. (1991). *Functional Consequences of a Transient Visual Processing Deficit in Reading Disabled Children.* Unpublished doctoral dissertation. University of Toronto.

Lawrence, D. (1971). Two studies of visual search for word targets with controlled rates of presentation. *Perception and Psychophysics,* 10, 85-89.

Lovegrove, W.J, Heddle, M., & Slaghuis, W. (1980). Reading disability: Spatial frequency specific deficits in visual information store. *Neuropsychologia,* 18, 111-115.

Lovegrove, W., Martin, F., & Slaghuis, W. (1986). A theoretical and experimental case for a visual deficit in specific reading disability. *Cognitive Neuropsychology,* 3, 225-267.

Lovegrove, W.J., & Slaghuis, W. (1989). How reliably are visual differences found in dyslexics? *The Irish Journal of Psychology, 10,* 542-550.

McGinitie, W.H. (1980). *Gates-McGinitie Reading Tests: Canadian Edition.* Toronto: Nelson.

Olson, R. (in press). Comparison of eye movements of disabled and normal readers. In D. Willows, R. Kruk, and E. Corcos (Eds.), *Visual Processes in Reading and Reading Disability.* Hillsdale, NJ: Erlbaum.

Pavlidis, G.T. (1983). How can dyslexia be objectively diagnosed? *Reading,* 13, 3-15.

Pollatsek, A. (1983). What can eye movements tell us about dyslexia? In K. Rayner (Ed.), *Eye Movements in Reading: Perceptual and Language Processes.* New York: Academic Press.

Potter, M.C. (1984). Rapid serial visual presentation (RSVP): A method for studying language processing. In D.E. Kieras and M.A. Just (Eds.), *New Methods in Reading Comprehension Research.* Hillsdale, NJ: Erlbaum.

Slaghuis, W.L., & Lovegrove, W.J. (1984). Flicker masking of spatial-frequency-dependent visible persistence and specific reading disability. *Perception, 13,* 527-534.

Stanley, G., & Hall, R. (1973). Short-term visual information processing in dyslexics. *Child Development,* 44, 841-844.

Stanovich, K.E. (1982). Individual differences in the cognitive processes of reading: I. Word decoding. *Journal of Learning Disabilities,* 15, 485-493.

Stanovich, K.E. (1985). Explaining the variance in terms of psychological processes: What have we learned? *Annals of Dyslexia,* 35, 67-96.

Vellutino, F.R. (1987). Dyslexia, *Scientific American,* 256, 34-41.

Williams, M.C., & LeCluyse, K. (1990). The perceptual consequences of a temporal processing deficit in reading disabled children. *Journal of the American Optometric Association,* 61, 111-121.

Williams, M.C., LeCluyse, K., & Bologna, N. (1990). Masking by light as a measure of visual integration time in normal and disabled readers. *Clinical Vision Sciences,* 5, 335-343.

Williams, M.C., Molinet, K., & LeCluyse, K. (1989). Visual masking as a measure of temporal processing in normal and disabled readers. *Clinical Vision Sciences,* 4, 137-144.

Willows, D.M. (1991). Visual processes in learning disabilities. In B.Y.L. Wong (Ed.), *Learning About Learning Disabilities.* New York: Academic Press.

Willows, D.M., Kruk, R.S., & Corcos, E. (in press). *Visual Processes in Reading and Reading Disability.* Hillsdale, NJ: Erlbaum.

Williamson, N.L., Muter, P., & Kruk, R.S. (1986). Computerized presentation of text for the visually handicapped. In E. Hjelmquist and L.-G. Nilsson (Eds.), *Communication and Handicap: Aspects of Psychological Compensation and Technical Aids.* Amsterdam: Elsevier Science Publishers (North-Holland).

Facets of Dyslexia and its Remediation
S.F. Wright and R. Groner (Editors)

SOME REFLECTIONS ON PSYCHOPHYSICAL MEASUREMENT WITH DYSLEXIC CHILDREN

P. U. Müller,
University of Bern,
Switzerland.

INTRODUCTION

Psychophysical experiments with adult observers is usually a very time-consuming and rather boring task for the subjects. They sit in a darkened room, and try for hours to detect small lines of light (Wilson, 1978), identify missing dots (DiLollo, 1977) or decide which way a circle appears to turn (Bischof & Groner, 1985). On the basis of these experiments psychophysical functions are determined, usually under different conditions. This requires a high degree of accuracy of the measurements and a large quantity of measurements. There are some differences in investigations with dyslexic children. Only some points on the functions relating probability of detection to physical stimulus intensity are tested and the accuracy has just to be as high, to permit the detection of differences between the groups. Therefore fewer measurements of lower accuracy are often sufficient. A similar situation exists for group comparisons with psychophysical measurements in adults (e. g. Groner et al., 1992), a growing field of research.

What are the Requirements for a 'Good' Subject in Psychophysical Research ?
1. Subjects must be motivated in some way to carry out the experiment. The usual means are either extrinsic motivation by rewards like money or credits, or intrinsic motivation of the observer by his interest in the results.
2. Subjects must be able to concentrate for some time on the tasks. Even adults experience difficulties, up to the point of falling asleep during the experiment.
3. Subjects must be able to translate their sensation or perception into the form, which is recorded by the experimenter. This task is usually so easy for adults that this step is not taken into account. However, there are difficulties in some experiments with respect to this point, as the paradigm for measuring visible persistence (e.g. Hogben & DiLollo, 1974).

What are the Special Problems with Children?
Motivation: For the children a financial attraction would probably work, but this might be unacceptable for their parents. More suitable are small gifts or credit notes. Parents must

be convinced of the usefulness of the investigation, otherwise they may not allow their children to take part. Since there is often no personal profit for the investigated child, one has to convince their parents of the scientific value of the research. It can be assumed that this is easier with more educated parents. In addition it can not be excluded that secondary behaviour problems of dyslexic children (Thomson, 1991) yield a further preselection.

Lasting Attention and Alertness: Amongst practitioners it is well known that children can not concentrate for as long as adults, especially in situations where they receive no feedback concerning their performance. This must be considered in the design of experiments for children. In addition there is a serious methodological problem in dyslexia research. It has been reported (Linder, 1951) that dyslexic children are less concentrated, have a tendency to hyperactivity and tire faster. Therefore, it should be checked, to what degree the results of an experiment could be explained by lower concentration or alertness in dyslexic children.

Cognitive Load: In most experiments a simple yes/no answer is required which should also be easy for school children. But sometimes even the translation of a simple yes/no response into the pressing of the adequate button seems to be rather difficult. This problem increases, if more complex answer-schemes are used, or if decisions about the position (especially right/left) are required. As with attention, the important question is, whether there are specific deficits, other than the one under question, which could account for differences between dyslexics and good readers in a psychophysical experiment. Probable candidates could be a short term memory deficit (e.g. Jorm, 1983) or a problem with sequencing (e.g. Shapiro, Ogden & Lind-Blad, 1990).

These problems make psychophysical measurements with children, and especially with dyslexic children, more difficult than with adults. Given that good readers do not differ from dyslexics on some variables which influence psychophysical measurements, the problems can be resolved with a careful experimental design. The most powerful method of avoiding a bias from differences in higher cognitive functions is a psychophysical task which is better solved by dyslexic than normal children using a forced choice paradigm.

In the next section some methods to estimate thresholds will be mentioned and the procedures used in investigations with dyslexic children will be outlined.

THRESHOLD MEASUREMENTS WITH DYSLEXIC CHILDREN

For a long time, the notion of a transient system deficit in dyslexia was based solely on psychophysical research. In more recent research, physiological and anatomical data (e.g. Livingstone, Rosen, Drislane & Galaburda, 1991) support this hypothesis. However, the question of how the psychophysical results were obtained remains interesting.

Psychophysics can be described as the study of the sensations evoked by physical stimuli. In the case of thresholds only one point on this function relating physics to 'psychics' is considered: the physical intensity at which a stimulus or a stimulus property can just be perceived. However there is a problem with this point - there is no discrete change between visibility and non-visibility but a gradual transition. Usually, the threshold is defined as the physical intensity where the property in question can be detected in 50% of the cases. Actually this definition is based on the method of constant stimuli: the stimuli are presented at different (but fixed) intensities and the frequency of detecting that property is determined. This process takes considerable time, because the stimulus has to presented frequently to obtain a accurate measurement of the probability For instance, in order to obtain a 95% confidence interval between 0.4 and 0.6 of the estimated probability at threshold 98 trials are needed (Hays, 1988). Only a few studies with dyslexic children have used the method of constant stimuli (Geiger & Lettvin, in this book; Willows, Corcos & Kershner, in this book) to determine one or a small number of psychometric functions. However, other more efficient procedures have been established for finding the threshold.

Two classical alternatives are the method of limits from which staircase methods were derived and the method of adjustment. The method of limits can be used either with a judgement or with a forced choice task, whereas the method of adjustment requires judgement. Judgements have the disadvantage, that they are more influenced by criteria and attitudes of the observer than forced choice tasks. This holds even more for untrained subjects, as is usually the case in dyslexia research. With trained observers, however the method of adjustment is one of the most efficient. With children it has the advantage, that they can actively do something and see the effect they cause which obviously increases motivation. With respect to criteria, the signal detection theory should be mentioned which allows for a differentiation between criteria and sensitivity. However, it is almost impossible to apply it in dyslexia research, because approximately 500 trials are needed to determine the decision criterion and the sensitivity of a subject at a given stimulus intensity (Green & Swets, 1974).

One of the first studies investigating visual temporal processing in dyslexics (Stanley & Hall, 1973) was done using the method of limits to measure visible persistence, Lovegrove et al. (1982) used the method of adjustment for obtaining the contrast sensitivity function, and Winters, Patterson & Shontz (1989) applied signal detection theory in a investigation of visible persistence with adult dyslexics. But the majority of studies was done using adaptive procedures.

Based on some assumptions and the availability of computers, several adaptive threshold tracking procedures have been developed. They are based on the fact, that maximum information is available around the threshold intensity itself (Smith, 1961) and on assumptions concerning the form of the function relating probability of detecting a stimulus to stimulus intensity.

The up-and-down methods as well as the staircase methods can be seen as generalizations of the method of limits: the run is not stopped after the first reversal but continues until a certain criterion is reached (Dixon & Mood, 1948; Brownlee, Hodges & Rosenblatt, 1953; Wetherill, 1963; Wetherill & Levitt, 1965; Jestaedt, 1980). These methods work without assumptions about the psychometric function and can be varied in many ways, as length of runs, required precision, and point estimated on the function. Wetherill & Levitt (1965) require different numbers of incorrect trials to decrease the intensity of the stimulus than they require correct trials to increase the intensity level. However, the same effect could be obtained by differentially increasing or decreasing the intensity. Some pre-assessment is also needed with naive subjects. If the stepsize of this method is constant, it takes a long time to reach threshold. This means that a lot of trials can be wasted in what is a completely uninteresting range.

Using a statistical decision procedures (Wald-Runs-Test) a decreasing step size and some other features, Taylor & Creelman (1967) described a highly elaborated tracking procedure. Findlay (1978) proposed two additions, which should make it even more efficient. Finally, Pentland (1980; an implementation is described by Liberman & Pentland, 1982) found the mathematically most efficient procedure: the best PEST. Based on an assumed psychometric function after every response the point of most information is determined based on all previous responses. On the one hand, no information at all is lost - which is the reason why it is the best procedure from a mathematical point of view - but this procedure has two (practical) disadvantages: every error is included in the computation of the threshold and after a few initial trials all stimuli are shown at threshold level which is very tiring - a similar procedure is proposed by Watson & Pelli (1988).

For computer sorting and comparable tasks binary search works very well. In psychophysical research it is not directly applicable, because an error has fatal consequences i.e: it cannot be corrected later. A modified binary search (MOBS) described by Tyrell & Owens (1988) avoids this problem elegantly and is suited for the measurement of fluctuating thresholds.

The staircase method in several variations was used by Cornelissen (this book) for 6 different contrast thresholds, Evans, Drasdo & Richards (this book) determined one flicker threshold. Lovegrove (this book) worked in investigations of flicker, uniform field flicker masking, contrast sensitivity functions and visible persistence and Williams (this book) in studies of temporal sequence and flicker. Arnett & DiLollo (1979) and DiLollo, Hanson & McIntyre (1983) used PEST for visible persistence and masking. None of the more sophisticated and mathematically most efficient procedures as BESTPEST or QUEST were used. Why? Before attempting to answer this question, some studies comparing different techniques will be reviewed.

There have been few studies comparing these different psychophysical techniques, and only some of them will be mentioned here. Blackwell (1952) compared phenomenal report with forced choice tasks and found that there was a bias in phenomenal report in the direction of 'yes' responses which he termed 'positive channelization'. He argued, that the difference between these two task types can be explained by the low quality of phenomenal report. Considerable differences in measures of sensitivity caused by different types of tasks were also found by Creelman & Macmillan (1979). An interesting comparison was made by Shelton, Picardi & Green (1982), who found that with the exception of minor differences the efficiency of PEST (accuracy after a comparable number of trials), the simple staircase and maximum-likelihood methods were the same. This is surprising since from a mathematical point of view, the maximum-likelihood procedure should be the most efficient. The result of Shelton et al. indicate that there are some effects on the threshold measure which are not considered in the mathematical model of maximum-likelihood methods. Evidence for further components in threshold tracking comes also from results of Taylor & Forbes (1983) comparing PEST and the method of constant stimuli. PEST seems to be more accurate, i.e. has a higher reliability which indicates a higher sensitivity and yields a shallower psychometric function. Telage & Fucci (1974) found no differences in accuracy between one- and three-trial thresholds, which could also be due to some additional factors not considered by conventional threshold-determination-modeling.

Not all procedures allow for the use of feedback. Only in forced choice procedures can correct and incorrect responses be distinguished. Constant stimuli are easy, because the difficulty remains constant. However in most adaptive tracking procedures the difficulty of a trial changes often and the meaning of the feedback changes accordingly. One possibility would be to indicate the actual stimulus intensity level to the subject and instruct the subject accordingly. This issue merits consideration because it includes possibilities of motivating children.

Some Critical Remarks

If the number of applications is taken as an indicator of the appropriateness of a procedure, it seems clear that the staircase methods are the most appropriate. There are clearly some advantages of these procedures:
- They are not very susceptible to errors of the subjects. Errors can be corrected and do not have much influence on the result.
- They are quite efficient, at least if several levels or step sizes are used.
- They can easily be used in a forced choice paradigm.
- Feedback can be given, although not as easily as with constant stimuli.

All of the other procedures mentioned have at least one severe disadvantage:
- The method of adjustment does not allow for forced choice tasks.
- The method of constant stimuli is very inefficient for determining just one threshold.

- Maximum-likelihood-methods are susceptible to errors and are very tiring for the subject.

There are also mixed form procedures as for example the PEST proposed by Findlay (1978) and specially tailored procedures as MOBS of Tyrell & Owens (1988). But it is doubtful whether they work better than the simple methods such as the ones proposed by Wetherill & Levitt (1965).

One could conclude that although the psychophysical procedure chosen can help greatly, the most critical variable is the motivation of the child to perform well in the task. One possibility of increasing the attractiveness of the task is a feedback. Good results were obtained with more or less direct feedback.

An empirical investigation of the components mentioned in psychophysical measurement with children would be highly desirable. One step in this direction could be the construction of models which take these components into account and could be simulated on a computer.

REFERENCES

Arnett, J. L. & DiLollo, V. (1979). Visual information processing in relation to age and to reading ability. *Journal of Experimental Child Psychology*, 27, 143-152.

Bischof, W.F. & Groner. M. (1985). Beyond the displacement-limit: an analysis of short-range processes in apparent motion. *Vision Research*, 25, 839-847.

Blackwell, H. R. (1952). The influence of data collection procedures upon psychophysical measurement of two sensory functions. *Journal of Experimental Psychology*, 44, 306-315.

Brownlee, K.A., Hodges, J.L. & Rosenblatt, H. (1953). The up and down method with small samples. *Journal of the American Statistical Society*, 48, 262-277.

Creelman, C. D. & Macmillan, N. A. (1979). Auditory phase and frequency discrimination: A comparison of nine procedures. *Journal of Experimental Psychology: Human Perception and Performance*, 5, 146-156.

DiLollo, V. (1977). Temporal Characteristics of iconic memory. *Nature*, 267, 241-243.

DiLollo, V., Hanson, D. & McIntyre, J. S. (1983). Initial stages of visual information processing in dyslexia. *Journal of Experimental Psychology: Human Perception and Performance*, 9, 923-935.

Dixon, W. J. & Mood, A. M. (1948). A method for obtaining and analyzing sensitivity data. *Journal of the American Statistical Association*, 43, 109-126.

Findlay, J. M. (1978). Estimates on probability functions: A more virulent PEST. *Perception & Psychophysics*, 23 , 181-185.

Green, D. M. & Swets, J. A. (1974). *Signal Detection Theory and Psychophysics*. New

Groner, M.T., Fisch, H.U., Walder, R., Groner, R., Hofer, D., Koelbing, U., Duss, I., Bianchi, R. & Bircher, B. (1992). Specific effects of midazolam on human visual receptive fields in light and dark adapted subjects. (submitted)

Hays, W. L. (1988). *Statistics* (4th edition). New York: Holt, Rinehart & Winston.

Hogben, J. H. & DiLollo, V. (1974). Perceptual integration and perceptual segregation of brief visual stimuli. *Vision Research*, 14, 1059-1069.

Jestaedt, W. (1980). An adaptive procedure for subjective judgements. *Perception and Psychophysics*, 28, 85-88.

Jorm, A. F. (1983). Specific reading retardation and working memory: A review. *British Journal of Psychology*, 74, 311-342.

Liberman, H. R. & Pentland, A. P. (1982). Microcomputer-based estimation of psychophysical thresholds: The Best PEST. *Behaviour Research Methods & Instrumentation*, 14, 21-25.

Linder, M. (1951). Ueber Legasthenie (spezielle Leseschwäche). *Zeitschrift für Kinderpsychiatrie*, 18, 97-143.

Livingstone, M.S., Rosen, G.D., Drislane, F.W. & Galaburda, A.M. (1991). Physiological evidence for a magnocellular defect in developmental dyslexia. *Academy of Sciences*, USA, 88, 7943-7947.

Lovegrove, W., Martin, F., Bowling, A., Blackwood, M., Badcock, D. & Paxton, S. (1982). Contrast sensitivity functions and specific reading disability. *Neuropsychologia*, 20, 309-315.

Pentland, A. (1980). Maximum likelihood estimation: The best PEST. *Perception and Psychophysics*, 28, 377-379.

Shapiro, K. L., Ogden, N. & Lind-Blad, F. (1990). Temporal processing in dyslexia. *Journal of Learning Disabilities*, 23, 99-107.

Shelton, B. R., Picardi, M. C. & Green, D. M. (1982). Comparison of three adaptive psychophysical procedures. *Journal of the Acoustical Society of America*, 71, 1527-1533.

Smith, J. E. K. (1961). Stimulus programming in psychophysics. *Psychometrica*, 26, 27-33.

Stanley, G. & Hall, R. (1973). Short-term-visual-information-processing in dyslexics. *Child Development*, 44, 841-844.

Taylor, M. M. & Creelman, C. D. (1967). PEST: Efficient Estimates on Probability Functions. *Journal of the Acoustical Society of America*, 41, 782-787.

Taylor, M. M. & Forbes, S. M. (1983). PEST reduces bias in forced choice psychophysics. *Journal of the Acoustical Society of America*, 74, 1367-1374.

Thomson, M. (1991). *Developmental Dyslexia* (3rd ed.). London: Whurr publishers.

Tyrell, R. A. & Owens, D. A. (1988). A rapid technique to assess the resting states of the eyes and other threshold phenomena: The Modified Binary Search (MOBS). *Behaviour Research Methods, Instruments, & Computers*, 20, 137-141.

Telage, K. M. & Fucci, D. J. (1974). Measurement of lingual vibrotactile sensitivity using one-trial and three-trial threshold criteria. *Bulletin of the Psychonomic Society*, 3, 373-374.

Watson, A. B. & Pelli, D. G. (1983). QUEST: A Bayesian adaptive psychometric method. *Perception & Psychophysics*, 33, 113-120.

Wetherill, G. B. (1963). Sequential estimation of Quantal Response Curves. *Journal of the Royal Statistical Society: Series B*, 25, 1-48.

Wetherill, G.B. & Levitt, H. (1965). Sequential estimation of points on a psychometric function. *The British Journal of Mathematical and Statistical Psychology*, 18, 1-10.

Wilson, H. R. (1978). Quantitative prediction of line spread function measurements: Implications for channel bandwidths. *Vision Research*, 18, 493-496.

Winters, R. L., Patterson, R. & Shontz, W. (1989). Visual persistence and adult dyslexia. *Journal of Learning Disabilities*, 22, 641-645.

II. EYE MOVEMENTS

Facets of Dyslexia and its Remediation
S.F. Wright and R. Groner (Editors)
© 1993 Elsevier Science Publishers B.V. All rights reserved.

SACCADIC EYE MOVEMENTS OF DYSLEXIC CHILDREN IN NON-COGNITIVE TASKS

M. Biscaldi & B. Fischer

Department of Neurophysiology

University of Freiburg, Germany

INTRODUCTION

"Do eye movements hold the key to dyslexia?" At present there are two extreme answers to this question. Some have argued that dyslexics make an usually high number of erratic eye movements when required to scan sequential visual stimuli (e.g. Pavlidis, 1981, 1985) and that their eye movements differ from those of non dyslexic readers (e.g. Romero, Estanol, Lee-Kim, Romero & Martinez, 1982). Others, such as Olson, Kliegl & Davidson (1983), who tried to replicate the results of Pavlidis, have found the eye movements of dyslexics to be normal. It has also been argued that any erratic eye movements seen in dyslexics are a peripheral component of reading and not a cause of dyslexia (Compagnoni, Falcone, Fiorita & Nappo, 1984; Rayner, 1985).

Fischer & Weber (1990) reported that the saccadic reaction times of dyslexic children (recruited from a special school class) in a task, which required saccades from a central fixation point to a single target occurring randomly to the right or left, were different from a control group of normal readers of the same age. In particular, they found two dyslexics who produced almost exclusively express saccades, i.e. saccades after extremely short reaction times in the order of 100 - 120 ms (Fischer & Ramsperger, 1984). It remained open whether or not other aspects of the eye movements of the dyslexics in a non-cognitive single target task would also be different from those of a control group of normally reading children.

In this study we again applied the same non-cognitive saccade tasks (single targets in gap and overlap trials) and analysed the spatial and temporal parameters of the eye movements, such as size and velocity of the primary saccades, the final eye position, the number of under- and overshoots, the reaction time, the time to corrective saccades, and the fixation quality. The data were analysed by group statistics as well as by comparing the individual with the group data.

The results show that in addition to the reaction times, the spatial parameters of the eye movements were different in dyslexics when compared to age matched controls. We found that the dyslexic children formed two groups according to their performance on a psychological test battery. Differential statistical analysis revealed that the eye movement parameters were differently affected in the two groups.

METHODS

Visual Stimulation and Eye Movement Recording
We applied two eye movement tasks. Only one target was used at a time in addition to a fixation point. In the GAP task the fixation point went off 200 ms before the target appeared, in the OVERLAP task it remained visible when the target appeared randomly 4 deg to the right or to the left (Fig. 1).

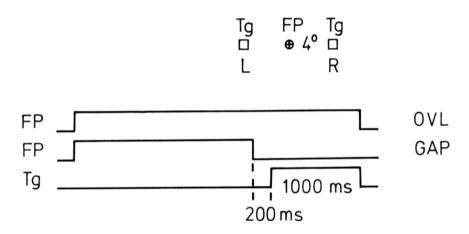

FIG. 1. Temporal and and spatial arrangements in the overlap (ovl) and gap tasks. FP= fixation point; Tg= target; L=left; R=right.

300 trials were applied with each child (75 trials for each target position in each task) and 48 parameters were analysed, which were extracted from the eye movement records. Eye movements were recorded by an infrared light reflexion method and stored on disc with a sample rate of 1 ms and a spatial resolution of 0.1 deg. The targets were computer generated white squares .2 by .2 deg in size presented on a green background at a viewing distance of 57 cm. The fixation point was a small red point, .1 by .1 deg in size. All stimuli were well above perceptual threshold.

Data Analysis

The analyses were done off-line by a computer program, which determined the beginning and the end of the first two saccades after target onset, as well as the other parameters listed below. Each eye movement trace was presented trial by trial on the computer screen together with the "results" of the computer program analysis. Mistakes made by this automatic analysis were immediately corrected by manual reanalysis of the trial or by aborting the trial altogether. For calibration the eye position signal sampled 700 ms after target onset, averaged over all 75 trials, was set to +- 4 deg.

The parameters analysed are given below:

Fix   = absolute value of eye velocity averaged over the last 200 ms before target
         onset
Amp   = Size of the first saccade after target onset (deg)
Sig   = Standard deviation of final eye position
Under = Number of undershoots as indicated by a secondary saccade in the same
         direction as the first saccade
Over = Number of overshoots as indicated by a secondary saccade in the direction
         opposite to the first saccade
VdA   = Ratio of maximal velocity and amp of the first saccade
Tcor = Time between first and second (corrective) saccade (minimum was 20 ms).
Anti   = Number of anticipated saccades, i.e. reaction times between 0 & 80 ms.
Srt   = Reaction time of the first saccade after target onset. The beginning of a
         saccade was detected by a velocity threshold of 30 deg/sec and a duration
         threshold of 15 ms.
Exp   = Number of express saccades, i.e. reaction times between 80 and 140 ms.
Freg = Number of fast regular saccades, i.e. reaction times between 140 & 200 ms.
Sreg = Number of slow regular saccades, i.e. reaction times above 200 ms.

These 12 parameters were determined for trials with the target to the right and left side and for the gap as well as for the overlap task. Altogether 48 parameters were analysed for each subject.

Subjects and Diagnostic Psychological Tests

Twenty four children, aged 9 to 11 years, were recruited from grammar schools in Freiburg. All subjects had normal visual acuity and no neurological impairments. All of them were subjected to a battery of psychological tests:

1. Handedness: tracing, dotting, tapping, H-T-D (Steingrüber, Lienert, 1976).
2. Non verbal intelligence: Raven coloured progressive matrices (Raven, 1986).

3. Attentional-concentration performance: Test d2 (Brickenkamp, 1981).

4. Serial number memory: Zahlenfolgen-Gedächtnis (ZFG) (Kubinger & Wurst,1988)

5. Serial syllables memory: Mottier test (Mottier, 1981).

6. Writing: Diagnostischer Rechtschreibtest (DRT 3) (Müller, 1983).

   Westermann Rechtschreibtest (WRT 4/5) (Rathenow, 1980).

7. Reading: Zürcher Lesetest (Grissemann & Linder, 1981).

Twelve children were accepted as members of the control group on the basis of their performance on the reading and writing tests. Each child had to score above 50% in both tests. In the other psychological tests they scored above 25%. The remaining 12 subjects did not reach the 30th percentile in reading and writing (the group means were 14% and 9%, respectively). Six of these subjects failed either Test d2, the ZFG or the Mottier (where criterion for failure was a score of less than 25%, or below 20 in the Mottier). They formed group 1 of the dyslexics in this population. The other 6 subjects had no impairment that could be detected by this test battery and were therefore grouped together as Group II dyslexics. Group means are given in Table 1.

TABLE 1

Mean Age and Scores on the 7 Diagnostic Tests

Group	Age	H-D-T	Raven	Test d2	ZFG	Mottier	Reading	Writing
Control	9 ; 9	61%	84%	61%	62%	25	83%	72%
Dyslexic I	10 ; 2	46%	67%	47%	33%	20	11%	5%
Dyslexic II	10 ; 5	28%	77%	60%	53%	24	17%	13%

RESULTS

Group Statistics: Mean Values

As a first step the pooled data of the control group with the pooled data of all 12 dyslexics were compared using the t-test. Of the 24 parameters obtained from the gap task, 12 were significantly different and from the 24 obtained from the overlap task, 17 were significantly different from the control group ($p < .0001$). Therefore, in total, 29 out of the 48 eye movement parameters (60%) differentiated normal from disabled readers.

If only the group I data were pooled and compared with the control data 21 (44%), parameters were different, 10 in the gap and 11 in the overlap data, respectively. Comparison of the group II data with the control data revealed 18 (gap task) and 15 (overlap task) different parameters. In all, 33 of the 48 (69%) parameters differentiated group II from the control group. These results are summarized in Table 2. The parameters which were statistically different ($p<.0001$) are indicated by a dot in the corresponding line.

TABLE 2

List of the 48 Parameters Obtained From the Two Tasks.

The Results of the Group Statistic are Reported: Each Dot Indicates if a Given Parameter was Equal or Significantly Larger or Smaller for the Test Groups: DYS = All Dyslexics D1 = Group I Dyslexics; D2 = Group II Dyslexics in Comparison to the Controls (CON). On the Right, the Look up Table is Shown.

Parameter		CON vs DYS N=12 N=12			D1vsCON		D2vsCON		D1 ⊙ D2 vs CON N=6 N=6					LOOK UP		
		=	>	<	>	<	>	<	==	>>	><	<>	<<	>	=	<
GAP fix	l		•		•		•			•					C	C
	r		•		•		•			•					C	C
sig	l		•		•		•			•					C	C
	r		•		•		•			•					C	C
amp	l			•				•								D2
	r			•				•								D2
v/a	l	•				•	•					•		D2	C	D1
	r		•				•							D2		
srt	l	•						•								D2
	r	•			•									D1		
corr	l	•						•								D2
	r	•						•								D2
under	l		•				•							D2		
	r		•				•							D2		
over	l			•		•		•					•	C	C	
	r			•		•		•					•	C	C	
anti	l	•							•					-	-	-
	r	•							•					-	-	-
exp	l	•					•							D2		
	r	•				•										D1
freg	l	•						•								D2
	r	•						•								D2
sreg	l	•							•					-	-	-
	r		•		•									D1		
			8	4	6	4	9	9								
OVL fix	l	•							•					-	-	-
	r		•		•		•			•					C	C
sig	l		•		•									D1		
	r		•		•									D1		
amp	l			•		•		•					•	C	C	
	r			•				•								D2
v/a	l		•				•							D2		
	r		•		•		•			•					C	C
srt	l			•				•								D2
	r	•			•			•			•			D1	C	D2
corr	l	•			•									D1		
	r	•							•					-	-	-
under	l		•		•		•			•					C	C
	r		•				•							D2		
over	l			•				•								D2
	r			•		•		•					•	C	C	
anti	l	•							•					-	-	-
	r	•							•					-	-	-
exp	l		•				•							D2		
	r		•				•							D2		
freg	l		•		•									D1		
	r	•							•					-	-	-
sreg	l			•		•		•					•	C	C	
	r			•				•								D2
			10	7	8	3	7	8								

The 16 parameters (8 for the left and 8 for the right target) that were consistently different in both tasks and in the same way (smaller or larger) are listed in Table 3 together with the corresponding mean values.

TABLE 3

Parameters that were Significantly Different in both Tasks and in the Same Way for one or the other Test Group as Compared to the Control Group. Mean Values of the Parameters in Each Group are Reported.

Filled Arrows Indicate Significant Differences in Both Tasks.

Parameter	Task	CON LE	CON RI	D1+D2 LE	D1+D2 RI	D1 LE	D1 RI	D2 LE	D2 RI	sig D1+D2 LE	sig D1+D2 RI	sig D1 LE	sig D1 RI	sig D2 LE	sig D2 RI
FIXATION QUALITY (DEG/SEC)	GAP	2.4	2.5	7.6	6.5	7.7	6.7	7.5	6.3	∧	▲	∧	▲	∧	▲
	OVL	2.9	2.3	4.5	5.4	5.8	6.3	3.3	4.4	‖	▲	‖	▲	‖	▲
SRT (MS)	GAP	161	159	155	168	165	186	144	159	‖	‖	‖	▲	▼	‖
	OVL	252	231	214	230	235	260	194	198	∨	‖	∨	▲	▼	∨
N-EXPRESS (%)	GAP	41	42	49	40	40	27	59	42	‖	‖	‖	∨	▲	‖
	OVL	8	12	24	20	10	10	37	30	∧	∧	‖	‖	▲	∧
SIZE OF THE FIRST SACCADE (DEG)	GAP	-3.7	3.7	-3.5	3.4	-3.7	3.6	-3.3	3.5	▼	▼	‖	‖	▼	▼
	OVL	-3.8	3.7	-3.5	3.4	-3.5	3.5	-3.2	3.2	▼	▼	∨	‖	▼	▼
UNDERSHOOTS (%)	GAP	41	35	55	49	41	41	70	35	▲	▲	‖	‖	▲	▲
	OVL	27	38	59	58	46	45	72	72	▲	▲	∧	‖	▲	▲
OVERSHOOTS (%)	GAP	6	7	1	1	1	1	1	1	▼	▼	∨	‖	▼	▼
	OVL	7	6	3	0	5	1	1	0	▼	▼	‖	‖	▼	▼
DEVIATION OF THE FINAL EYE POSITION (DEG)	GAP	0.5	0.5	0.8	0.7	0.8	0.8	0.7	0.7	▲	▲	▲	▼	∧	∧
	OVL	0.5	0.5	0.7	0.7	0.7	0.8	0.6	0.6	▲	▲	▲	▼	‖	‖
VEL/AMP (1/SEC)	GAP	50	50	51	53	49	51	55	54	‖	▲	∨	‖	▲	▲
	OVL	50	50	52	53	50	52	55	53	∧	▲	‖	∧	▲	▲
										4	6	1	4	6	5

First of all saccadic reaction times are among these differentiating parameters supporting the earlier observation (Fischer &Weber, 1990). For the group I subjects the reaction times were longer to the right, for the group II subjects they were shorter to the left. The group II subjects produced more express saccades to the left. Both groups had a decreased fixation stability (increased eye velocity) and they produced smaller amounts of overshoots than the control group. Only group II subjects had smaller amplitudes of their primary saccades, more undershoots, and an increased ratio of the maximal velocity versus amplitude. Only group I subjects reached the target inconsistently even after a second corrective saccade, i.e. the standard deviation of the eye position is increased. (An alternative explanation is that the eyes of the group I subjects did not really fixate at the centre of the screen at the time of target onset and therefore the measured total amplitude of the two saccades scatters more than in the normally fixating subjects.)

The difficulties the dyslexics have in hitting the target become even clearer if one looks at their corrective saccades (Fig. 2).

In group I, of 900 saccades, 401 (290 for the controls) were undershoots and only 25 (55 for the controls) were overshoots. This deviation from the control group is even more drastic in the group II subjects, where 639 saccades out of 900 were undershoots and a minimum of only 4 saccades were overshoots. The subjects in all three groups made reasonable amounts of anticipations, but the number of anticipations is a parameter which never showed a significant difference among the groups.

From the statistical group comparison of the data from the single target tasks we conclude, that dyslexics differ from normals not only in their saccadic reaction time but also in their fixation stability and in their ability to reach a parafoveal visual target by one or by two saccades.

Group Statistics: Consistency

In addition to the differences in the mean values of several parameters of their eye movements we noticed that the dyslexics' data show larger amounts of scatter in two ways: first, the mean values of the subjects scatter over a larger range, and second, the single values contributing to a mean value for a given subject scatter more. This second fact indicates that the dyslexic children produce a given parameter of their eye movements more inconsistently from trial to trial than do the control children. To quantify this observation we computed a relative inconsistency factor RIF defined by s/m divided by MRIF, where s is the standard deviation and m is the mean value of a given parameter of a given subject, and MRIF is the mean of the m/s-values averaged over the parameters of the control group.

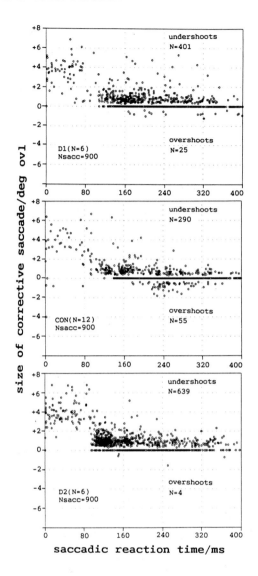

FIG. 2. Size of corrective saccades versus reaction time. Each dot on the figure represents a saccade in the overlap task. The dots along the horizontal midline correspond to the saccades not followed by correction. Group II subjects undershot the target altogether twice as often as the control group and almost never overshot it. CON= controls; D1= dyslexics group I; D2=dyslexics group II.

Table 4 shows the RIFactors for the two groups of dyslexics and the control group. The asterisks indicate significant deviations of the RIFactors (p < 1 %) from the mean of the control and the dots indicate significant deviations (p < .01 %) of the corresponding mean values (see also Table 2). One clearly sees that in group I the saccadic reaction times differ in their mean value as well as in their consistency, whereas the size of the first saccade and the ratio v/a show an increased inconsistency but no difference in their mean values. In group II deviations of the mean values as well as those of consistency are observed in the number of undershoots, size of the first saccade, and the ratio v/a. Note that the number of undershoots and the time to corrective saccade are produced by the group II subjects with a higher consistency than by the control children.

Altogether group I produced 7 and group II produced 12 significant deviations from the control data. This emphasizes again that group II differs more clearly in their eye movement performance from the controls than does group I. The important message from this analysis is that consistency is a variable, which - in addition to differences of certain mean values - differentiates the eye movements of the dyslexics from those of the controls.

TABLE 4

RIFactors for the Three Groups of Controls, Dyslexic I & Dyslexic II.
Asterisks Indicate Significant Difference of the RIFactor Compared to the Control Group. Dots Indicate Significant Difference of the Corresponding Mean Value.

PARAMETERS	CON	D1	D2
FIXATION QUALITY	1.00	1.27●	1.04●
SRT	0.99	* 1.33●	1.15●
N-EXPRESS	1.00	1.05	0.61●
SIZE OF THE FIRST SACCADE	0.99	* 1.52	* 1.48●
T-CORRECTIONS	1.00	1.06	* 0.75
UNDERSHOOTS	0.99	0.79	* 0.45●
OVERSHOOTS	0.99	1.03●	0.69●
DEVIATION OF THE FINAL EYE POSITION	0.99	1.10●	* 1.18
VEL/AMP	1.00	* 1.28	* 1.25●

Figure 3 shows the distributions of the RIFactor values for the three groups of subjects including only five parameters that were statistically different in both tasks (parameters of Table 2 except n-express, undershoots and overshoots).

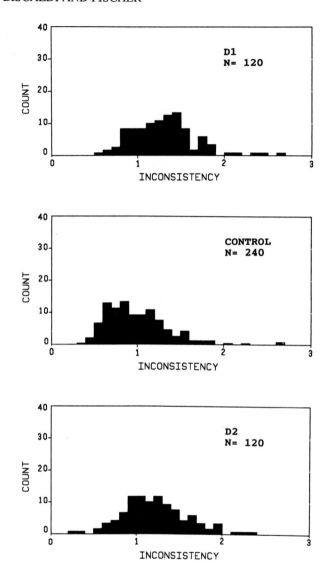

FIG. 3. Distribution of the relative inconsistency factors

Individual Data Analysis

The comparison of the group data has revealed statistically significant differences of certain mean values and certain differences in the consistency of the single values that contributed to these mean values. However such an analysis tells little or nothing about the individual members of the different groups. To learn more about the individuals on the basis of their eye movements the data of each single subject were reevaluated. Using the group statistic, a look up table of the eye movement data was made available in the following way: the data of each child were compared with the control group data and for each parameter it was determined whether it was equal, significantly smaller or larger. By comparing the result with that of the group comparison it was determined whether a parameter differed in the same or in another way from the group mean value. Thus each comparison gave a hint to which of the groups the child should belong. The table of these 'hints' or 'votes' constitute the look up table which is shown to the right of Table 1. Note that only 21 of the gap and 19 of the overlap parameters contribute to the look up table, because the remaining parameters were statistically equal.

In this way 21 votes were collected for each child from the gap data and 19 from the overlap data, each vote falling into one of six categories: for or against the control group (C or C̲), for or against group I (D1 or D̲1̲), and for or against group II (D2 or D̲2̲). As result of this procedure six numbers were assigned to each child. A factor analysis on these numbers for all 24 subjects suggested that two combinations carried the most relevant information. These two combinations were: (C-C̲) - (D2-D̲2̲) and D1-D̲1̲. They can be represented by a point in a plane for each child. The resulting graph is shown in Fig. 4, with the abbreviations Con for C-C̲, D2 for D2-D̲2̲, and D1 for D1-D̲1̲. Part A depicts the results if one uses the gap data only, part B if one uses the overlap data only, and part C, if one uses both sets of data together.

The circles represent the control subjects, the "ones" the group I subjects, and the "twos" the group II subjects. The centre of gravity of the single points belonging to each group is indicated by the black dots. The three lines represent the points of equal distances in this plane. They divide the plane in three "territories", one for controls, one for group I, and one for group II. One can see that indeed most, but not all, of the children land in their territory regardless of the data pool used (gap, overlap, or both). Table 5 shows the placements of the individual children according to Fig. 4.

It becomes clear that the gap task produced 29% false placements, the overlap task 21%, and both tasks together only 12%. Furthermore, the table emphasizes again that the group II subjects are more easily identified than the group I or the control group subjects: in the group II only 11% are misplaced as compared to 28% and 22% in group I and in the control group, respectively. Using the gap and overlap data together produced no mistakes at all in the group II.

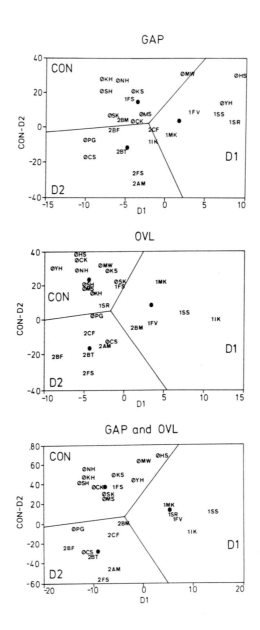

FIG. 4. Placement of the single subjects in the control (CON), dyslexic I (D1) or dyslexic II (D2) territory on the basis of their results in the eye movement task. The number on the left of each subject indicate to which group he belonged (0= control, 1= dyslexic group I, 2=dyslexic group II). (See also text).

TABLE 5

Placement of each Subject According to Fig. 4

	Subject	Gap	Ovl	Gap+Ovl	False
	CK	C	C	C	
	KH	C	C	C	
C	KS	C	C	C	
O	MS	C	C	C	
N	MW	C	C	C	
T	NH	C	C	C	
R	SH	C	C	C	
O	SK	C	C	C	
L	HS	**D1**	C	C	
	YH	**D1**	C	C	
	CS	**D2**	**D2**	**D2**	
	PG	**D2**	**D2**	**D2**	
	False	4	2	2	8 (22%)
	FV	D1	D1	D1	
D	MK	D1	D1	D1	
Y	SS	D1	D1	D1	
S	IK	**D2**	D1	D1	
	SR	D1	**C**	D1	
I	FS	**C**	**C**	**C**	
	False	2	2	1	5 (28%)
	AM	D2	D2	D2	
D	BF	D2	D2	D2	
Y	BT	D2	D2	D2	
S	CF	D2	D2	D2	
	FS	D2	D2	D2	
II	BM	**C**	**D1**	D2	
	False	1	1	0	2 (11%)
	Total False	7 (29%)	5 (21%)	3 (12%)	

DISCUSSION

This study has shown that sets of parameters of the eye movements of two different groups of dyslexic children are statistically different from those of aged matched normal readers. As has been reported earlier (Raymond, Ogden, Fagan & Kaplan, 1988) fixation stability was also found to be decreased in the dyslexics. This finding can be interpreted as a deficit in active inhibition of the eye movement generating system during directed visual attention, fixation being a situation where attention is directed to the foveal stimulus. From stimulation experiments in monkeys we know that the frontal eye fields (Goldberg, Bushnell & Bruce, 1986) as well as the parietal cortex (Mountcastle, Andersen & Motter, 1981; Sakata, Shibutani & Kawano, 1980; Shibutani, Sakata & Hyvarinen, 1984) exert such inhibitory actions during active and attentive fixation.

This interpretation would explain the shorter reaction times of the group II dyslexics and in particular of those, who make excessive numbers of express saccades: in order to generate a saccade they do not have to disinhibit the saccade system thus saving the corresponding time. This saccade, however, will occur "too soon", without previous proper fixation, it will reach a high velocity and end too short with respect to the target. Group I dyslexics have also a decreased fixation stability and a decreased ability to hit the target, but altogether their eye movement parameters are less affected than those of group II subjects. Because they failed in one of the psychological tests their reading and writing problems may be the consequence of other deficits, so that one would not really expect their eye movements to be very different from those of normal readers.

If one assumes that indeed a deficit in maintaining foveal fixation is a major problem in dyslexia one wonders whether other quantitative abnormalities can be of the same origin. For example problems in macular perception leading to improper vergence movements for macular sized fusion targets (Stein, Riddell & Fowler, 1988), or an increased number of saccades during smooth pursuit eye movements (Black, Collins, De Roach & Zubrick, 1984). This then would explain, why the training of oculomotor skills can reduce the rate of saccades, decrease the number of regressions, increase the fixation durations, which in turn improve the reading performance (Punnett & Steinhauer, 1984).

According to the three loop model for the generation of saccadic eye movements (Fischer, 1987), which links the processes of visual attention (Fischer & Breitmeyer, 1987), decision making and the computation of the metrics of the saccades, the abnormalities of the saccadic eye movements in non-cognitive tasks reported in this paper would occur as a consequence of a central deficit which also causes problems in reading. The present data, therefore, do not necessarily lead to the conclusion that abnormal eye movements cause dyslexia. But one would certainly maintain the view that the inconsistency in the generation of saccades decreases the performance rate in any task which requires both proper visual attention and proper timing of saccadic eye movements.

We developed a procedure which results in a point on a plane, which indicates for an individual child its chances of being a member of the control group or of belonging to one or the other group of dyslexics. Of course, each subject has contributed to the group mean values. For a real test of the procedure one would have to use 'new' subjects and compare their placement in one group with their results on the psychological tests. Therefore Fig. 4 could be said to represent a kind of control of how reliably the procedure works for the subjects that constitute the data base. Given this caveat, the best identification was obtained for the children in dyslexic group II. However, two subjects from the control group were placed in the dyslexic group II on the basis of their results in the eye movement tasks. We are now looking at the performance of all these subjects in other tasks such as point sequences (like those of Pavlidis and Olson). From the point sequences further eye movement parameters can be collected, which will be analysed together with the parameters from the gap and overlap tasks. We hope this will succeed in a better classification of single subjects. Other mathematical procedures can then be developed in order to identify a dyslexic child on the basis of his/her eye movements only.

ACKNOWLEDGMENTS

This work was supported by the Deutsche Forschungsgemeinschaft (SFB 325, Tp C5 and C7). The computer assistance of Dr. F. Aiple is greatly acknowledged.

REFERENCES

Black, J.L., Collins, D.W., De Roach, J.N. & Zubrick, S.R. (1984). Smooth pursuit eye movements in normal and dyslexic children. *Perceptual Motor Skills* 59, 91-100.

Brickenkamp, R. (1981). Test d2, Aufmerksamkeit-Belastung-Test. Göttingen: Verlag für Psychologie-J. Hogrefe.

Compagnoni L., Falcone, N., Fiorita, G.F. & Nappo, A. (1984). Sequential eye movements in developmental dyslexia (electrooculographic study). *Rivista di Neurobiologia*, 30, 40-46.

Fischer, B. (1987). The preparation of visually guided saccades. *Reviews in Physiology, Biochemistry and Pharmacology*, 106, 1-35.

Fischer, B. & Ramsperger, E. (1984). Human express sacades: extremely short reaction times of goal directed eye movements. Experimental Brain Research, 57, 191-195.

Fischer, B. & Breitmeyer, B. (1987). Mechanisms of visual attention revealed by saccadic eye movements. *Neuropsychologia* 25, 73-83.

Fischer, B. & Weber, H. (1990). Saccadic reaction times of dyslexic and age-matched normal subjects. *Perception*, 19, 73-83.

Goldberg, M.E., Bushnell, M.C. & Bruce, C.J. (1986). The effect of attentive fixation on eye movements evoked by electrical stimulation of the frontal eye fields. *Experimental Brain Research*, 61, 579-584.

Grissemann, H. & Linder, M. (1981). Zürcher Lesetest. Bern: H. Huber Verlag.

Kubinger, K.D. & Wurst, E. (1988). Adaptive Intelligenz Diagnostik, AID. Weinheim: Belz Verlag.

Mountcastle, V.B., Andersen, R.A. & Motter, B.C. (1981). The influence of attentive fixation upon the excitability of the light-sensitive neurons of the posterior parietal cortex. *Journal of Neuroscience* , 1, 1218-1225.

Mottier, G. (1981). *Mottier-Test in Handanweisung zum Zürcher Lesetest Förderdiagnostik der Legasthenie*. Bern: H. Huber Verlag, p. 19.

Müller, R. (1983). *Diagnostischer Rechtschreibtest für 3. Klassen*, DRT3. Weinheim: Belz Verlag.

Olson, R.K., Kliegl, R. & Davidson, B.J. (1983). Dyslexic and normal readers' eye movements. *Journal of Experimental Psychology [Hum Percept]* 9, 816-825.

Pavlidis, G.T. (1981). Do eye movements hold the key to dyslexia? *Neuropsychologia* 19, 57-64.

Pavlidis, G.T. (1985). Eye movement differences between dyslexics, normal, and retarded readers while sequentially fixating digits. *American Journal of Optometry & Physiological Optics*, 62, 820-832.

Punnett, A.F. & Steinhauer, G.D. (1984). Relationship between reinforcement and eye movements during ocular motor training with learning disabled children. *Journal of Learning Disabilities,* 17, 16-19.

Rathenow, P. (1980). *Westermann Rechtschreibtest 4/5*. Braunschweig: Westermann Verlag.

Raven, J.C. (1986). *The Coloured Progressive Matrices*. London: Lewis.

Raymond, J.E., Ogden N.A., Fagan, J.E. & Kaplan, B.J. (1988). Fixational instability and saccadic eye movements of dyslexic children with subtle cerebellar dysfunction. *American Journal of Optometry & Physiological Optics*, 65, 174-181.

Rayner, K. (1985). Do faulty eye movements cause dyslexia? *Developmental Neuropsychology*, 1, 3-15.

Romero, R., Estanol, B., Lee-Kim, M., Romero, O. & Martinez, A. (1982). Analysis of sequential saccadic eye movements during the act of reading: comparison between normal and dyslexic subjects. *Arch Invest Med* (Mex), 13, 185-9 passim.

Sakata H., Shibutani, H. & Kawano, K. (1980). Spatial properties of visual fixation neurons in posterior parietal association cortex of the monkey. *Journal of Neurophysiology* , 43, 1654-1672

Shibutani, H., Sakata, H. & Hyvarinen, J. (1984). Saccade and blinking evoked by microstimulation of the posterior parietal association cortex of the monkey. *Experimental Brain Research* , 55, 1-8.

Stein, J.F., Riddell, P.M. & Fowler, S. (1988). Disordered vergence control in dyslexic children. *British Journal of Ophthalmology* , 72, 162-166.

Steingrüber, H.J. & Lienert, G.A. (1976). *Hand-Dominanz-Test*, H-D-T. Göttigen: Verlag für Psychologie-J.Hogrefe.

Facets of Dyslexia and its Remediation
S.F. Wright and R. Groner (Editors)
235

SACCADIC EYE MOVEMENTS IN DYSLEXICS, LOW ACHIEVERS, AND COMPETENT READERS.

H. Fields, S. Newman & S.F. Wright.

University College & Middlesex School of Medicine,

London

INTRODUCTION

The fact that dyslexics show distinctive eye movements while reading is well established. Compared to competent readers, they are reported to make longer or more variable fixations (Adler-Grinberg & Stark, 1978) , more and smaller saccades (Adler-Grinberg & Stark, 1978; Martos & Vila, 1990), and more regressions (Rubino & Minden, 1973; Martos & Vila, 1990). There is, however, controversy about the precise relationship between abnormal eye movements and developmental dyslexia. Few now consider eye movement abnormalities to be a major cause of dyslexia. An alternative view is that they are more likely to be a result of problems experienced by the reader with the decoding and comprehension of the material being read (Tinker, 1958, Olson, Kliegl & Davidson, 1983). Another theory is that the unusual patterns of eye movements and the reading problems are together symptomatic of an underlying abnormality - in visual perception or visuospatial function, for example - which is the real cause of some dyslexics' reading problems (Pirozzolo & Rayner, 1979). Finally, it has been suggested that, whether or not they have a causal role in dyslexia, the presence of abnormal eye movements is sufficiently reliable to serve as a useful diagnostic indicator of dyslexia (Pavlidis, 1985).

In order to discover whether there is anything distinctive about the eye movements of dyslexics which is not simply a reflection of linguistic processing difficulties, researchers have monitored their eye movements in non-linguistic tasks. A typical procedure is one in which the subject is required to track, visually, a horizontal array of sequentially illuminated lights. The idea is that such tasks elicit saccadic eye movements similar to those made in reading.

The evidence from studies comparing the saccadic eye movements made by groups of dyslexics and non-dyslexics, whilst performing tasks without a linguistic component, has been inconsistent. In some cases there have been clear differences between the groups. For example, Pavlidis (1985) found that a dyslexic group made more regressions than did a group of controls. In a study carried out by Martos & Vila (1990) dyslexics made more saccades and regressions, and shorter fixations. Some differences between dyslexics and controls were also reported by Lesevre (1968). In Lesevre's study, however, distinctive eye movements were found only in a subgroup of dyslexics who were characterised by spatial problems, as opposed to those dyslexics whose problems were of an auditory-linguistic nature. In other studies such as those reported by Brown et al. (1983) Stanley et al. (1983) and Olson et al. (1991), no notable group differences were apparent.

Any attempt to interpret these contradictory findings should take into account methodological differences between the various studies. For example, subject selection has varied considerably from one study to another. In some cases subjects have been drawn from an ordinary population of schoolchildren, but in many they have been exclusively referrals from clinics and dyslexia institutions, and therefore subject to preselection by teachers or parents. Definitions of dyslexia have also varied. In another part of the Linbury Study, Fields, Wright & Newman (1990) have shown that the various criteria which have been used to define spelling disability result in quite different sets of subjects being selected. Similarly, in the case of reading disability, a group of children identified as being dyslexic because their prose comprehension is poor relative to their age may differ in composition from a group so identified because their comprehension is poor relative to that of their classmates, or a group whose single word decoding falls below the level predicted by their age and I.Q.

Another variable has been the nature of control groups to which the dyslexics have been compared. These have sometimes been matched to the dyslexics in reading ability but not age, or in reading ability but not IQ, or in age alone. Clearly, the type of group used for comparison has important implications for interpretation of the results. For example, if dyslexics are compared to a control group of better readers, it may be difficult to determine whether any differences found in their eye movements are actually related to the cause of their dyslexia, or merely a result of their having less reading experience. Similarly, if there are age or IQ discrepancies between dyslexics and control groups, these variables could be responsible for any eye movement differences found.

Finally, the nature of the experimental task must also be considered. In some studies comparing the reading eye movements of dyslexics and controls, all subjects have been required to read the same material, with the inevitable result that the poorer readers are faced with a relatively more difficult task. In such cases, therefore, group differences in eye movements may not be directly related to dyslexia, but instead reflect the fact that the text is too advanced for one group and/or too easy for another. Non-linguistic tracking tasks have also varied from one study to another, requiring saccades of anything from 1 to 8 degrees or more, with some involving an unpredictability factor in the task, and others not.

THE LINBURY STUDY

With these methodological issues in mind, one objective of our study of dyslexics' eye movements was to be particularly rigorous in the selection of dyslexic subjects, the groups to which they would be compared, and the tasks they were to perform. This study forms a part of the Linbury Research Project. The Linbury subjects were a large sample of boys and girls attending normal schools in various regions of England. They ranged in age from 6-10 years, and covered a wide range of academic ability. Over 500 of these children underwent a comprehensive investigation which included assessments of reading, spelling, IQ, and other cognitive measures, as well as various neuro-psychological, social, and emotional assessments.

The method for identifying children who had unexpectedly poor reading was as follows. Using the data of over 800 Linbury Study subjects for whom full reading, spelling and IQ assessments were available, Pearson correlations and a stepwise multiple regression analysis were carried out in order to discover which other variables were related to reading achievement. The regression equation derived from this data was:

Predicted Schonell Reading Raw score= (.770*Verbal IQ) + (.785* age) - 119.96

Using this regression equation, it was possible to identify those children whose reading achievement as measured by the Schonell Graded Word Reading Test (Schonell & Schonell, 1963) fell well below the level predicted by a combination of their age and WISC-R Verbal IQ (Wechsler, 1974). The standard error of prediction was 14.02 and the equation accounted for 44% of the variance in reading scores. A score falling at or below the 5% point on the distribution of residual reading scores was adopted as the criterion for underachievement. The 5% cut-off is equivalent, in terms of the more traditional standard error cut off, to -1.65.

In the eye movement study described here, 3 groups of children were compared. The first consisted of 12 dyslexics: individuals of various age and IQ, but all reading well below expectation. These children were selected from the underachievers identified by the regression method described above. They were regarded as dyslexic because there was no apparent reason for their reading underachievement. Having identified a group of dyslexics, it was possible to draw upon the remaining large sample for groups of contrasting abilities with which to compare the dyslexics. In the second group were 12 competent readers, individually matched with the dyslexics in IQ, but with reading as good as, or better than, predicted by IQ and age. The third group was comprised of 12 low achievers: individually and closely matched in reading ability to the dyslexics. The low achievers were also poor readers, therefore, but in contrast to the dyslexics, their reading performance was consistent with that predicted by the formula derived from their IQ and age. None of these children were known to have any organic or behavioural problems.

Selection of these particular children enabled two comparisons to be made: first, of 2 groups similar in IQ but very different in reading achievement, and second, of 2 groups identical in their reading achievement despite considerable differences in IQ. Using this design to compare the groups on other measures, it can be established whether any differences which emerge are specific to dyslexia or associated with low reading achievement in general. Table 1 shows the respective ages, IQs, and reading ages, of the 3 groups.

TABLE 1

Age, IQ and Reading of Dyslexics and Control Groups (SD's in brackets)

	Dyslexics	Competent Readers	Low Achievers	F	p
Age (Months)	101.33	101.5	101.83	< 1	ns
	(8.8)	(8.7)	(11.2)		
WISC-R Verbal IQ	109.2	110.1	92.8	9.77	<.001
	(12.9)	(9.7)	(9.3)		
Schonell Reading Age	83.6	110.8	83.3	18.7	<.001
	(8.3)	(18.4)	(8.6)		

The infra-red limbus tracking technique was used to monitor eye movements while the children carried out a number of tasks. One task consisted of each child reading a passage of prose of a standard appropriate to his or her reading ability. In addition, there were visual tracking tasks in which the child was required to follow a horizontal display of 7 light-emitting diodes illuminated sequentially. Duration and direction were each varied so that there were 4 tracking tasks: from left to right at intervals of 1 second and 1/3 second, and right to left at those two speeds. The display included a total arc of 28 degrees, thus involving saccades of about 4.5 degrees. For each of the 5 tasks, 7 eye movement variables were examined in a series of 3-way analyses of variance. These variables were: the number and size of saccades, duration of fixations, number and size of regressive saccades, duration of regressive fixations, and regressions as a proportion of all saccades. The results are shown in Table 2.

For reasons of clarity the dyslexic group, which is the focus of this study, will be discussed relative to the other two groups in turn.

TABLE 2

Group Comparisons of Eye Movement Measures in Reading and Tracking

(SD's in brackets)

	Dyslexics (D)	Competent Readers (C)	Low Achievers (L)	F	p	Duncan Tests
Reading						
Saccades / second	1.6 (0.5)	1.3 (0.4)	1.1 (0.5)	3.9	<.05	C,L < D
Saccade Size	5.2 (1.5)	9.8 (2.9)	4.7 (0.8)	24.5	<.001	D,L< C
Fixation Duration	367.1 (96.4)	387.8 (87.9)	450.4 (104.2)	2.2	ns	-
Regressions / second	0.6 (0.3)	0.3 (0.3)	0.4 (0.3)	2.1	ns	-
Regression Size	4.7 (2.4)	7.6 (2.8)	4.2 (1.8)	7.0	<.01	D,L< C
Regression Fixation Duration	364.0 (94.8)	396.8 (115.5)	529.7 (180.3)	5.1	<.05	D,C < L
% of Regressions	23.8 (12.1)	16.2 (9.3)	17.8 (12.9)	1.4	ns	-
Tracking	No	significant	differences			

Dyslexics Compared to Competent Readers.

First, the comparison of the Dyslexics and Competent Readers, the two groups who were of similar IQ but significantly different reading achievement. Analysis of their reading eye movements showed a significant difference between the two groups on 3 of the 7 measures. The dyslexics made more saccades per second, smaller saccades, and smaller regressives. This is consistent with some of the findings previously reported for dyslexics when compared to control groups. In this study, however, there was no indication that the dyslexics made more frequent regressions than the Competent Readers.

If basic visuomotor factors were responsible for the eye movement differences between these two groups, those differences ought also to occur in the tasks involving a reading-like pattern of saccades but without the linguistic component of reading. An examination of the dyslexics' and the Competent Readers' tracking eye movements showed that this was not the case. None of the 28 comparisons between the groups revealed significant differences.

The discovery that the differences between the reading eye movements of the dyslexics and the Competent Readers did not persist in the tracking comparison strongly suggests that the reading eye movement differences were associated with some aspect of reading ability which was independent of basic oculomotor skills or IQ.

Dyslexics Compared to Low Achievers.

The other important comparison in this study was between the Dyslexics and the Low Achievers, the two groups who were identical in their reading achievement, despite having widely differing IQs. When comparing their reading eye movements, two significant differences emerged. The Dyslexics made more saccades per second, and shorter regressive fixations, than did the Low Achievers.

A comparison of the tracking eye movements of the Dyslexics and Low Achievers revealed no differences between the two groups. The similarity of the Dyslexics and Low Achievers' eye movements during the tracking tasks appears to rule out the existence of a difference between the groups in basic oculomotor skills. The possibility remains that the differences between their reading eye movements are related to the IQ discrepancy which exists between the groups.

DISCUSSION

When an overall assessment is made of the group differences in reading eye movements, it is evident that a tendency to make significantly smaller saccades and regressions is common to both groups of poor readers. Both the Dyslexics and the Low Achievers made saccades with a mean size of only 5 character spaces, whereas the saccades of the Competent Readers travelled approximately 10 character spaces. A similar pattern was apparent in the regressive saccades, although the difference was not so great. Since all children in this study read texts which were matched to their individual levels of reading achievement, the relatively short saccade length of the poor readers appears to be a function of their absolute reading ability and/or experience, rather than any difficulties they are experiencing with the particular text. It must be borne in mind, however, that where attempts are made to equate the relative difficulty of the texts, the better readers are likely to be reading longer words and sentences with more complex structure than those read by the poorer readers. The effect of this on saccade size in particular is difficult to gauge. The range in average word length between the easiest and hardest of the 6 texts was only just over 2 characters, however, so it seems unlikely that the highly significant difference in this variable could be accounted for simply by physical characteristics of the text.

Only one result distinguished the Low Achievers from the other groups. The Low Achievers made longer regressive fixations, implying that they were taking longer to process information when they looked back to examine previously scanned text. Although the Low Achievers and Dyslexics were equal in their scores on the reading assessment, was this difference in their regressive fixation durations indicative of some other discrepancy in their reading abilities? The two groups of poor readers were matched on single word decoding thus allowing the possibility of differences between them in reading comprehension. But a comparison of the Dyslexics and Low Achievers' comprehension showed that there was no significant difference between the groups. The possibility remains that the longer fixation durations of the Low Achievers may be associated in some other way with their relatively low IQ.

The reading eye movements of the Dyslexic group were distinguishable from those of both other groups only in that the Dyslexics made significantly more saccades. Because the texts were varied according to the subject's reading ability, it was not meaningful to compare absolute numbers of saccades, so the measure employed was in terms of saccades per second. From the finding that Dyslexics made significantly more saccades per second than did both other groups, two other results might logically be expected. First, that the fixation durations of the Dyslexics would be shorter, and second, that the Dyslexics would read their passages in a shorter time than the Low Achievers. In fact, although the fixation durations of the Dyslexics were somewhat shorter than the other two groups, this difference failed to reach significance. Neither were any significant differences found between the Dyslexics and Low Achievers in the time taken to read the passage. Overall, therefore, it seems that although the Dyslexics had a slight tendency to move more quickly through the text, this effect was marginal.

Individual Differences.

Not only were the group means similar for most of the tracking eye movement variables, but the distributions were highly overlapping. An analysis was made of individual performance within the groups with a view to determining how accurate each subject was in terms of the task requirements. This included variables such as the number and size of saccades, including return sweeps, and the duration of fixations. The analysis revealed that each of the groups contained individuals who were notably inaccurate in these variables. For example, 3 of the Dyslexics consistently made dysmetric saccades, either too large or too small, as did 4 of the Low Achievers and 3 of the Competent Readers. Of those whose return sweeps repeatedly consisted of 2 or more saccades, 4 were in the

Dyslexic group, 3 in the Low Achievers, and 6 in the Competent Readers. Equally, there were individuals in each group whose tracking was remarkably accurate. These children made single-saccade return sweeps, they made approximately the right number of progressive saccades, and fixations of the correct duration. Other researchers (e.g. Lesevre, 1968; Pirozzolo & Rayner, 1979) have identified a subgroup of dyslexics suffering from visual-spatial problems as opposed to auditory-linguistic, and have suggested that eye movement abnormalities are likely to be confined to the former. It may be that our Dyslexic group contained a mixture of these two types, so that eye movement abnormalities would be expected to occur in only a subset of the dyslexics studied here. Pirozzolo & Rayner concluded that apparent eye movement deficits were actually the result of dyslexia in cases of auditory-linguistic dyslexia, but in visual-spatial dyslexia eye movement deficits were the result of a spatial dysfunction. But the fact that inaccurate tracking and dysmetric return sweeps were found in all the groups in roughly equal proportions suggests that these abnormalities are not peculiar to dyslexics or even a small group of dyslexics. These conclusions need to be qualified, however, by the small number of subjects studied.

CONCLUSION

The results of this study indicate that, with careful selection of subjects from a large sample pool of children in normal schools, matching to groups of comparable reading ability and IQ, and ensuring that reading materials are of the appropriate level for each subject's ability, there is little that is distinctive about the saccadic eye movements of dyslexics during reading and tracking.

ACKNOWLEDGEMENTS

The authors would like to thank: The Linbury Trust and the Middlesex Hospital Trustees for their generous support; Hellmut Karle, Rosemarie Archer, and Jane Wadsworth for their contributions to the Linbury Study; John Redburn and Jim Shand for their respective roles in developing the eye movement analysis system; and the children, teachers, and parents for their participation.

REFERENCES

Adler-Grinberg, D. & Stark, L. (1978). Eye movements, scanpaths, and dyslexia. *American Journal of Optometry and Physiological Optics,* 55(8), 557-570.

Brown, B., Haegerstrom-Portnoy, G., Yingling, C. D., Herron, J., Galin, D. & Marcus, M. (1983). Tracking eye movements are normal in dyslexic children. *American Journal of Optometry and Physiological Optics*, 60 (5), 376-383.

Fields, H., Wright, S.F. & Newman, S. (1989). Techniques for identifying a group with spelling difficulties in the absence of reading difficulties. *Irish Journal of Psychology*, 10, 657-663.

Lesevre, N. (1968). L'organisation du regard chez des enfants d'age scolaire, lecteurs normaux et dyslexiques. *Revue de Neuropsychiatrie Infantile et d'Hygiene Mentale de l'Enfance*, 16, 323-349.

Martos, F.J. & Vila, J. (1990). Differences in Eye Movement Control Among Dyslexic, Retarded and Normal Readers in the Spanish Population. *Reading and Writing*, 2, 175-188.

Olson, R. K., Kliegl, R. & Davidson, B. J. (1983). Eye movements in reading disability. In K. Rayner (Ed.) *Eye Movements in Reading: Perceptual and Language Processes*. New York: Academic Press.

Pavlidis, G. Th. (1985). Eye movement differences between dyslexics, normal and retarded readers while sequentially fixating digits. *American Journal of Optometry and Physiological Optics*, 62 (12), 820-832.

Pirozzolo, F. J. & Rayner, K. (1979). The neural control of eye movements in acquired and developmental reading disorders. In H. Whitaker & H.A. Whitaker (Eds.), *Studies in Neurolinguistics, Vol. 4*. New York: Academic Press.

Schonell, F. J. & Schonell, F. E. (1963). *Diagnostic Attainment Testing*. Edinburgh: Oliver & Boyd.

Stanley, G., Smith, G. A. & Howell, E. A. (1983). Eye movements and sequential tracking in dyslexic and control children. *British Journal of Psychology*, 74, 181-187.

Tinker, M. A. (1958) Recent studies of eye movements in reading. *Psychological Bulletin*, 55, 215-231.

Wechsler, D. (1974). *Wechsler Intelligence Scale for Children - Revised*. Windsor: NFER-Nelson.

Facets of Dyslexia and its Remediation
S.F. Wright and R. Groner (Editors)
245

EYE MOVEMENTS IN READING CHINESE: PARAGRAPHS, SINGLE CHARACTERS AND PINYIN

F. Sun

Shanghai Institute of Physiology,
Chinese Academy of Sciences, Shanghai, China

INTRODUCTION

Chinese written language differs greatly from Western languages both in linguistic and in visual information aspects. The ideographic Chinese written language consists mainly of invariable monosyllabic morphemes called "characters" or "Hanzi". All the characters are constructed from two types of smaller units called radicals and strokes. There are about 20 different strokes and 189 radicals in Chinese (Wang, 1973, Huang, 1983). A stroke is a line or a curve which is completed every time the writing pen leaves the paper and a radical is built up by one or several strokes. Each Chinese character contains one radical with or without remainder[1]. Figure 1 gives examples of the stroke and the radical.

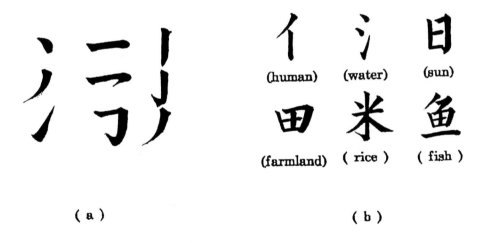

(human)   (water)   (sun)

(farmland) (rice) (fish)

(a)   (b)

FIG. 1 Examples of Chinese strokes (a) and radicals (b). The English translation of the radicals are in the parentheses.

---

[1] Each Chinese character contains only one radical. In some cases a character itself is a radical in which case it has no remainder. But in most cases, a character is built up with one radical and an additional component, the remainder.

All Chinese characters can be divided into six categories according to their formation. The main category is the "pictophonetic category", which contains about 80% of all the characters. For a pictophonetic character, the radical part represents the ideographic meaning of the character and the remainder offers a hint of the pronunciation of the character. Some examples are shown in Fig. 2.

FIG. 2. Two examples of Chinese characters from the pictophonetic category. (a) The character "rice" and four characters, which are related to food, with radical "rice". (b) The character "fish" and 6 characters, which are different fishes, with radical "fish". English translations are in parentheses.

There are also two other categories of characters which are used frequently at present. One is called the "associative compounds" character, the meaning of which is explicitly related to its radical and construction.

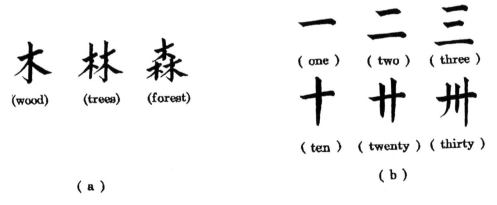

FIG. 3. Examples of Chinese characters from the "Associative Compound" category (a) or from the "Self-explanatory" category (b). Their meanings are directly connected with their forms with English translation in parentheses.

Another is called the "self-explanatory" character, the meaning of which is directly expressed by the form of the character (Huang, 1983). Examples of these are given in Fig. 3. Since there is a close connection between meaning and form in Chinese characters, a native Chinese can often guess roughly the meaning and/or the pronunciation of a character which he has never seen before (Wang, 1973).

From a visual information aspect, Chinese characters with their boxlike shapes might be better adapted to exploiting the small disc fovea in the retina than alphabetic language words with their linear array outlines. Most Chinese Hanzi are monocharacter words, but there are double-character words and a few triple character words. Statistically, we found 1.5 Chinese characters were equivalent to one English word.

In addition to the traditional character "Hanzi", a new system of Chinese written language called "Pinyin" (means "spell sound") has been created in China since 1956. "Pinyin" is an alphabetic phonetic language, in which the Chinese words are built up phonetically by 26 Latin letters. Figure 4 shows an example of Chinese Pinyin in comparison with Chinese Hanzi.

视 觉 信 息 处 理 研 究

Shi Jue Xin Xi Chu Li Yan Jiu

(Studies in Visual Information Processing)

FIG. 4. An example of Chinese pinyin in comparison with Chinese Hanzi. The English translation in parenthesis.

It has been generally supported that eye movements reflect the human perceptual and cognitive processes (Yarbus, 1967; Noton & Stark, 1971; Stark & Ellis, 1981; Menz & Groner, 1982; O'Regan & Levy-Schoen, 1985). So in comparison with the large number of research data for reading Western languages, the eye movement data for reading non-alphabetic Chinese could provide a fresh insight into the understanding of language processing (Holender, 1987).

METHOD

Reading materials consisted of: short reading paragraphs selected from popular scientific articles which were written both in Chinese characters and in English; reading texts selected from text books for Chinese primary schools, which were written both in

Chinese characters and in Chinese Pinyin; Single Chinese characters; Sets of "number matrix" for visual search experiments (Fig.7), each matrix contains 5 x 5 elements, which are numbers selected randomly from one to ten, written in Chinese Hanzi or in Chinese Pinyin. The reading materials were presented in CRT of a IBM/AT computer with a VGA graphic adapter. The stimulus display area was about 20 x 20 degrees visual angle.

Eye movements were measured by an infrared TV eye tracker PUP1 and infrared LED eye glasses. The output of measurement was recorded and processed by the computer. At the beginning of each run of the experiment, calibrations were performed to minimize the non-linearity of measurements.

Subjects consisted of three groups of Chinese graduate students, Chinese high school students and primary school students. There were a small group of native English speakers and Chinese-English bilinguals for comparison.

RESULTS AND DISCUSSION

Chinese Subjects Reading in Chinese Hanzi
The average duration of the fixation pause was 257 ± 63 ms. The span of recognition was 2.57 ± 0.5 characters/fixation (or $1.7\pm$ equivalent words/fixation, since 1.5 Chinese characters are equivalent to one English word for the same contents). The reading rate was 580 characters/min (or 380 equivalent words/min).

Native English Speakers Reading English (same contents as in Experiment 1)
The average duration of fixation pause was 265 ± 72 ms. The recognition span was 1.75 ± 0.5 words/fixation. The reading rate was 380 words/min.

The two eye-movement patterns (for reading Chinese and English) appeared very similar. The small amplitude saccades move the eyes right following the line, and the large, left return saccades move the eyes from the end of one line to the beginning of the next line. The means for fixation duration, recognition span and reading rate for reading Chinese and for reading English were very similar. The data obtained from experiments 1 and 2 can be summarized in Table 1. The results indicated that there were no significant differences between the fixation durations ($t(26) = 0.365$, $p>.05$), the recognition spans ($t(26) = 0.229$, $p>0.5$) or the reading rate ($t(26) = 0.087$, $p>.05$) for the two groups. These experimental data confirmed the results reported by Sun, Morita & Stark, (1985) and Sun & Stark (1988) using similar experimental method but different subjects.

It is quite reasonable to regard the reading text as a visual pattern. According to the aspects of "scanpaths" (Noton & Stark, 1971; Stark & Ellis, 1981), we could consider the reading eye-movement pattern as "scanpaths" of the subject for that reading text.

TABLE 1

Eye-movement Data for Reading Chinese and English

	Chinese Reading Chinese	English Reading English
Fixation Duration (ms)	257 (63)	265 (72)
Recognition Span (characters / fix.) (words / fix.)	2.57 (0.51) 1.71 eq.	1.75 (0.58)
Reading Rate (characters / fix.) (words / fix.)	580 (155) 386 eq.	382 (127)

Noton & Stark (1971) have argued that scanpaths are not "simply resulted from low level control of the eye by peripheral visual feature detectors". Groner, Walder & Groner (1984) suggested that the "global scanpaths are assumed to operate in a top-down fashion". The results, obtained from the present experiments, confirm their viewpoints. In our experiments the two reading eye-movement patterns were very similar given that the contents of two reading paragraphs were the same, although their visual forms were strikingly different. Resemblances between eye-movement patterns for reading these two languages indicate that reading eye movements are controlled mainly by the higher level centre of brain, which decodes the meaning of languages. The "local scanpaths analysis" (Groner, Walder & Groner, 1984; Koga & Groner, 1989) could be a useful method to process these eye-movement data in local detail. But further work is required.

Chinese-English Bilingual Reading the Same Paragraph in Chinese and English.

The results obtained from the 10 subjects show that 6 of the bilinguals read Chinese much faster than English (380 equivalent words/min for Chinese and 255 words/min for English). These 6 performed similarly to native Chinese readers in reading Chinese (386 equivalent words/min - see Experiment 1), but were much slower than native English readers reading English (382 words/min - see Experiment 2). However, the other 4 bilingual subjects read Chinese much slower than they read English and were similar to native English readers in reading English but much slower than Chinese readers reading Chinese (233 equivalent words/min for Chinese and 385 words/min for English). One can conclude from these results that each of the 10 bilinguals has a distinct bias in reading skill and none has equal reading skill in both languages. Present eye-movement data suggests that no real Chinese-English bilinguals exist.

Reading a Single Chinese Character Hanzi

(1) The single character is well-known to the subjects

All subjects made only one fixation during recognition of the character. This result leads us to assume that one fixation pause is enough for recognition of a well-known character, because the reading practice of the subject has built an efficient model in his brain. In one fixation pause, the fovea and peripheral vision have acquired sufficient information for recognizing the well-known character, and it is unnecessary to make another fixation.

Recognizing a well-known visual image pattern should be similar to reading a well-known character. Groner et al. (1984) have described a case in their experiments: "one of seven subjects was excluded from further analysis because in the recognition session he mostly spent only one single fixation per picture and was still able to give a high percentage of correct responses". This phenomenon in our single character reading experiments was very common. Possibly the learning capability of the subject who was excluded from their experiments was very high, so experimental pictures became well-known to him already.

(2) The single character is unknown to the subject

For pictophonetic characters, most fixations are concentrated on the remainder part (which is usually the phonetic part and on right side) since the meaning of the radical is known already. According to the formation of Chinese, the phonetic remainder has more linguistic information for distinguishing a character from others with the same radical (Fig. 5). The results from this experiment are in accordance with those of Koga & Groner (1989) from European subjects in a post-learning phase, during the recognition of Japanese "Kanji", which are derived from Chinese Hanzi in history.

When the unknown Chinese character is a complex one which can be divided into several components, each of which if separate from others, is well-known as a simple character, the eye-movement pattern demonstrates a "scanpaths" trajectory scanning the components of the character to guess the meaning. Some of the data are similar to those obtained in previous experiments (Sun & Stark in preparation for submission). Figure 6 illustrates some examples. The eye-movement pattern was a set of random fixations for a subject with no knowledge of Chinese.

Chinese Primary School Students, High School Students and Graduate Students Reading Primary School Texts in Hanzi and Pinyin.

Data from these experiments are briefly summarized in Table 2. Two-factor analyses of variance were calculated for these data. One factor represented orthography difference

FIG. 5. Eye-movement patterns for a subject reading single pictophonetic characters. The 6 characters on the left are not well-known to the subject. Most fixations are concentrated on the remainder part (the phonetic part). After the subject has had enough practice, the recognition eye movement consists of a single fixation. An example is shown on the right side of this figure.

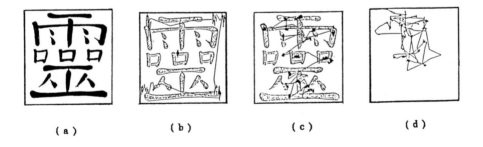

(a)   (b)   (c)   (d)

FIG. 6. Eye movement patterns from 3 different subjects reading a complex character: (a) The form of a complex character (means "soul"); (b) The character is well-known to the subject the eye movement for reading the character is a single fixation. The 4 fixations on the corners are calibration eye movements. (c) The character is unknown to the subject, but he/she is familiar with components of the character. The eye movement pattern is scanning components. (d) Eye movement patterns from a non-Chinese subject ignorant of the Chinese language.

(Chinese Hanzi and Chinese Pinyin) and the other factor represented education levels in the 3 groups.

The difference between fixation duration for reading between the two Chinese orthgraphies was highly significant (F (1,66)=26.47, p<.001) but the fixation duration did not differ significantly among the 3 levels (F (2,66)=0.015, p>0.5) nor was there a significant interaction (F (2,66) = 1.45, p>0.2). Similarly for recognition span and reading rate, the difference between reading Hanzi and reading Pinyin was highly significant (F (1,66) = 188.0, p<.001 and F (1,66) = 127.0, p<.001, respectively). The difference between education levels was not significant for either recognition span (F (2,66) F= 2.01, p>0.2), nor reading rate (F (2,66) = 1.10, p>0.5). No interaction effects were found.

TABLE 2

Eye-movement Data for Reading Chinese Hanzi and Pinyin for 3 Groups of Subjects
(SD's in parenthesis)

	GRP. 1 Primary School	GRP. 2 High School	GRP. 3 Graduate
Fixation Duration			
Hanzi (ms)	268 (50)	266 (61)	240 (54)
Pinyin (ms)	326 (65)	326 (70)	353 (77)
Recognition Span			
Hanzi (chars / fix.)	2.08 (0.62)	2.11 (0.59)	2.61 (0.56)
Pinyin (chars / fix.)	0.89 (0.22)	0.79 (0.26)	0.77 (0.21)
Reading Rate			
Hanzi (chars / min.)	459 (168)	491 (190)	598 (197)
Pinyin (chars / min.)	171 (59)	145 (60)	137 (55)

This suggests that for all subjects, regardless of their level of education, the average fixation duration for reading Hanzi is much shorter than that for reading Pinyin. Similarly the average recognition span and reading rate are also faster for Hanzi.

There are at least two possible explantions for these results. Firstly, given the large number of homophone characters in Chinese, the alphabetic phonetic Pinyin may be more difficult to comprehend than ideographic Hanzi. Secondly, most of the books in China are printed in Hanzi, so the greater difficulties associated with reading Pinyin may be due to a lack of practice. Thus, it is not possible from these data to conclude that Pinyin orthography is less suited to the Chinese language. The reason that education failed to play a role was probably due to the ease of the reading texts - these were primary school texts and easy for all groups.

Visual Search Experiments

Subjects were required to find an assigned number from a matrix presented on a screen. The matrix consisted of 25 numbers, selected randomly from 1-10 written in Hanzi or in Pinyin. Experimental results show a clear difference in search and fixation times depending on whether the target number was written in Hanzi or in Pinyin (Fig. 7).

The search time for finding a Pinyin number was about five times longer than finding a Hanzi number; the fixation times for Pinyin were also about five times more than Hanzi; there was no big difference in fixation durations. The data is summarized in Table 3. The results obtained from these experiments indicate that numbers in Hanzi are better than Pinyin for visual search tasks. Probably the form of Hanzi has a direct connection with its meaning and the form of Pinyin is connected with pronunciation. Of course, it should be noticed that the numbers are only a small part in language, and the mental processing in visual search tasks is not the same as that in normal reading. Also, as Per Udden suggested (personal communication) subjects involved in such visual search experiments should be selected from different areas of China to diminish the influence of dialect.

TABLE 3

Eye-movement Data for Searching Hanzi and Pinyin Numbers

Ratio of Search Time	(Pinyin / Hanzi)	5.4 (4.5)
Ratio of Fixation Times	(Pinyin / Hanzi)	5.0 (3.8)
Average Fixation Pause	for Hanzi (ms)	249 (30)
	for Pinyin (ms)	232 (21)

CONCLUSION

The similarities of reading eye movements for two very different languages, Chinese character and alphabetic English, indicate that reading eye movements are controlled by the high level centre of the human brain and not by peripheral visual feature detectors and that the recognition spans are determined by the linguistic information, not by the visual geometric form of the reading text. The language information processing capability is limited to the brain, not by the peripheral system of vision.

Eye-movement patterns in reading single Chinese characters are determined by the subject's familiarity with the character.

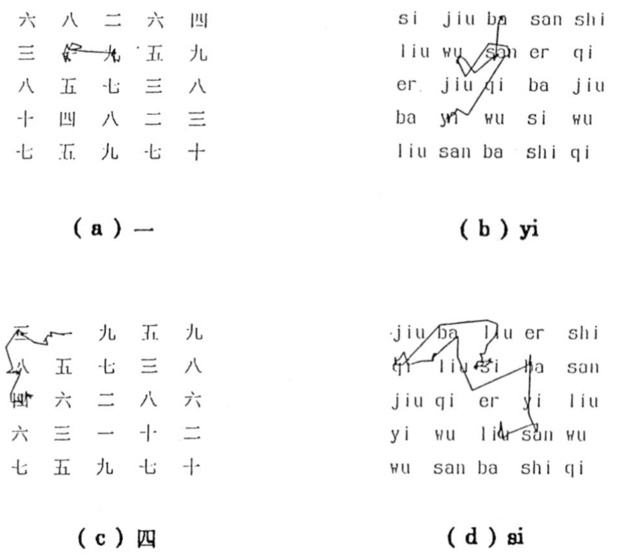

```
六  八  二  六  四          si  jiu ba  san shi
三  九  九  五  九          liu wu  san  er  qi
八  五  七  三  八          er  jiu qi  ba  jiu
十  四  八  二  三          ba  yi  wu  si  wu
七  五  九  七  十          liu san ba  shi qi
```

（ a ）一   （ b ）yi


```
五  一  九  五  九          jiu ba  liu er  shi
八  五  七  三  八          qi  liu si  ba  san
十  六  二  八  六          jiu qi  er  yi  liu
六  三  一  十  二          yi  wu  liu san wu
七  五  九  七  十          wu  san ba  shi qi
```

（ c ）四   （ d ）si

FIG. 7. Four examples of eye movement patterns during visual search. The matrix is constructed of 25 numbers selected randomly from 1 -10 written in Hanzi (a & c), or in Pinyin (b & d). The subject was required to find an assigned number from each matrice. The subject made 2 fixations to find Hanzi "1", in matrix (a); 5 fixations to find Pinyin "1" in matrix (b); 4 fixations to find Hanzi "4" in matrix (c); 7 fixations to find Pinyin " 4" in matrix (d).

The eye movements demonstrated three types of patterns as follows: single fixation for a well-known character, more fixations concentrated on the unknown part for a not so well-known character, fixations scanning components of the character for an unknown complex character, components of which are familiar to the subject.

The reading eye movement data from bilingual subjects showed there are no real Chinese-English bilinguals. No one really possesses equal skill for reading Chinese and English quantitatively. For each bilingual subject, only one language is dominant, and the other has been suppressed.

For all the Chinese subjects in the present investigation, the reading rate for Chinese Hanzi is much faster than that for alphabetic Pinyin. This indicates that either the "Pinyin" orthography is not suitable to Chinese language, or more efforts are required to master Pinyin language for the Chinese. However, it is obvious that data obtained from visual search experiments show that Hanzi have certain advantages over Pinyin.

ACKNOWLEDGMENTS

This work was partially supported by the National Science Foundation of China. The author also gratefully acknowledges the help of Professor Lawrence Stark, Professor William S-Y Wang, Professor Rudolf Groner and Dr.Per Udden, and the assistance of Xinzhen Zhao for experiments.

REFERENCES

Groner, R., Walder, F., & Groner, M. (1984). Looking at faces: Local and Global aspects of scanpaths. In A.G Gale and F.Johnson (Eds.), *Theoretical and Applied Aspects of Eye-Movement Research.* Amsterdam: North-Holland.

Holender, D. (1987). Synchronic description of present-day writing systems: some implications for reading research. In J.K.O'Regan and A.Levy-Schoen (Eds.), *Eye Movements from Physiology to Cognition.* Amsterdam: North-Holland.

Huang, B. (1983). *Modern Hanyu.* China Gansu: Gansu People's Publisher Company.

Koga, K., & Groner, R. (1989). Intercultural experiments as a research tool in the study of cognitive skill acquisition: Japanese character recognition and eye movements in non Japanese subjects. In H. Mandl, J.R. Levin (Eds.), *Knowledge and Acquisition from Text and Pictures.* Amsterdam:North-Holland.

Menz, C., & Groner, R., (1982). The analysis of some componental skills of reading acquisition. In R. Groner and P. Fraisse (Eds.), *Cognition and Eye Move-ments.* Amsterdam: North Holland.

Noton, D., & Stark, L. (1971). Scanpaths in eye movements during pattern perception *Science*, 171, 308-311.

O'Regan, J.K. & Levy-Schoen, A. (1987). Eye movements from physiology to cognition Amsterdam: North-Holland, 273-274.

Sun, F., Morita, M., & Stark, L. (1985). Comparative patterns of reading eye movement in Chinese and English. *Perception and Psychophysics*, 37, 502-506.

Sun, F., & Stark, L. (1988). Visual information processing eye movements during reading Chinese and English. *Acta Biophysica Sinica*, 4, 1-6.

Stark, L., & Ellis, S.R. (1981). Scanpaths revisited: Cognitive models direct active looking. In D.F. Fisher, R.A. Monty & J.W. Senders (Eds.) *Eye Movements: Cognition and Visual Perception.* Hillside N.J.: Lawrence Erlbaum.

Wang, W., S-Y. (1973). The Chinese Language, *Scientific American*, Feb.1973, 50-60.

Yarbus, A. (1967). *Eye movements and Vision.* New York: Plenum Press

READING VERTICALLY WITHOUT A FOVEA

N. Osaka

Department of Psychology, Faculty of Letters

Kyoto University, Kyoto 606, Japan

INTRODUCTION

The importance of parafoveal vision during normal reading was investigated for the first time by McConkie & Rayner (1975) in their sophisticated computer-controlled experiments. It was shown that a line of text projected onto the reader's retina could be divided into three major regions based on the retinal cone/rod distribution density: the foveal, parafoveal, and peripheral (Rayner & Bertera, 1979). The highest visual acuity was found in the foveal region within 2.5 deg visual angle from the fixation point. Outside this region, the parafoveal region extends up to 9 deg, and the remainder, the peripheral region, extends up to the Ora Serrata at which the retina ends (Osaka, 1983). Visual acuity drops off sharply as the distance from the fovea increases: the difficulty in identifying a word or character increases as a function of distance from the fovea. Saccades have an important role in bringing words or characters into the foveal region for detailed analysis as we read.

The present experiment examines the extent to which readers acquire different types of information from the foveal and parafoveal regions. If the information necessary for visual identification is limited to foveal vision, reading rate would be much more limited than if visual information were obtained from a larger region extending well into the parafoveal region (Rayner & Bertera, 1979).

To estimate the size of the effective visual field during reading, it is necessary to introduce a moving-window technique in order to control the amount of information available to the reader during each fixation (McConkie & Rayner, 1975; Osaka & Oda, 1992; Rayner & Bertera, 1979). In these studies, observer's eye movements were monitored by a computer system and the information about current eye position was fed into a computer system controlling the cathode-ray tube (CRT) from which the observer was reading. Changes were made on the basis of the position of the observer's fixation. McConkie & Rayner (1975) employed a passage of mutilated text which was initially presented on the CRT with every character from the original text replaced by character x, then wherever the observer fixated, a region around the current fixation point changed into readable text. This window region moved in synchrony with the observer's eye movements so that wherever the observer fixated, the normal text was exposed, however everywhere outside the window region the mutilated text remained. By changing the size of the window region, the effective visual field size necessary for reading could be estimated (Rayner & Bertera, 1979).

The present study employed a method similar to that of McConkie & Rayner (1975) and Rayner & Bertera (1979). The purpose was to estimate the extent to which observers can get sufficient information from outside the foveal region to identify the meaning of words. As Rayner & Bertera, we excluded foveal as well as parafoveal information from the observer at each fixation thereby simulating the artificial central scotoma.

METHOD

Apparatus

Eye movements were recorded with an eye-mark recording system (NAC Corporation, type V), in which eye position data could be recorded in 33-ms frames by a corneal reflected infrared light. The eye position data were analyzed with a connected module (NAC model V-99B) coupled with a GPIB-based interface connected to PC system (Epson Corporation, PC286VE).

The size of the effective visual field has traditionally been estimated with the use of the moving-window technique (McConkie & Rayner, 1975), however, it can also estimated with the use of the mask-window technique. The mask-window technique involves highly sophisticated real-time feedback and provides definitive information about the size of the effective visual field. It employs on-line recording of the observer's eye positions to present text on a computer-controlled screen, contingent on where the observer is fixating. This technique is the same as the moving-window technique; a visible window area in the moving-window is replaced with a visual mask-window area. Characters of the text are replaced with grey-mask patches, except in an experimenter-defined region outside the observer's current point of fixation. Wherever the observer looks, the original text is obliterated by the mask, while outside this window region the text is visible. When the observer moves his or her fixation point, a new region of text is obscured, and the previously obscured region reappears. The observer's impression of reading with this mask-window is that he or she seems to see the text with artificial central scotoma (simulated central scotoma).

Using the mask-window technique, we can estimate the effective visual field size necessary for normal reading.

As Fig. 1 indicates, the current eye position is fed back by a fast 12-bit analog-to digital converter (Analog Pro Corporation, 100 kHz sampling rate) to control the mask-window. Figure 1 shows the block diagram of the moving mask generator system.

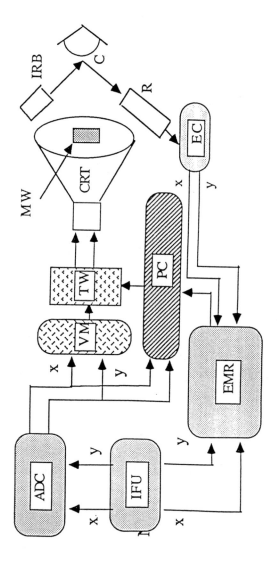

FIG. 1. Moving-mask generator system: EMR, Eye movement recording system; EC, Eye camera; IRB. Infrared beam; C, cornea; R, Receptor-array; IFU, Interfacing unit; ADC, Analog-to-digital converter; VM, Video memory; TW, Text window; CRT, Cathode ray tube; MW, Mask-window; PC, Personal computer system.

Four different mask-window sizes were used: 1, 2, 4 and 6 characters (corresponding to 1 to 6 deg.). In addition, two reading directions were introduced: vertical and horizontal. The mask-window length changed along the vertical column or horizontal line but had a fixed width of 32 pixels. The separation between character columns or lines was kept at 12 pixels and the vertical or horizontal inter-character separation was kept to 5 pixels. The position of the current fixation utilized for window control was the centre of the window. Figure 2 shows example screen views with mask-window for vertical (left) and horizontal (right) reading conditions.

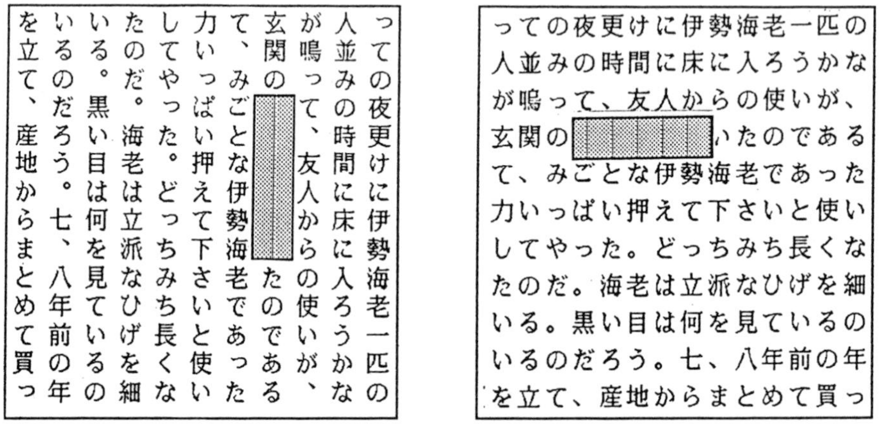
っての夜更けに伊勢海老一匹の
人並みの時間に床に入ろうかな
が鳴って、友人からの使いが、
玄関の
たのである
て、みごとな伊勢海老であった
力いっぱい押えて下さいと使い
してやった。どっちみち長くな
たのだ。海老は立派なひげを細
いる。黒い目は何を見ているの
いるのだろう。七、八年前の年
を立て、産地からまとめて買っ
っての夜更けに伊勢海老一匹の
人並みの時間に床に入ろうかな
が鳴って、友人からの使いが、
玄関の
いたのである
て、みごとな伊勢海老であった
力いっぱい押えて下さいと使い
してやった。どっちみち長くな
たのだ。海老は立派なひげを細
いる。黒い目は何を見ているの
いるのだろう。七、八年前の年
を立て、産地からまとめて買っ

FIG. 2. Two example screen views. Mask-window superimposed on the text for vertical (left) and horizontal (right) reading (part view). Six-character vertical mask and five-character horizontal mask is shown here. Text type is kanji-hirakana-mixed.

Text

Sixteen different sets of written Japanese text, each of which contained 170 characters, were used. They consisted of normal Japanese text, composed of both kanji and hirakana, and equivalent hirakana-only text. One character subtended a visual angle of about 1 deg including inter-character separation. They were selected from high school texts. The kanji contribution factor (Osaka, 1989) of the composite texts was about 0.3, and their familiarity level was controlled.

Procedure

After a ten minute calibration period for saccade and fixation measurements, each subject was asked to read the text from top to bottom silently, as it appeared on the screen. The eye movement system sampled the position of the eye every 33 ms, and a visual mask was superimposed over the text to obliterate the text in the foveal and parafoveal regions. The size of the mask-window was changed but always moved in synchrony with the eye. The measured delay between saccadic initiation and mask movement was, on average, 18.3 ms. Reading rate was computed for each mask-window condition based on the time needed to read the whole text. Fixation within the area of 1-deg of visual angle and longer than four frames were selected for the data. Saccades, including regressive saccades, within each column and line, were collected, but the regressive movements due to return sweeps were excluded. Each observer received three sessions for each of the four mask-window conditions, in random order. The first session was the training period; measurements were made during the remaining sessions. Eye movements were recorded from the left eye of the observers.

Observers

Five undergraduate students who had either normal or corrected-to-normal vision were chosen as subjects. They had a 20/20 or better acuity of the left eye.

RESULTS

Figure 3 shows reading rate as a function of mask-window size.

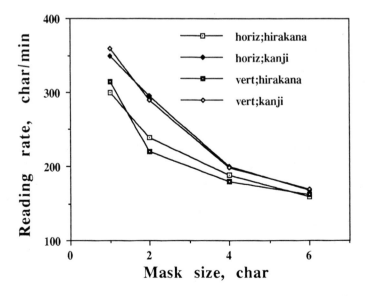

FIG. 3. Reading rate (character/min) as a function of mask-window size during reading. Parameter is reading direction and text type.

Parameters were reading direction and text type. Note that the reading rate defined here is not in word/min but in character/min due to the characteristics of Japanese text: there are no blank spaces between words in normal Japanese text. One data point is the average of 25 trials across subjects. As the mask size increased reading rate decreased (F(3,12)=32.56,p<.01), however no significant difference was found between vertical and horizontal reading. It was noted that the reading rate dropped sharply beyond the 4-character mask-window.

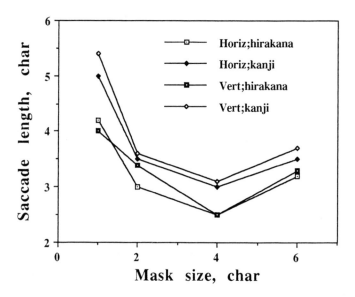

FIG. 4. Saccade length as a function of mask-window size during reading. Parameter is reading direction and text type.

Figure 4 shows saccade length (in character) as a function of mask-window size. Parameters are reading direction and text type. As the mask size increased saccade length decreased, but increased beyond the 4-character mask-window (F(3,12) =12.30,p<.01). No significant differences were found between reading directions. The saccade length appears to be long beyond the 4-character mask.

Figure 5 shows fixation duration as a function of mask-window size. As mask size increased fixation duration tended to increase gradually up to the 4-character mask-window, then increased sharply beyond this condition (F(3,12)=19.52, p<.01). Multiple range t-tests revealed significant differences between the 6-character mask compared with the other mask conditions (p<.01) which, in themselves, did not appear significantly different from one another. Again, no significant differences were found between reading directions.

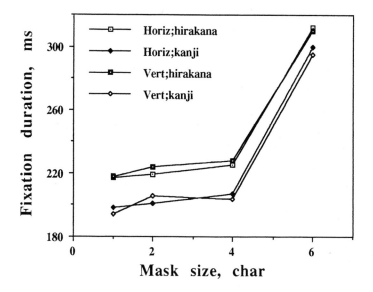

FIG. 5. Fixation duration as a function of mask-window size during reading. Parameter is reading direction and text type.

Due to the finding of no significant differences between horizontal and vertical reading, the data were collapsed across conditions. Significant differences were found between text types. Kanji-based text showed an overall higher reading rate (Fig. 3), longer saccade length (Fig. 4), and shorter fixation duration (Fig. 5) (F (1,4)=11.32, p<.05). This suggests that kanji tended to facilitate text reading to a larger extent than hirakana.

DISCUSSION

Visual Mask

As mask-window size increased, the mean reading rate decreased. Similarly, as mask-size increased, the mean fixation duration increased, while the mean saccade length decreased up to 4-character window. These data indicate that as the mask size increased, reading became more difficult. Observers had little difficulty when they read the text with the 1-character mask. However, when the mask increased up to 4-6 characters the observer had more difficulty in identifying which words were present in the parafovea and near periphery. With the use of the mask-window Rayner & Bertera (1979), using the percentage of words correctly reported and the effective reading rate, found that the 1-character mask decreased effective reading rate by a half to 165 words/min. Furthermore, they found that the 3- and 5-character masks decreased reading rate by 55 and 42 words/min, respectively. Although the estimated reading rate index used here is not the effective rate, the decreasing function, in general, appears similar to that obtained by Rayner and Bertera. It is noted that masks having 4-6 characters covered the most effective visual field for reading (Osaka, 1991; Osaka & Oda, 1991) which is also the useful visual field in terms of visual angle for learning to discriminate hirakana symbols (Koga & Groner, 1989). Significant decrease of the reading rate between 4 and 6 characters is likely to correspond to the limit of the effective visual field. The measured lag time between the saccade and the mask movement appears to place some constraints on the type of research that can be explored. However, the system appears sufficient for measuring the approximate size of the effective visual field during attention-demanding Japanese text reading, in which the saccade size, as well as the velocity, are relatively small compared with English text reading (Osaka, 1992).

Reading Direction

The direction of reading varies between languages: horizontal reading is the major way of reading in the present world. English as well as other modern European languages are read from left to right but some Semitic languages are read from right to left. There appear to be no clear conclusions that can be drawn about the effect of direction of reading on eye movements (Rayner & Pollatsek, 1989). As Figs. 3 to 5 show, the examination of this issue with regard to vertical reading revealed that the effect of reading direction is negligible for Japanese text processing although some advantages for horizontal over vertical reading have been reported during Chinese text reading (Sun, Morita & Stark, 1985). These findings are in agreement with the data by Osaka & Oda (1991): they found that the effective visual field size necessary for vertical and horizontal reading was about the same.

The effect of text type on reading performance is clearly observed. The results showed that the kanji component facilitated the reading during Japanese text processing. This could be explained in terms of dual strategy model of Japanese text reading (e.g. Osaka, 1991) in which kanji logograph is supposed to have an advantage over phonogram-based hirakana script in terms of lexical access properties.

As the present study on the simulated central scotoma shows masking the fovea resulted in more severe problems in text reading than when the parafovea was masked. The results are consistent with previous findings that information necessary for semantic processing is obtained from the fovea and parafoveal region close to the fovea (McConkie & Rayner, 1975; Rayner & Bertera, 1979; Osaka, 1987, 1990, 1991), while more global information is obtained from the parafovea and near periphery (Rayner & Bertera, 1979).

ACKNOWLEDGEMENTS

Thanks are due to Koichi Oda, Hiroshi Ashida, and Hitoshi Tsuji for their help in making the moving window software. This research was supported in part by grants #02801014, #03801010, and #03401003 from the Ministry of Education, Japan.

REFERENCES

Koga, K. & Groner, R. (1989). Intercultural experiments as a research tool in the study of cognitive skill acquisition: Japanese character recognition and eye movements in non-Japanese subjects. In. H. Mandl & J. Levin (Eds), *Knowledge Acquisition from Texts and Pictures* (pp. 279-291). Amsterdam: North Holland.

McConkie, G.W., & Rayner, K. (1975). The span of the effective stimulus during a fixation in reading. *Perception and Psychophysics*, 17, 578-586.

Osaka, N. (1983). *Psychophysical analysis of peripheral vision. Tokyo: Kazamashobo* (In Japanese).

Osaka, N. (1987). Effect of peripheral visual field size upon eye movements during Japanese text processing. In J.K.O'Regan & A.Levy-Schoen (Eds.), *Eye movements: From physiology to cognition* (pp.421-429). Amsterdam:North Holland.

Osaka, N. (1989). Eye fixation and saccade during kana and kanji text reading: Comparison of English and Japanese text processing. *Bulletin of the Psychonomic Society*, 27, 548-550.

Osaka, N. (1990). Spread of visual attention during fixation while reading Japanese text. In R.Groner, G. D'Ydewalle & R. Parham (Eds.). *From Eye to Mind* (pp.167-178). Amsterdam: North Holland.

Osaka, N. (1992). Size of saccade and fixation duration of eye movements during reading: Psychophysics of Japanese text processing. *Journal of the Optical Society of America,* 9, 5-13.

Osaka, N., & Oda, K. (1991). Effective visual field size necessary for vertical reading during Japanese text processing. *Bulletin of the Psychonomic Society,* 29, 345-347.

Osaka, N., & Oda, K. (1992). Moving window for reading and related psychophysical experiments. Manuscript submitted for publication.

Rayner, K., & Bertera, J.H. (1979). Reading without a fovea. *Science*, 206, 468-469.

Rayner, K., & Pollatsek, A. (1989). *The psychology of reading.* New Jersey: Prentice-Hall.

Sun, F., Morita, M. & Stark, L.W. (1985). Comparative patterns of reading eye movement in Chinese and English. *Perception & Psychophysics,* 37, 502-506

Facets of Dyslexia and its Remediation
S.F. Wright and R. Groner (Editors)
© 1993 Elsevier Science Publishers B.V. All rights reserved.

EYE AND HEAD READING PATH IN HEMIANOPIC PATIENTS

D. Schoepf and W. H. Zangemeister

Department of Neurology, University of Hamburg, Germany

INTRODUCTION

In 1881, Mauthner first described reading disorders in hemianopic patients. He found that patients with left homonymous hemianopia had difficulties in finding the beginning of a line. Syllables and whole words were omitted while the following part of the lines were read correctly. Patients with right homonymous hemianopia were not able to read continuously. These findings were confirmed by Willbrandt in 1907. Additionally, he reported similar deficits in patients with paracentral scotomas, i.e. scotomas that occur next to the fovea. Both authors pointed out that a right hemifield loss disturbs the reading behaviour more than a left hemifield loss. Poppelreuter (1917) and Mackensen (1962) supported the theory that a disturbed reading coordination was responsible for these reading disorders. Poppelreuter emphazised the value of compensatory strategies for guiding visual behaviour in everyday situations. Mackensen (1962) and Zihl, Krischer & Meissen (1984) stated clearly that an adequate level of reading may be achieved, when a residual hemifield of between three and five degree on the horizontal meridian remains. Global reading time, however, remains higher in these cases. Eye movements of patients with a residual hemifield of less than three degrees are mostly disorganized. Here, the classical staircase pattern, characterized by a relatively symmetrical exchange between fixation pauses and saccades from left to right, appears to be disintegrated (Gassel & Williams, 1963, Eber, Metz-Lutz, Bataillard & Collard, 1986). With perimetric saccade training Zihl (1981) and Zihl & von Cramon (1985, 1986) obtained a persistent relocation of the visual field border. There remains, however, an ongoing controversy as to whether specific training can restore vision (Campion, Latto & Smith, 1983, Balliet, Blood & Bach-y-Rita, 1985, 1986, Zihl & von Cramon, 1986).

In hemianopic patients visual information is available only in one hemifield, thus to compensate for hemianopia it is necessary to have appropriate ocular motor strategies for efficient use of the remaining half of the visual field (Poppelreuter, 1917; Mackensen, 1962; Meienberg, Zangemeister, Rosenberg, Hoyt & Stark, 1981, 1988, Zangemeister et al., 1982, 1983, 1986). To some extent, it is possible to learn to compensate for this visual handicap. Zangemeister, Meienberg, Stark & Hoyt (1981, 1982, 1983) found distinct adaptive ocular motor strategies in hemianopic patients to search for objects in their blind hemifield. They used a horizontal target, a small green light spot, the appearance of which was predictable or randomized in amplitude and time. In this artificial experimental situation the subjects employed a consistent set of presumably unconscious compensatory eye movement strategies. So far no study has reported on adaptive ocular motor strategies used in reading by hemianopic patients, including short adaptation and transient learning. With respect to a possible positive effect of head movements Zangemeister et al., (1982) pointed out that hemianopic patients often simplify search and fixation strategies by eliminating head movements.

The purpose of the present study was to extend these previous studies and to analyse the qualitative and quantitative changes of the ocular motor scanning behaviour during reading i.e. the reading path. Our hemianopic patients had to read aloud four different short texts as accurately and quickly as possible. First, in an experiment where the head was fixed and second, in a more natural head-free-to-move condition. In one of the four texts presented the distances between the letters and words were decreased. The question was whether this would have a positive or a negative effect on reading behaviour and comprehension. So far we have analysed the reading path for different strategies with respect to the efficiency of reading time and accuracy.

METHODS

Subjects

The experimental group was composed of four male patients with homonymous hemianopia due to ischemic strokes in the occipital region (mean age: 57.5 years, range 42 to 73) and four young (26 +/-3 years) healthy male subjects: two patients with incomplete homonymous hemianopia (Figures 1a,b) and one patient with a complete, dense right and left homonymous hemianopia. All subjects had undergone a complete neuropsychological assessment, which included the following tests: Mosaic and Figure Test (Hawie), Line Bisection Test, Cross Out test, Benton test (form c, instruction a), several option vocabulary Test, Body Scheme Test, remembering and recalling of twelve different geometrical figures, spontaneous drawing of a face, a sunflower and a clock. In these tests no patient showed any sign of inattention that could be attributed to neglect phenomena.

The intelligence of all patients was at an average level. Visual acuity examination, perimetry and CCT scanning confirmed the clinical investigations, with CCTs showing discrete occipital lesions in all patients. With respect to acuity, all patients were fully sight corrected. Time from the acute event to the day of EOG recordings varied between twelve days to one year. The patients did not participate in any special visual rehabilitation training. The four young normal subjects without any neurological diseases or reading disorders were recorded as a comparison group.

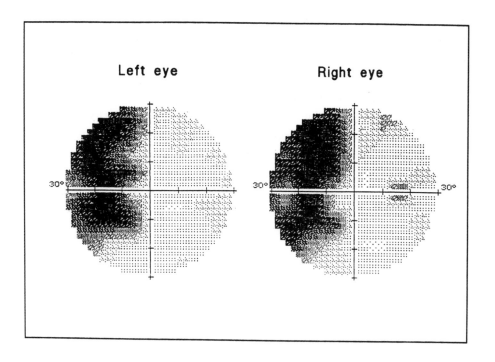

FIG. 1. Visual fields of left & right eyes of two well adapted patients with incomplete left hemianopia with five degree sparing (1a) & incomplete right hemianopia (1b).

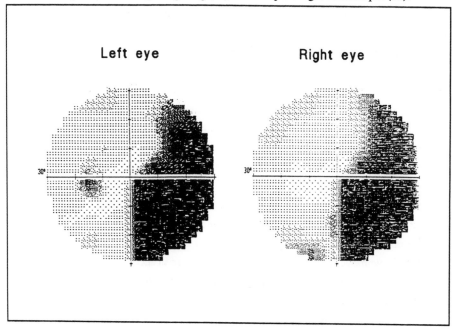

Apparatus

Subjects were seated on a dental chair in front of a white screen. The room was darkened and at least ten minutes were permitted for eye adaptation before any recording started. Horizontal binocular and vertical monocular DC EOG signals (analogue filtered at 10 Hz) were recorded on-line with an 80286 12 CPU computer, running at 12 Mhz. Horizontal recordings were consistently linear within the range of eye movements examined, that were always coordinated, i.e. showed the same velocities and position. Additionally, a high resolution ceramic potentiometer and a high resolution torquemeter for horizontal head rotations and torque recordings were used. The distance between the screen and the head rotation axis was 1.2 m. Sampling rate of the system was 25 Hz DC. Texts were projected on a white screen in front of the subjects. Text borders varied between 24 to 26 degrees to the left, and 26 to 28 degrees to the right. Size of characters and spaces within words was 1.5 degrees, between words 4 degrees. The vertical size of the text was ± 15 degrees.

Several calibrations were made before and after each reading set. With a specially developed software the text and frame calibrations were presented immediately on-line on the computer screen, such that the electrodes could be replaced if necessary. Overall accuracy of the EOG was estimated to be ± 1 degree horizontally and vertically, ± 0.25 degree for the potentiometer, ± 0.01 relative torque units for the torquemeter. One relative torque unit equals 70 newton micrometer.

Subjects had to read the texts aloud and they were asked to read them as accurately and as quickly as possible. Additionally, an acoustical back up was recorded for control reasons. We used eight different texts, that were not simple but also not too easy to comprehend. In a low letter density mode they had 28 letters per line, in high density mode they had 46 letters per line. The following text gives an example of a typical short text:

> Es treibt mich aus dem Zimmer hinaus,
> ich muß in den Straßen schlendern:
> die Seele sucht eine Seele und späht
> nach zärtlich weißen Gewändern.

RESULTS

Normal Subjects

The reading eye movements of the normal subjects were in accordance with the classical `staircase pattern', as demonstrated in Fig. 2 by time functions of the vertical and horizontal EOG (upper) and head horizontal torque (lower). In Fig. 2 the eye trace and

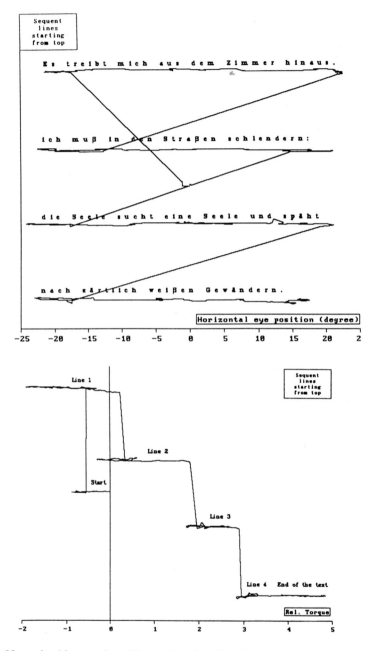

FIG. 2. Normal subject reading 4 lines of text in 2 fixed head condition, starting at the centre (dot). Upper: Left to right reading path: Vertical and horizontal eye movements (time functions in degree) plotted successively. Lower: Horizontal head torque during reading path. Ordinate: Lines from upper to lower. Abscissa: Relative horizontal torque.

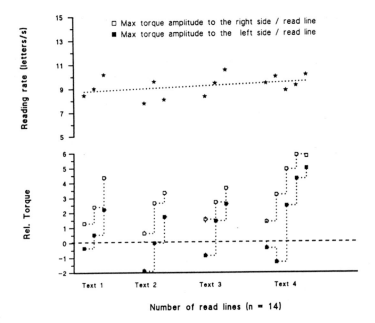

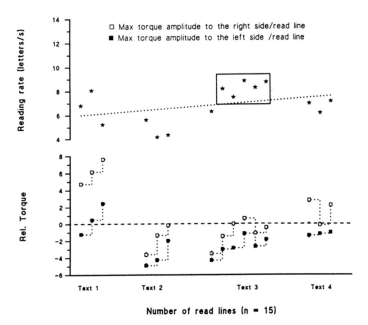

FIG. 3a. Upper: Head fixed condition of a normal subject reading different texts successively. Lower: Same procedure and description for the well adapted patient with right homonymous hemianopia. Note: In text three the distances between letters are reduced. The reading rate is increased.

typical corresponding horizontal torque pattern of one of the normal subjects reading in the head fixed condition is shown. The subject had an individual touque accentuation to the right side and performed in all texts the same `staircase' torque pattern to the right-hand side. It was characterized by decreasing maximal torque amplitudes to the left side from line to line, and synchronously increasing amplitudes to the right-hand side. The duration of torque to the right-hand side predominated in three out of four texts.

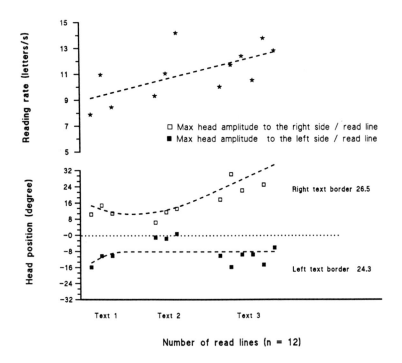

FIG. 3b. Head free condition for a normal subject with reading rate on the upper part and head position in degree as time function in the lower part. Reading rate is increased when the reader employs active and forced horizontal head movements during reading shown through the correlation of increase of head position to the right together with elevated reading rate.

This torque pattern appeared in all texts the same way without any adaptation (Fig. 3a upper). In the more natural head free-to-move situation the healthy readers were able to reduce their global reading time with active and forced head employment (Fig. 3b). Therefore, the mean number of positional fixation pauses (a fixation pause of 60 to 130 msec) was increased, whereas the mean number of lexical fixation pauses (a fixational pause of 200-300 msec) was decreased. Because of this the mean reading rate became faster. In general the mean duration of lexical fixation pauses did not change in texts with decreased spaces between letters and words.

In the head fixed condition, the mean reading rate was rather constant and fluctuated much less than in the hemianopic patients (Fig. 3a upper & lower). This was accompanied by a low number of regressions and a comparatively low global reading time. The constant reading rate generated a synchronous torque pattern (Fig. 3a upper) that appeared in all tested normal subjects to be individually accentuated.

Hemianopic Patients: Ocular Motor Strategies

No patient had participated in any special visual rehabilitation training. The two patients with incomplete homonymous hemianopia (Figs. 1a, b) did not produce any acoustical reading errors. As these two out of the four patients appeared to be better adapted, we describe their ocular motor patterns of adaptation in more detail.

The patient with incomplete right homonymous hemianopia (Fig. 1b) was a 56 years old former policeman, who came along to the investigation with his own car against medical instruction. For two weeks he had regarded his visual field defect to be no longer as disabling. He was the most agile and intelligent subject in the experimental group. Recordings were made three months after his acute event. The subject with incomplete left homonymous hemianopia (Fig. 1a) was a 73 year old self-employed industrialist, who appeared to be very active and motivated in all kinds of rehabilitation training. Recordings were made three weeks after his acute event. Both subjects had developed particular adaptive ocular motor strategies to compensate for their visual handicap.

The patient with right homonymous hemianopia (Figs. 1b, 3a lower, 4) often overshot especially the end of the first line of a text as he looked for the general end of the line and text. In the following line he often changed his strategy and approached the end of the line with small, staircase-like saccades followed by small regressions. We called this the blind hemifield `overshooting' strategy, and the `end of line' detective strategy. In the seeing hemifield the saccadic amplitudes were most often smaller than in the blind hemifield: Because of this we called it 'the saccadic resolution' strategy. The application of the three different reading strategies appeared to be completely interchangeable.

The other right homonymous patient presented a complete, dense right homonymous hemianopia. He employed these strategies only rudimentarily. The `end of line' detective strategy was often employed but much slower in this patient, that is more regressions occurred that were most often inaccurate. Correspondingly, the global reading time was significantly prolonged. Sometimes this patient overshot the end of a line, but obviously he did so unsystematically. The increased number of regressions and the frequent acoustical reading errors mirrored the uncertainty of this patient.

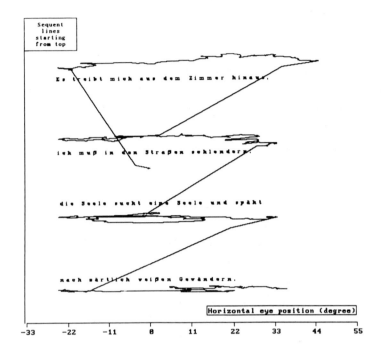

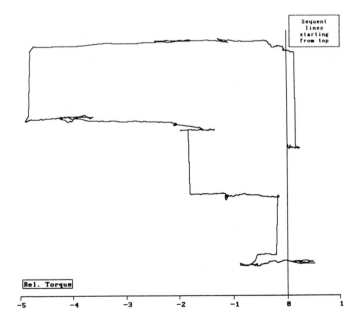

FIG. 4. Patient with right hemianopia in head fixed condition. Upper part: Reading path as in Fig. 2. Lower part: Horizontal head torque during reading path.

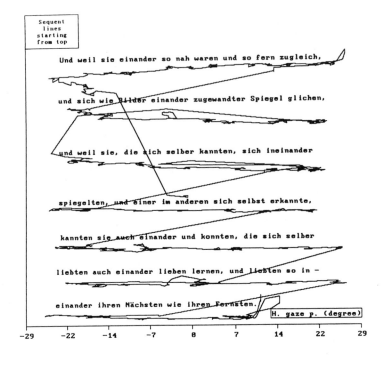

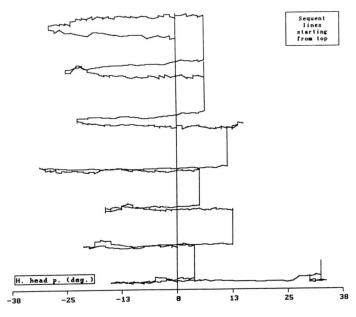

FIG. 5. Patient with left hemianopia in head free-to-move condition. Reading path as in Fig. 2 (gaze defined as sum of horizontal eye movement and horizontal head movement). Lower: Horizontal head movement.

The patient with incomplete left homonymous hemianopia (Fig. 1a) generated a high number of small saccades that were sometimes followed by regressions at the beginning of a line (Fig. 5). Because of this we defined the `beginning of line' detective strategy for patients with left homonymous hemianopia. In some lines there was an accumulation of saccades without regressions at the beginning. The other part of the eye movement pattern was in accordance with the classical `staircase pattern'.

The second left hemianopic patient had a more complete, dense homonymous hemianopia. He was much more insecure and produced multiple acoustical reading errors. During the whole reading path there was an increase of progressive and regressive saccades in this patient. The global reading time was therefore significantly prolonged.

Short Adaptation in Horizontal Head Torque and Free Head Rotations

Both hemianopic subjects with incomplete hemianopia (Figs. 1a, b) and *without* acoustical reading errors demonstrated a high short adaptation that appeared side inverted with respect to the right- and left hemifield loss. In the subject with left homonymous hemianopia (Figs. 1a, 6a) this was most clearly marked. A head accentuation to the blind hemifield predominated in the first two texts. The maximal torque and head amplitudes to the blind left hemifield were mostly enlarged compared with those to the seeing right hemifield. Correspondingly, the duration spent in a left side head position predominated. The relative zero position of the head was shifted into the blind hemifield.

In the texts the maximal amplitudes to the blind hemifield were small compared to amplitudes with direction to the seeing hemifield (Fig. 6a, text 3). The leftwards shifted zero position was changed into primary zero position (Fig. 6a, text 4). In the head fixed condition the percent time of horizontal head torque (Fig. 7) to the seeing hemifield was 95% in the fourth text. In the head-free-to-move condition (Fig. 7) the percentage time of head movements to the right-hand side was 48% in the last text. Evidently, the head movement component was reduced and minimized (Fig. 6a, 8), and the side of attention had changed from the blind hemifield to the seeing hemifield with increased reading time.

The same adaptive behaviour occurred in the well adapted right homonymous patient with incomplete homonymous hemianopia (Fig. 1b, 6b). In both subjects a distinct accentuation of the blind hemifield through the reading stimulus obviously caused a neglect-like phenomenon of the seeing hemifield. After all the reading rate did not increase and global reading time could not be reduced.

The other two less adapted patients did not show this kind of short term adaptation. The patient with the complete, dense right homonymous hemianopia that produced acoustical reading errors did not use his head at all. The less adapted left homonymous

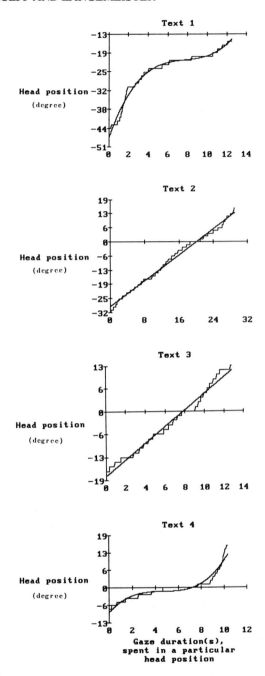

FIG. 6a. Well adapted patient with left homonymous hemianopia reading texts successively in the head free condition. Abscissa: Head position as function of gaze duration spent in a particular head position; head position to the left is negatively displayed, to the right positively displayed.

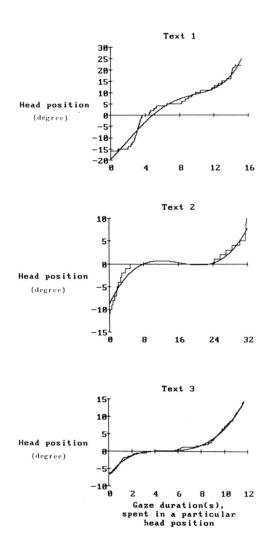

FIG. 6b. Side inverted the same kind of short adaption as in Fig. 6a for the well adapted patient with right homonymous hemianopia.

patient showed a distinct blind hemifield accentuation for all reading paths. The duration from the acute event to the day EOG recordings were made was twelve days in this patient, so it was a very recent lesion. The active head accentuation of the blind hemifield in the head-free-to-move situation had no positive effect on reading accuracy of this patient. On the contrary, the mean number of regressions in the seeing hemifield and the acoustical reading errors were increased.

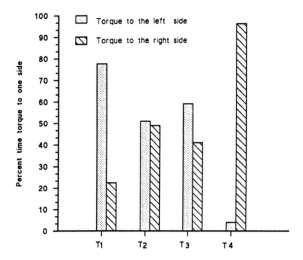

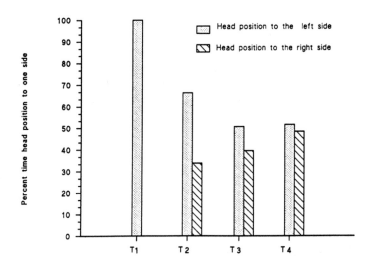

FIG. 7. Statistical distribution of percent time of relative torque and percent time of horizontal head position to one side for four texts read successively.

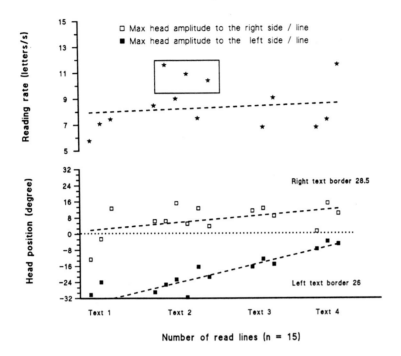

FIG. 8. Reading rate and horizontal head position as a function of numbers of read lines in a patient with left homonymous hemianopia. Note the short adaptation of head movements and reading rate with successively decreasing accentuation from the left to right, from text to text with reduced head movements at the last text. Note: In text two the distances between letters are reduced. The reading rate is increased.

We concluded that hemianopic patients at first had the intention of compensating their hemifield loss with active and forced head employment to the affected side. However, in doing so they could not gain an advantage in such a complex, high level task as reading. Successively, they reduced their head movements and relied on eye movements instead.

Influences of Decreased Spaces Between Letters and Words

Change of distances between letters had no significant effect on the mean number of fixation pauses during reading. Generally, the mean number of positional fixation pauses was increased with active and forced head employment. Correspondingly, the ratio between mean number of lexical fixation pauses and mean number of positional fixation pauses was reduced in the head-free-to-move situation (Table 1). In the two hemianopic subjects that produced no acoustical reading errors (Figs. 1a,1b) the global reading time was increased with reduced spaces between letters. Simultaneously, the reading rate was enhanced in some lines. The increase of the global reading time correlated with an increased duration (Table 2) of lexical fixation pauses for both head fixed and head free conditions.

TABLE 1

Ratio of Lexical Fixations Divided by Positional Fixations in Hemianopic Patients without Acoustical Reading Errors.

	LHH	RHH	NOR
Head fixed	5.2	4.0	4.8
Head free	1.8	3.0	2.6

TABLE 2

Lexical Fixation Durations of Well Adapted Hemianopic Patients & a Normal Subject for a Low Letter Density Mode (28 letters/line) & High Letter Density Mode (46 letters/line).

	LHH	RHH	NOR
Mean letters / line	28 (+18=) 46	28 (+ 18=) 46	28 (+18= 46)
Head fixed mean duration of lexical fixations	272 ms + 49 ms	253 + 42 ms	250 ms + (=) ms
Head free mean duration of lexical fixations	253 + 24 ms	254 ms + 29 ms	268 ms - 11ms

Reduced spaces had the most distinct effect in the subject with left homonymous hemianopia that was recorded twelve days after the acute event. The global reading time of a text with 48 letters per line was nearly the same as in texts with 28 letters per line. Correspondingly, the mean reading rate was significantly higher in texts with reduced spaces between letters and words (see Table 3). Obviously, this hemianopic subject had more trouble in understanding the meaning of especially long words. Also, it was difficult for him to get the idea of a sentence with enlarged spaces between letters and words. This correlated with an increase in acoustical reading errors and of global reading times.

TABLE 3

Mean Reading Rate Change of Less Adapted Left Hemianopic Subject with Acoustical Reading Errors. Time from the Acute Event to Day of EOG Recordings was 12 Days.

Less adapted patient with acoustical reading errors	MEAN READING RATE	
	Head fixed	Head free
28 letters per line	5.4 1 / s	5.7 1 / s
46 letters per line	7.5 1 / s	8.2 1 / s

DISCUSSION

None of our four patients showed any sign of inattention that could be attributed to neglect phenomena. Two of them did not produce any acoustical reading errors (Figs. 1a,b). Characteristically, they were the best adapted patients. The patient with right homonymous hemianopia had a complete dense right lower quadrantic hemifield loss that reached far over the horizontal meridian into the right upper quadrant. The macula was not affected in the upper quadrant (Fig. 1b).

The patient with left homonymous hemianopia had an incomplete left hemifield loss with macular sparing (Fig. 1a). The residual hemifield on the horizontal meridian was about five degree. In this patient our results were in accordance with earlier findings. He was able to read without acoustical reading errors, whereas the global reading time was increased. To compensate for the blind hemifield, the patient had a 'beginning of line' detective strategy while the other part of the eye movement patterns resembled normal behaviour.

The patient with incomplete right homonymous hemianopia (Fig. 1b) could read the texts immediately, without any acoustical reading errors. Of course, the global reading time was increased compared with a normal subject. The eye movement pattern did not appear to be disorganized in this patient, although ocassionally the classical hemianopic staircase pattern was evident.

This subject had adopted three different reading strategies to compensate for the hemifield loss. The application of these strategies appeared completely interchangeable. Pommerenke & Markowitsch (1989) concluded that a specific and systematic exploration training can be of significant help for improving visuo-spatial behaviour in hemianopic patients. Both of our subjects were highly motivated in any training and did not appear to be satisfied with the present situation. As the application and efficient use of different ocular motor strategies during reading represents the highest level of visual control (Stark et al., 1991), high motivation of the patients is essential for a successful rehabilitation.

Both of our well adapted hemianopic patients demonstrated the similar kind of short term adaptation in horizontal head torque and horizontal head rotations: The side of attention changed from the blind hemifield into the seeing hemifield. The reduction of head movements with increased reading time were in accordance with earlier results of Zangemeister et al., (1982). Consequently, with this reduction of especially large head amplitudes to the blind hemifield the seeing hemifield was more and more accentuated. Obviously, both patients had at last a subjective advantage in using head movements of an equal amount to both sides. It is well known that hemianopic patients often try to assume an oblique head position to the blind hemifield that shifts the gaze direction additionally into the seeing hemifield (Zihl et al., 1984). This kind of adaptation is inefficient, because the gaze direction is shifted into the opposite direction. Nevertheless, our results demonstrate that in an advanced stage it may be useful to train the same rehabilitation tasks in a head-free-to-move condition. A well trained and highly motivated patient should be able to profit from a well accentuated horizontal head employment to both sides.

Our two other hemianopic subjects displayed simpler adaptive mechanisms. They produced the typical acoustical reading errors that have been known since Mauthner (1881). While the right hemianopic patient did not use his head at all, the left homonymous patient displayed a distinct blind hemifield accentuation in all texts. As he had a very recent lesion, this patient did not have much experience in compensatory strategies. His behaviour was not in accordance with our previous results (Zangemeister et al., 1982, 1986).

With enlarged spaces between letters and words hemianopic patients had more problems in understanding especially long words. In the two well adapted subjects we found the duration of lexical fixation pauses increased with decreased spaces between letters and words. This confirms the results of Eber et al., (1988), who suggested that ocular motor behaviour is related to altered sensorial input such as size of words or extension of spaces between words.

CONCLUSION

To compensate for hemianopia it is necessary to have appropriate ocular motor strategies for efficient use of the remaining half of the visual field. Our study indicates that patients with pure hemianopia and foveal sparing optimally learn to compensate their visual handicap by active and motivated visual training. Characteristically, the two of our four hemianopic subjects that did <u>not</u> produce any acoustical reading errors were the best adapted ones. In daily life they had developed a consistent set of ocular motor strategies to compensate for their specific visual handicap.

ACKNOWLEDGEMENTS

We would like to thank Dirk Christian Ernst for his intense technical help, and also Mrs. Jutta Kleim for doing the psychological tests. Also we would like to thank Prof. Dannheim and Priv.-Doz. Dr.Kai Uwe Hamann for referring patients to us.

REFERENCES

Balliet, R., Blood, K.M.T. & Bach-y-Rita, P. (1985). Visual field rehabilitation in the cortically   blind ? *Journal of Neurology Neuroscience and Psychiatry*, 49, 1113-1124.

Balliet, R., Blood, K.M.T. & Bach-y-Rita, P. (1986). Visual field rehabilitation in the cortically blind? (Reply). *Journal of Neurology Neuroscience and Psychiatry*, 49, 966-967.

Benton, A.L. (1972). Der Benton Test. Bern: Huber Verlag.

Campion, J., Latto, R. & Smith,Y.M. (1983). Is blindsight an effect of scattered light, spared cortex, and near threshold vision? *Behavioural Brain Science*, 6, 423-486.

Eber, A.M., Metz-Lutz, M.N., Strubel, D., Veterano, E. & Copllard, M. (1988). Electro-oculographic study of reading in hemianopic patients. Paris: *Revue Neurologie*, 144 (2-9), 515-8.

Eber, A.M., Metz-Lutz, M.N., Bataillard, M. & Collard, M. (1986). Reading eye movements of patients with homonymous hemianopia. In J. K. O`Reagan and A. Levy-Schoen (Eds.). *Eye movements: From physiology to cognition*. Amsterdam: Elsevier North Holland (pp. 544-545).

Gassel, M.M. & Williams, D. (1963). Visual function in patients with homonymous hemianopia. Part 2. Oculomotor mechanisms. *Brain*, 86, 1-36.

Gassel, M.M., Williams, D. (1963). Visual function in patients with homonymous hemianopia. Part 3. The completion phenomenon; insight and attitude to the defect and visual functional efficiency. *Brain*, 86, 229-260.

Mackensen, G. (1962). Examination of reading ability as a clinical function test. Germany: Bibliothek *Ophthalmologie*, 59, 344-379.

Mauthner, L. (1881). Vorträge aus dem Gesamtgebiete der Augenheilkunde für Studierende und Aerzte. In J.F. Bergmann (Ed.). *Gehirn und Auge*. Wiesbaden: Hefte 6-8 (pp. 345-600), 8°.

Meienberg, O., Zangemeister, W.H., Rosenberg, M., Hoyt, W.F. & Stark, L. (1981). Saccadic eye movement strategies in patients with homonymous hemianopia. *Annals of Neurology*, 9, 537-544.

Meienberg, O. (1983). Clinical examination of saccadic eye movements in hemianopia. *Neurology*, 33, 1311-1315.

Meienberg, O. (1988). Augenbewegungsmuster bei homonymer Hemianopsie und visuellem Hemineglekt. Okulographische Abgrenzungs-Kriterien anhand von neunzehn Fällen und diagnostische Bedeutung. Klinische Monatsblätter, *Augenheilkunde*, 192, 108-112.

Pommerenke, K. & Markowitsch, H.J. (1989). Rehabilitation training of homonymous visual field defects in patients with postgeniculate damage on the visual system. Elsevier, *Restorative Neurology and Neuroscience*, 1, 47-63.

Poppelreuter, W. (1917). Die psychischen Schädigungen durch Kopfschuß im Kriege 1914-1916. In L.Voss (Ed.). *Band 1: Die Störungen der niederen und höreren Sehleistungen durch Verletzungen des Occipitalhirns*. Leipzig: 8.

Stark, L.W., Giveen, S.C. & Terdiman, J.F. (1991). Specific dyslexia and eye movements. In J.F. Stein (Ed.).*Vision and Visual Dyslexia*. London: Macmillan, (pp. 203-232).

Wechsler, D. (1964). Die Messung der Intelligenz Erwachsener. Bern: Huber Verlag.

Willbrandt, H. (1907). Ueber die makulär-hemianopische Lesestörung und die von Monakowsche Projection der Makula auf die Sehsphäre.Stuttgart: *Klinisches Monatsblatt für Augenheilkunde*, 16, 1-39.

Zangemeister, W.H., Meienberg, O., Stark, L. & Hoyt, W.F. (1982). Eye head coordination in homonymous hemianopia. *Journal of Neurology*, 226, 243-254.

Zangemeister, W.H. & Stark, L. (1983). Pathological types of eye-head coordination in neurological disorders. *Neuro-Ophthalmology*, 3, 259-276.

Zangemeister, W.H., Dannheim, F. & Kunze, K. (1986). Adaptation of gaze to eccentric fixation in homonymous hemianopia. In E. L. Keller and D.S. Zee (Eds.). *Adaptive processes in visual and oculomotor systems*. Pergamon Press.

Zihl, J. (1981). Recovery of visual functions in patients with cerebral blindness: Effect of specific practise with saccadic localization. *Experimental Brain Research*, 44, 159-169.

Zihl, J. & von Cramon, D. (1985). Visual field recovery from scotoma in patients with postgeniculate damage. A review of 55 cases. *Brain*, 108, 335-365.

Zihl, J. & von Cramon, D. (1986). Recovery of visual field in patients with postgeniculate damage. In K. Poeck, H. Freund and H. Gänshirt (Eds.). *Neurology*. Berlin: Springer, 188-194.

Zihl, J., Krischer, C. & Meißen, R. (1984). Die hemianopische Lesestörung und ihre Behandlung. *Nervenarzt,* 55, 317-323.

III. LANGUAGE PROCESSING

Facets of Dyslexia and its Remediation
S.F. Wright and R. Groner (Editors)

A NEW THEORETICAL FRAMEWORK FOR UNDERSTANDING READING AND SPELLING TASKS

E. Graf

University of York, York, U.K.[*]

INTRODUCTION

Currently there are two important theoretical approaches to developmental dyslexia. The developmental models for reading and spelling, aim to describe mainly the normal development of reading and spelling. The different developmental stages denote a specified set of skills acquired by the individual. Most of these models have in common a similar taxonomy of skills for each stage of development.

Only recently have there been some attempts to apply the processing models of cognitive neuropsychology to developmental dyslexia. Basically the same material that has been used in studies of patients with an acquired reading and/or spelling disorder, such as words differing in length, frequency, age of acquisition, regularity and non-words, is employed in studies of developmental dyslexia. By an analysis of errors one can find out what type of strategy (direct/indirect route) an individual uses. Therefore the focus of this paper lies on the underlying processes used in reading and spelling.

Besides these theoretical approaches (which still lack clear empirical evidence for developmental dyslexics) there are a lot of empirical studies. These focus on the relationship between the development of reading and spelling and various cognitive skills, such as short-term memory and 'phonological awareness'. Unfortunately in most of these studies little interest is shown in the nature of the underlying processes that are involved in solving a specific task.

Baddeley's model of short-term memory (1986) is an exception as the development of short-term memory is empirically well examined and documented in normal children. The author claims that developmental dyslexics are not lacking the articulatory loop which is the most important part of the model but that the capacity of the system seems to be insufficient. Evidence for the important role of short-term memory also comes from very specific findings in a brain-damaged patient. Consequently the short-term memory will be integrated in the proposed generic cognitive neuropsychological model.

---

[*] This paper was supported by a grant of the Swiss National Research Foundation which allowed the author to stay at the Department of Psychology, University of York. I am grateful to Andy Ellis and Sue Franklin for discussions and to Judy Turner for revising the paper.

In this paper an attempt will be made to analyze the underlying processes of specific tasks, such as rhyme decision and the blending of phonemes to make a word. A generic processing model of cognitive neuropsychology, into which short-term memory is integrated, will provide the theoretical basis for this.

Developmental Models for the Acquisition of Reading and Spelling

At present there are several models which aim to describe the normal development of reading and spelling (Marsh, Friedman, Welch & Desberg, 1981; Frith, 1985, 1986; Ehri, 1986). The different developmental stages denote a specified set of skills acquired by the individual. Most of these models have in common a similar taxonomy of skills for each stage of development.

Frith's approach (1985, 1986) seems particularly interesting because she postulates a single model for reading and spelling, as well as posing some issues in considering "classic developmental dyslexia". According to Frith (1985, p. 311) writing and reading skills will sometimes show dissociations in development. Consequently each phase is divided into two steps with either reading or writing as the pacemaker of the strategy that identifies the phase. Reading is the pacemaker for the logographic strategy, writing for the alphabetic strategy and reading again for the orthographic one.

The logographic strategy is characterized by use of a sight vocabulary. According to Eichler (1986) the visual representation of a word need not to be very detailed in this case. Words are recognised by salient features, such as the first letter, the last letter and word length. This technique quite often leads to reading errors. During the alphabetic stage the child learns about the relationship between letters and sounds. The orthographic strategy is characterized by using orthographic units and morphological knowledge. Unfortunately the stages in Frith's model (1985, p.306) are not clearly defined (Stuart & Coltheart, 1988).

Frith (1986) assumes that classic developmental dyslexia describes problems in acquiring the alphabetic stage: there are great difficulties in the reading of new words which are not already present in the sight vocabulary because of difficulties in blending the individual phonemes of a word. Spelling performance is even more compromised. The further development of reading and spelling is claimed to be abnormal because a breakdown at a certain stage affects all other stages. However by using compensatory strategies, a developmental dyslexic can develop an approach which resembles the orthographic strategy.

Marsh et al.,'s (1981) and Ehri's models (1986) are similar and indeed Snowling (1987) proposes that any differences between them are mainly terminological. Both imply that there is a strong connection between reading and spelling. They also show that the crucial points in the acquisition of reading are grapheme/phoneme-conversion and the blending of phonemes to form a word, whereas the important points for spelling are the segmentation of phonemes and subsequent phoneme/ grapheme-correspondence. It is important in both reading and spelling that a child gives up the one-to-one correspondence (in spelling this is the one-letter-for-every-sound principle).

There are several hypotheses for how reading and spelling are connected but very few empirical studies focus on this point. In a long-term study Cataldo & Ellis (1989) found that the transition from the pre-alphabetic stage (which seems to be similar to Frith's logographic stage) to the alphabetic stage of reading is clearly facilitated by the acquisition of spelling. This finding supports Frith's hypothesis that the alphabetic stage is firstly acquired in writing. As reading and spelling are closely related it is assumed that dyslexic children not only have reading but also spelling problems. In the literature the focus clearly lies on reading problems.

Each stage is typically defined by the skills acquired during the development of reading and spelling. This approach seems to be useful for providing an assessment of reading and spelling skills, but does not offer an account of the processes which lead to the achievement of a particular stage. It remains an open question as to why a child becomes able to blend phonemes after initially relying on sight vocabulary.

Having reached the highest level of reading and spelling, people are supposed to work with phonemes, morphemes, orthographic principles and analogies. Reading by analogies sometimes is thought to be acquired at a late stage of reading and writing development (Marsh et al., 1981). According to Goswami (1988a, 1988b) children of 6 to 7 years already know that using analogies does not always work (it is fine with 'peak'/'leak', but it doesn't work with 'peak'/'steak'). As sight vocabulary grows, children seem to be less dependant on the analogy strategy. It is claimed that the many irregularly spelled words in English have to be stored as a whole in order to be read or spelled correctly (e.g. Leicester) and therefore it is important to have a big sight vocabulary.

There are a few other drawbacks to these models: most of the previously mentioned authors have presented the stages of reading/spelling as being invariant in order. These, however, have not yet been empirically demonstrated. Ellis & Large (1988) state that these sequences are not necessarily invariant, and that the strategy used by an individual is influenced by the material read. Furthermore, the choice of reading and spelling method seems to have a major influence on the acquisition of these skills. Ehri (1987) and Seymour & Elder (1986) have assumed that in the case of a synthetic method the logographic stage might be skipped entirely, and the strategy which represents the alphabetic stage will be directly acquired.

There is another recent approach to developmental dyslexia, which is regarded as an alternative to stage models. According to Colley & Beech (1989) it is called 'componential approach'. It focuses on the relationship between the development of reading and spelling and various cognitive skills, such as short-term memory and phonological awareness (Goswami & Bryant, 1990). Typically comparisons are made between a group of dyslexic children and a group of normal children matched for same reading/spelling age or chronological age. Unfortunately most of these studies are atheoretical and little interest is shown in the underlying skills and processes that are involved in solving a specific task.

Process Models of Single Words of the Cognitive Neuropsychology

Cognitive neuropsychology represents a convergence of cognitive psychology and neuropsychology. According to Ellis & Young (1988, p.4) cognitive neuropsychology has two basic aims: "The first is *to explain the patterns of impaired and intact cognitive performance seen in brain injured patients in terms of damage to one or more of the components of a theory or model of normal cognitive functioning. (...) The second aim is (...) to draw conclusions about normal, intact cognitive processes from the patterns of impaired and intact capabilities seen in brain-injured patients.*" (italic by the authors).

In 1980, Morton published the so called 'logogen-model'. In the literature there are a large number of models which are related to the logogen model although the authors tend to use different terminology. A generic process model of reading might somehow overcome these terminological difficulties.

In reading there are two different routes, a direct route and an indirect route (Fig. 1). Patients with an impaired direct route are usually called surface dyslexics, whereas patients whose indirect route is impaired are called deep dyslexics. The same distinction is also made with dyslexic children, but different terms are used. If the direct route is not available or is temporarily inaccessible, Seymour (1986) speaks of morphemic developmental dyslexia, whereas when the indirect route cannot be used he refers to phonological developmental dyslexia.

Despite numerous attempts to subtype learning disabilities (see Hooper & Willis, 1989 for a review) there is, at present, no clear evidence for the existence of two (or more) subgroups in developmental dyslexia. Seymour (1986) argues, there are only a few developmental dyslexics who can clearly be identified as suffering from morphemic dyslexia. In most of these cases there seem several components which are undeveloped or not functioning normally. It might be the case that originally one of the routes was weaker, but the child has now developed some kind of compensatory strategy (phonological developmental dyslexics are quite often taught 'phonics'). It has been assumed that patients with an acquired dyslexia will generally show a clearer pattern of impairment and therefore be more readily assigned to one of the groups.

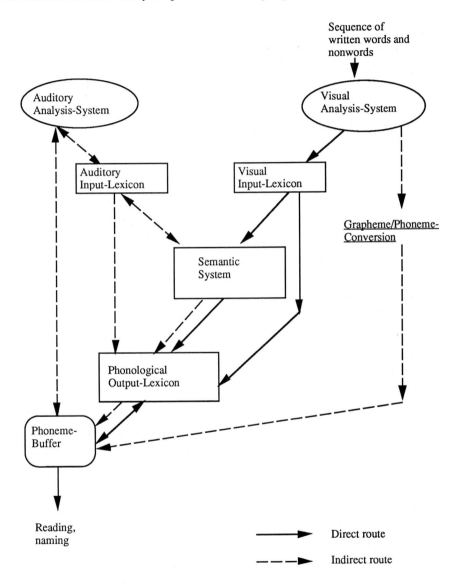

FIG. 1 Generic cognitive neuropsychological model of reading.

Three different types of 'boxes' are seen in Fig. 1 representing: (1) Visual and phonological identification and discrimination and visual and auditory analysis system; (2) Different lexicons and the semantic system which are all linked to the long-term memory (the type of representation of information for the different components is still open) and (3) Phoneme-buffer (Patterson & Shewell, 1987) which seems to be closely related to the short-term memory. The same component is called 'phonological short-term memory system' by Miceli (1989) or simply 'phoneme level' by Ellis & Young (1988).

In fact a reading problem might be caused by any of the components involved. Impaired visual processing (visual analysis-system), a restricted sight vocabulary (visual input-lexicon), a generally restricted vocabulary (semantic system), problems with phonological processing (auditory analysis-system, auditory input-lexicon) and blending of phonemes to a word (supposedly phoneme-buffer) are usually meant to cause developmental dyslexia individually, or in combination.

Using the indirect grapheme/phoneme-conversion route in reading quite often does not cause any difficulties, but the blending of phonemes appears to be a problem. In spelling, the segmentation of words into phonemes is crucial in use of the indirect route. There is some evidence that the short-term memory is involved in blending and segmenting.

The big difference between patients with an acquired dyslexia and children with a developmental dyslexia is that the former group's problems occurred after they had mastered reading and spelling skills. Therefore, the question arises whether cognitive neuropsychological models can be adopted for developmental dyslexia. Here it will be suggested that in normal acquisition of written language the different components of the generic model are acquired or developed at different instants. Those components being used for the naming of objects, such as the semantic system, the phonological output-lexicon and the phoneme buffer will be acquired very early, before school starts. Recognition, comprehension and repetition of words are also skills that are shown early (additionally involving the auditory analysis-system and auditory input-lexicon). The specific components for written language usually are developed after school starts. For reading these are the visual input-lexicon, the phoneme buffer and grapheme/phoneme-conversion. (For spelling the graphemic output-lexicon, the grapheme buffer and the phoneme/grapheme-conversion have to be developed and accessible.)

It is debatable to what extent the developmental models and the cognitive neuropsychological approach are related. The critical comparison being whether the highest level of reading and spelling in developmental models can be explained within the cognitive neuropsychological models which were specifically designed to describe the processes of good readers and spellers. Seymour & Elder (1986) suggested that good readers use an 'adult logogen system' which developed from the sight vocabulary used at the logographic stage but contains a very precise representation of the individual word. These authors also suggest that different visual processing styles are necessary in order to use the direct and indirect route in reading: a wholistic visual processing (as a strategy of the logographic and the orthographic stage) will be used for the direct route and an analytic visual processing (as a strategy of the alphabetic and the orthographic stage) will be used for the indirect route.

The cognitive neuropsychological models are restricted to the processing of single words which means that context is not integrated. It is evident that context influences the reading of beginners both at the sentence and text level (Marsh et al. 1981). The model also suggests that a reader/speller must use either the direct or the indirect strategy but there is evidence that reality is much more complex.

It seems to be important that a good reader/ speller shows a great flexibility of strategy and is able to use different approaches, according to the words or nonwords to be read or spelt. It is even likely that in reading one single word there will be two different strategies used simultaneously.

Recently there has been a great deal of interest in parallel processes as proposed in connectionist models. Interestingly there are some attempts at simulating acquired dyslexia by lesioning specific components of the network in such a model (e.g. Hinton & Shallice, 1989). Seidenberg & McClelland (1989) taught their model to read and tried to simulate developmental dyslexia. Unfortunately these first attempts at relating cognitive neuropsychological models to the connectionist approach are not too convincing but there is hope that with a greater vocabulary and with more effective connectionist models the simulation will be more adequate.

Task Analysis: Towards a New Theoretically Based Approach

For further empirical and theoretical work in the field of developmental dyslexia it seems appropriate to try to combine the different approaches mentioned above. On the basis of a cognitive neuropsychological model in which short-term memory is integrated one can analyse the tasks usually presented, such as reading and spelling of words and nonwords or phonological awareness. This seems to be an important step towards theoretically guided empirical studies of developmental dyslexics and controls.

Baddeley's model of short-term memory (1986) consists of a 'central executive' which integrates the information coming from the visuo-spatial sketch-pad and the articulatory loop, the latter is divided into the phonological input-system and the articulatory rehearsal process. Empirical studies (e.g. Hitch et al. 1988) clearly show that between six and ten years the children become gradually able to use rehearsal as a specific memory strategy. In the beginning this strategy will only be used with auditorily presented material but as the child develops it will also be used for visually presented stimuli.

On the basis of this model Gathercole & Baddeley (1990) assume that an impairment in the functioning of the articulatory loop might cause developmental dyslexia. They also claim that phonological storage skills play an important role in the development of language skills, such as reading, vocabulary and word comprehension. In fact developmental dyslexics quite often show an impaired memory span. The authors suggest that the information is phonologically encoded and even rehearsed in order to maintain it. Nevertheless the lack of efficiency of the processes which are related to phonological representation seems to cause the poor memory performance.

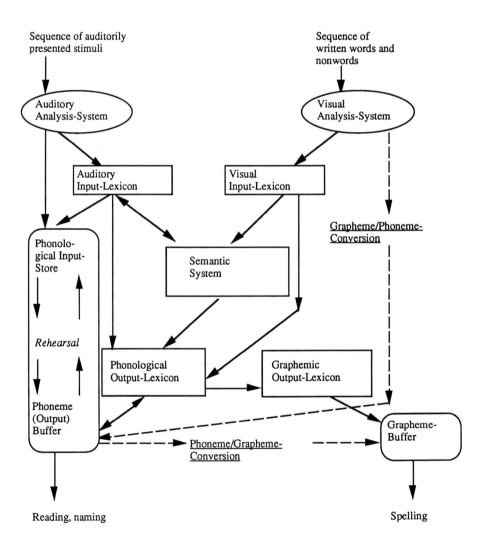

FIG. 2. Generic cognitive neuropsychological model with integrated short-term memory (articulatory loop)

Franklin (1989) suggested the integration of short-term memory into the models of cognitive neuropsychology in order to explain a very specific finding in her patient MK. Howard & Franklin (1990) propose a 3 component model of short term memory which looks quite similar to Baddeley's model of working memory (1986). In Fig. 2 the short-term memory (represented by the articulatory loop) is integrated into a generic cognitive neuropsychological model. In this model the articulatory loop contains a phonological input-store (items are phonologically stored), a phoneme output-store (storage of sequences of phonemes; rhyme judgement, blending) and the rehearsal which links both stores.

It is often claimed that developmental dyslexics show a lack of phonological awareness. One task frequently used to examine this skill is the detection of rhymes. Goswami & Bryant (1990) suggest that pre-school rhyming skills are quite a good indicator for reading in the third grade. In this paper the attempt is made to demonstrate which processes are being used in rhyme decision through use of a generic cognitive neuropsychological model in which the short-term memory is integrated (Fig. 3).

Rhyme decision can be presented in several different ways: spoken words, pictures and written words. Obviously short-term memory plays a crucial role. In the case of auditory presentation, the information passes to the articulatory loop directly. In the case of pictures, which are recognised by the object recognition-system access to the semantic system (meaning) and to the phonological output-lexicon (naming) are necessary. Subsequently the information will be passed on to the articulatory loop. In the case of written words, there is also no direct access to the articulatory loop. Words and nonwords firstly have to be read either by the direct or by the indirect route. Whatever the form of the stimulus, the rhyme decision seems to take place in the articulatory loop and the words are compared by segmenting them into onset and rime (beginning and end of the word).

Blending of phonemes to make a word seems to cause big problems for phonological developmental dyslexics. Again there are two different methods of presentation (Fig.4). In the case of auditorily presented phonemes the information need not undergo a grapheme/phoneme conversion, but directly passes to the articulatory loop. Therefore, it should be easier to blend auditorily presented phonemes than ones presented visually. This hypothesis still has to be empirically tested.

The same theoretical framework can thus be applied to rhyme decision, segmenting words into phonemes and to manipulation of letters in words or nonwords. This might be an important step towards theoretically guided studies of specific aspects of developmental dyslexia. It would be of interest to investigate which component(s) is (are) lacking or malfunctioning in a specific individual. These results should show if there is any evidence for the existence of the two subgroups of developmental dyslexics distinguished above. More importantly, however, the use of a model should assist in gaining knowledge of weaknesses and strengths of a specific individual in respect of his/her specific training or remediation. Testing should contain not only specific reading and spelling tasks but also visual processing (recognition of abstract forms, naming and categorising of objects and pictures, visual short-term memory, rhyme decision using visual stimuli), phonological processing (acoustic differentiation, repetition of words and nonwords, repetition of sentences, auditory short-term memory, rhyme decision and manipulation of letters in a word or nonword).

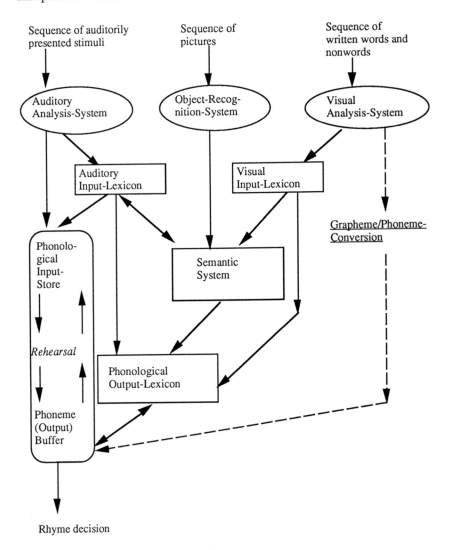

FIG. 3. Different processes in rhyme decision depending on the input

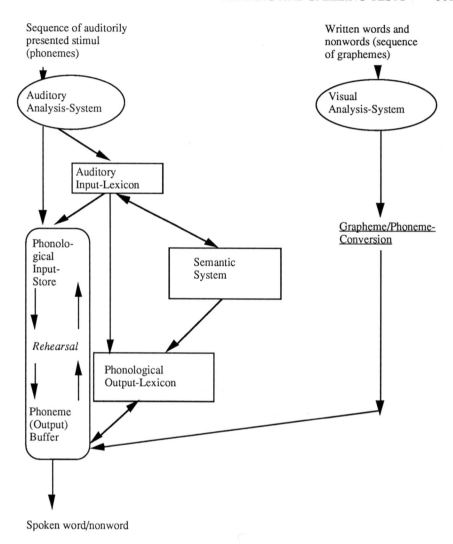

Sequence of auditorily presented stimul (phonemes)

Written words and nonwords (sequence of graphemes)

Auditory Analysis-System

Visual Analysis-System

Auditory Input-Lexicon

Grapheme/Phoneme-Conversion

Phonological Input-Store

Semantic System

Rehearsal

Phonological Output-Lexicon

Phoneme (Output) Buffer

Spoken word/nonword

FIG. 4. Blending of phonemes or graphemes to a word/nonword

CONCLUSION

The explanations given by the two most important theoretical approaches to developmental dyslexia are rather restricted. Developmental models demonstrate which level of reading and spelling an individual has reached, whereas process models of cognitive neuropsychology emphasize the availability and functioning of the direct and indirect routes of reading and spelling. The subsequent subtyping of dyslexic children into two groups (morphemic or phonological developmental dyslexics) is in all likelihood not an adequate solution.

This paper advocates a more individual approach that focuses on the individual strengths and weaknesses of dyslexics and which moves away from such a dichotomous classification. This approach is based on the analysis of common tasks related to reading and spelling (eg. phonological awareness) which entails the breaking down of tasks into their (sub)components, the tesing of which provides the basis for individual diagnosis. As a result of such detailed testing the individual dyslexic may be shown to have a very specific pattern of problems with components of visual and phonological processing.

REFERENCES

Baddeley, A.D. (1986). *Working memory.* Oxford: Clarendon Press.

Cataldo, S. & Ellis, N. (1989). Learning to spell, learning to read. In P. Pumphrey & C. Elliott (Eds.). *Primary school pupil's reading and spelling difficulties: Current research and practice.* Baisingstone: Falmer Press.

Colley, A. & Beech, J. (1989). *Acquisition and performance of cognitive skills.* New York: Wiley.

Ehri, L. (1986). Sources of difficulty in learning to spell and read. *Advances in Developmental and Behavioural Pediatrics, 7,* 121-195.

Ehri, L. (1987). Learning to read and spell words. *Journal of Reading Behaviour,* XIX, 5-31.

Eichler, W. (1986). Zu Uta Frith's Dreiphasenmodell des Lesen- (und Schreiben-) Lernens. In G. Augst, (Hrsg.): *New trends in graphemics and orthography* (pp. 234-247). Berlin: De Gruyter.

Ellis, A. & Young, A. (1988). *Human Cognitive Neuropsychology.* London: Erlbaum.

Ellis, N. & Large, B. (1988). The early strategies of reading: A longitudinal study. *Applied Cognitive Psychology, 2,* 47-76.

Franklin, S. (1989). *Understanding and repeating words: Evidence from aphasia.* Unpublished doctoral thesis. City University, London.

Frith, U. (1985). Beneath the surface of developmental dyslexia. In K. Patterson, J. Marshall, & M. Coltheart, (Eds.) *Surface Dyslexia* (pp. 301-330). London: Erlbaum.

Frith, U. (1986). Psychologische Aspekte des orthographischen Wissens: Entwicklung und Entwicklungsstörung. In G. Augst, (Ed.). *New Trends in Graphemics and Orthography* (pp. 218-233). New York: De Gruyter.

Gathercole, S. & Baddeley, A. (1990). Phonological memory deficits in language disordered children: Is there a causal connection? *Journal of Memory and Language*, 29, 336-360.

Goswami, U. (1988a). Orthographic analogies and reading development. *Quarterly Journal of Experimental Psychology*, 40A, 239-268.

Goswami, U. (1988b). Children's use of analogy in learning to read. *British Journal of Developmental Psychology*, 6, 21-34.

Goswami, U. & Bryant, P. (1990). *Phonological Skills and Learning to Read*. Hillsdale: Erlbaum.

Hinton, G. & Shallice, T. (1989). *Lesioning a connectionist network: Investigations of acquired dyslexia*. Unpublished technical report. Department of Computer Science, University of Toronto, Canada.

Hooper, S. & Willis, W.G. (1989). *Learning Disability Subtyping. Neuropsychological Foundations, Conceptual Models and Issues in Clinical Differentation*. New York: Springer.

Howard, D. & Franklin, S. (1990). Memory without rehearsal. In G.Vallar & T. Shallice (Eds.) *Neuropsychological Impairments of Short-Term Memory* (pp. 287-318). Cambridge: Cambridge University Press.

Marsh, G., Friedman, M., Welch, V. & Desberg, P. (1981). A cognitive-developmental theory of reading acquisition. *Reading Research: Advances in Theory and Practice*, 3, 199-221.

Miceli, G. (1989). A model of the spelling process: Evidence from cognitively impaired subjects. In P. Aaron, & R. Joshi (Eds.) *Reading and Writing Disorders in Different Orthographic Systems* (pp. 305-328). Dodrecht: Kluwer Academic Publishers.

Morton, J. (1980). The logogen model and orthographic structure. In U. Frith (Ed.) *Cognitive Processes in Spelling*. London: Academic Press.

Patterson, K. & Shewell, Ch. (1987). Speak and spell: Dissociations and word-class effects. In M. Coltheart, G. Sartori, & R. Job, (Eds.). *The Cognitive Neuropsychology of Language* (pp. 237-294). London: Erlbaum.

Seymour, P. (1986). *Cognitive Analysis of Dyslexia*. London: Routledge and Kegan Paul.

Seymour, P. & Elder, L. (1986). Beginning reading without phonology. *Cognitive Neuropsychology*, 3, 1-36.

Seymour, P. & MacGregor, C. (1984). Developmental dyslexia: a cognitive experimental analysis of phonological, morphemic and visual impairments. *Cognitive Neuropsychology*, 1, 43-82.

Snowling, M. (1987). *Dyslexia. A Cognitive Developmental Perspective.* New York: Blackwell.

Stuart, M. (1990). Processing strategies in a phoneme deletion task. *The Quarterly Journal of Experimental Psychology,* 42A, (2), 305-327.

Stuart, M. & Coltheart, M. (1988). Does reading develop in a sequence of stages? *Cognition,* 30, 139-181.

Facets of Dyslexia and its Remediation
S.F. Wright and R. Groner (Editors)

INFORMATION INTEGRATION AND READING DISABILITIES

K.L. Holtz,

Pädagogische Hochschule, Heidelberg

Germany

Discussing the subject of "verbal comprehension" (Sternberg & Powell, 1983) as well as "reading disabilities" Sternberg used an analogy in order to stress the need for a comprehensive theoretical framework: in an old Indian folktale some blind men want to know, what an elephant 'looks' like. The first man proclaims that the elephant is just like a snake, the second asserts that the animal is like a wall and so on, with each blind man contributing a different, only partially correct opinion on the nature of the elephant" (Spear & Sternberg, 1987, p.4).

In our research project "Information Integration and Reading Disabilities" carried out over the last six years, we very often felt ourselves to be in the position of these blind men. Contradictory results from our own research and from those of other investigators led us to the conclusion, that it would be necessary to rely on a comprehensive model, that could incorporate at least the following features:

- it should be componential and allow a process-evaluation of different strategies
- strategies should be related to different task requirements and developmental aspects of the reading process and should therefore allow the determination of different subtypes of reading disability.

Especially in the field of reading and writing, researchers become aware that the analysis of the basic skills of these abilities must be a highly theory-driven enterprise. The paradox arises, from the fact that we need a theory-based model of reading, "before we can model the reading system" (Carr, Brown, Vavrus & Evans, 1990, p.7). Especially in the componential analyses of complex processes the iterative character of this kind of research has become evident (Carr & Levy, 1990; Frederiksen, 1982). So we started to look for a comprehensive model of information integration, that could serve as a heuristic model and could integrate relevant findings to explain problems of reading development.

In this paper I will present first the model based on Sternberg's triarchic theory of intelligence (e.g. 1985). In a second part I would like to show that this model may help to generate further hypotheses for empirical research. It will be made clear, that a "unified approach" to intelligence and learning disability (Das 1987), a synthesis between "intelligence, information processing, and specific learning disabilities" (Kolligian & Sternberg, 1987) is both possible and necessary.

A MODEL OF INFORMATION INTEGRATION

We will use the model presented below as a "causal" model (cf. Cavanaugh, Kramer, Sinnott, Camp & Markley, 1985), which is composed of a finite set of independent "abilities" integrated into a more global system that specifies the relationship between parts (cf. Detterman, 1987, p.8). The "wholeness" of the system is a function of the level of measurement and it is possible to develop hypotheses about the interrelations of different aspects in separate parts of the system.

In this sense the model may serve for "causal modeling" (Cavanaugh et al. 1985), also in the sense of "componential skills analysis" (Frederiksen, 1982), where connections between constructs or components (e.g. of the reading process) will be made explicit, as well as the relationships between hypothetical constructs, and their measures (for further advantages of causal modeling in task analysis see Cavanaugh et al., 1985). Thus, this approach can make a theory explicit and testable and allows for the integration of different research studies, as Spear & Sternberg (1987) have suggested. As Detterman (1987) has pointed out, each construct (or a special combination of components) may be the subject of an entire "causal model", where in addition to componential analysis it may lead to further testable hypotheses in an experimental or (individual) quasi-experimental process. Our intention here is to present at first a more general "heuristic" model relating high-level constructs and then to give an example of hypotheses-generation under process-orientated and developmental aspects. Methodological problems of these approaches cannot be discussed here in detail.

THE TRIARCHIC MIND

Sternberg (e.g. 1985) has proposed a triarchic model of human intelligence which is comprised of three subtheories.

1. The **contextual part** which relates intelligence to the external world of the individual (adapting, shaping, selecting).

2. **The experiential part** which relates intelligence to experience with both the internal and external world of the individual (familiarity of tasks and situations to the child: novelty and automatization).

3. **The componential part**, which relates intelligence to the internal world of the individual (meta-, performance- and knowledge-acquisition-components, see Fig.1).

Hypotheses about school failure (as a transactional aspect of information integration) may be deduced on every level of subtheory or level of components, or on their respective interactions.

The componential part is briefly discussed below.

Metacomponents are a central source of individual differences in general intelligence and consist of higher order, executive processes that are used to plan, monitor and evaluate one's task performance. They are the only kind of component that can act upon other components and include: definition of the nature of the task; selection of lower order processes to accomplish the task; formation or selection of one or more strategies into which to combine the lower order processes; formation or selection of a mental representation upon which the lower order processes act; allocation of mental resources in task performance; monitoring of one's task performance and the evaluation of one's task performance.

Performance components are lower order components that execute the plans and strategies designed by the metacomponents ("metacomponents operate upon components, whereas performance components operate upon data", Kolligian & Sternberg, 1987, p.10). Sternberg differentiates three types of performance components.

1. General components necessary for every type of task-solution (e.g. decoding).

2. Specific components only activated for specific tasks, (e.g. transformation in spatial rotation-tasks).

3. Class components which have, in different types of task, a differing number of common components (e.g. categorization).

Knowledge-acquisition components are also nonexecutive processes that are used in learning new information. They consist of three different types of components.

1. Selective encoding, information which is relevant for one's problem solving purposes is distinguished from not relevant information.

2. Selective combination, selective encoded information is assimilated and organized into a more or less coherent and usable cognitive structure.

3. Selective comparison, the process by which the newly encoded and combined information is compared to old information and antecedent cognitive structures.

The model also includes working memory and the knowledge base. Interaction between the components is here seen to take place in the working memory.

A strategy may be defined as the organized collection of component processes, where metacomponents are higher order control processes for selecting and monitoring strategies (cf. Sternberg & Salter, 1982). Strategic deficits may therefore be due to many different conditions, e.g. to deficits of any one of the components or to a "production deficit" because of insufficient interaction of the components (cf. the problem of flexibility, Sternberg, 1981; Sternberg & Frensch, 1989). Jorm (1983) has shown that also deficits of the working memory may be responsible for incomplete, incorrect or inflexible performance of metacomponents.

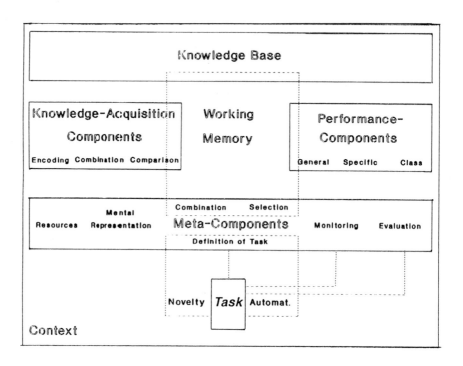

FIG. 1. Information-Processing sensu Sternberg

As Kolligian & Sternberg (1987), referring to Glaser (1984) and Chi & Koeske (1981), have pointed out, the "richness of knowledge base" may also be responsible for strategy choice. The transactional nature of the model is also stressed by the authors: "If, either for reasons of inadequate environmental opportunities or for reasons of componential inadequacies (especially inadequacies in the functioning of knowledge-acquisition components), a child's knowledge base is limited in a specific domain, then componential processing in that domain will be impaired (1987, p.12)."

This transaction also includes the relationship between the knowledge base and the automatization of information processing and coping with novelty (on the level of the experiential subtheory).Information processing and reading development

As we have pointed out (Holtz, 1990) our research project began with a replication of the findings by Das et al. (1982) concerning simultaneous and successive information processing in reading. One of our intentions was to find out, whether some kind of cultural bias would lead to different results, given it could be that different degrees of grapheme-phoneme irregularities in different languages (e.g. German and English) may be responsible for different reading strategies and problems (Venezky, 1976; Wimmer, Hartl & Moser, 1990).

Das, Snart & Mulcahy (1982) have described a model of information integration based on the empirical work of Luria (1973) on the neuropsychological theory of brain functioning (see Naglieri & Das, 1988 for a description). This model is linked to language and school learning and consists mainly of the following variables:

Simultaneous processing - "refers to the synthesis of separate elements of information into a holistic unitary representation, in which any portion of the synthesis is immediately surveyable without dependence upon its position in the whole" (Kirby & Robinson, 1987, p.243). The Marker tests of this process are:

Raven Coloured Progressive Matrices (Raven, 1956): The subject must indicate which of six alternatives correctly completes a figure. The subjects score is the number of correct items.
Memory for Designs (MFD: Graham & Kendall, 1960): The test consists of 15 simple designs which are individually inspected by the subject for 5 seconds and then reproduced from memory.
Figure Copying (FIGCOP: Ilg & Ames, 1965): This test involves the copying of figures that are always visible to the child and are scored 0, 1, 2, or 3 according to the correctness of the geometric proportions and relations.

Successive processing "refers to the synthesis of separate elements of information into a sequential, temporally dependent ordering. The entire synthesis in not surveyable at any one time and individual elements are independent of one another. Elements can be assessed only from their preceding elements in temporal order" (Kirby & Robinson, loc.cit). Marker tests of this dimension are:

Digit Span Forward (DIGIT: taken from the WISC for children):
Serial Order Recall (SERR): Auditory recall of words.

Visual Short-term Memory (VISHOT, Das et al, 1982). Subjects are presented with two-dimensional digit grid patterns for 5 seconds after which the digits have to be reproduced on a blank copy of the grid. There are 15 grid patterns with the number of digits in each grid increasing from 5 to 9 along with a corresponding increase in grid complexity. test score is the total number of digits correctly positioned in the grid cells.

Planning: refers to "a host of varied activities" (Das, 1987, p.107), which bear some resemblance to the metacomponents in the Sternberg model already cited. As Das states, these include " setting up goals, selection of strategies, performance monitoring, evaluating ones own and others behaviour, as well as the three planning functions: generation, selection and execution of plans" (Das et al, op. cit.). The marker tests include:

Trail Making (TRAILS): Numbers which are quasi-randomly distributed on a page have to be connected in the correct numerical order.
Visual Search (VIS): The subject is required to find objects or letters which have been shown on a separate piece of paper (a modified version of Das, 1984).

Re-test reliability and interrater reliability of the above tests was examined. The former ranged from .79 (VISHOT) to .89 (TRAILS) and the latter from .91 (SERR) to .93 (FIGCOP).

METHODS

Subjects
The sample consisted of 106 Lernbehinderte[1]: 57 male and 49 female. Mean IQ as measured by the CFT was 80, SD 10.8. Mean chronological age was 10 years 2 months, with a SD of 8.7 months. A group of 104 children with no reading problems gave the data for an additional factor analysis which is not discussed here.

Using the same marker tests we confirmed in general the same factorial dimensions of simultaneous/successive coding and of a planning strategy. Similar results have been reported with samples of different cognitive levels and within different cultures (e.g. U.S., Canadian, Canadian Natives, Indian [Orissa, India], Australian and Chinese children; Naglieri et al. 1989).

---

[1]Lernbehinderte are those students attending special schools, who have shown school failure in normal schools and have an IQ (as measured by a standardised IQ test) of between 1 and 2 SDs *below* the mean. This represents quite a heterogeneous group with many different reasons for school failure.

Our sample of reading disabled students however showed a factorial structure, where one marker test for successive coding, "visual short term memory" (VISHOT) had no significant loadings (and no high communalities) on the "successive" dimension, supporting studies of Leong (1974) and Kirby & Robinson (1987) with similar populations. Leong as well as Kirby and Robinson interpret their findings as a sign for a specific difficulty of the reading disabled child:

"They employ simultaneous processing in the early stage of reading for both word recognition and syntactic analysis, tasks which are more appropriately handled with successive processing (cf. Leong, 1976/1977). Simultaneous processing is less than optimal for syntactic analysis, and is only appropriate for word recognition when the child can recognize a considerable number of words by sight. In the early stages of reading a successive analysis of sound-symbol relationships should bring more success. Two possible explanations for the use of a suboptimal mode of processing can be suggested: a deficit in the skill of successive processing or a poor strategic choice of processing modes (Kirby & Robinson, 1987, p 249)".

As several authors have pointed out (e.g. Das et al., 1982, Torgeson, 1980), if this inability is seen in a kind of production deficit, strategic training would be adequate, especially when it uses decoding and syntax tasks (cf. Krywaniuk & Das, 1976; Kaufman & Kaufman, 1979), that means "ecologically valid" procedures.

Planning and Transformation as Possible Sources of Production Deficits

In order to clarify the dimensional structure of some other marker tests relevant for the reading process, we added those subtests that are marker tests for "planning" and some other tests we included in order to control, whether the simultaneous factor only represented the visual dimension compared to the acoustic dimension of successive coding (for a detailed criticism of factor analyses and this kind of interpretation see Carroll, 1979). Another reason for the extension of the battery was to find out, whether the "noted relationships are not a function of idiosyncrasies in the defining battery" (Ryckman, 1981, p.75). We, therefore added tests for rhythms and tonal memory, very similar to those Ryckman had proposed (Seashore-Test), and a subtest, where chords had to be compared with a sequence of tones. And we added a subtest of the HSET (Grimm & Schöler, 1978), where a sequentially arranged story (an Indian fairy tale) had to be remembered, using pictures for memorial help. We found typical subtests for successive coding, but there was an another subtest (CHORDS) besides VISHOT that only contributed to the factor solution of the not disabled population.

If we look at both tests, they seem, at the first sight, to have at least one common feature. Both offer at first a simultaneous structured task that has to be transformed into a sequential (possibly acoustic coded) order to be solved. This transformation may require a special mental representation (one aspect of the metacomponents in Fig.1), that may lead to a more or less optimal storing in the working memory according to the requirements of the task (Stanovich, 1982, Vellutino, 1979).

We therefore computed an extended factorial design with four dimensions (eigenvalues > 1). The results suggest that some marker tests which were originally designed as successive tasks are in fact better conceptualised as tests of a special form of transformation process in reading (eg. VISHOT, CHORDS, HSET). This can be seen in Table 1.

TABLE 1

Principal Components Analysis of Coding and Planing Measures

FACTOR

Variable	I (SIM.)	II (TRANS.)	III (PLAN.)	IV (SUCC.)	h
RAVEN (percept)	.75	.03	.24	.02	.65
RAVEN (concept)	.24	.24	.67	.15	.42
FIGCOP	.67	.12	.29	.24	.61
MFD	.97	.10	.13	.15	.99
VISHOT	.04	.62	.10	.23	.28
VIS	.14	.02	.51	.04	.31
MELODY	.11	.44	.51	.07	.47
RHYTHM	.06	.28	.15	.07	.10
CHORDS	.03	.73	.08	.01	.54
TRAILS	.08	.20	.76	.01	.62
DIGIT	.12	.03	.01	.96	.93
SERR	.26	.06	.02	.38	.22
HSET	.08	.71	.25	.20	.61
Total Variance	52.0	21.9	15.7	10.4	
% Total Var.	28.9%	13.8%	11.4%	8.2%	62.4%

This hypothesis, supported by analyses of variance of our samples (Holtz, 1990) was consistent with results reported by Das et al.

In discussing the relation of coding processes and reading, Das, Snart & Mulcahy (1982) as well as Das (1987) mention some special problems of reading disabled students. They point out, that disabled readers not only have difficulties in the "rapid coding of linguistic information in short-term memory" (Das et al., 1982), citing Vellutino (1977) and Perfetti & Goldman (1976), they also suffer "from a defective control process in terms of memory" and they have deficits in phonological coding strategies (Das et al., 1982, p.86). And because the "most obvious use of phonological coding is that it converts print which is visual and spatial, to something that can be articulated sequentially (in other words, can be spoken).... is perhaps this demand of conversion to a sequential order that is beyond the capacity of the poor reader" (Das, 1987, p.104). If we consider the requirements in solving the VISHOT task, it may represent this kind of spatial to sequential (phonological) transformation. This task like the other tasks of the "successive" and "transformation"-dimensions stresses coding in short term memory as well (Torgeson & Houck, 1980, for example use DIGITS as a test to measure these aspects).

Research has also concentrated on transformation processes and more comprehensive models have been proposed (Ellis, 1984; Scheerer-Neumann, 1977, 1981; Spear & Sternberg, 1987; Vellutino, 1979), although the exact nature of these processes "is a matter of some dispute" (Spear & Sternberg, 1987, p.18). Nevertheless, besides the problems of defining subtypes of reading disorders, it seems to be out of dispute, that the main cognitive aspects of delayed reading acquisition are problems of phonological conversion and phonetic coding in short term memory (e.g. Perfetti & McCutchen, 1982; Spear & Sternberg, 1987; Stanovich, 1986). As Stanovich (1982) has pointed out, there is evidence, that the most stable short-term memory code is a phonetic code, in part, because during reading "sequences of words must be held in short-term memory, while comprehension processes operate on the words to integrate them into a meaningful conceptual structure that can be stored in long-term memory" (Stanovich, 1982, p.488).

In respect of this discussion and in view of our results of factor analysis and analyses of variance, we have interpreted the fourth factor as a sign of an important dimension of the reading process, that is independent from successive/ simultaneous coding and the "strategic" planning factor. This factor may represent the dimension of task related representation and transformation of visual-holistic information into sequentially (and/or phonetically) adequate units.

But it has to be taken into account, that simultaneous/successive coding as well as planning and transformation strategies may play a different part at different levels of reading development and with subtypes of reading disorders related to these levels as for example Frith (1985) has pointed out in her developmental model.

We therefore computed analyses of variance with stages of reading development (logographic, alphabetic, orthographic, Frith, 1985) and subtypes of reading disorders (dyseidetics and dysphonetics, Bayliss & Livesey, 1985) as independent variables. As dependent variables we used Visual Short Term Memory in different variations, varying the amount of visual and phonetic storing possibilities (e.g.pictures and nonsense-symbols, letters and digits with different phonetic coding possibilities, see Fig.2).

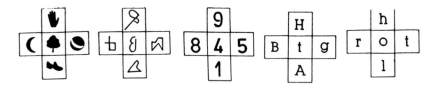

FIG.2. Examples of variations on Visual Short Term Memory

Seventy-two students in our sample were classified by teacher ratings and reading tests as having severe reading problems. Table 2 shows their classification according to Frith's (1986) groups (I=logographic, II=alphabetic and III=orthographic) and Bayliss & Livesey (1985) groups (dyseidetic, dysphonetic). Twenty-two students could not be reliably classified into one of the above categories.

TABLE 2

Classification of Reading Disorders

	Dyseidetic	Dysphonetic	Not classified	No problems	N
I	-	9	6	-	15
II	12	6	14	2	34
III	4	19	2	32	57
N	16	34	22	34	106

Analyses of variance for the three groups of reading level and the marker tests gave the following results (Table 3). These results and further analyses of variance led us to the following conclusions:

On the logographic reading level a holistic (simultaneous) strategy is predominant. (Simultaneous coding, as measured by the marker tests, e.g. Raven, has to be divided into a "holistic" and an "analytic" strategy (Perceptual and Conceptual in our factor solution, see Table 1, cf. Hunt, 1975, Carlson & Wiedl, 1976).

TABLE 3

ANOVA of Different Reading Levels (logographic, alphabetic and orthographic).

	Logographic N=15	Alphabetic N=34	Orthographic N=57	F	p
FIGCOP	7.87	12.83	<17.57	13.29	<.001
RAVEN (Percept.)	45.0	50.54	57.5	1.60	.24
RAVEN (Concept.)	<<36.4	50.74	59.65	3.65	.02
TRAILS	<<34.68	24.16	19.13	8.55	<.001
MFD	8.27	6.57	<3.42	28.56	<.001
VIS	<<16.84	13.23	9.73	6.82	.03
DIGITS	2.25	3.24	3.84	1.22	.17
SERR	15.25	<<12.22	18.24	13.01	<.001
VISHOT	33.37	<<26.16	37.31	4.55	<.01
MELOD	5.99	5.76	<7.83	6.82	.03
CHORDS	45.0	48.44	60.09	4.08	.09
RHYTHM	5.62	6.11	<7.63	5.23	.04
HSET	<<19.12	22.15	31.37	7.42	.02
CFT1	<29.07	<45.31	64.82	19.63	<.001

Note: < indicates that the marked group is significantly different from both other groups; << indicates that the marked group is significantly different from the Orthographic group.

Compared to all other groups the dysphonetics have significantly lower scores on simultaneously (holistic) tasks (F=8.55, p<.001).

As can be seen from Fig. 3, the reading disordered students, especially the dysphonetics, score significantly lower on phonetic transformation tasks. In contrast, this seems to be no problem for the dyseidetics at the orthographic level. Analyses of variance showed a main effect of disorder (F=9.23, p<.01) and reading level (F=3.11, p<.01) but no interaction.

At all levels the dysphonetics had significantly lower scores in the transformation tasks but not in the holistic coding and planning tasks when compared to normal readers.

Given that reading level was correlated with age (r=.68) analyses of covariance with age and IQ were computed. Age as a covariate had no effect although IQ covaried with planning (F=6.08, p<.05) and with the transformation task (F=13.65, p<.01).

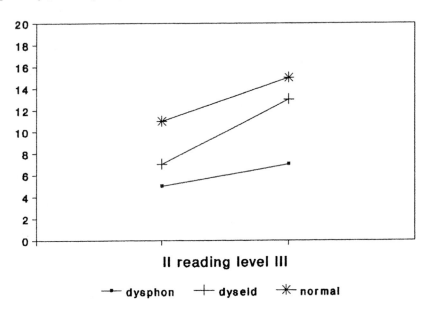

II reading level III

── **dysphon**   ─┼─ **dyseid**   ─✱─ **normal**

FIG. 3. Phonetic Transformation

CONCLUSIONS

The results cited above are at the moment mainly of heuristic value within the model presented at the beginning. It leads us to hypotheses for analysing and modifying the tasks, that show a relation to our subgroups of reading levels and give some information about the interaction of specific components. We therefore have to refine our methodology concerning the "causal modeling" (Cavanaugh et al., 1985) and the componential analysis (Carr & Levy, 1990). The interactive nature of this process will also lead to a refinement of our model and/ or a specification of the interactive nature of components. As was first stated, our model is seen at first as a general survey, where subparts and their interactions may be "blown up" for more detailed transactional processes (Cavanaugh et al., 1985, Detterman, 1987).

Due to the complex and interactive nature of the process it is hard to decide, whether the reading disorders mentioned above are the result of a deficit of specific components or rather the result of a strategic deficit in the sense of Sternberg, i.e. one caused by an inadequate selection or combination of componential parts. It is also possible, that the inability to transform, store and retrieve phonologically relevant units may lead to an inadequate selection of only successive or only holistic coding strategies.

Our aim is to specify and test further hypotheses about the reading (and writing) process by more empirical work on the basis of a componential analysis. This may lead to a refinement of the model and to a selection of some other tasks reported in the literature about empirical research in reading. Up to now it seems plausible, that one problem of retarded readers either on a logographic or on a successive (alphabetic) level is their (strategic ?) difficulty in transforming (eventually by insufficient mental representation) "holistic" presented tasks into sequential units, to store them and/or to combine them again into a unitary representation. This interpretation is not new and in variations one main dispute in reading research, but we hope to clarify the complex process of reading by further componential analyses on the basis of our proposed model.

On the other side, our model offers a lot of other interpretations about reading problems and it seems to be possible, that one can find subgroups of students with the same surface reading problems but with different interacting processes and compensating strategies. And it might be possible as well, that the reported problems on different reading levels may be influenced by other metacomponential or even contextual aspects, e.g. phonological awareness (this construct has to be reformulated in terms of the model) or motivational aspects. The individual assumes because of past failure that it has no chance in decoding and therefore has a rigid strategy selection. Spear & Sternberg (1987) for example argue that context difficulties - in adapting, changing or selecting - may as well be responsible for an inadequate interaction of bottom-up and top-down processes. The developmental and transactional nature of the model (including aspects of task and context) will make it possible to attribute failure in reading not only to "inner" deficits but to recognize experience and instruction as important variables in learning behaviour.

REFERENCES

Bayliss, J. & Livesey, P.J. (1985). Cognitive strategies of children with reading disability and normal readers in visual sequential memory. *Journal of Learning Disabilities*, 18(6), 326-332.

Carlson, J.S. & Wiedl, K.H. (1976). Modes of information integration and Piagetian measures of concrete operational thought. *Intelligence*, 1, 335-343.

Carr, Th.H. & Levy, B.H. (1990). *Reading and its Development: Componential Skill Approaches*. San Diego: Academic Press.

Carr, Th.H., Brown, T.L., Vavrus, L.G. & Evans, M.A. (1990). Cognitive skill maps and cognitive skill profiles: componential analysis of individual differences in children's reading efficiency. In Th. H. Carr, & B.H. Levy, (Eds.), *Reading and its Development: Componential Skill Approaches*. San Diego: Academic Press.

Carroll, J.B. (1979). How shall we study individual differences in cognitive abilities? - Methodological and theoretical perspectives. In R.J. Sternberg, & D.K. Detterman (Eds.), *Human Intelligence*. Norwood, N.J.: Ablex.

Cavanaugh, J.C., Kramer, D.A., Sinnott, J.D., Camp, C.J. & Markley, R.P. (1985). On missing links and such: Interfaces between cognitive research and everyday problem-solving. *Human Development*, 28, 146-168.

Das, J. P. (1987). Intelligence and learning disability: A unified approach. *Mental Retardation and Learning Disability Bulletin*, 15(2), 103-113.

Das, J.P., Snart, F., & Mulcahy, R.F. (1982). Reading disability and its relation to information integration. In J.P. Das, R.F. Mulcahy, & A.E. Wall (Eds.), *Theory and Research in Learning Disabilities* (pp. 85-110). New York: Plenum.

Detterman, D. K. (1987). Theoretical notions of intelligence and mental retardation. *American Journal of Mental Deficiency*, 92,2-11.

Ellis, A.W. (1984). *Reading, Writing and Dyslexia - a Cognitive Analysis*. London: Lawrence Erlbaum.

Frederiksen, J.R. (1982). A componential theory of reading skills and their interactions. In R.J. Sternberg (Ed.). *Advances in the Psychology of Human Intelligence*, Vol.1. Hillsdale, N.J.: Erlbaum

Frith, U. (1986). Psychologische Aspekte des orthographischen Wissens: Entwicklung und Entwicklungsstörung (Psychological aspects of orthographic skills: development and disorder). In G. Augst (Ed.) *New Trends in Graphemics and Orthography*. Berlin: Walter de Gruyter, 218-233.

Glaser, R. (1984). Education and thinking. American Psychologist, 39(2), 93-104.

Grimm, H. & Schöler, H. (1978). *Sprachentwicklungsdiagnostik. Was leistet der Heidelberger Sprachentwicklungstest?* Göttingen: Hogrefe.

Graham, F.K. & Kendall, B.S. (1960). Memory-for-Designs-Test: Revised general manual. *Perceptual and Motor Skills*, 11 147-188.

Holtz, K.L. (1990). Information Processing and Reading Disabilities . *Paper presented at the Meeting of the International Association for Cognitive Education.* University of Mons, Belgium. Arbeitsbericht Nr.3.

Hunt, E. (1975). Quote the Raven? Nevermore! In Gregg, L. (Ed.) *Knowledge and Cognition*. Potomac, Md: Lawrence Erlbaum.

Ilg, F.L. & Ames, L.B. (1964). *School Readiness: Behaviour Tests used at the Gesell Institute*. New York: Harper & Row.

Jorm, A. F. (1983). Specific reading retardation and working memory: A review. *British Journal of Psychology*, 74, 311-342.

Kaufman, D. & Kaufman, P. (1979). Strategy training and remedial techniques. *Journal of Learning Disabilities*, 12, 416-419.

Kirby, J. R., & Robinson, G. L. W. (1987). Simultaneous and successive processing in reading disabled children. *Journal of Learning Disabilities*, 20(4), 243-251.

Kolligian, J. & Sternberg, R. J. (1987). Intelligence, information processing, and specific learning disabilities: A triarchic synthesis.*Journal of Learning Disabilities*, 20(1), 8-17.

Krywaniuk, L.W. & Das, J.P. (1976). Cognitive strategies in native children: Analysis and intervention. *Alberta Journal of Educational Research*, 22, 271-280.

Leong, C. K. (1974). Spatial-temporal information processing in disabled readers. *Unpublished Doctoral Dissertation, Department of Educational Psychology*, University of Alberta, Edmonton, Canada.

Naglieri, J. A., Prewett, P. N., & Bardos, A. N. (1989). An exploratory study of planning, attention, simultaneous, and successive cognitive processes. *Journal of School Psychology*, 27, 347-364.

Naglieri, J.A. & Das, J.P. (1988). Planning-Arousal-Simultaneous-Successive (PASS): a model for assessment. *Journal of School Psychology*, 26, 35-48.

Perfetti, C. & McCutchen, D. (1982). Speech processes in reading. In N. Lass (Ed.) *Speech and Language: Advances in Basic Research and Practice*. Vol. 7. New York: Academic Press, 237-269.

Raven, J.C. (1956). *Coloured Progressive Matrices: Sets A, Ab, B*. London: H.K. Lewis.

Ryckman, D.B. (1981). Reading achievement, IQ and simultaneous-successive processing among normal and learning-disabled children. *The Alberta Journal of Educational Research*, XXVII, 74-83.

Scheerer-Neumann, G. (1977). Funktionsanalyse des Lesens. *Psychologie in Erziehung und Unterricht*, 24,125-135.

Spear, L. C., & Sternberg, R. J. (1986). An information processing framework for understanding learning disabilities. In S. Ceci (Ed.), *Handbook of Cognitive, Social and Neuropsychological Aspects of Learning Disabilities* (Vol. 2) (pp. 2-30). Hillsdale, New Jersey: Erlbaum.

Stanovich, K.E. (1982). Individual differences in the cognitive processes of reading I: Word decoding. *Journal of Learning Disabilities*, 15, 485-493.

Stanovich, K.E. (1986). Cognitive processes and the reading problems of learning disabled children: evaluating the assumption of specificity. In J.K.Torgesen & B.Y.L. Wong (Eds.) *Psychological and Educational Perspectives on Learning Disabilities*. Orlando, Fl.

Sternberg, R. J., & Frensch, P. A. (1989). A balance-level theory of intelligent thinking. *Zeitschrift für Pädagogische Psychologie*, 3(2), 79-96.

Sternberg, R. J., & Salter, W. (1982). Conceptions of intelligence. In R.J. Sternberg (Ed.), *Handbook of Human Intelligence* (pp. 3-28). New York: Cambridge University Press.

Sternberg, R. J. (1981). Toward a unified componential theory of human intelligence: I. Fluid ability. In M. P. Friedman, J. P. Das, & N. O'Conner (Eds.), *Intelligence and Learning* (pp. 327-344). New York: Plenum Press.

Sternberg, R. J. (1985). *Beyond IQ: A Triarchic Theory of Human Intelligence*. New York: Cambridge University Press.

Torgeson, J.K. (1980). Conceptual and educational implications of the use of efficient task strategies by learning disabled children. *Journal of Learning Disabilities*, 13, 19-26.

Torgeson, J.K. & Houck, D. (1980). Processing deficiencies of learning disabled children who perform poorly on the digit span test. *Journal of Educational Psychology*, 72, 141-160.

Vellutino, F. (1979). *Dyslexia: Theory and Research*. Cambridge, Mass: MIT Press.

Venezky, R. (1976). *Theoretical and Experimental Base for Teaching Reading*. The Hague: Mouton.

Wimmer, H., Hartl, M., & Moser, E. (1990). Passen "englische" Modelle des Schriftspracherwerbs auf "deutsche" Kinder? Zweifel an der Bedeutsamkeit der logographischen Stufe. *Zeitschrift für Entwicklungspsychologie und Pädagogische Psychologie*, 22(2), 136-154.

Facets of Dyslexia and its Remediation
S.F. Wright and R. Groner (Editors)

VISUOSPATIAL ABILITY AND LANGUAGE PROCESSING IN READING DISABLED AND NORMAL CHILDREN

G.F. Eden, J.F. Stein

University Laboratory of Physiology,

Oxford, England

and

F.B. Wood

Bowman Gray School of Medicine,

Winston Salem, North Carolina, USA

INTRODUCTION

Stein & Fowler (1980, 1985, 1989) have suggested that many dyslexic children are not only poor at phonological but also at visual analysis. They have particular difficulties localising small targets, such as letters; and they often complain that letters appear to blur or move around during reading. Stein and colleagues have shown that these symptoms are often associated with impaired vergence control, unstable binocular fixation and inaccurate visual direction sense.

There is good evidence that the phonological problems in these reading disabled children are the result of disordered development of the specialised linguistic functions of the left hemisphere (Liberman, Cooper, Shankweiler & Studdert Kennedy, 1967). Their visual sequencing problems, on the other hand, may be the result of abnormal development of the visuospatial functions of the right hemisphere (Stein & Fowler, 1989).

The aim of the study reported below was therefore to compare the phonological and visuospatial abilities of a group of thoroughly studied, normal and reading disabled children. These were drawn from the 'Learning Disability' population which Wood and colleagues have studied in great detail at the Bowman Gray School of Medicine, North Carolina (Felton & Wood, 1989). We were able to make use of the large variety of test results on these children and compare these results with the visuospatial tests that we had been using in Oxford.

METHOD

Subjects

The subjects were 38 children from the Bowman Gray learning disability project. If a child's reading standardised score at 5th grade was between 85 and 115 (± 1 SD from the mean) on the Woodcock Johnson he or she was classified as a good reader. Those below this range were classified as poor readers. Twenty one good readers and 17 poor readers were studied (see Table 1).

TABLE 1

Profile of the 38 Subjects Studied

Group	N	Age range/mean	Reading Age range/mean	Full IQ range/mean	Male	Female
Normal	21	10.5 - 13.4 10.9	9.5 - 18.0 11.7	85 - 115 103.0	11	10
Reading Disabled	17	11.4 - 13.4 11.5	7.2 - 9.4 8.5	85 - 115 94	9	8
All	38	10.5 - 13.4 11.2	7.2 - 18.0 10.1	85 - 115 99.0	20	18

The children in the Bowman Gray Project have been followed for 5 years; and 450 variables have been recorded on each. The 38 children reported here were chosen to have normal IQ and good or poor reading as described above. In addition, both the good and poor readers were selected to display a wide range of phonological and visual abilities on the basis of composite visual scores (based on Complex Figure of Rey, Judgement of Lines and Meier's Concentric Circles) and phonological scores (based on Word Attack, Test of Auditory Analysis and Pig Latin) derived from tests administered earlier. Details of these tests are given later.

A standard eye examination was administered before the children entered the study. All were normal.

Tests

Phonological Tests: We recorded the children's performance in a wide variety of phonological tests; but here only those which were found to be related to reading ability are described in detail.

1. Test of Auditory Analysis Skills: In this test the child has to break words down to either the syllabic or phonemic level. For example the child is asked to say 'cowboy'. Then he is asked to say it again, but without saying 'boy'. The correct response is thus 'cow'. Similarly say 'smack'; now say it again but don't say 'm', the correct response is 'sack'.

2. Pig Latin: Here the order of sounds in words are reassembled by the child by deleting the initial phoneme from the word and placing this at the end of the word, followed by 'ay'. So 'pig' is spoken as 'igpay'.

3. The Lindamood Auditory Conceptualisation Test: In this test of phonemic awareness the child is taught to associate phonemes with coloured blocks and then reassemble them for isolated sounds in sequence, and sounds within a syllable pattern.

Visual Tests: The visual tests used in the Bowman Gray learning disability project were the following:

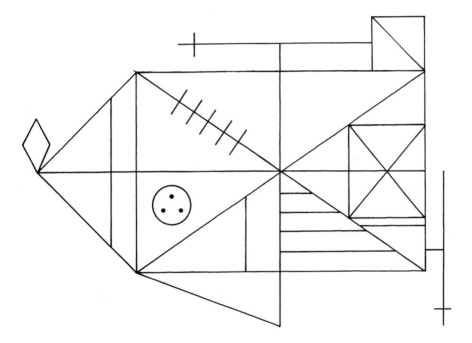

FIG. 1. The Complex Figure of Rey.

1. The Complex Figure of Rey: The child has to copy this figure (Fig.1) first from the original (copy) and then again from memory (immediate recall). Completeness of their drawings is scored.

2. Judgement of Lines: The child is shown a single line with a particular orientation to the horizontal. From a fan of sample lines the child is asked to pick that which matches its orientation.

3. Meier's Concentric Circles: Here the child has to spot which one of the four circles has the odd pattern.

Additional visual tests introduced from Oxford were the following:

Left-Right Differences. In several of our tests we separated the children's errors on the left and right sides, because our earlier studies had suggested that dyslexic children with visual problems such as unstable binocular control, show a mild tendency to neglect objects on their left side (Stein & Fowler, 1989). This abnormality is also characteristic of patients with lesions of the posterior parietal cortex. Like these patients some of the dyslexic children we have seen show left neglect and inaccurate localisation on the left side. Detection of this left neglect could be an indication of disordered development of the visuomotor processing functions of the right hemisphere. Therefore we went back to some of the above visual tests and looked for a left/right difference.

1. The Complex Figure of Rey: The scores for copying and immediate recall of the figure were rescored in terms of left and right performance down the midline.

2. Judgement of Lines: Here those lines picked out by the child to match the orientation of the sample were analysed in terms of left and right side.

3. Dot Localisation (Riddell, Fowler & Stein, 1990): This is a computer game designed to assess the child's ability to localise small targets on the left and right sides (Fig. 2). The child has to report whether the second spot on the screen appeared to the left or right of the first. The subject was given 2 seconds to fixate the target spot so that any eye movement instability that might occur would show up in the child's answer to where the second spot seemed to appear.

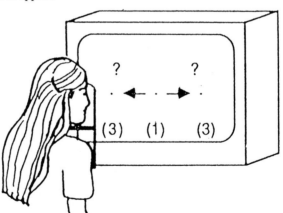

(1) Priming Spot

(2) Delay: 500ms

(3) Second spot: 200ms

(4) Subject makes a choice whether second spot appeared on the left or on the right of the priming spot.

FIG. 2. Dot Localisation

Mesulam: 1. Ordered letters   3. Ordered circles
          2. Random letters   4. Random circles

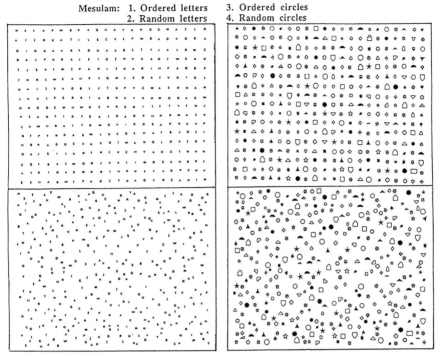

Halligan : Random Star Cancellation

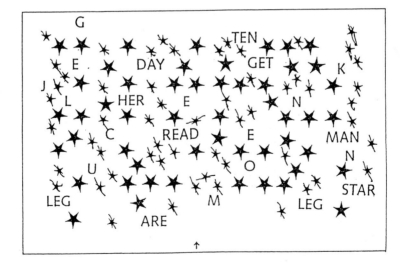

FIG. 3. Five different cancellation charts: total errors on the left and right sides for all 5 charts are added together.

4. Cancellation Tasks: Five charts showing large and small stars, letters or starlike circles were presented to the child (Fig.3). Such charts are used with stroke patients in order to assess 'neglect' (Halligan, Marshall & Wade, 1989; Mesulam, 1985). Particular letters or stars have to be crossed out. We have presented the results as the total objects missed on the left or right side of all five charts.

5. Clocks: We simply asked the children to draw a clock with its numbers.

Tests of Binocular Stability

1. Vertical Tracking Task: The children's ability to follow a line down a page just using their eyes, beginning at a box at the top and ending at a number at the bottom (Fig. 4) was tested. If they had poor eye movement control they were more likely to skip over to another line and name the wrong number at the bottom.

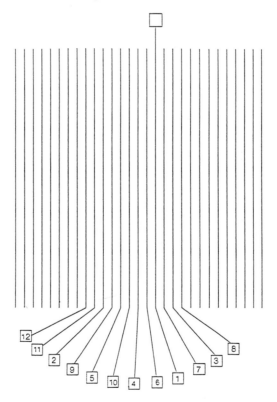

FIG. 4. Vertical tracking task.

2. Fixation During Vergence: The children's eye movements were recorded when they were fixating a small target at three levels of convergence. The target consisted of a star of < 0.5° of visual angle presented to the subject using a synoptophore (amblyoscope). Eye

movements were measured using an infrared limbus tracking system. Fig. 5a shows an example of well controlled fixation compared with the poor fixation shown in Fig. 5b.

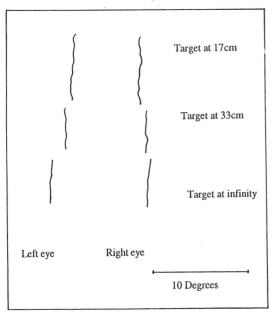

FIG. 5a. Well stabilised binocular fixation in a normal child.

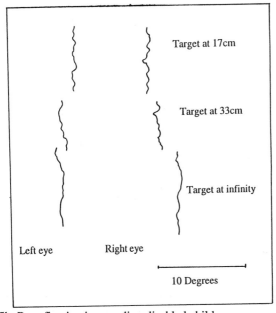

FIG. 5b. Poor fixation in a reading disabled child

3. Using the same technique we also measured the amplitude of divergent or convergent eye movements that the children could make, also using the synoptophore (Stein, 1989).

RESULTS

We wished to see whether the visual tests could be used to identify differences between good and poor readers. Therefore we will present the significant differences we found between the normal and poor reading groups.

In the phonological tests there were significant differences between the normal and reading disabled children both for the amount of time taken to complete Pig Latin and for the Test of Auditory Analysis. We concentrated upon the Pig Latin Completion Time as this correlated most strongly with reading ability.

The visual tests offered by Bowman Gray showed significant differences for the Complex Figure of Rey on the immediate copy: normals: 26.5 ± 1.2 (SE); RDs: 23.6 ± 1.3), but copying directly from the example did not reach significance (normals: 29.9 ± 0.53; RDs: 28.2 ± 1.0). In both cases the reading disabled group performed on average worse than the normal readers as they reproduced less of the original figure.

For the Judgement of Lines test the normals scored significantly higher (14.5 ± 0.6) than the reading disabled group (12.9 ± 0.7). Meier's concentric circles did not prove to be so useful in differentiating between the two groups as they scored equally well (normals = $13.4, \pm 0.3$; RDs = 13.5 ± 0.5).

In the visual tests designed to compare performance on the left with that on the right there were clear differences between normal and reading disabled children. In the dot localisation computer game in which the child had to report if the second dot appeared to the left or right of the first, normals scored more points (122 ± 6.3) than the reading disabled group (114 ± 6.4). Also the normals scored better on their left side (left: 130 ± 6.9; right: 115 ± 5.7) whilst the reading disabled group scored less on their left side 112 ± 5.6 than on their right (117 ± 7.2).

In the cancellation tasks although the total number of errors made by normal and dyslexic children was similar, normals again made less errors in the left side (left: 3.7 ± 1.1; right: 4.9 ± 1.3) whereas the reading disabled children made many more in the left (left: 4.9 ± 1.1; right: 3.5 ± 0.7) so that their left minus right scores were again significantly different from those of normals, as seen in Fig. 6.

In the clock drawing test many reading disabled children squashed all the numbers into the right side and neglected the left (e.g. see Fig. 7). There was a significantly greater tendency for the reading disabled group to draw clocks with these errors than the normal readers (chi-square test: $p < 0.05$).

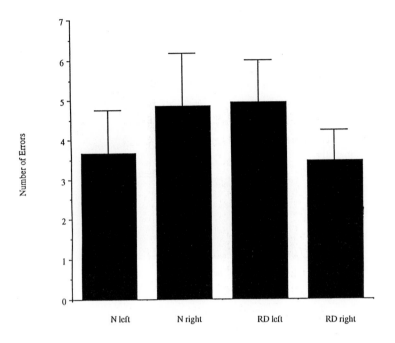

FIG. 6. Total number of errors on the left & right sides in the 5 cancellation tasks.

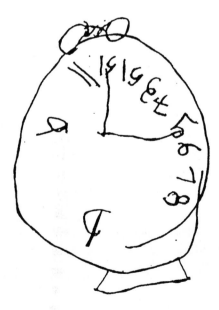

FIG. 7. Clock drawn by a reading disabled child: figures squashed into the right side

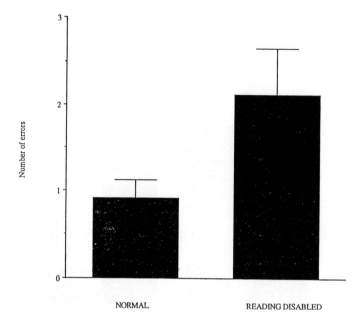

FIG. 8. Vertical tracking task

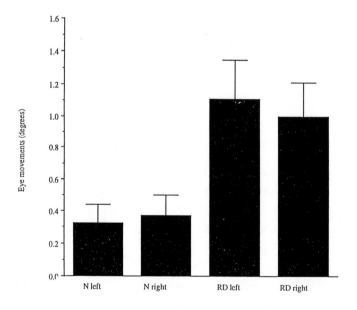

FIG. 9. Fixation Stability in Normal and Reading Disabled children.

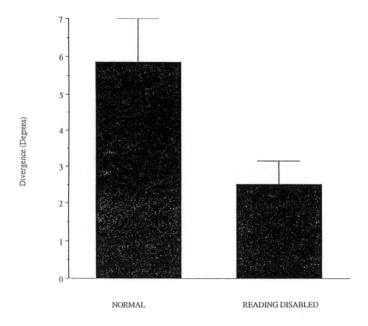

FIG. 10. Divergence amplitudes measured in normal and Reading Disabled children

In the tests designed to assess the children's binocular control the reading disabled children made significantly more mistakes in the vertical tracking task (Fig.8). Likewise the reading disabled children had significantly less stable fixation than the normal children particularly in their left eyes (Fig. 9). As reported by Stein et al. (1988) the amplitude of divergence of the reading disabled children was significantly smaller than that of the normal children, as seen in Fig. 10.

Simple Correlations

Another way of looking at how useful these tests are for understanding these children's reading problems is to calculate how well they correlate with reading in both normal and reading disabled children. Reading ability was assessed using the Woodcock Johnson standardised score. We also gave each child the British Ability Scales (BAS) reading test. Thankfully these two are highly correlated (r = 0.78)!

Figure 11 shows the percentage of the total variance in reading scores for all 38 children which can be explained by the various variables. The strongest single predictor was the time in which the Pig Latin Test was completed (r-squared = 36%). We found that in the 5th Grade children that we tested, Pig Latin Time was a more powerful predictor of reading than the Pig Latin errors themselves. In the 3rd Grade Pig Latin errors had been a stronger predictor however.

It was noticeable that many of the visual tests we employed were almost as good predictors of reading ability as the phonological tests. Examples are shown in Fig. 11. Significantly the scores on the left hand side in the visual tests which could be separated into left and right, were almost as good predictors as the total scores.

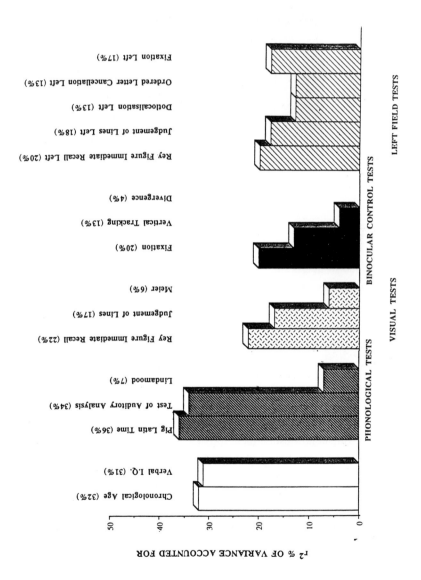

FIG. 11. Percentage of total variance (R^2) in reading scores

Multiple Correlations

Obviously many of the variables shown in Fig.11 correlate with each other. So we calculated the correlations between all seventeen. Chronological Age and Verbal IQ correlated strongly with most of the other variables. So it was important to allow for these first in multiple correlations. Also it was clear that many of the visual tests, apart from those indexing binocular stability, correlated significantly with the phonological tests. We therefore ran several multiple correlations to see which combination of the variables, which we knew individually were strongly correlated with reading ability, gave the best overall predictions. We took out Verbal IQ and Chronological Age first, and then added the different variables. Figure 12 shows the results of some of our models. All seven of these were statistically significant and the top three were significantly different from the model which included only Chronological Age and Verbal IQ. The latter accounted for 38% of the variance of reading ability in this group of children.

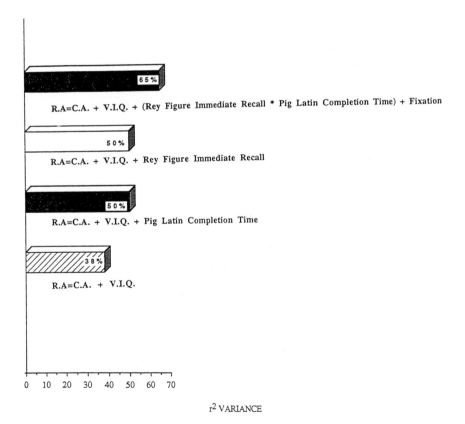

FIG. 12. Models Predicting Reading Age

When the strongest phonemic variable, Pig Latin Completion Time, was added the model accounted for 50% of the variance. The same procedure was then used for the visual tests which previously had been shown to be well correlated with reading ability. When the Rey figure immediate recall score was added to Chronological Age and Verbal IQ this model was as good as that involving Pig Latin Time. It, too, accounted for 50% of the variance in reading ability in this group of children.

Since we had found that Pig Latin Completion Time and the Rey Figure scores were correlated strongly with each other in the correlation matrix, but that neither of these correlated with our binocular fixation index, after taking out Chronological Age and Verbal IQ we entered the covariable (Rey Figure Immediate Recall x Pig Latin Completion Time) and the binocular Fixation score to generate the top model shown in Fig. 12. This accounted for 65% of the variance of reading ability. It was significantly different from the other models and the best we found.

DISCUSSION

These results demonstrate that visual tasks which test children's ability to localise and orientate small targets, particularly on the left hand side, together with those which index binocular stability were almost as useful in discriminating between good and bad readers as were the phonological tests. Likewise they correlated as well with reading ability. A model which allowed for Chronological Age and Verbal IQ, the shared variance between a phonological task, the Pig Latin Completion Time, and our visual localisation test, together with our binocular stability score, accounted for a very high proportion of the total variance in reading ability (65%). These results therefore support Stein and Fowler's hypothesis that children with reading problems often suffer disordered development of the specialised processing functions of both the left and the right hemispheres i.e. they may have both phonological and visuospatial difficulties.

It is important to bear in mind however that these 38 children were not selected randomly. They were chosen because they performed poorly on either reading, composite visual or phonological scores. So we need to see whether these relationships hold true for a truly randomly selected sample. Although Felton & Wood (1989) have not found evidence for a clearly separate visual subtype amongst the dyslexics in the Bowman Gray Learning Disability population, there are in fact strong correlations between reading ability and performance in the Complex Figure of Rey ($r = 0.42$) and the Judgement of Lines Test ($r = 0.35$) in their large reading disabled population, just as we report here. Therefore it is likely that in an unselected sample of reading disabled children the same relationships would hold true; i.e. most dyslexics probably suffer both phonological and visual deficits.

ACKNOWLEDGEMENTS

We wish to express our thanks to Dr. William McKinney, the Neurology Department of the Bowman Gray School of Medicine, the Guarantors of Brain and the McDonell Pew Foundation. Thanks also to Dr. Robert Hiorns and Alison Margetts in the Department of Statistics, University of Oxford and to Dr. Grey Weaver, Department of Ophthalmology, Bowman Gray School of Medicine.

REFERENCES

Felton, R.H. & Wood, F.B. (1989). Cognitive deficits in reading disability and attention deficit disorder. *Journal of Learning Disabilities.*, 3-13.

Halligan, P.W., Marshall, J.C. & Wade, D.T. (1989). Visuospatial neglect: under-lying factors and test sensitivity. *The Lancet*, 14 October.

Liberman, A.M., Cooper, F.S., Shankweiler, D.P. & Studdert Kennedy, M. (1967). A motor theory of speech perception. *Physiological Review*, 431-492.

Mesulam, M-M. (1985). *Principles of Behavioural Neurology.* Contemporary Neurology Series, F.A. Davis.

Riddell, P.M., Fowler, M.S. & Stein, J.F. (1990). Spatial discrimination in children with poor vergence control. *Perceptual and Motor Skills*, 707-718.

Stein, J.F. (1989). Unstable vergence control and dyslexia. *British Journal of Ophthalmology*, 73, 319-320.

Stein, J.F. & Fowler, M.S. (1980). Visual dyslexia. *British Orthoptic Journal*, 11-17.

Stein, J.F. & Fowler, M.S. (1985). Effect of monocular occlusion on visuomotor perception and reading in dyslexic children. *The Lancet*, 69-73.

Stein, J.F. & Fowler, M.S. (1989). Disordered right hemisphere function in dyslexic children. In C. von Euler, I. Lundberg & G. Lennerstrand (Eds), *Brain and Reading* pp.139-158. London, Macmillan. 10

Facets of Dyslexia and its Remediation
S.F. Wright and R. Groner (Editors)

RATE OF ELEMENTARY SYMBOL PROCESSING IN DYSLEXICS

R. Yap and A. van der Leij

Vrije Universiteit, Amsterdam,

The Netherlands

INTRODUCTION

There are a large number of studies which point to a phonological deficit in dyslexics, indicating that dyslexics have a disturbance in the mental representation of speech, (see e.g. Olson, Kliegl, Davidson & Foltz, 1985; Snowling, 1980, 1981). In particular, the rate of access to and retrieval of, phonological properties of the language, seems to discriminate poor from good readers. Impairments in the rate of phonological processing do not only include the name retrieval of printed words (e.g Ehri &Wilce, 1983; Manis, 1985) but also the name retrieval of single symbols, such as letters and digits. Some authors argue that dyslexics suffer from a general speed deficit in retrieving names and that this deficit affects the lower-level aspects of reading, tapped by word recognition and oral reading speed (Bowers & Swanson, 1991; Denckla & Rudel, 1974; Wolf, Bally, & Morris, 1986).

Impaired speed of processing in dyslexics is not merely related to the phonological nature of the code to be processed, it also seems to be influenced by the temporal separation between stimuli. Bowers & Swanson (1991) demonstrated that the difference in naming speed between poor and average readers is greater when the inter-stimulus interval between digits presented is smaller. Furthermore, poor readers were more affected by the temporal separation between items than average readers of the same age. Tallal (1980) found similar results in the processing of nonlinguistic auditory stimuli. Research of Lovegrove and colleagues (1989) showed that dyslexics have a deficit in the visual transient system, a system which is involved in fast processing of visual temporal information. These results raise the possibility that, irrespective of phonological processing, dyslexics may have a general difficulty with temporal discrimination across senses and perceptions.

The aim of the present study was to find out whether dyslexics have a deficit in the rate of symbol processing and whether this is related to temporal separation between stimuli and different demands on name retrieval ability. Since the material used had an alphanumeric character (digits 1 to 9), name retrieval ability is always involved to a greater or lesser degree. However, the demand placed on it was varied by changing the task demands. Stress on rapid processing was made by masking the stimulus after 200 msec. of exposure duration. The assumption is that impaired performance in the masked condition is the result of insufficient time to process the symbol before the arrival of the mask.

A reading-level-match design was used to decide whether the problems found in dyslexics were subject to a lag or a deficit. In this design, dyslexics were matched with normal readers of the same age and younger readers of the same reading level. Because of the aberrant reading profile of dyslexics, it is important to decide on which aspect of reading dyslexics are matched with younger readers (see for discussions on this design Backman, Mamen & Ferguson, 1984; Goswami & Bryant, 1989; Jackson & Butterfield, 1989; Vellutino & Scanlon, 1989). In the present experiment, the match was on oral reading speed because of its close relationship with name retrieval ability. The argument in favour of a deficit will be strongest if the dyslexics perform lower or have a different pattern of symbol processing than the younger readers. In that case, a developmental lag is most unlikely because the problem is not observed in a population who is, because of a younger age, less developed. If the younger reading-age group has the same low level of performance as the dyslexic group, it can be argued these levels, that is the level of symbol speed and reading speed, reflect a lag in development in both groups. Notice here that the argument in favour of a lag explanation of dyslexia is based on some resemblances with younger children, namely the resemblances on symbol speed and reading speed. However, the hypothesis in favour of a deficit in dyslexics cannot be fully rejected on the basis of some resemblances with younger children. The fact remains that dyslexics are older and still have problems that have to be overcome by that age. Therefore, an interpretation of the results towards a lag or deficit is most unequivocal when differential performances appeared between the dyslexic group and the reading-age group, in favour of the latter.

METHOD

Subjects

All groups were tested in May/June, at the end of the school year. Twenty-one dyslexics in the age range of 9 to 11 years, who were at least 2 years below grade level in reading, were selected from a special school for Primary Learning Disabled Children (mean age= 10;2 years, SD = 0.67, reading grade equivalent= 2.3, PPVT IQ= 109). The normal readers in the experiment were recruited from two normal primary schools. The reading-age controls (RA group) were 15 average readers of grade one (mean age= 7;1 years, SD = 0.33, reading grade equivalent= 2.1, PPVT IQ= 96.4). The chronological-age controls (CA group) were 15 average readers from fourth and fifth grade (mean age=10;1 years, SD. = 0.72, reading grade equivalent= 4.6, PPVT IQ= 109). The groups did not differ significantly in PPVT IQ.

A standardized Dutch test of speeded word reading (EMT; Brus & Voeten, 1973) was used to determine the reading grade levels of the groups. The same test was used to match the dyslexic group with the RA group for equal reading speed. In this way, the reading rates on the EMT were 24 words per minute (wpm) in the normal RA group, 28 wpm in the dyslexic group, and 64 wpm in the CA group. The EMT test score encompasses both accuracy and rate of word-decoding processes.

Apparatus, Materials, and Overall procedure

The stimuli consisted of the overlearned digits 1 to 9. These were presented randomly in point size 48, in discrete-trial format in the centre of a microcomputer screen, and at a distance of 40 cm from the subject, subtending at a visual angle of approximately 2°. The computer automatically recorded response latencies and number of correct scores. The tasks were administered in separate sessions, spread over separate days. Subjects were instructed to work as quickly as possible without making errors.

The onset of each trial was preceded by a warning tone of 600 Hz. with a duration of 200 msec., followed by a one-second interstimulus interval after which a frame appeared in which the digit was displayed. In the baseline condition, frame plus digit were removed from the monitor by response onset. In the masked condition, the visual digit disappeared after 200 msec. and was replaced by a 200 msec. lasting backward-masking stimulus, a nonsense pattern that was made up of partial features of digits (i.e. circles and lines). Afterwards, the masking stimulus disappeared and an empty frame was left that was not removed until the onset of the subject's response. The subjects were instructed that the onset of their own responses terminated the frame display. After a response was made, a 3-second delay followed until the warning tone announcing the next stimulus appeared.

The Tasks

1. Visual -Visual Matching Task.

Subjects were instructed to press the YES button if two visual presented digits were similar and the NO button if they were dissimilar. The V-V matching task makes least demand on name retrieval ability because decisions can be made by comparing the digits visually, without using a phonological representation. The proportion of signal and noise trials was equal. Response corrections were permitted, as many and as often as was possible within 3 seconds after the first press; the computer registered the last press as the final answer. There were three conditions in which temporal separation between stimuli was varied. Each condition consisted of 20 trials and was separated from each other by a ten-second rest.

a. Parallel condition. Two visual digits were presented simultaneously, one above the other, and terminated by a mask after an exposure time of 200 msec. So the interstimulus interval (ISI) was zero seconds.
b. Serial condition. Two visual digits, each masked after 200 msec., were presented one after the other with an ISI of one second.
c. Baseline condition. Analogous to the parallel condition, except that the digits remained on the screen until the subject responded (ISI= 0 sec.).

2. Auditory-Visual Matching Task.

Subjects had to compare a spoken digit with a visual digit. They were instructed to press the YES button if the two digits were similar and the NO button if they were dissimilar. The A-V matching task requires a print-sound translation or vice versa and hence made an implicit demand on name retrieval. The auditory digits were stored on harddisk by using an analog-to-digital converter which sampled the naturally spoken digits at a rate of 22 KHz. Subjects heard the digits through a head phone. The proportion of signal and noise trials was equal. The procedure for response corrections was similar to that in the visual-visual matching task.

a. Parallel condition. Subjects heard a spoken digit through a head phone, while simultaneously they saw a masked visual digit on the screen (ISI= 0 sec.).
b. Serial condition. The auditory digit was presented first, followed after one second by the visual masked digit (ISI= 1 sec.).
c. Baseline condition. Analogous to the parallel condition, except that the visual digit remained on the screen until the subject responded (ISI= 0 sec.).

3. The Naming Task.
This task requires explicit demand on name retrieval because the name of the symbol had to be pronounced. Twenty digits had to be named in the baseline condition and, after a ten-second rest, twenty digits in the masked condition. A microphone was attached to a voice-activated relay, which was triggered by the onset of vocalization. Simultaneously a digital timer started and registered response latency with millisecond accuracy. The response latency was indicated by the interval between the onset of the stimulus and the onset of vocalization. As soon as the subject responded, the experimenter recorded response errors by hand. Vocalizations other than a word were scored invalid; in that case another digit was presented until a valid response was made.

RESULTS

Only response latencies were included in the following analyses because in none of the other tasks did the groups differ significantly in accuracy of performance, neither did they show differential interactions in error scores; they all performed near the 100 accuracy level. In each task, 1% of the response latency data was excluded from analyses because of too extreme scores (2.5 SD. from the mean of the group). The effect of masking in each group was indicated by simple effects, calculated in a 3 (group) x 2 (baseline vs. masked condition) ANOVA. The effect of temporal separation in each group was calculated in a separate 3 (group) x 2 (parallel vs. serial) ANOVA.

The Visual-Visual Matching Task
There was a differential group effect in the baseline condition, F (2,48)= 11.8, p ≤ .001. Newman-Keuls post hoc test (alpha=.05) showed that the RA group performed slower than both the dyslexic and the CA groups, who did not differ significantly from each other. The same pattern was found in the serial condition, F(2,48)= 8.43, p ≤ .001. In the parallel condition, the dyslexic group performed slower than the CA group, F (1, 34)= 6.69, p ≤ .01, but remained better than the RA group, F (1, 34)= 4.08, p ≤ .05. There was a group x masking interaction at the 10% significance level, F(2,48)= 3.05, p=.06.

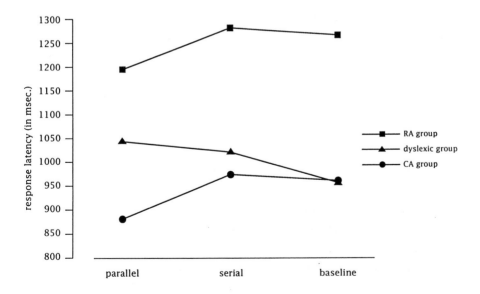

FIG. 1. V-V Matching Task

Simple effects showed that only the dyslexic group was sensitive to the parallel masking effect, $F(1,48)= 3.27$, p= .08, indicating better performances in the baseline condition than in the parallel condition (Fig. 1). Both the RA group and the CA group were not impaired in the parallel condition. The effect of temporal separation indicated slightly better performances in the serial condition than in the parallel condition, $F(1,48)= 2.82$, p=.10. The interaction between group and temporal separation was not significant.

Auditory-Visual Matching Task

In the baseline condition, there was one significant group difference, which indicated that the RA group was slower than the CA group, $F (2, 48)= 3.40$, $p \leq .05$. The dyslexic group fell somewhere between the CA group and the RA group and did not differ significantly from both groups. This same pattern was found in the serial condition, in which again a significant difference was found between the RA group and the CA group, $F (1, 28)= 8.48$, $p \leq .01$. In the parallel condition, the pattern was somewhat different; the RA group remained slower than the CA group, $F (1,28)= 4.31$, $p \leq .05$; the dyslexic group, however, performed also slower than the CA group, $F (1,34)= 6.37$, $p \leq .05$; no differences were found between the dyslexic group and the RA group.

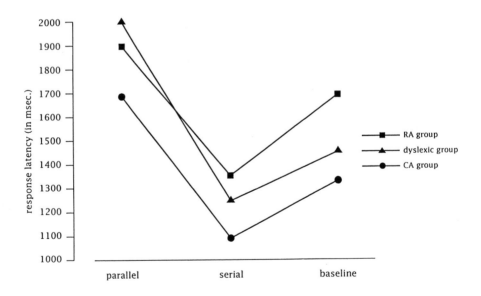

FIG. 2. A-V Matching Task

There was a significant group x masking interaction, F (2,48)= 4.02, p ≤ .05. All groups were impaired in the masked parallel condition as opposed to the baseline condition. The impairment was, however, largest in the dyslexic group (see Fig. 2); simple effects showed that F= 54.9 in the dyslexic group, whereas F= 6.01 in the RA group and F= 18.6 in the CA group. There was a main effect of temporal separation, F(1,48)= 212.5, p ≤ .001, indicating that all groups performed faster in the serial condition than in the parallel condition (Fig. 2).

The Naming Task

A 3 (group) x 2 (baseline vs. masked) ANOVA was run. There was a main effect of group, F(2,48)= 21.9, p ≤ .001. Newman-Keuls post hoc analyses (alpha= .05) showed significant differences between all groups, with the RA group performing the slowest, followed, in turn, by the dyslexic group and the CA group (Fig. 3). There was no significant effect of masking. Nor was there an interaction between group and masking.

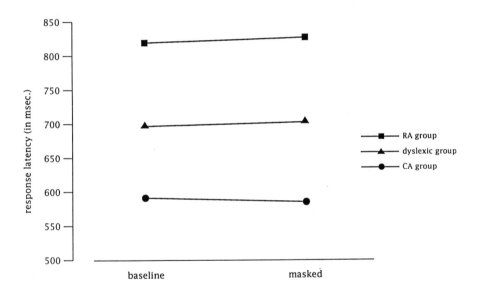

FIG. 3. Naming Task.

DISCUSSION

The aim of the present study was to find out whether dyslexics have a deficit in the rate of elementary symbol processing and whether this is related to temporal separation between stimuli and different demands on name retrieval ability. The results showed that on the naming task, which makes strongest demand on name retrieval ability, dyslexics performed slower than their peers but not slower than the younger readers. When demand on name retrieval diminishes, in the A-V matching task and the V-V matching task, respectively, the same pattern of group performance was found but only in the parallel condition. Thus, similar to the naming task, in the parallel condition of both the V-V and the A-V matching task dyslexics were slower than the peers but not slower than the younger readers. In the naming task and the parallel condition of the VV matching task they were even faster than the reading-age group.

It is worth mentioning that in the parallel condition of both matching tasks, two stimuli had to be processed fast, within 200 msec. This notion of speed in parallel processing is very important because dyslexics did not perform below age level when parallel processing was combined with loss of time pressure (baseline condition). This appeared in both the AV and the VV matching task. Furthermore, dyslexics were more sensitive than the younger and same age normal readers to the effect of masking, again in both the V-V and the A-V matching task.

Another aspect of parallel processing, as defined in this experiment, was the proximity of stimuli in the temporal dimension. This aspect also differentiated dyslexics from normal reading peers. For, when the temporal separation between stimuli was made larger (serial condition) dyslexics performed as well as their peers. Performances were also better in the serial condition than in the parallel condition. However, this was not unique for the dyslexic group because both groups of normal readers showed this pattern as well. It is worth mentioning that the present interstimulus interval of one second was quite long. Therefore, it is possible that dyslexics will show a different pattern than normal readers when the interstimulus interval is made smaller and smaller. It is interesting to investigate at which point of temporal separation, other than the zero second interval, the differences will occur.

The extent of the parallel processing problem in dyslexics is greater when demand on name retrieval increases from V-V matching to A-V matching. This was indicated by two findings. First, the effect of time pressure on parallel processing, which differentiated dyslexics from normal readers, was larger in the A-V matching task (F= 54.9) than in the V-V matching task (F=3.27). Second, dyslexics fell down to the level of the younger reading-age controls in the A-V matching task, whereas they exceeded the younger readers in the V-V matching task. However, these patterns of contrasts between dyslexics and reading-age controls were not restricted to the parallel condition but occurred in the other conditions as well. Thus, as was the case in the parallel, baseline and serial conditions, dyslexics performed as slowly as the younger readers in the AV matching task and faster than the younger readers in the VV matching task. This suggests that, irrespective of whether parallel processing is involved or not, dyslexics have more problems when the demand on name retrieval increases. However, this holds only for the process of stimuli comparison because in the naming task, dyslexics did not perform at the level of the younger readers but instead performed faster.

In conclusion, strongest evidence for a deficit in dyslexics was found in rapid processing of stimuli that are proximal in temporal space and the deficit in parallel processing is greater when demand on name retrieval increases. However, the argument in favour of a deficit is not based on mean group performances. For, although dyslexics did perform below age level in the parallel condition, they did not perform below the level of younger readers, who are generally behind in development. Hence, the possibility of a developmental lag can still be argued. The presence of a deficit is, however, more convincingly demonstrated by the fact that dyslexics were more impaired by the effect of time pressure in the parallel condition than the younger readers. This suggests that dyslexics have a defective mechanism that is responsible for speed and parallel processing. If there is an underlying defective mechanism dyslexics may use other routes than the younger readers to obtain the same level of reading speed. No such strong evidence for a deficit was found on the naming task. The fact that dyslexics did perform slower than their peers but faster than the younger readers on this task suggests that naming is partly subject to developmental trends.

The results of the present experiment imply that the primary deficit of dyslexics is not so much the inability of rapid name retrieval as the inability to process temporal-proximal information rapidly. For, when name retrieval ability is the underlying defective mechanism, there would be evidence for a deficit on the naming task, which taps name retrieval ability explicitly. However, more convincing evidence for a deficit was found on parallel processing ability. Presumably, the deficit appears in an early stage of information processing because time pressure and temporal separation were manipulated at the perceptual level, before the execution of the subject's response.

Parallel processing as defined here is not the same as parallel processing of serial information which has to be processed from left to right. The parallel nature of the latter kind of processing refers to the ability of processing the target information while partially processing the information right next to it. It is sometimes said that dyslexics have this kind of parallelity problem when they have to name words and symbols in a continuous list (e.g. Spring & Davis, 1988). Recently, Bowers & Swanson (1991) rejected this left-to-right parallel processing hypothesis. They, instead, argue that it is the shortness of the interstimulus interval between to-be-named items that accounts for the differential performance of poor readers on continuous and discrete naming tasks. Our results support their opinion but, in addition, suggest that the influence of temporal separation between items already operates at the perceptual input level and not yet at the level of naming-output processing.

Although the processing of symbols in the experiment involves the use of perceptual-cognitive resources rather than mere sensory activation, the present notion of parallel processing may have implications for sensory processing as well. Actually, the parallel processing problem of dyslexics occurs irrespective of which task is used. This may reflect a general parallel processing deficit across senses. However, any generalization from this experiment must be considered tentatively because we did not include a matching task in which comparisons were based on auditory aspects alone. Another particular point concerning sense activation is that dyslexics show greatest problems with parallel processing in the A-V matching task. Because of the requirement of print-sound translation on this task, the problem can be described as a phonological problem. However, it can also be argued that dyslexics have a slowness of the system to activate two different senses simultaneously: the visual and the auditory sense.

Further research is needed to clarify the mechanism by which parallel processing affects the reading process. Of particular importance is the issue of how and which aspects of temporal-proximal information are related to the automaticity in lower-level aspects of reading.

REFERENCES

Backman, J. E., Mamen, M., & Ferguson, H. B. (1984). Reading level design: Conceptual and methodological issues in reading research. *Psychological Bulletin*, 96, 560-568.

Bowers, P. G., & Swanson, L. B. (1991). Naming speed deficits in reading disability: multiple measures of single processes. *Journal of Experimental Child Psychology*, 51, 195-219.

Brus, B. T. & Voeten, M. J. M. (1973). *Een-minuuttest, vorm A en B*. Nijmegen: Berkhout.

Denckla, M. & Rudel, R. (1976). Rapid 'automatized' naming R.A.N: Dyslexia differentiated from other learning disabilities. *Neuropsychologia*, 14, 471-479.

Ehri, L. C. & Wilce, L. S. (1983). Development of word identification speed in skilled and less skilled beginning readers. *Journal of Educational Psychology*, 75, 3-18.

Goswami, U. & Bryant, P. E. (1989). The interpretation of studies using the reading level design. *Journal of Reading Behaviour*, 21(4), 413-424.

Jackson, N. E. & Butterfield, E. C. (1989). Reading-level-match designs: Myths and realities. *Journal of Reading Behaviour*, 21(4), 387-412.

Lovegrove, W., Pepper, K., Martin, F., Mackenzie, B. & McNicol, D. (1989). Phonological recoding, memory processing and visual deficits in specific reading disability. In D. Vickers & P. L. Smith (Eds.), *Human Information Processing:Measures, Mechanisms, and Models*. Amsterdam: Elsevier Science Publishers B. V.

Manis, F. R. (1985). Acquisition of word identification skills in normal and disabled readers. *Journal of Educational Psychology*, 77, 78-90.

Olson, R., Kliegl, R., Davidson, B., & Foltz, G. (1985). Individual and developmental differences in reading disability. In T. Waller (Ed.), *Reading Research: Advances in Theory and Practice*. 4 (pp. pp 1-64). London: Academic Press.

Snowling, M. (1980). The development of grapheme-phoneme correspondence in normal and dyslexic readers. *Journal of Experimental Child Psychology*, 29, 294-305.

Snowling, M. (1981). Phonemic deficits in developmental dyslexia. *Psychological Research*, 43, 219-234.

Spring, C., & Davis, J. M. (1988). Relations of digit naming speed with three components of reading. *Applied Psycholinguistics*, 9, 315-334.

Tallal, P. (1980). Auditory temporal perception, phonics, and reading disabilities in children. *Brain and Language*, 9, 182-198.

Wolf, M., Bally, H. & Morris, R. (1986). Automaticity, retrieval processes, and reading: A longitudinal study in average and impaired readers. , 57, 988-1000.

Facets of Dyslexia and its Remediation
S.F. Wright and R. Groner (Editors)
349

THE DEVELOPMENT OF SYMBOLIC - MOTOR PERFORMANCE IN
MINIMAL BRAIN DYSFUNCTION BOYS

P. Cakirpaloglu and T. Radil

Institute of Physiology, Czechoslovak Academy of Science, Prague

One of the causes of Developmental Learning Disability, within the Minimal Brain Dysfunction syndrome, (MBD) lies in disorders of visual symbolic motor processing. These disorders are a part of wider perceptual motor dysfunctions (Cakirpaloglu & Radil, 1991a) complicating child development from the cognitive, behavioural and social points of view. Dyslexia, Dysgraphia and Dyscalculia being the most frequent problems. Even when learning disabilities are not as clearly expressed, teacher ratings of such children's achievements in reading, writing and calculating usually reflect poorer performance relative to their peers.

The basis of MBD relates to problems with cognitive manipulations with specific visual symbols and programming corresponding motor responses (Cakirpaloglu & Radil, 1990). Disorders in symbolic motor processing of letters might be related therefore to problems with reading and writing. On the other hand, visual number miss-processing causes severe dyscalculic problems.

The aim of this paper was to evaluate symbolic motor processes in Minimal Brain Dysfunction and healthy control children of different ages and cultural and social backgrounds (Prague, Czechoslovakia and Skopje, Yugoslavia) and to correlate the results with achievements in the school rated by teachers.

The analysis was carried out on 240 right-handed boys with the diagnosis of Minimal Brain Dysfunction syndrome (120 in Prague and 120 in Skopje, 30 boys in each class, from the first to the fourth class of elementary school). Their performance and achievements were compared with age (class) matched groups of 191 healthy boys in Prague and 181 in Skopje. Highly anxious children and those with trained finger movements (playing piano, experienced with computer keyboards or type-writer, etc.) were not included in the sample.

A special technique for analyzing simple visual symbolic-motor performance which might reflect basic cognitive manipulations with symbols such as letters, numbers and meaningless patterns was developed (Cakirpaloglu & Radil, 1991b). The task of boys was to reproduce certain combinations of symbols (letters, numbers or meaningless patterns) generated on a computer display by pushing with the corresponding fingers (without the thumb) of the right (dominant) hand the proper key on the keyboard.

The final selection of the sets of symbols was made according psychometric criteria usually adopted for creating new tests, (see Fig. 1). The sequence of visual stimuli consisted of one to three symbols switched on successively with time intervals lasting 500 msec. each. The display remained on until the boys responded. Visual control of fingers was prevented by covering the hand.

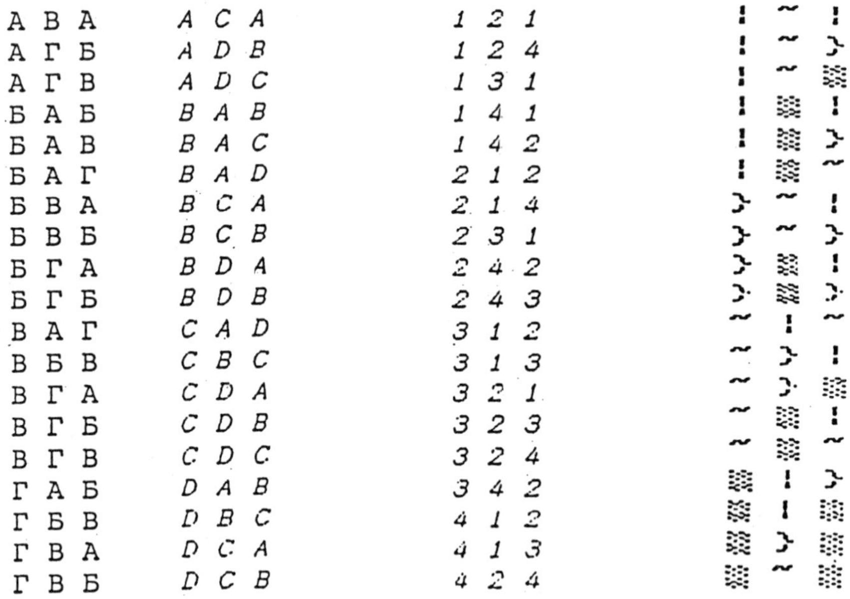

A	B	A		A	C	A		1	2	1			
A	Г	Б		A	D	B		1	2	4			
A	Г	B		A	D	C		1	3	1			
Б	A	Б		B	A	B		1	4	1			
Б	A	B		B	A	C		1	4	2			
Б	A	Г		B	A	D		2	1	2			
Б	B	A		B	C	A		2	1	4			
Б	B	Б		B	C	B		2	3	1			
Б	Г	A		B	D	A		2	4	2			
Б	Г	Б		B	D	B		2	4	3			
B	A	Г		C	A	D		3	1	2			
B	Б	B		C	B	C		3	1	3			
B	Г	A		C	D	A		3	2	1			
B	Г	Б		C	D	B		3	2	3			
B	Г	B		C	D	C		3	2	4			
Г	A	Б		D	A	B		3	4	2			
Г	Б	B		D	B	C		4	1	2			
Г	B	A		D	C	A		4	1	3			
Г	B	Б		D	C	B		4	2	4			

FIG. 1. Sets of stimulus symbols

The average reaction time (RT) in seconds for sets with one, two and three symbols (letters, numbers and meaningless patterns), mean RT for each finger, the total RT and the number of errors were evaluated statistically by a computer program. In the present paper we concentrate mainly upon the values of total RT-s.

The achievement of boys in reading, writing and calculation were also evaluated by their teachers on the five point scale usual in European schools. Both types of average individual values were correlated and evaluated through T-test values.

In general, the total RT was the longest for letters and shortest for meaningless symbols, (Fig. 2). Speed in manipulation with letters might be lower (in comparison with numbers) due to more elaborate cognitive operations and the larger cognitive space

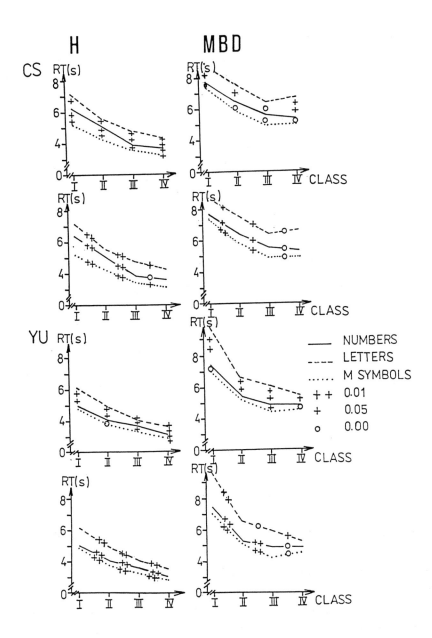

FIG. 2. Reaction times for Czechoslovakian (CS) and Yugoslavian (YU), Healthy (H) and Minimal Brain Dysfunction (MBD) children, from 1st to IVth class for numbers, letters and meaningless symbols

involved. For similar reasons, the RT-s for meaningless symbols were the shortest, probably because of the lack of preceding experience of the subjects with this type of pattern, i.e. the limited cognitive space involved.

Additionally, the RT-s for Minimal Brain Dysfunction children were longer in comparison with RT's in the control subjects which could reflect the circumstance that even very elementary visual motor performance is changed in Minimal Brain Dysfunction children in contrast with healthy ones.

The performance of the healthy children significantly improved during their development from the first class (6-7 years) to the fourth (10-11 years), with the only exception being with the numbers between the third and fourth class in Prague. However, it seems to be important that no further improvement could be seen within the Minimal Brain Dysfunction children when passing from the third to the fourth class (with the only exception for letters in Skopje). That points to the seriousness of limitations in plasticity of brain processes underlying cognitive mechanisms involved in symbolic motor performance and/or to the inadequacy of actual teaching procedures adopted.

Differences in RT-s between healthy groups from Prague and Skopje were not significant. For the Minimal Brain Dysfunction group the RT-s did not differ for numbers or meaningless symbols. However, significant differences were obtained for letters between Prague and Skopje children in the second and fourth classes. These results point to the relative independence of visual motor processing of symbolic stimuli of the social and cultural background of the children.

The results of correlations between the total individual RT-s for letters and numbers on one side, and school ratings in reading-writing and calculating on the other, indicate that the speed of symbolic motor manipulations with letters correlated positively with school grades for reading-writing for all groups of Minimal Brain Dysfunction children both in Prague and Skopje (see Table 1). The same was true for symbolic motor performance with numbers and calculation. However, no significant correlations were found when performance with letters was related to the school achievement in calculations or performance with numbers to school achievements in reading-writing. Thus basic manipulations with symbols (Letters and Numbers) seem to be separately and in a specific way related to reading-writing and numerical skills, respectively.

Similar results were found for symbolic motor performance with numbers in all group of normal boys. However, for normal children the correlations disappear between symbolic motor performance with letters and reading-writing by grades third and fourth (ages 9 and 10 years), both in Prague and Skopje. This might be because in healthy children the cognitive skills that are involved in reading and writing become organized more or less independently of the basic cognitive manipulations reflected in the symbolic motor task for letters by this age, whereas there is apparently a specific developmental retardation of these skills in Minimal Brain Dysfunction children. No differences have been found in this respect comparing Prague and Skopje boys.

TABLE 1

Pearson Correlation Coefficients between Average Individual Reaction Times for Symbolic Motor Performance (SMP), with Letters (L), Numbers (N), Calculation (C) and School Grades in Reading and Writing (R-W) in Minimal Brain Dysfunction (MBD) and Control Boys (C) in 2 Areas in Grades 1 to IV.

	PRAGUE				SKOPJE			
	L/R-W		N/C		L/R-W		N/C	
	MBD	H	MBD	H	MBD	H	MBD	H
Grade								
I	.56 **	.51**	.53**	.58**	.63**	.46**	.61**	.50**
II	.42**	.40**	.56**	.48**	.44**	.41*	.51**	.52**
III	.57**	.24	.51**	.44**	.56**	.33	.48**	.55**
IV	.51**	.20	.44**	.54**	.55**	.29	.53**	.49**

** p<.01; * p<.05

REFERENCES

Cakirpaloglu, P. & Radil, T. (1990). On development of visual motor performance in normal boys and boys with Minimal Brain Dysfunction. *Perceptual and Motor Skills*, 70, 426.

Cakirpaloglu, P. & Radil, T. (1991a). Some neuropsychological and psychological parameters in children with Minimal Brain dysfunction: A transcultural comparison. *Perceptual and Motor Skills*, 72, 80-82.

Cakirpaloglu, P. & Radil, T. (1991b). Minimal Brain Dysfunction children: Correlations between symbolic motor performance and achievements in school. *Perceptual and Motor Skills*, (in press).

IV. ATTENTIONAL CORRELATES OF DYSLEXIA

Facets of Dyslexia and its Remediation
S.F. Wright and R. Groner (Editors)
© 1993 Elsevier Science Publishers B.V. All rights reserved.

POSSIBLE ATTENTIONAL ORIGINS OF WORD DECODING DEFICITS IN DYSLEXIA[1]

R. Stringer and J. Kershner

Ontario Institute for Studies in Education

University of Toronto, Canada

INTRODUCTION

In 1937 Samuel Orton posed a question that has shaped the course of dyslexia research for over half of a century. Given the accepted belief that in right-handed adults language function was lateralized to the left cerebral hemisphere (Broca, 1861) and his own astute observations of the nature of single word decoding deficits in dyslexia, Orton was prompted to ask whether dyslexics might be suffering from a developmental delay in language lateralization (Orton, 1937). When research methods such as dichotic listening became available to test Orton's proposition, however, the results were bewildering. Some experimenters found support for Orton's view; but others reported that dyslexics actually were more lateralized than good readers; and others failed to detect any language laterality difference at all between their dyslexic and normal subjects. Because all three findings were replicated on a number of occasions, these conflicting results proved as difficult to interpret as they were to ignore.

This impasse in the interpretation of laterality research in dyslexia remains (Bryden, 1988). Recently, however, insights from the attentional literature, together with some intriguing empirical findings with dyslexic children have led a number of researchers to regard lateralization in a new way (Obrzut, 1991). It appears that the poor lateralization hypothesis of dyslexia was not as much correct or wrong as it was misinformed about background assumptions. As Churchland (1986) has argued in defending the Quine-Duhem thesis on epistemology in science, there may be no crucial experiment for the falsification of a theoretical claim that can be interpreted independently of background assumptions. If key background assumptions are erroneous, it is highly likely that negative experimental results may lead to the false rejection of true hypotheses. Experiments test whole networks of interrelated ideas, not isolated hypotheses. Indeed, the mounting dissatisfaction with Orton's original conceptualization linking lateralization

---

[1] Portions of this chapter are reprinted with permission from J. Kershner and R. Stringer (1991). Effects of reading and writing on cerebral laterality in good readers and children with dyslexia. *Journal of Learning Disabilities,* 24, 560-567.

very specifically to word decoding failures in dyslexia may be paradigmatic of the Quine-Duhem error in scientific inference.

For instance, new research findings are at odds with two of Orton's principle background assumptions. First, the assumption of progressive lateralization has been proven to be erroneous. Left hemisphere specialization for language in the broad sense implied by Orton does not develop progressively in early childhood (Hiscock, 1988). Rather, a related skill which does appear to develop is children's control over the allocation of attention (Kershner & Chyczij, 1991; Pearson & Lane, 1991; Roeltgen & Roeltgen, 1990). Because contralateral hemisphere-specific mechanisms are involved in such attentional control, we use the term attentional lateralization to describe it; but it is equally important to distinguish it clearly from the more traditional cerebral asymmetries, i.e., left hemisphere dominance for speech vs. right hemisphere superiority for aspects of visuo-spatial processing (Hellige, 1990). Attentional lateralization refers to the dynamic interaction of hemispheric attention/ arousal mechanisms with age-invariant asymmetries for processing cognitive content (Kinsbourne, 1977; Posner & Peterson, 1990). Cerebral asymmetry remains constant across chronological ages and tasks; while children's cognitive performance varies with their time-sharing efficiency in simultaneously processing cognitive content and controlling the direction of attention (Hugdahl & Andersson, 1987; Kershner & Morton, 1990).

Second, the assumption of situational stability in lateralized processing has been replaced by the view of lateralization as a situationally variable phenomenon (Kershner, 1985; Kinsbourne, 1977). This key notion was first demonstrated with dyslexic children in a pace-setting study by Obrzut, Hynd, Obrzut, and Pirozzolo (1981). In dichotic listening, these researchers observed that dyslexic children were more lateralized than normals when they were forced to attend to the right ear; but they were less lateralized than normals when they were forced to attend to the left ear. Overall, the dichotic performance of the nondisabled children was relatively uninfluenced by the attentional manipulations. Age-invariant cerebral asymmetries for cognitive content appears to interact with fluid attentional demands and processes to produce task-specific patterns of lateralization (Kershner, 1988; Obrzut, 1988). Thus, it is not surprising at all that past studies have found dyslexic children to be both less and more lateralized in comparison to average readers. The outcome that emerges from any given experiment depends to some extent on the attentional demands of the task (Kershner & Morton, 1990). The important point to be made is that each of these more recent inferences discredits background assumptions, leaving the quintessential claim untouched for a fundamental relationship in dyslexia between aspects of language lateralization and failures in single word decoding.

This neo-Ortonian perspective, however, substantially changes the nature of the central questions in laterality dyslexia research. As opposed to asking simply whether children with dyslexia differ from normal readers in the physical lateralization of their language functions, whether they are more or less lateralized, we now ask what are the causes of situational variability in attentional lateralization, and more pointedly, how task variations in attentional lateralization might produce the symptoms of dyslexia? In this chapter, we present a theoretical perspective and supporting data to suggest the hypothesis that both phonological and visual word decoding failures in dyslexia may be caused by an impairment in attentional lateralization.

Serendipitous Dichotic Findings

In the data presented here we used the dichotic listening test as devised by Kimura (1961). Kimura noted that when different verbal stimuli (digits) were presented simultaneously to the two ears in a free-recall paradigm, listeners were more accurate in identifying stimuli presented to the right ear. This typical effect was replicated many times and was referred to as a right ear advantage (REA). She theorized that an REA occurred because the left hemisphere was dominant for language in most right-handed individuals and the contralateral auditory pathways were prepotent over the ipsilateral pathways. Thus, with qualifications for the potential effects of attention, an REA reflects the superiority of the left hemisphere for language.

This test is given usually in individual presentations in which the experimenter records the child's spoken responses. However, to use the dichotic listening test in the usual manner is a drawn-out process. Many trials are required to get reliable results and instructions must be repeated for each child. In 1984, Polly Henninger, Wally Cooke, and one of us (Kershner, Henninger, & Cooke, 1984) were seeking a short-cut to perform the dichotic digits task. With this objective in mind we developed a group testing technique which allowed from five to eight subjects to be tested at the same time. All subjects were connected in parallel by stereo headsets to a central tape deck and heard the same stimulus presentations. Instead of giving an oral report, however, they all wrote their responses silently with a paper and pencil. When this group test was performed on dyslexics and control children, we were not surprised (Orton, 1937) to find that the dyslexics demonstrated a reversed pattern of linguistic dominance, that is, a left ear (right hemisphere) advantage; whereas age-matched good readers displayed the normal right ear (left hemisphere) advantage.

This result was a nice fit with the influential Ortonian model of dyslexia. What happened next, however, transformed our fundamental assumptions about what the dichotic test was measuring. Several of the dyslexics were selected for more careful individual study as part of a music discrimination experiment. A second dichotic listening test, using the typical oral report procedure, was administered individually as part of the new protocol. Now, a "normal" REA was revealed in the same dyslexics who several weeks earlier had shown the reversed LEA pattern! According to the Ortonian viewpoint, the dichotic test was supposed to reflect the physical localization of the language functions. Because to the best of anyone's knowledge the dyslexic children had not undergone brain surgery nor been completely remediated of their disability in the time between testings, quite an Ortonian to Neo-Ortonian stir was created. Notably, post-hoc tests indicated that the only significant within-ear difference was in right ear performance which had been depressed during written recall relative to the oral recall condition.

CONTEXT EFFECTS ON LATERALIZATION: A REPLICATION

Purpose

To attempt a better understanding of what may have taken place in the dyslexics to alter their dichotic performance over the two testing sessions, we designed another similar experiment with better control procedures. We added a concurrent reading task in addition to the written and oral dichotic response conditions. We also added a reading-level control group of good readers and a variety of school achievement measures. Our purpose was twofold: to try to replicate the original study, and to investigate the practical educational implications of those findings. We approached this study with three questions:

1. Do the cognitive demands of reading and writing produce changes in language lateralization?
2. Do dyslexics and good readers differ in such task-induced patterns of lateralization?
3. Are task-induced patterns of language lateralization educationally significant, and specifically, do they have an impact on word decoding skills in dyslexia?

Sample and Academic Measures

Thirty-six public school children from an upper-middle class area of a suburb of Toronto participated in the study. Table 1 lists the defining attributes of the dyslexics and the two groups of chronological-age and reading-level matched good readers.

The generalizability of our results is limited by our selection of dyslexic children who were disabled severely in both pseudoword decoding (a recognized measure of phonologically decoding real words) on the Woodcock Word Attack Test-Revised and in reading comprehension on the Woodcock Test of Passage Comprehension-Revised, (Woodcock, 1987). This sampling strategy reflects our agreement with the growing consensus that phonological coding deficits may constitute a primary symptom and final common pathway for dyslexia (Pennington, Van Orden, Smith, Green, & Haith, 1990; Vellutino, et al., 1991). In addition to the Woodcock Tests, all of the children were tested on the Block Design and Vocabulary subtests from the Wechsler Intelligence Scale for Children-Revised (WISC-R) (Wechsler, 1974); and in Reading (visual word recognition) from the Wide Range Achievement Test-Revised (Jastak & Wilkinson, 1984).

TABLE 1

Subject Characteristics of the Dyslexic, Age-Matched and Reading-Matched Groups.

A.   **Dyslexic N = 12**
1. Right-handed males between 8-12 years old
2. No apparent problems with visual acuity or hearing
3. No neurological impairment
4   No evidence of a primary emotional problem
5. not attention deficit disordered or hyperactive
6. Caucasian, English spoken at home
7. identified as reading disabled with normal IQ by a Special Education Identification, Placement and Review Committee
8. teacher identified phonological decoding problem
9. $\leq$ the 17th percentile in reading comprehension and pseudoword decoding on the Woodcock Tests-Revised
10. .$\geq$ 90 on Performance IQ estimates from the WISC-R Block Design

B.   **Reading Matched N = 12**
1. Right-handed Caucasian males between 6-8 years old
2. average achievement in grade two
3. $\geq$ 90 on Performance IQ estimates from the WISC-R Block Design

C.   **Age Matched N = 12**
1. Right-handed Caucasian males between 8-12 years old
2. average achievement in grades four and five
3. $\geq$ 90 on Performance IQ estimate for the WISC-R Block Design

Dichotic Methods

The dichotic stimuli were pairs of monosyllable digits (one, two, three, four, five, six, eight, nine, ten, twelve). Each trial consisted of four digit pairs, and each condition consisted of 16 trials with a 14 sec. interval between trials. The children were asked to report all eight numbers on each trial which produced a maximum correct of 32 for each ear by condition. Free-recall dichotic testing was done individually in three conditions: Baseline; Reading; Writing. The Baseline condition consisted of simple oral reports recorded by the experimenter after each trial. The Writing condition required the children to record their own responses using a paper and pencil. The Reading condition used an oral report, but during the dichotic 14 sec. intertrial interval the children were required to read aloud and for comprehension from a graded passage (Spache, 1981) that was slightly above each child's individual reading level. Two children were assigned to each of six condition orders to counterbalance for any order effects.

RESULTS AND DISCUSSION

No differences were found among the groups on the IQ measures, and the dyslexics were matched successfully with the younger good readers in raw score Passage Comprehension.

TABLE 2

Means and Standard deviations for the Age-Matched, Reading-Matched and Dyslexics on the Academic Measures and F-tests comparing the Groups.

	Dyslexics	Age-Matched	Reading-Matched
Vocabulary [a]	9.5 (3.6)	11.7 (2.4)	11.6 (1.8)
Block Design[a]	14.4 (3.3)	12.3 (2.8)	13.4 (2.5)
Word Attack[b]			
raw score	13.0* (6.7)	33.2 (4.9)	23.0 (7.9)
standard score	71.7* (13.8)	104.1 (10.9)	106.4 (10.2)
Passage Comp.[b]			
raw score	29.0 (4.9)	41.6* (4.8)	30.7 (3.7)
standard score	78.2* (7.1)	100.3 (10.7)	110.6 (7.4)
WRAT Reading [c]	71.0* (8.9)	103.2 (9.9)	108.1 (11.2)

Note: * significantly different, p<.05 from the other groups

    [a] Vocabulary and Block Design subtest of the WISC-R

    [b] Subtest of the Woodcock Mastery Test-Revised

    [c] Subtest of the Wide Range Achievement Test-Revised.

On all standard score measures (Woodcock Passage Comprehension; Woodcock Word Attack; WRAT Reading) the dyslexics were significantly poorer in comparison to both control groups. These results are displayed in Table 2.

The results of the Baseline, Reading, and Writing dichotic conditions are plotted in Fig. 1. In Baseline, the dyslexic and older controls (age-matched) demonstrated a similar mean REA, but the younger controls (reading-matched) failed to produce an ear advantage. This unequivocal lack of an REA for the younger children in Baseline performance suggests that relatively younger children may need more demanding tasks to engage the attentional mechanisms necessary for lateralized processing and, indeed, under the increased cognitive requirements of the Reading and Writing condition, they did show evidence of a left hemisphere advantage. This transient REA in the younger children could explain some previous "developmental" lateralization effects which were interpreted in error as supporting the idea of progressive lateralization.

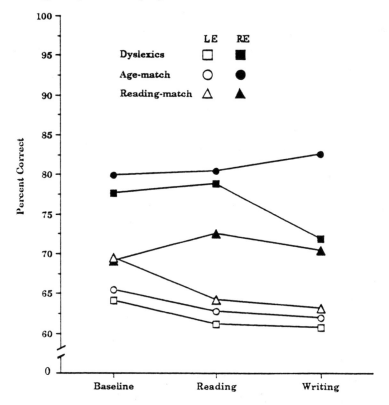

FIG. 1. Mean Percentage Correct Dichotic Scores for the Left Ear (LE) and Right Ear (RE) for the 3 Groups in Baseline, Reading, and Writing.

Reading, as a concurrent task, produced an equal increase in the REA in all groups, and did so by depressing performance at the left ear. This result is consistent with other research showing that the attentional/arousal processes underlying the REA may be heightened as a task-specific response to increased functional demands (Wexler & Heninger, 1980).

When the response mode was changed from oral to written in the Writing condition, the REA was increased again, by a depression in left ear scores, but only for the two groups of good readers. Although the dyslexics also reacted with a depression in left ear performance, they demonstrated a much larger decrement in right ear recall which lowered their REA. The attentionally-mediated inhibition in right ear performance in the Writing condition was unique to the dyslexic pattern of results. This effect replicates our earlier work (Kershner et al., 1984).

Therefore the first two questions of the study were answered. The information processing requirements of reading and writing did produce shifts in attentional lateralization, for good and for poor readers. Moreover, we can see a neuropsychological difference in how the dyslexics and good readers responded to these modifications in task demands, particularly when making written responses to the orally presented dichotic task. This takes us to the third question. Are task-induced patterns of lateralization significant educationally?

To answer the third questions, the two measures of word decoding (phonological and visual) and passage comprehension were correlated in each dichotic condition with the children's laterality scores. For a single measure of lateralization corresponding to each condition, we calculated lambda scores (Bryden & Sprott, 1981) which normalize the difference scores for the two ears ($1 = \text{Ln} (R/L)$. Pooling the lambda scores over the three conditions revealed that 90% of the children in all three groups produced an REA (positive mean lambda) which compares favourably with results from more direct, intrusive measures of language lateralization (Wada & Rasmussen, 1960). This result adds a degree of credibility to our dichotic procedures. The correlations controlled for the variability due to chronological age and IQ, and produced some fascinating results.

In the Baseline condition, only one significant correlational effect was found and that was for the older good readers between dichotic lambda and reading comprehension, $r_x = .73$, p<.01. This shows that the fluent readers' inclination, in the unprimed Baseline state, to direct more attention to left lateralized processes during receptive speech was related to their ability to comprehend written language. We interpret this as a context-specific window into the dynamic relationship between attentional lateralization and reading comprehension processes. It was not observed in the Baseline performance of the beginning good readers or the dyslexics, nor in the same children when they were given the Reading and Writing conditions.

When the neural processes underlying the REA were attentionally primed by the concurrent Reading condition (which increased the REA), the same older good readers produced significant lambda correlations with both phonological decoding, $r_{xy} = .64$, p<.05, and visual word recognition, r_{xy} = .59, p<.05. In this condition, lateralization was not related to reading comprehension. It seems that reading aloud in the advanced readers activated the lateralized left hemisphere processes that are predominantly involved in single word reading (Peterson, Fox, Posner, Mintun, & Raichle, 1989). Additionally, because the increase in lateralization in the Reading condition was purchased by depressing left ear scores we infer that the content processes that are engaged in reading single words by fluent readers are somehow released through concurrent inhibition of the right hemisphere. Reading inhibited the right hemisphere equally in all groups; but lateralization/reading correlations were found only in the older good readers. Thus the effect appears to be a correlate of advanced reading skills.

This result is consistent with brain imaging, positron emission tomographic (PET) studies which have identified a common left-lateralized area in posterior, extrastriate cortex that activates when adult good readers decode pseudowords and real words (Peterson, Fox, Snyder, & Raichle, 1990). This area does not activate when letter clusters and visual features are processed which do not conform to the rules of English spelling. These PET results suggest strongly that this left-lateralized orthographic processor for words may only become task engaged as a function of experience with printed words as children gain in literacy competence. Hence, these results suggest that dyslexic children may suffer from a developmental delay or arrest in activating this left-lateralized mechanism when they read passages for comprehension. Presumably such localized attentional activation is a correlate of decoding visual stimuli that have been learned to be recognized as words. Consequently, this problem qualifies as a correlated rather than as a primary symptom in that it results secondarily as a by-product of their low level of reading. The younger beginning readers showed the same dichotic processing characteristic.

It is in the Writing condition that the dyslexics showed a unique pattern of performance and unique lambda-reading correlations suggesting a primary-symptom attentional dysfunction. In the Writing condition, the dyslexics were the only group to show significant lambda/reading correlations. In Writing, which attentionally decreased the REA in the dyslexics, the resulting REA was related positively to their phonological decoding, r_{xy} = .84, p<.01, and to their visual word recognition, r_{xy} = .73, p<.01. Somehow the written response induced an REA that promoted selective interference with left hemisphere word decoding processes in the dyslexics. This effect appears to be symptomatic of dyslexia and may provide insights into the primary underlying deficit of the disorder.

This result suggests that efficient single word decoding skills, which are heavily dependent on left hemisphere processes, may fail to develop in children with dyslexia partly because of a vulnerability to attentional interference from the cognitive processing dynamics of other selective tasks and task components. It is important to emphasize the specificity of both cause and effect. Left hemisphere interference was produced by writing and not by reading. Additionally, it was seen to have a linear disruptive effect on two measures of word decoding but to have no effect on reading comprehension. Clearly, it is the specificity of the results that may give us some clues to the possible etiology of dyslexia.

One possible theoretical explanation comes from the cognitive-anatomical theory developed by Kinsbourne and Hicks (1978) and supported by PET blood flow studies (Posner, Sandson, Dhawan, & Shulman, 1989). The essence of the theory can be expressed in a single proposition: that two cognitive operations will interfere to the the extent that they share components in an interconnected anatomical system. Furthermore, Posner and Peterson (1990) have identified an anatomically separate and highly interlinked attentional system in the brain that subserves the widely distributed systems that are dedicated to processing cognitive content, that is, semantics, phonology, and so forth. The attentional system includes thalamus, anterior cingulate, dorsolateral prefrontal cortex, posterior parietal lobe, and interhemispheric connections. Thus multiple anatomical sites are candidates for potential cognitive interference. Indeed, task interference has been shown to occur in normal adults between lateral shifts in attention and both phoneme monitoring and digit recall (Posner & Peterson, 1990).

From the perspective of this cognitive-anatomical theory, the writing effect suggests two things. First, the hemisphere-specific and task-specificity of the interference supports the view that a deficit in single word decoding may be a final common pathway and marker for dyslexia. Second, the fact that writing was the only task to produce such left hemisphere attentional interference suggests that writing and single word decoding may share common attentional pathways in dyslexia. We presume that the attentional system is the source of such interference because the interference was produced as a result of an attentional shift in lateralization.

Just where the neuronal overlap may occur is not known; but there are some promising possibilities. One candidate is the anterior cingulate which is thought to monitor left and right posterior parietal lobes when they engage attention to cues in contralateral space (Posner et al., 1989). Actual and imagined right hand movement activates the supplementary motor area (SMA) which is in close proximity to cingulate cortex (Fox, Pardo, Peterson, & Raichle, 1987). Thus the attentional interference that was observed with word decoding in the written dichotic response condition may have been mediated by a pathological dependence on common attentional pathways in or near the anterior cingulate. A second possibility is that the interference may have been caused more directly by an overlap between the posterior cortical areas involved in lateral attentional control and word decoding. For instance, PET studies of single word reading in adult dyslexics have shown localized activation in the same posterior, extrastriate region that activates when normal adults read single words (Gross-Glenn et al., 1991). However, the dyslexics in comparison produced an excessive amount of activation bilaterally with asymmetrically greater hyperperfusion in the right hemisphere. Such excessive neuronal activity apparently is characteristic of inefficient processing (Parks et al., 1988) and the diffuse spread of neuronal activity suggests a potential for interference from posterior parietal lobe attentional functions. Writing but not speaking requires careful left-to-right attentional control implicating posterior parietal lobe engagement. Moreover, in the same study, the adult dyslexics showed abnormal prefrontal activation, an area that has been implicated in higher-level attentional control (Posner & Peterson, 1990; LaBerge, 1990). It is important to point out that the physical location for word decoding is not shifting from the right hemisphere in the dyslexics to the left hemisphere in advanced readers as Orton's theory would have predicted. Rather, it appears that adult dyslexics engage in excessive attentional activation bilaterally during single word reading. Writing and word decoding in dyslexics therefore may share aberrant attentional pathways in the posterior parietal area of the brain. Indeed, there is additional support for this intriguing possibility. In a task that required both an orthographic and an auditory letter-by-letter analysis of single words (a central process that would be required in writing dichotic digits and in word decoding), blood flow measurements (rCBF) using the 133-Xenon method with dyslexic adults found (1) excessive left hemisphere activation and (2) the activation was displaced posteriorly from its normal temporal location to an area encroaching on the posterior parietal region (Wood, Felton, Flowers, & Naylor, 1991).

These are, of course, just two theoretical possibilities out of many potential explanations for the pattern of attentional interference with single word decoding in dyslexia that we found in the present study. Our results suggest that phonological and visual word processing may both be compromised by a high-order attentional impairment. The hypothesis that an attentional lateralization impairment may be a primary causal factor in dyslexia is relatively new. Nonetheless, we believe that the theoretical and experimental evidence is substantial enough to warrant its careful consideration. Orton's speculation of long ago may have been reasonably close to the truth after all.

REFERENCES

Broca, P. (1861). Remarques sur le siege de la faculte du language articule, suivies d'une observation d'aphemie. *Bulletin of the Society of Anatomy,* Paris, Series 2, 6, 398.407.

Bryden, P. (1988). Does laterality make any difference? Thoughts on the relation between cerebral asymmetry and reading. In D. Molfese and S. Segalowitz

(Eds.), *Brain Lateralization in children* (pp. 509-525). New York: Guilford Press.

Bryden, M., & Sprott, D. (1981). Statistical determination of degree of laterality. *Neuropsychologia*, 19, 517-581.

Churchland, P. (1986). *Neurophilosophy*. Cambridge, MS: MIT Press.

Fox, P., Pardo, J., Peterson, S., & Raichle, M. (1987). Supplementary motor and premotor responses to actual and imagined hand movement with positron emission tomography. *Society for Neuroscience*: Abstract, 13, 1433.

Gross-Glenn, K., Duara, R., Barker, W., Loewenstein, D., Chang, J., Yoshii, F., Apicella, A., Pascal, S., Boother, T., Sevush, S., Jallad, B., Novoa, L., & Lubs, H. (1991). Positron emission tomographic studies during serial word-reading by normal and dyslexia adults. *Journal of Clinical and Experimental Neuropsychology*, 13, 531-544.

Hellige, J. (1990). Hemispheric asymmetry. *Annual Review of Psychology*, 41, 55-80.

Hiscock, M. (1988). Behavioural asymmetries in normal children. In D. Molfese & S. Segalowitz (Eds.), *Brain Lateralization in Children* (pp. 84-169). New York: Guilford Press.

Hugdahl, K., & Andersson, B. (1987). Dichotic listening and reading acquisition in children: A one-year follow-up. *Journal of Clinical and Experimental Neuropsychology*, 9, 631-649.

Jastak, J., & Wilkinson, S. (1984). *Wide Range Achievement Test-Revised*. Wilmington, DL: Jastak Associates.

Kershner, J. (1985). Ontogeny of hemispheric specialization and relationship of developmental patterns to complex reasoning skills and academic achievement. In C. Best (Ed.), *Hemispheric \unction and Collaboration in the Child* (pp. 327-360). New York: Academic Press.

Kershner, J. (1988). Dual processing models of learning disability. In D. Molfese & S. Segalowitz (Eds.), *Brain Lateralization in Children* (pp. 527-546). New York: Guilford Press.

Kershner, J., Henninger, P., & Cooke, W. (1984). Writing induces a right hemisphere linguistic advantage in dysphonetic dyslexic children: Implications for attention and capacity models of laterality. *Brain and Language*, 21, 105-122.

Kershner, J., & Chyczij, M. (1991). The development of lateralized attention and its relationship to intellectual ability and school achievement: Abstract. *Journal of Clinical and Experimental Neuropsychology*, 13, 416.

Kershner, J., & Morton, L. (1990). Directed attention dichotic listening in reading disabled children: A test of four models of maladaptive lateralization. *Neuropsychologia*, 28, 181-198.

Kimura, D. (1961). Cerebral dominance and the perception of verbal stimuli. *Canadian Journal of Psychology,* 15, 166-171.

Kinsbourne, M. (1977). Hemi-neglect and hemisphere rivalry. In E. Weinstein & R. Friedland (Eds.), *Advances in Neurology* (vol. 18, pp. 41-49). New York: Raven.

Kinsbourne, M., & Hicks, H. (1978). Functional cerebral space: A model for overflow, transfer and interference effects in human performance. In J. Requin (Ed.), *Attention and Performance* VII (pp. 345-363). Toronto: Wiley.

LaBerge, D. (1990). Thalamic and cortical mechanisms of attention suggested by recent positron emission tomographic experiments. *Journal of Cognitive Neuroscience,* 2, 358-372.

Obrzut, J. (1988). Deficient lateralization in learning disabled children. In D. Molfese and S. Segalowitz (Eds.),*Brain Lateralization in Children* (pp. 567-589). New York: Guilford Press.

Obrzut, J. (1991). Hemispheric activation and arousal asymmetry in learning disabled children. In J. Obrzut & G. Hynd (Eds.), *Neuropsychological Foundations of Learning Disabilities: A handbook of Issues, Methods and Practice* (pp. 179-198). New York: Academic Press.

Obrzut, J., Hynd, G., Obrzut, A., & Pirozzolo, F. (1981). Effect of directed attention on cerebral asymmetries in normal and learning disabled children. *Developmental Psychology,* 17, 118-125.

Orton, S. (1937). *Reading, Writing and Speech Problems in Children..* New York: Norton.

Parks, R., Loewenstein, D., Dodrill, K., Barker, W., Yoshii, F., Chang, J., Emron, A., Apicella, A., Sheremata, W., & Duara, R. (1988). Cerebral metabolic affects of a verbal fluency test: A PET scan study. *Journal of Clinical and Experimental Neuro-psychology,* 10, 565-575.

Pearson, D., & Lane, D. (1991). Auditory attention switching: A developmental study. *Journal of Experimental Child Psychology,* 51, 320-334.

Pennington, B., Van Orden, G., Smith, S., Green, P., & Haith, M. (1990). Phonological processing skills and deficits in adult dyslexics.*Child Development,* 61, 1753-1778.

Peterson, S., Fox, P., Posner, M., Mintun, M., & Raichle, M.(1989). Positron emission tomographic studies of the processing of single words. *Journal of Cognitive Neuroscience,* 1, 153-170.

Peterson, S., Fox, P., Snyder, A., & Raichle, M. (1990). Activation of extrastriate and frontal cortical areas by visual words and work-like stimuli. *Science,* 249, 1041-1043.

Posner, M., & Peterson, S. (1990). The attention system of the human brain. *Annual Review of Neuroscience*, 13, 25-42.

Posner, M., Sandson, J., Dhawan, M., & Shulman, G. (1989). Is word recognition automatic? A cognitive-anatomical approach. *Journal of Cognitive Neuroscience*. 1, 50-60.

Roeltgen, M., & Roeltgen, D. (1990). Asymmetric lateralized attention in children. *Developmental Neuropsychology*, 6, 25-37.

Spache, G. (1981). *DRS-81:Diagnostic Reading Scales*. Monterey, CA: CBT/ MacMillan.

Vellutino, F., Scanlon, D., & Tanzman, M. (1991). Differential sensitivity to the meaning and structural attributes of printed words in poor and normal readers. *Learning and Individual Differences*, 2, 19-43.

Wechsler, D. (1974). *Manual for the Wechsler Intelligence Scale for Children-Revised*. San Antonio, TX: Psychological Corp.

Wexler, B., & Heninger, G. (1980). Effects on concurrent administration of verbal and spatial visual tasks on a language related dichotic listening measure of perceptual asymmetry. *Neuropsychologia*, 18, 379-382.

Wood, F., Felton, R., Flowers, L., & Naylor, C. (1991). Neurobehavioral definition of dyslexia. In D. Duane & D. Gray (Eds.), *The Reading Brain* (pp. 1-25). Parkton, ML: York Press.

Woodcock, R. (1987). *Woodcock Reading Mastery Tests-Revised*. Circle Pines, MN: American Guidance Service.

Facets of Dyslexia and its Remediation
S.F. Wright and R. Groner (Editors)

TOWARDS THE ORIGINS OF DYSLEXIA

R. I Nicolson and A. J. Fawcett
Department of Psychology
University of Sheffield, UK.

INTRODUCTION

The past decade has seen continuing and substantial progress in both the understanding and the remediation of dyslexia. The 'dyslexia community' — theorists, practitioners, administrators and supporters — have much to be proud of. In many ways, dyslexia research epitomises good science. The problems of dyslexia have been addressed by theorists from a bewildering variety of backgrounds — experimental psychologists, educationalists, geneticists, neurophysiologists, neuroanatomists, statisticians, and now cognitive scientists. Each discipline has brought a different perspective and contributed a range of intriguing findings that should eventually fit together to complete the jigsaw connecting the genes with the brain with the mind. International cooperation on an impressive scale, inspired by voluntary associations including the Orton Society in America, the British Dyslexia Association in Britain and the Rodin Remediation Foundation in Europe, has led to a fruitful cross-fertilisation of ideas, which in turn has led to further progress. Dyslexia conferences provide a stimulating mixture not just of different disciplines but also of different perspectives, from that of a dyslexic person, through those of teachers of dyslexia and administrators of dyslexia support systems, through to researchers from many disciplines.

It is beyond the scope of this chapter to review all the theoretical advances that have been made during the past decade, and the advances in teaching method that now promise to alleviate very significantly the reading problems of dyslexic children. We shall concentrate upon cognitive analyses of the deficits. Theoretical hypotheses for the cognitive origins of dyslexia have been rigorously tested in the laboratory and the classroom, leading to the emergence of the 'phonological deficit' hypothesis (PDH) which was proposed by the Haskins Lab in the 1970's (e.g. Liberman, 1973), and has been refined by many researchers over the next decade, with Stanovich giving arguably the clearest description in 1988. The PDH theory predicted that early phonological problems should precede the emergence of dyslexia, and a series of experiments by Bradley & Bryant, (e.g. 1983) in England; Lundberg and his colleagues in Denmark (e.g. Lundberg & Høien, 1989); and Olson and his team in the USA (e.g. Olson, Wise & Rack, 1989) have confirmed that those young pre-reading children who show phonological problems in terms of lack of sensitivity to rhyme and alliteration and general phonemic segmentation are likely to show typical dyslexic problems when they try to learn to read. Furthermore, early instruction in phonemic segmentation seems to allow the children to progress at a more normal rate through the early stages of reading, thus mitigating the effects of the underlying deficit, thus providing strong evidence for phonological segmentation performing a key role in learning to read.

In related work, following Miles (1983) and Jorm (1983) it is well known that reduced memory span is a common weakness of dyslexic children, and recent research (Gathercole & Baddeley, 1990) has both demonstrated its occurrence early in dyslexia, and devised a technique (non-word repetition) which promises to be valuable in the early diagnosis of dyslexia. The PDH is a theoretically plausible hypothesis which accounts for the reading-related problems of dyslexia; has received support from direct longitudinal tests in young children, and has led to direct benefits in remediation. In short, it appears that the PDH has been supported by converging operations over the past decade and should provide a solid basis for this decade's research and practice.

The main burden of this paper is a direct challenge to the causal role of the PDH. We argue, on the contrary, that despite the undoubted success of the PDH, it is not able to account for the full range of findings related to dyslexia, and that an analysis of the likely causes for the PDH suggests that the fundamental problem may lie somewhere within the general learning process. We support this bold assertion by data collected from our ongoing research programme in which we have explicitly attempted to fit dyslexia within a wider framework in which phonological deficits are seen as one of many symptoms of a deeper underlying deficit. Unfortunately (from this perspective, but understandably in view of the key importance of reading) research on dyslexia has tended to concentrate entirely on components of, or contributors to, the reading process and so evidence of deficit in areas unrelated to reading is surprisingly meagre. In terms of spatial skills Ellis & Miles (1978) found no evidence of memory impairment for spatial tasks (using a checkerboard pattern). Carpenter, Just & McDonald (1984) demonstrated that dyslexic students performed normally on mental rotation of complex spatial stimuli. However, no-one appears to have undertaken a systematic exploration of spatial capabilities, and there is always the nagging concern that to meet the normal criteria of dyslexia, children normally have to compensate for their relatively poor verbal performance by relatively good spatial performance. There is of course considerable evidence of deficit in visual skills, and it is particularly apt that this issue should be raised at a Rodin conference, which is notable for the breadth of its research into visual deficits (see Euler, 1990, for an overview of the area).

There is already persistent evidence from a number of sources that dyslexic children suffer problems in skills quite independent of phonological processing. Anecdotal evidence suggests forgetfulness, distractibility, and clumsiness all tend to accompany dyslexia, and these impressions have been substantiated by careful research. Augur (1985) summarised a good deal of clinical evidence, deriving 21 classic problems for dyslexic children. The majority of them were related to reading and phonological skills but a surprisingly large number (clumsiness, difficulties in hopping and skipping, clapping in rhythm, throwing and catching a ball) indicated problems in motor skill, and another set (carrying out several instructions simultaneously, high distractibility, and rapid tiring under continuous load) were perhaps more indicative of central attentional problems. Haslum's (1989) recent analysis of the National Cohort study data for 17,000 British schoolchildren born in 1970 investigated those characteristics which correlated highly with the established symptoms of dyslexia at age 10. She found that, after partialling out the effects of related characteristics, 4 primary independent factors emerged with significance at the .01 level, and these factors included catching and clapping, and also walking backwards. Even the well-established symptom of atrocious hand writing suggests a problem with motor skill rather than phonological skill. There is, then, quite a large body of evidence against the predictions of the PDH.

Karl Popper, the influential philosopher, has argued at length (e.g. 1968) that most scientific research may be characterised as attempting to confirm a hypothesis (usually by testing whether predicted effects can be found), whereas the preferred scientific method should be to attempt to falsify hypotheses (for instance by testing whether predicted non-effects can be found). If one evaluates the theoretical research on dyslexia from this critical viewpoint, one is forced to the conclusion that the PDH may suffer from the problem of premature specificity, in that the research to date is insufficient to discriminate between the predictions of the PDH and of a range of alternative theories.

OVERVIEW OF THE RESEARCH PROGRAMME

In the interests of clarity of exposition we shall report the results of our research programme within the framework established above, presenting the findings from a series of experiments, all performed within the past two years. The questions addressed were as follows:

Question 1. Are there deficits in performance in tasks independent of phonological processing? These experiments involved an investigation of whether there was a deficit in an area far removed from phonological processing, namely motor balance.

Question 2. Can we pinpoint the stage at which basic skills become impaired? This represents a line of investigation aimed at pinpointing the likely cause of the effect. The final set of experiments investigated speed of reaction, using a range of tasks from complex to the simplest possible.

Space precludes a detailed description of all the experiments, so we shall offer an abbreviated summary of the key points in each of the experiments, followed by a more speculative analysis of the likely explanations of our results. First we explain the methodological issues that led us to use 5 groups of children for the studies, and then, given the problems of differing interpretations of dyslexia, it is necessary to explain our criteria for subject selection.

OVERALL DESIGN OF THE STUDIES

Although it is valuable to identify whether dyslexic children perform significantly worse than their same-age controls, one of the key discriminants between theories is a test of performance of dyslexic children against reading age controls, since a significant impairment compared with reading age controls is indicative of developmental disorder rather than just a developmental lag (cf. Bryant & Goswami, 1986). Since the specific nature of dyslexic children's deficits may also change with age, it is important to examine the effects of age separately. These considerations suggest an experiment with at least six groups of subjects: two groups of dyslexic children of different mean ages; two groups of non-dyslexic children matched to the dyslexic children on chronological age; and two groups of non-dyslexic children matched to the dyslexic children on reading age.

Three separate issues are of interest in the statistical analyses for each experiment. First, whether there are any between-group differences at all. This involves a design which treats all the six groups within one factor, irrespective of age and presence/absence of dyslexia. We refer to this as the 'Overall Analysis'. A lack of a significant effect here would suggest that the variable under investigation was unaffected by either age or dyslexia. Second, it is important to identify whether dyslexic children perform worse than their same-age controls. This design has the two level factor age and the two level factor presence/absence of dyslexia. We refer to this as 'CA & Dyslexia'. A main effect of age would indicate a developmental trend in the variable in question, while a main effect of dyslexia would suggest a reliable difference between dyslexic and control subjects of equivalent age. Such a difference may, however, be attributable either to a fundamental difference, or to a developmental lag. Deciding between these requires a third analysis, one involving a comparison with reading age controls; this also has two factors, namely a two level factor reading age, and a two level factor presence/absence of dyslexia. We refer to this as 'RA & Dyslexia'. A negative effect of dyslexia on this analysis would indicate that dyslexic subjects are performing more poorly than younger children of equivalent reading age, and would argue against a developmental lag interpretation.

Note that the latter two analyses are based on only four of the experimental groups. Fortunately, as described in the next section we were able to select two groups of dyslexic children who were sufficiently similar in IQ to allow a single control group to be used both as RA control for the older dyslexics and CA control for the younger dyslexics, thus leading to a total of only five groups.

Subjects

Five groups of subjects participated. The groups were: 14 dyslexic children around 15 years old ('Dys 15'); 11 dyslexic children around 11 years old ('Dys 11'); a group of 12 non-dyslexic children matched to the older dyslexics for age and full IQ ('Cont 15'); a group of 12 non-dyslexic children of similar IQ to the two dyslexic groups, matched for chronological age with the younger dyslexics and for reading age with the older dyslexics ('Cont 11'); and a fifth group of 9 non-dyslexic children around 8 years old matched for reading age and full IQ with the younger dyslexics ('Cont 8').

TABLE 1

Psychometric Details for the 5 groups

Group	N	Mean CA	Mean IQ	Mean RA	CA Range	IQ Range	RA Range
Cont 15	12	15.4	106.2	14.9*	15.1 to 16.0	92 to 130	14.4 to 15.0
Dys 15	14	15.5	105	11.8	14.4 to 16.2	88 to 126	8.1 to 14.4
Cont 11	12	11.4	107.7	11.4	10.6 to 12.2	94 to 126	9.8 to 12.7
Dys 11	11	11.3	111.5	8.6	9.7 to 12.9	101 to 128	7.0 to 10.8
Cont 8	9	8.3	112.0	8.8	7.9 to 8.8	101 to 133	7.7 to 10.6

* 15.0 represents ceiling on the Schonell test of reading age used. Most of this group were reading at this level

All dyslexic children had been diagnosed between the ages of 7 and 10, based on discrepancies of at least 18 months between chronological and reading age, and the standard 'exclusionary criteria' — namely that their IQ levels fell in the normal to superior range on the Wechsler Intelligence Scale for Children (Wechsler, 1976), and they had no known neurological deficit or primary emotional difficulty. The dyslexic children were recruited via the local British Dyslexia Association and the Dyslexia Institute[1], and the

---

[1] We are grateful to Dr. Harry Chasty and to Mrs Jean Walker of the Dyslexia Institute for allowing us to approach the parents of the children involved and for permitting access (with parental consent) to the original diagnostic data.

non-dyslexic controls were recruited from local schools. The overall psychometric details are shown in Table 1 above.

Experiment 1: Motor Balance

The experiments reported here were extensions and replications of earlier experiments we had performed on a larger group of dyslexic children which had included a subset who had improved their reading to the extent that they were technically 'remediated' in that their RA was within 1 year of their age (Nicolson & Fawcett, 1990). Interestingly, although the vocabulary of the remediated group of older dyslexic children was at least as good as normal children, any more sensitive analyses of reading suggested a lack of fluency — their reading was more laboured, more prone to error, and more susceptible to interference from other tasks. In fact this incomplete mastery characterises many features of dyslexic performance, sometimes established difficulties, such as problems in tying shoelaces, riding a bicycle (Augur, 1985) or walking backwards (Haslum, 1989), but more often anecdotal reports such as tendencies to being accident-prone, distractible, or absent-minded.

Theoretical accounts of the acquisition of skill (Fitts & Posner, 1967; Anderson, 1982) attribute fluency in performance to 'automatisation' of the skill — the gradual reduction in the need for conscious attentive control of performance, with a concomitant increase in speed and efficiency together with a decreased likelihood of breakdown under stress. There is no established procedure for testing for automaticity, but the general rule of thumb (Shiffrin, Dumais & Schneider, 1981) is to see whether performance breaks down under stress — typically by introducing a secondary task which has to be performed concurrently, thus distracting conscious attention from performance of the primary task. If performance on both tasks is unaffected, then the primary task skill is clearly automatic. If performance on the primary task degrades, and the tasks do not interfere explicitly (e.g. by both requiring visual input, or both requiring manual output) then it is probable that the primary skill is not fully automated. This analysis provided the rationale for the series of experiments described below.

In a series of studies reported recently (Fawcett, 1990; Nicolson & Fawcett, 1990) we examined a range of motor skills, from fine motor skills such as fitting pegs into holes to motor skills as gross as standing on two feet without falling over, and we used the 'classic' distracting task of counting backwards as the secondary task. Unfortunately, the dyslexic subjects were worse at the counting task under 'just counting' conditions, and so it was adjusted in objective difficulty for each subject to ensure equivalent performance under 'just counting' conditions. Most of the control subjects ended up counting backwards in threes, whereas the dyslexics counted backwards (or even forwards) in

unimpaired by the addition of the dual task, whereas both groups of dyslexics showed significant deterioration when the secondary task was introduced.

Since the key issue in this study was whether the children's balance was disrupted by the addition of a secondary task, the balance data were then collapsed by subtracting the errors in the single task condition from those in the dual task condition, thus deriving the 'Dual Task Balance Deficit' (DTBD). One factor analyses of variance were performed on the DTBD data and showed a significant main effect of group for all four tasks [$F(4,46)=$ 8.0, $p<.001$; $F(4,46)= 18.4$, $p<.001$; $F(4,50)= 5.6$, $p<.001$; $F(4,50)= 8.1$, $p<.001$ respectively]. Pairwise comparisons indicated that the DTDB was significantly worse for the 11 year old and 15 year old dyslexics than even the 8 year old controls.

Comparisons with CA Controls. A second analysis of variance was then undertaken limited to the four older groups, taking age and presence/absence of dyslexia as two factors and difference condition as the third. On all four tasks, the main effect of dyslexia was highly significant [$F(1,38)=20.6$, $p<.001$; $F(1,38)=7.8$, $p<.01$; $F(1,42)=15.2$, $p<.001$; $F(1,42)=19.2$, $p<.001$ respectively], whereas the effect of age did not reach significance.

Comparisons with RA Controls. A third analysis of variance was conducted on the DTDB data, omitting the oldest controls and comparing the two dyslexic groups with their reading age controls. The main effect of dyslexia was again highly significant, [$F(1,39)=28.0$, $p<.001$; $F(1,39)=59.4$, $p<.001$; $F(1,40)=11.4$, $p<.01$; $F(1,40)=19.2$, $p<.001$ respectively], whereas that of reading age was not significant .

The effects of the secondary task are summarised in Table 2 below. Under the dual task conditions the counting accuracy declined significantly for the two groups of dyslexics but not for the controls, and the SCRT latency increased significantly for the two groups of dyslexics and the 8 year old controls.

Discussion of the Dual Task Balance Results

For the two dyslexic groups all the comparisons of dual task vs single task performance were significant (with 7 out of the 8 at $p<.001$), whereas none of the 8 comparisons for the control groups approached significance. The two groups of dyslexic children were significantly impaired in their balance under dual task conditions, whereas the three control groups showed no such impairment. The two groups of dyslexic children showed significant impairments on the secondary task in most of the conditions, whereas the three groups of control children showed almost no significant impairment on secondary task performance. Furthermore, these group results also applied to almost all the individuals involved. It appears extremely difficult to explain these results in terms of impairment in phonological processing, given the complete absence of any speech input or output in the balance/SCRT conditions.

TABLE 2

Summary of the Dual Task Impairments for the Five Groups

	Counting		Selective CRT		
	bal^{ce}	count %	bal^{ce}	SCRT %	SCRT
Cont 15					
2 foot	↔	↔	↔	↔	↔ *
1 foot	↔	↔	↔	↔	↔
Dys 15					
2 foot	↓	↓	↓	↓	↓
1 foot	↓	↓	↓	↓	↓
Cont 11					
2 foot	↔	↔	↔	↔	↔
1 foot	↔	↔	↔	↔	↔ *
Dys 11					
2 foot	↓	↓	↓	↔	↔ *
1 foot	↓	↓	↓	↔	↓
Cont 8					
2 foot	↔	↔	↔	↔	↔
1 foot	↑	↔	↔	↔	↓

Key

↔	NS
↔ *	NS but p<.10
↓	sig worse
↑	sig better

To summarise, for the two dyslexic groups all the comparisons of dual task vs single task performance were significant (with 7 out of the 8 at p<.001), whereas none of the 8 comparisons for the control groups approached significance. These results indicate that dyslexic children show a greater dual task deficit than reading age controls, symptomatic of a developmental disorder in balance.

In our earlier research we have suggested that dyslexic people may suffer from a deficit in the ability to automatise skills, our Dyslexic Automatisation Deficit (DAD) hypothesis. Automatisation is the process by which well-learned skills slowly become more and more fluent to the extent that one no longer needs to think about them, and is thought to be a key component of the development of any skill, whether cognitive (such as say knowing that 6x7=42, or how to spell biscuit) or motor (how to write the letter a, how to catch a ball etc.), and is an especially important requirement for reading, which depends critically on the automatisation of a range of sub-skills (Laberge & Samuels, 1974). Two requirements for automatic processing are that it is fast, and that it needs little conscious effort. A standard test for automaticity, therefore, is to see whether performance degrades under dual task conditions (Shiffrin, Dumais & Schneider, 1981). This is of course exactly what we found with our dual task balance results. For the two older non-dyslexic groups there was no decrement in balance under the dual-task conditions, and so their balancing was automatic, as one would expect for such a highly overlearned skill. The youngest controls' performance on the single task was poor but even so they showed no further impairment under dual task conditions. By contrast, however, there was a clear decrement under dual task conditions for the two dyslexic groups, and so their balance was not automatic. Although dual task deficits are difficult to interpret, the blindfold balance results confirmed an underlying deficit in even the most basic skill, simply standing on both feet without wobbling.

The balance results overall therefore provide strong support for the DAD hypothesis (and, incidentally, suggest that even under normal conditions, the dyslexic children have to invest more conscious attention to keep performance at normal levels).

Blindfold Balance
Nonetheless, dual task deficits are notoriously difficult to interpret, reflecting either an attentional deficit or time sharing between the two concurrent tasks. Therefore we presented a simplified version of the balance task in an attempt to circumvent these problems in interpretation. In this experiment, subjects were blindfolded to ensure that they were unable to use cues from the environment as 'conscious compensation' for any underlying balance deficit. The same method was used as in the previous balance tasks, but subjects balanced on the floor on this occasion, and were steadied by the experimenter at the start of the test (in the one foot condition only).

Results

The overall results are shown in Fig. 3. Naturally enough, all subjects found the one foot blindfold balance particularly difficult, but the performance of the dyslexic groups was more severely affected than that of the controls. Most strikingly of all, the performance of the dyslexic children was significantly impaired while balancing on both feet blindfold, whereas the non-dyslexic children, including the eight year old controls, were unaffected.

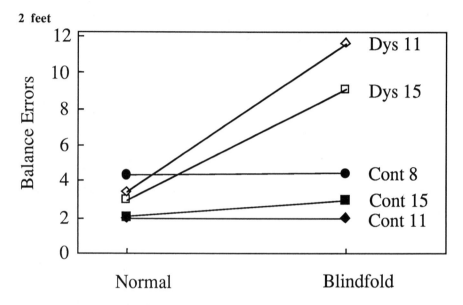

FIG. 3. Blindfold Balance

Analysis of the dual task balance deficits for both feet balance showed a highly significant effect of group,[F(4,49)= 7.2] Pairwise comparisons indicated that the older dyslexic groups were significantly worse than the 11 year old and 8 year old controls (p<.05) and the 11 year old dyslexics were significantly worse than all three control groups at the .01 level. Both the chronological age and the reading age analyses showed a highly significant effect of dyslexia [F(1,42)= 20.93, p<.001; F(1,40)= 21.71, p<.001] and no effect of age[F(1,42)= 0 15, NS; F(1,40)= 0 5, NS for chronological and reading age respectively].

Results for the one foot balance were equally significant, but pairwise comparisons here showed that both dyslexic groups were significantly worse than all three control groups at the .01 level.

Interpretation of the Balance Results

The blindfold balance results allow us to interpret our earlier balance findings with some confidence in that the latter can no longer be explained purely in terms of some attentional deficit which makes any combination of tasks difficult to carry out simultaneously. There is evidence of a *developmental disorder* in balance which is normally masked by the input of greater attentional resources. These results provide strong evidence for both DAD and for Conscious Compensation (CC), moreover the blindfold balance effect suggests that the deficit is not caused by the dual task nature of the situation, but by the removal of the opportunity for CC.

Clearly the phonological deficit hypothesis is unable to explain this pattern of results, but can the DAD hypothesis cope with the existing findings? Let us consider these in turn: phonological deficits are entirely to be expected, since a child has to *learn* to discriminate the phonology of his/her culture; reading deficits act as a litmus test for automatisation since reading involves the smooth interplay of a number of skills; motor skill deficits (in both balance and visual control) are again, as expected, especially when conscious attention is distracted from the task and finally articulation rate deficits are a combined motor/phonological skill which needs to be learned, and so one would again expect deficits.

Thus the DAD copes comfortably with the findings to emerge so far. Even so, it is by no means the only explanation, and in reality only describes yet another *symptom* of dyslexia, rather than being a causal explanation. So we need to analyse the components of 'automatisation' in an attempt to identify which one(s) appear to be causing the impairment.

One key aspect of automatised performance is that it is faster than consciously controlled performance, and slower speed of information processing appears to be a recurring symptom of dyslexia. Many researchers and practitioners have noted that dyslexic children appear to show more marked deficits under paced or timed tests than under more relaxed conditions (e.g. Ellis & Miles, 1981; Seymour, 1986). There is also a substantial literature on deficits in speed of access to the spoken word, initially discovered using the 'rapid automatised naming' test (e.g. Denckla & Rudel, 1976). This test involves the rapid sequential naming of a series of 50 familiar stimuli presented simultaneously on a card. The authors showed that dyslexic children were slower to name colours, pictures, digits and letters than slow learners matched for reading age. The naming deficit has been consistently replicated (e.g. Swanson, 1987), and is evident not only with visual stimuli but also with auditory presentations and the naming of palpated objects (Rudel, Denckla & Broman, 1981). It therefore seems well established that any task which demands continuous speeded access to spoken language highlights deficiencies in speed of information processing in dyslexic subjects. Unfortunately, these and similar demonstrations of reduced speed of information processing have involved linguistic stimuli, and it is therefore not clear whether the deficit represents another correlate of the known linguistic abnormalities of dyslexic children or whether reduced speed of information processing reflects some deeper problem. It is particularly surprising in view of the central role of information processing speed in cognitive skills that there appear to be no reports of direct investigations of information processing speed in the literature. These issues prompted us to examine the role of information processing speed in dyslexia more exhaustively.

SPEED OF PROCESSING AND DYSLEXIA.

In an effort to trace the dyslexic deficit back to its source, we decided to investigate speed of information processing, using a variety of reaction time tasks in the hope that at some point we would find a cut off where tasks of lesser complexity would show no deficit, whereas more complex tasks would result in a deficit. We tried a variety of simpler and simpler tasks — 'Coding', lexical access, choice reaction, and finally simple reaction to a tone in an attempt to find normal performance. Deficits were found all the way down to 2-choice reactions, and so we present the data for the two simplest conceivable tasks, a simple reaction and an selective choice reaction task (SCRT). The study is presented in detail in Nicolson & Fawcett (note 2) and so a summary should suffice here.

In both tasks, subjects sat with a single button in their preferred hand, and their task was to press it as quickly as possible whenever they heard a low tone. In the simple reaction task, no other tone was ever presented, but in the SCRT task, there was an equal probability of a high tone being presented. These tasks were introduced by Donders well over a century ago. His rationale was that the only difference between the tasks was the need to classify the stimulus before responding in the SCRT trials, and he argued that subtracting the simple reaction time (SRT) from the SCRT time gave an estimate of 'stimulus classification' time. The experiment was computer-controlled, and 100 stimuli were presented at an average rate of 1 per 2 seconds. The results are shown in Fig. 4, plotted on the same graph to facilitate comparison. Note that median reaction times have been used to avoid the danger of outliers biasing the results.

It is clear that, although the dyslexic subjects performed at the same level as their same-age controls on the simple reactions [$F(4,52)=17.3$, $p<.0001$], they were slowed down more by the need to make an SCRT, [$F(4,52)=8.0$, $p<.0001$]. Analysis of the simple reactions indicated that both groups of older subjects were significantly faster than all three groups of younger subjects.

Comparisons with CA Controls. Analysis of the SRT's in terms of age and dyslexia indicated a highly significant effect of age ($p<.0001$), whereas there was no effect of dyslexia at all. Analysis of the SCRTs indicated a significant effect of both age and dyslexia at the .01 level. Pairwise comparisons indicated that performance of the older controls was significantly faster than for the other four groups.

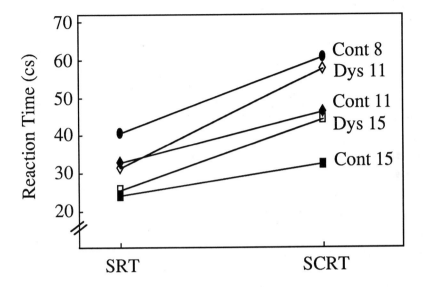

FIG. 4. Simple Reactions and Selective Choice Reactions

Comparisons with RA Controls. A third analysis of variance was conducted on the SRT's, omitting the oldest controls and comparing the two dyslexic groups with their reading age controls. The main effects of both age and dyslexia were highly significant at the .001 level. Analysis of the SCRTs indicated a significant effect of age (p<.01) but no effect of dyslexia.

Discussion of the Reaction Time Results
It would appear that we have at last found a critical pair of tasks, which together capture the transition point from no deficit to deficit, thus pinpointing the area for subsequent exploration. Both groups of dyslexic subjects performed at the appropriate level for their age on the simple reaction. Furthermore, the older dyslexic subjects significantly outperformed the younger controls (p<.01) for the first time in this suite of experiments. However, for the SCRT condition, the dyslexic subjects slipped back, to the extent that there was a significant main effect of dyslexia.

OVERALL DISCUSSION

It is appropriate to start the discussion by reviewing the questions which guided our research programme:

Question 1. Are there deficits in performance in tasks independent of phonological processing? The first set of experiments involved an investigation of whether there was a deficit in an area far removed from phonological processing, namely motor balance. The experiments demonstrated unequivocally that there was indeed a deficit (though it showed up only when the children were prevented from consciously compensating), thus embarrassing the PDH, and suggesting that a more general explanation was required.

Question 2. Can we pinpoint the stage at which basic skills become impaired? The second set of experiments investigated reaction time performance. Here it was found that deficits were found in all complex reaction time tasks, including even the simplest of choice reactions, but that no deficit was found on simple reactions. It would seem, therefore, that the deficit appears as soon as a not completely predictable decision must be made.

Any viable theory of the origins of dyslexia must account for the above set of findings. Before going on to consider specific theories, however, it is valuable to consider what kind of theory is needed — is dyslexia merely a developmental lag, merely slower development than normal, or is it a developmental disorder, a qualitatively different developmental sequence? On the one hand not only the dual task balance but also the blindfold balance results, in which the older dyslexics were significantly worse even than the youngest controls, provide clear evidence of disorder. By contrast, the reaction time results, in which there appears to be a change from normal speed to impaired speed as one adds a simple decision, provide evidence only of a lag in reaction speed somewhere in the system. Furthermore, in a series of experiments investigating working memory performance with these groups of children, we found evidence of at most a lag in memory span, paralleled by a lag in speed of articulation (see Nicolson, Fawcett & Baddeley, note 3).

It would appear that the phonological deficit hypothesis is unable to account for either the balance results or the reaction time results, since deficits were found in the absence of any phonological component of the stimulus or the response. The possibility that impaired working memory might underlie the full range of deficits (Gathercole & Baddeley, 1990) appears also to be insufficient, since it is hard to see why it should predict a blindfold balance deficit or a reaction time deficit with only two stimuli. Consequently it would appear that a more general explanation is required.

The DAD hypothesis provides a natural explanation of both the dual task balance deficits and the blindfold balance deficits for the dyslexic children. It also able to explain the working memory deficit naturally in terms of a reduced (less fluent) articulation rate (Nicolson, Fawcett & Baddeley, note 3). Consequently the only real issue is whether it is able to give a principled account of the reason for the dichotomy between SRTs and SCRTs. Here the argument is less convincing. Why, for instance, is there no deficit in SRTs, but a deficit in SCRT and other more complex choice reactions? Presumably the reason lies in the need to make a decision in the latter cases, rather than merely execute a pre-determined response. The slowed cognitive decision is quite consistent with the framework of automatisation — for the same amount of practice the decision is less automatic — but unfortunately for DAD, it can offer no principled reason why automatisation should be difficult only for decision-based tasks. Once one is attempting to explore central processes, the DAD hypothesis offers too coarse grain an analysis.

A number of possible explanations can be offered for the slower decision problem — presumably the problem must lie either in the time taken to analyse the stimuli, the time taken for the 'central executive' to 'notice' that the stimulus has been classified, the time taken to determine the appropriate motor response, or the time taken to 'load' it ready to despatch the neural impulses to the finger. Further research would be needed to distinguish between these possibilities, but our guess is that the problem is not a perceptual one — the time taken to classify the input — since this is not sufficiently general to explain the range of deficits suffered. Consequently it seems inevitable that the problem lies somewhere within the central processing system.

In their recent connectionist/control architecture for working memory, Schneider and his colleagues (e.g. Schneider & Detweiler, 1988) have modelled these central processes as a set of modules (audition, vision, semantics etc) all of which are linked via an 'inner loop', which have an output link to the other modules, and input links from each of the other modules. In the case of interference between messages a central control structure forces all the messages into sequential order. As tasks become more automatic, the messages involved may pass directly between the appropriate modules (for instance, the auditory module could send the message 'tone' directly to the motor output module) without the need for it to be routed through the central controller, and thus the link becomes less susceptible to interference from other tasks. Almost any problem within this system would lead to effects qualitatively similar to those we have obtained. It might be, for instance, that the control system has a relatively low memory capacity, and so some messages decay before they can routed to the appropriate module; it might be that the amount of interference between messages within the inner loop is higher than normal;

or maybe that the ambient level of noise within the system is critically higher. All of these features would lead to less reliable data transfer, requiring the input of greater attentional resources, and would very likely lead to a more variable mapping between effective stimulus and response — conditions which have been shown to militate against the development of automaticity (e.g. Shiffrin & Schneider, 1977).

Interestingly, Galaburda and his colleagues have identified both "*a uniform absence of left-right asymmetry in the language area and focal dysgenesis referrable to midgestation ...possibly having widespread cytoarchitectonic and connectional repercussions. ...Both types of changes in the male brains are associated with increased numbers of neurons and connections and qualitatively different patterns of cellular architecture and connections*" (Galaburda, Rosen & Sherman, 1989, p 383). Presumably the cortical dysplasias would result in noisier perceptual processes, whereas the left-right abnormalities would result in noisier central processes.

Consider now the implications of these findings for the phonological deficit hypothesis and the automatisation deficit hypothesis. The deficit in selective choice reaction to pure tones occurred with the simplest possible auditory stimuli, and therefore at the very least would require that the phonological deficit be broadened to an 'auditory deficit' hypothesis. The auditory deficit hypothesis would therefore require that the underlying cause of the SCRT impairment was slower perceptual processing. Both slower perceptual processing and slower central decision making are consistent with an automatisation deficit. The automatisation deficit hypothesis handles the lack of deficit on simple reactions as deriving from the simplified central decision process, or from the simplified perceptual route, deriving from stimulus onset detection rather than stimulus classification.

In a parallel investigation as part of our overall research programme (Nicolson, Fawcett, & Baddeley, note 3) we established that working memory problems of dyslexic children appear to diminish over time, with marked problems around 8 years, slight problems at 11 years, and no identifiable problems around 15 years. It appears therefore that the dyslexic children's deficit should be seen as an initial disorder which is gradually overcome. We speculated that the cause of the disorder might be 'noisy' neural networks, quite probably arising in response to the neuroanatomical abnormalities identified by Galaburda and his colleagues. Unfortunately, attempts to model reaction time processes using connectionist techniques are not yet well developed, but we hope to investigate the issue further in subsequent research. We believe that the research issues raised could provide fruitful agenda spanning cognitive psychology, neuroscience and connectionist modelling, capable not only of illuminating the causes of abnormal information processing, but also providing a much deeper understanding of the bases of normal cognition.

It is too early to tell which of these factors is or are responsible for the deficits found in dyslexic children. Techniques for exploration of the issues involved are still under development. It seems likely that resolution of the issues will require an unprecedented collaboration between psychologists, neuroscientists and cognitive scientists, an enterprise which will not only illuminate the origins of dyslexia, but also the foundations of normal cognitive functioning.

CONCLUSIONS

To summarise our research findings, we established that a phonological deficit alone was not capable of accounting for the range of deficits found. There was strong evidence of an automatisation deficit on motor balance, and the automatisation deficit hypothesis appeared capable of explaining not only the balance deficits, but also the phonological deficit and the established problems of working memory. However, the automatisation deficits are best seen also as a symptom of some deeper underlying cause. The lack of a deficit for simple reactions, taken together with the appearance of a deficit in even the simplest choice reactions, suggest that the most likely cause is some problem within the central brain processes, most probably in the hypothetical inner loop for information transmission between different brain modules. Our best guess at the complex causes underlying the deficits suffered by dyslexia people is shown in Fig. 5. Note the complex interaction between automatisation, working memory and phonological skill in determining the likely deficits on any given task.

It is necessary to include a note of caution in generalising from the results from two small groups of dyslexic children, but, if our findings are replicated on a wide range of dyslexic children, we believe that this analysis sets new and exciting agenda for dyslexia research for the next decade. The research agenda are clear: inter-disciplinary collaboration towards a clearly identified target that promises to disperse some of the mists which presently shroud the relationship between mind, brain and behaviour. The identification of the problem of automatisation promises the development of remediation techniques which will complement those already developed. Discovery of the increased range of deficits facilitates the development of sensitive screening tests that should allow the diagnosis of dyslexia well before the child reaches school age. Early diagnosis allows the use of effective remedial support. Maybe within a few years, by early screening, effective prophylactic remedial support, and the better understanding of optimal methods for supporting dyslexic learning, dyslexic children will be able to enjoy their undoubted strengths without suffering from their concomitant weaknesses.

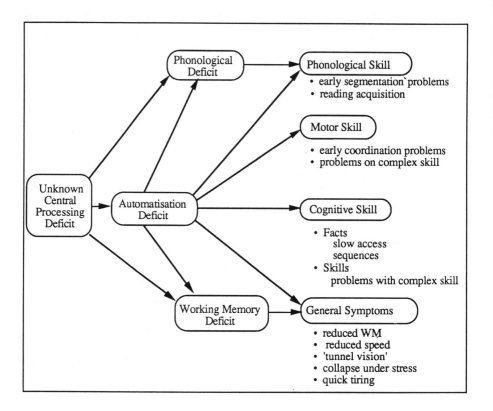

FIG. 5. A Causal Analysis of the Dyslexic Deficits

ACKNOWLEDGEMENTS

The research reported in this chapter is supported by a grant from the Leverhulme Trust to the first author. We acknowledge gratefully the support of the participants in this study, and the comments of Prof. Tim Miles and Dr Paul Dean on an early draft.

Note 1: Fawcett, A.J. & Nicolson, R.I. (submitted). Automatisation Deficits in Balance for dyslexic children. Currently available as Report LRG 12/90, Department of Psychology, University of Sheffield.
Note 2: Nicolson, R.I. & Fawcett, A.J. (submitted). Reaction Times and Dyslexia. Currently available as Report LRG 4/91, Department of Psychology, University of Sheffield .

Note 3: Nicolson, R.I, Fawcett, A.J & Baddeley, A.D (submitted). Working Memory and Dyslexia. Currently available as Report LRG 3/91, Department of Psychology, University of Sheffield.

REFERENCES

Anderson, J.R. (1982). Acquisition of cognitive skill. *Psychological Review,* 89, 369-406.

Augur, J. (1985). Guidelines for teachers, parents and learners. In M. Snowling (Ed.). *Children's written language difficulties.* NFER Nelson, Windsor.

Bradley, L. & Bryant, P.E. (1983). Categorising sounds and learning to read: A causal connection. *Nature*, 301, 419-421.

Bryant, P. & Goswami, U. (1986). Strengths and weaknesses of the reading level design. *Psychological Bulletin* , 100, 101-103.

Carpenter, P.A., Just, M.A. & McDonald, J.L. (1984). *Reading and processing differences between dyslexic and normal college students.* (Technical report). Pittsburgh: Carnegie Mellon University, Department of Psychology.

Donders, F.C. (1869). Over de snelheid van psychische processen. Onderzoekingen gedaan in het Psysiologish Laboratorium der Utrechtsche Hoogeschool: 1868-69. Tweede Reeks, II, 92-120. Translated by G. Koster in W.G. Koster (Ed.). Attention and Performance II. *Acta Psychologica,* 1969, 30, 412-431.

Ellis, N.C. & Miles, T.R. (1978). Visual Information processing in dyslexic children, in M.M. Gruneberg, P.E. Morris and R.N. Sykes (Eds.). *Practical aspects of memory*, Academic Press: London.

Fawcett, A.J. (1990). *A cognitive architecture of dyslexia.* Unpublished Ph.D. thesis, University of Sheffield.

Fitts, P.M. & Posner, M.I. (1967). *Human Performance.* Brooks Cole, Belmont CA.

Galaburda, A.M. (1989). Learning disability: biological, societal or both? *Journal of Learning Disabilities,* 22, 278-282.

Galaburda, A.M., Rosen, G.D., & Sherman, G.F. (1989). The neural origin of developmental dyslexia: Implications for medicine, neurology and cognition. In A.M. Galaburda (Ed.) *From Reading to Neurons.* Cambridge, MA: MIT Press.

Gathercole, S.E. & Baddeley, A.D. (1990). Phonological memory deficits in language disordered children: Is there a causal connection? *Journal of Memory and Language*, 29, 336-360.

Haslum, M.N. (1989). Predictors of dyslexia? *Irish Journal of Psychology,* 10, 622-630.

Jorm, A.F. (1983). Specific reading retardation and working memory: a review. *British Journal of Psychology,* 74, 311-342.

Laberge, D. & Samuels, S.J. (1974). Toward a theory of automatic information processing in reading. *Cognitive Psychology,* 6, 293-323.

Liberman, I.Y. (1973). Segmentation of the spoken word and reading acquisition. *Bulletin of the Orton Society,* 23, 65-77.

Lundberg, I. & Høien, T. (1989). Phonemic deficits: A core symptom of developmental dyslexia? *Irish Journal of Psychology,* 10, 579-592.

Miles, T.R. (1983). *Dyslexia: the pattern of difficulties.* London: Granada.

Nicolson, R.I. & Fawcett, A.J. (1990). Automaticity: a new framework for dyslexia research? *Cognition,* 30, 159-182.

Olson, R.K., Wise, B.W. & Rack, J.P. (1989). Dyslexia: Deficits, genetic aetiology and computer based remediation. *Irish Journal of Psychology*, 10, 594-508.

Popper, K.R. (1968). *The logic of scientific discovery.* Hutchinson: London.

Schneider, W. & Detweiler, M. (1988). The role of practice in dual task performance: toward workload modeling in a connectionist /control architecture. *Human Factors,* 30, 539-566.

Shiffrin, R.M., Dumais, S.T. & Schneider, W. (1981). Characteristics of automatism. In J. Long and A. Baddeley (Eds.) *Attention and Performance IX.* Erlbaum, London.

Shiffrin, R.M. & Schneider, W. (1977). Controlled and automatic human information processing: II. Perceptual learning, automatic attending and general theory. *Psychological Review,* 84, 127-190.

Stanovich, K.E. (1988). The right and wrong places to look for the cognitive locus of reading disability. *Annals of Dyslexia,* 38, 154-177.

Wechsler, D. (1976). Wechsler Intelligence Scale for Children-Revised (WISC-R). Windsor: NFER.

Facets of Dyslexia and its Remediation
S.F. Wright and R. Groner (Editors)
393

THE DEVELOPMENT OF THE AUTONOMOUS LEXICON OF READING
DISABLED STUDENTS.

A. van der Leij,

Vrije Universeit, Amsterdam

INTRODUCTION

Learning to read can be described as the strenghtening of connections between the relevant aspects of words. These aspects involve information about orthography, phonology, meaning, syntax, articulation and grapho-motor production of words. It is clear that when students start to learn to read they possess knowledge about meaning and articulation of words that are used frequently in their every-day life. They also have knowledge about the syntactical use of words, when they understand or express them within the context of sentences, albeit, at that stage, quite implicit knowledge (e.g. they cannot label words as nouns or adjectives but are capable of proper use). However, they lack orthographical, phonological and grapho-motor knowledge, the aspects of words that are the main subject of early reading instruction.

In the course of the process of learning to read, the awareness of links between grapheme-phoneme correspondencies is stimulated. As a result of practice, the student is able to exploit an increasing span of visual analysis (i.e. grapheme/phonemes, grapheme/ phoneme clusters, words). Words that have been read repeatedly, will be processed accurately and with increasing speed. New or unknown words will be processed at a lower level of visual analysis, which is indicated by a slower speed and possibly, a lower degree of accuracy. Of course, when repeatedly practised, these words are known better, so the response rate goes down and the accuracy increases.

At any stage of the learning process, 'well-connected' and 'less-connected' words can be differentiated by detection of gradual differences in speed and accuracy. According to Perfetti (in press), one may label the difference in information processing as 'autonomous' versus 'functional' processing. Within the autonomous lexicon, all relevant aspects of words are 'bonded', which facilitates accurate and fast processing, indicated by repeated correct responses over different testing situations, and by reduced effort. Interfering stimuli or change of stimulus or response conditions have no effect, indicating that word recognition may be called 'automatic'. Once the task is operationalised, it does not matter whether the words are presented auditorily or visually, nor whether the response involves oral reading, matching or spelling. When a word is processed autonomously, the result will be an accurate and fast response, time and again. If this is not yet the case, the word is still part of the functional lexicon.

With regard to reading disabled students, it is evident that they have to put much effort into the process of word recognition and that their reading development is very slow. The general research question that is raised in this chapter is whether they eventually are capable of building up an autonomous lexicon, comparable to normal readers of the same general reading level. The relevance of this question is that it relates to the developmental lag versus specific deficit discussion. Some authors (e.g. Treiman & Hirsh-Pasek, 1985) argue that reading disability is best explained by the developmental lag hypothesis, while other claim that it is a specific deficit (e.g. Olson, Kliegl, Davidson & Foltz, 1985). Operationalised within the concept of the autonomous lexicon, the idea of a developmental lag would mean that, when stimulus and response conditions are varied, no significant differences between reading disabled students and younger reading-age controls show up. In contrast, when differences are apparent, this result may be considered as evidence in favour of a specific deficit.

Because learning to read is a developmental process, we decided to answer the question in a longitudinal design to be able to detect whether cross-sectional differences at one point in time may be affected in the course of time. A group of 20 reading disabled students was matched to 20 reading-age controls twice, i.e. at both moments of testing the general reading level was comparable. It is important to note that the relative slowness of reading development of reading disabled students is indicated by the fact that the period between assessments had to be larger for the reading disabled group than for the reading-age controls. We will return to this issue in a later section.

METHODS

Design
The research questions related to two issues: the description of cross-sectional and longitudinal differences between reading disabled students and reading-age controls in the development of the autonomous lexicon, and the explanation of differences in terms of a developmental lag or a specific deficit.

Subjects
From a larger sample, 20 reading disabled students were matched to 20 reading age controls on general reading level and general IQ. The reading disabled students were selected from schools for Special Education (schools for Primary Learning Disabled children; for a description of the Dutch school system see Van der Leij, 1987). They had no gross neurological or physical handicaps. The reading-age controls were selected from

regular schools. As noted before, general reading level was equivalent at both testings. As a consequence of the slow development of the reading disabled students, the time interval was 18.8 months, in contrast to 9.2 months for the reading-age controls. In Table 1, relevant statistics are presented. To interpret the statistics at an international level, it is important to note that in the Dutch school system, formal instruction of reading starts at the beginning of grade 1, when the students are 6 years of age.

TABLE 1

Number, Chronological and Reading Ages of the Subjects

	n	Age (Months)		Reading Age (grade)	
First Testing		Mean	SD	Mean	SD
Reading Disabled (RD)	20	121.6	12.0	2.1[1]	0.4
Reading-age Controls (N)	20	88.2	3.1	2.1	0.4
Second Testing					
Reading Disabled (RD)	20	140.4[2]	12.0	2.8[3]	0.6
Reading-age Controls (N)	20	97.4[2]	3.1	2.9[3]	0.6

1 Comparable to the reading level of the average reader after one month in grade 2.

2 The interval between assessments was 18.8 months for the RD-group and 9.2 for the N-group.

3 Comparable to the reading level of the average reader in April (RD) or May (N) in grade 2.

Tasks

Two series of tasks were used: tasks to indicate general reading level and experimental tasks for specific reading processes.

Tests for General Reading Level (matching tasks)

EMT (one-minute test of reading, Brus & Voeten, 1973). In this oral reading test a list of words of increasing orthographic difficulty is presented. All words are real words. The score is the amount of correctly read words in one minute, so the score indicates the accuracy/speed trade-off at word level.

AVI (Van den Berg & Te Lintelo, 1977). In this oral reading test, pages of text of increasing complexity are presented. The score is the level of text that is read within accuracy and speed criteria, so the score indicates the accuracy/speed trade-off at text level.

Experimental Tasks

To operationalise the concept of autonomous processing in tasks, a computer-assisted device was developed. The test, called COTAL (computer-assisted test for automaticity of reading, Van der Leij & Smeets, in press), aims for precise diagnosis of reading disabilities. This device consists of a series of tasks with varied stimulus and response conditions. Furthermore, six word categories of one- and two-syllable high frequency words are used. In this chapter, the following tasks have been selected from the battery.

Auditory-visual matching (A-V). A word was presented by headphones. After a short interval, a word was presented on the screen. The students had to decide whether the words were the same or different by using the 'mouse.' The time interval between the stimuli extended from 1 to 3 to 5 seconds to enlarge the load on repetition processes in working memory.

Oral reading (R). A word was presented on the screen. The student had to read the word aloud. The experimenter used the key-board to note correct or incorrect responses.

Oral reading of flashed words (FL). This task was comparable to R, but the stimulus was flashed and masked afterwards by a nonsense design of lines and circles (to frustrate the use of the after-image on the retina). Exposure time decreased from 200 to 160 to 100 msec. to enlarge the pressure on fast visual analysis.

In the three experimental tasks, six categories of high-frequent words of increasing phonological and orthographical complexity were used: CVC ('jas' = coat), CVCC ('kalf' = calf), CCVC ('vloer' = floor), two-syllables ('poeder' = powder), 'closed' two syllables ('appel' = apple) and 'open' two syllables ('vogel' = bird). The general idea was that if students processed high frequency words autonomously at the level of a certain word category (starting with CVC), they would master all subtasks that contained that word category, independent of stimulus and response conditions. As a consequence, processing of that word category (and the easier word categories below that) was labelled as 'autonomous processing'.

Procedure

Presentation of the tasks was in blockwise order from easy to difficult (6 word categories; 3 time intervals in A-V; 3 exposure times in Fl.). Thus, A-V and Fl. consisted of 18 subtasks each, and R contained 6 subtasks. Per subtask, the number of items was 10. The criterion for mastery was set at 90 % correct answers per subtask.

Scoring

Per word category, 5 was the maximum score (A-V and Fl.: 0 when the subtasks were not mastered at all; 1 at the smallest time interval or largest exposure time; 3 at the intermediate level; 5 at the highest level of difficulty. R: 0 or 5). Summed over word categories per subtask, the scores ranged from 0 to 30.

Assessment

First testing date was at the end of grade 1 (June) for the N group, and November-December for the RD group. Second testing date was after 9 months in April (grade 2) for the N group, and after 18 months in May-June for the RD group.

RESULTS

Auditory-visual matching.

Within the A-V task, differences between RD and N were not significant at the first or second testing. However, RD performed better at the first and N scored better at the second testing. Progress of both groups was significant (p<.001) but larger in N, indicated by a repeated measures effect (repeated F * Group = 5.26, p<.003) (Fig. 1).

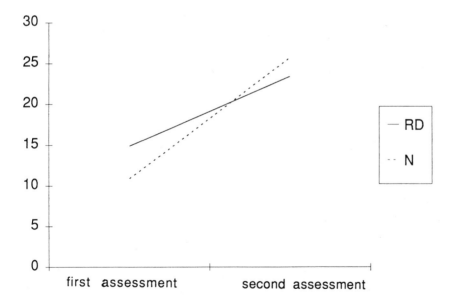

FIG. 1. Auditory-visual matching (A-V)

When the levels of difficulty were taken into account, the difference between RD and N was significant only at the shortest 1" interval, in favour of the RD students (F = 9.03, p<.01) (first assessment). The second time no differences showed up at any interval.

Oral Reading

RD performed significantly worse than N on the oral reading task at both assessments (first: F= 5.68, p<.03; second: F= 5.63, p<.03). However, the progression was comparable (p < .001) (Fig. 2). Further analysis showed that the groups differed in scores on the word categories (see for a detailed description the sections on phonological and orthographical complexity).

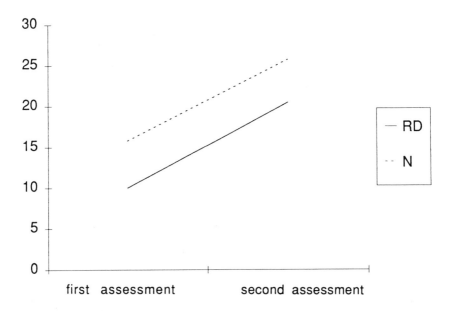

FIG. 2. Oral reading task (R)

Oral Reading of Flashed Words.

RD and N did not differ in Fl. scores at the first testing, but N performed better than RD the second time (F = 11.59, p<.002). Progression of both groups was significant (p< .001), but smaller for RD (Fig. 3). When reduction of exposure time was taken into account, it was clear that this factor troubled both RD and N. Interestingly, a marginally significant difference in favour of the N group showed up in the 100 msec. condition (F= 3.17, p=.08; first assessment). The second time, differences showed up at all levels of difficulty (200, 160 and 100 msec.), in favour of the normal group (resp. $F = 8.81$; $F = 10.27$; $F = 11.49$; all p's<.01).

The Effects of Phonological and Orthographical Complexity.
Because the experimental tasks included subtasks of increasing difficulty, we decided to explore the results in greater detail. First, we were interested in the effects of phonological and orthographical complexity, indicated by the six word categories. The task R was selected for further analysis because this task is a simple oral reading task, without demands on working memory or fast processing.

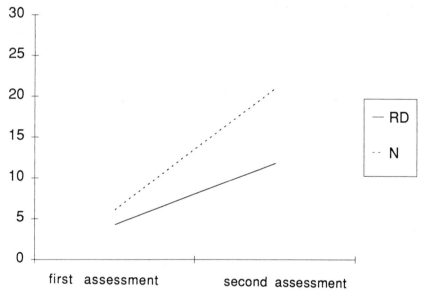

FIG. 3. Oral reading of flashed words (FL).

In Dutch orthography, most of the one-syllable words do not represent particular orthographical difficulties (the correspondence between orthography and phonology is quite regular). However, double (or even triple) consonants at the beginning or end of the word contribute to phonological complexity.

As is indicated by the results of Table 2, the effect of increasing phonological complexity of one-syllable words was more dramatic in the RD group than in the N group at both assessments (word categories a, b and c). At the first assessment, both groups mastered the words with the least phonological complexity (CVC). However, only 50 % of the RD students were competent in the accurate processing of double consonants at the end of the word (CVCC), and a mere 25 % could handle CCVC-words. In comparison, the N group performed better (resp. 75 and 65 % competent readers of CVCC and

CCVC). At the second assessment, all N students mastered the one-syllable words, independent of phonological complexity, whereas 30 % of the RD students still were not able to cope with CCVC-words.

TABLE 2

Percentages of Competent Readers of the Oral Reading Task (R).

GROUP WORD CATEGORIES							
	a CVC	b CVCC	c CCVC	d 2-syll.	e 'closed'	f 'open'	av. a-f [1]
First Assessment							
N	100	75	65	35	35	5	53
RD	100	50	25	15	10	0	33.3
Second Assessment							
N	100	100	100	90	85	40	85.8
RD	100	95	70	65	50	30	68.3

1 Averaged percentage over all word categories

The effect of orthographical complexity is indicated by the results of the two-syllable words. To explain this, another remark must be made concerning Dutch orthography. The word categories d, e and f represent different levels of orthographic complexity. The d category ('poeder') does not contain any orthographic ambiguity other than the pronunciation of the accentless 'e' at the end, but the categories e and f do ('appel', 'vogel'). In the e and f categories, the pronunciation of the first vowel has to be detected from the number of written consonants that follow. For example, in 'bommen' ('bombs'), the vowel is short (as in the English word 'rotten') because of the double 'm'; in 'bomen' ('trees), the pronunciation of the 'o' is long (as in the English word 'open'), because of the single 'm'. So the orthographical complexity is greater in e and f than in d.

To interpret the findings of Table 2, it is important to note that two-syllable words are more difficult to read than one-syllable words (word-length effect). Because they were less competent at the one-syllable level, the RD students were less able to read the two-syllable words than the N students. To detect an extra effect of orthographical complexity, comparison of the results of the two-syllable words will be taken as indication. The first time, most of the students of both groups did not master the two-syllable words. The second time, it was clear that the category of 'open' syllables caused both groups trouble.

When categories d and f are compared, only 40 % of the 90 % N students who mastered category d, could cope with the open-syllable words (f). Of the RD students, the proportion was 30 % of 65 %. So, the drop in competency, indicated by comparison of d and f, was proportionally about the same. This result indicates that the effect of orthographical complexity was comparable in the two groups.

The Effect of Exposure Time.

Another point of analysis was the effect of increasing difficulty by shortening of the exposure time. In Table 3 the results are shown, averaged over word categories.

TABLE 3

Percentages of Competent Readers in the Flashed Oral Reading Task (200 & 100 msec).

Group	Exposure time	
	200 msec.	100 msec.
First Assessment		
N	25.0	14.2
RD	28.2	4.2
Second Assessment		
N	80.0	60.0
RD	55.0	25.0

The first time, RD and N students were comparable at the 200 msec. level: most of them failed. A slight indication of a differentiating effect was apparent at the 100 msec. level (as mentioned before, the difference was marginally significant). The second time, the difference between RD and N was significant at all levels. At this point, it is relevant to take the averaged results of Table 2 into account, because it clearly demonstrates the effect of shortening of exposure time. The stimulus and response conditions of the oral reading task R (Table 2) and the flashed word reading task Fl. (Table 3) were only different with respect to exposure time, ranging from 'unlimited' (R) to 200 and 100 msec. (Fl.). Averaged over word categories, the proportion of competent normal readers dropped from 85.8 % (unlimited condition) to 80 % (200 msec.) to 60 % (100 msec.). From the RD group, the proportion decreased from 68.3 % to 55 % to 25 %. So the difference between RD and N accumulatively increased. These figures demonstrate not only the worse performance of RD in the 'unlimited' condition, due to the aforementioned effect of phonological complexity, but also an extra problem of RD with shortening of exposure time, i.e. a problem with speed of processing.

The overall conclusion of the frequency analyses is that at both assessments, RD and N showed differences in mastery. Most of the N students progressed quite regularly, i.e.

at an increasingly higher level of phonological/orthographical complexity and speed of processing. When task demands increased by complexity of the presented words, 85 % of N were able to process one- and two-syllable words accurately, except the 'open' two-syllable words (unlimited condition, see Table 2). In addition, 60 % was able to cope with the shortest exposure time across all word categories (se Table 3). In contrast, RD students were more affected by phonological complexity, and by the need for fast processing. At the age of 12 (in comparison, the end of grade 6), 50 % were able to read one- and two-syllable words except the 'open' category when exposure time was not limited, and only 25 % could cope with the 100 msec. condition across all word categories.

Regression Analysis.

Before turning to interpretation of the results, one question needs to be answered by additional analysis. It was not expected that the groups would differ on R, a simple oral reading task that involved stimulus and response conditions comparable to the matching tasks with the exception of the presentation of words in lists or texts (the matching tasks) or isolated (the experimental task). To be able to interpret the differences, a stepwise regression analysis was run. The question was to what extent the reading processes at the single word level would predict general reading level and which differences between the two groups would show up. Table 4 shows the results of the regression analysis.

TABLE 4

Results of the Stepwise Regression Analysis (Explained Variance, R^2).

Group: Predicted variables	EMT		AVI	
Predictor	task	R^2	task	R^2
First Assessment				
N	FL.	68[1]	FL	66[1]
RD	FL.	46[1]	FL.	53[1]
Second Assessment				
N	FL.	51[1]	FL.	45[2]
RD	-	ns	--	ns

1 $p < .001$; 2 $p < .01$

From these results, two conclusions may be drawn. First, the amount of variance explained by the three experimental tasks was far from perfect. Indeed, flashed word reading was the only task that contributed at the level of $p < .05$. To reach a certain general reading level, other processes are involved than processing of isolated words. Second and more importantly, the amount of variance predicted by flashed word reading was increasingly different. In the N group, the amount decreased over the two assessments, but was still about 50 % at the second assessment. However, in the RD group, it dropped below significance at the second assessment (i.e. below about 10 %). This result indicates that the RD group increasingly depended on other processes than triggered by the experimental subtasks to progress in general reading achievement.

DISCUSSION AND CONCLUSIONS

Although the general reading level was comparable at both assessments, reading-age controls cross-sectionally and longitudinally showed significant differences in single word processing in comparison to reading disabled students. Moreover, the general reading level of reading-age controls was for a greater part predicted at both assessments by accurate and fast processing of single words. In contrast, the prediction of the general reading level of reading disabled decreased below the level of significance, indicating that their (slow) progress in general reading level was increasingly dependent on other processes.

To return to our theoretical concept, the reading disabled students from our sample seemed to show very limited ability to develop an autonomous lexicon, indicated by the fact that they were more affected than their reading age controls by increasing task demands. In our experiment, three task demands were important: word categories, exposure time and interval between stimuli (triggered by the R, Fl. and A-V task, resp.). We will discuss the effect of the task demands in turn.

With respect to word categories, the results of the oral reading task R showed in Fig. 2 and Table 2 are indicative. At both assessments, the reading disabled students performed worse than the reading-age controls. Specifically, the phonological complexity of one-syllable words containing double consonants caused a drop in mastery in more RD than in N students. Developmentally the problem experienced by RD with double consonants moved from the end and beginning of the word (CVCC and CCVC) to the beginning of the word (CCVC). However, at the time of the second assessment the N students performed perfectly. The fact that RD students were to a greater proportion affected by phonological complexity, can be considered as support for the hypothesis that 'dyslexia' is caused by a phonological deficit (e.g. Snowling, 1987). There is little sign that two-syllable words which, as we argued before, relate to differences in orthographical complexity, caused reading disabled students extra trouble, apart from the observation that they performed worse in all word categories than reading-age controls.

With respect to the effect of exposure time (Table 3), only 25 % of the reading disabled students at the age of 12 could process the words fast enough in the 100 msec. flashed oral reading condition. The percentages of the normal readers at the age of 8 was 60. This finding is in agreement with conclusions of other authors that dyslexic students are troubled by short presentations (e.g. Bouma & Legein, 1980). However, our experiment adds a developmental aspect to the facts that are already known. Indeed, the results summed over word categories (Fig. 3 and Table 3) show that the difference between RD and N in flashed word reading increased with time. At the time of the first assessment, only a slight difference could be detected, but at the moment of the second assessment, it was very clear. In other words, in normal reading development, the student is able to process shortly presented words with increasing speed, whereas in abnormal reading development, the progress is relatively small. As we argued before, the results indicate that the effects of phonological complexity and short exposure time are accumulative and therefore, to some extent, independent.

The effect of an increasing time interval between stimuli is shown in Fig. 1. In the beginning, the RD students were slightly better than the N students, due to a better performance with the shortest time interval (1"). This could be the result of an age factor (maturity) because of the age difference between RD and N (about 33 months). However, the reading disabled students were not performing at the level of students of the same chronological age. Data, not presented in this chapter, showed that at the age of about 10, students normally had maximum scores on the A-V task, independent of word categories and time interval. If a maturity factor was operative, it only caused a very small advantage for the RD students. Moreover, the advantage disappeared when the students were retested, indicating that the use of repetition in working memory in a relatively simple matching task did not progress as much in RD as in N. Although our experimental task was not designed as a typical working memory task, this result gives some support to the idea that reading disabled students lag behind in development of working memory abilities (see for example, Baddeley, 1987).

With respect to the 'deficit' - 'lag' discussion, our results make it hard to accept that reading disability can be explained in terms of a developmental lag. For one thing, the results on the matching tasks (EMT and AVI) indicate that it is at least an increasing lag, because the interval between assessments had to be twice as large for RD than for N to get to an equivalent general level (18.8 versus 9.2 months). In other words, the RD group only progressed at half the speed of normal students. However, the results of the experimental tasks were not in favour of an increasing developmental lag. Theoretically, both groups would have to perform equally well (or badly) on the experimental tasks at both assessments. Instead, the differences between RD and N were significant at the first assessment (oral reading) and, more dramatically, at the second assessment (oral reading and flashed word reading). Besides, progression between assessments differed (auditory-visual matching and flashed word reading). These differences are better explained in terms of a deficit or, possibly, multiple deficits, than in terms of a developmental lag. The only result that could be supportive to the hypothesis of a developmental lag, was the equivalent progression in the oral reading task (Fig. 2). However, at both assessments, the RD group performed significantly less well than the N group. These findings do not support the idea of a developmental lag, because that would predict no differences at all between RD and N on this task.

At this point, the question may be forwarded why the differences in the oral reading task R appeared. How can the results on this task, when the groups were matched on two general reading tasks that resembled the experimental task in stimulus and response conditions (unlimited exposure time and oral reading, resp.), be explained? It is important to note that we were confronted with exactly the same phenomenon in other experiments as well (Yap & Van der Leij, submitted), which demonstrates that it is a reliable finding. In our opinion, the only way to explain why RD students did relatively better on the general reading tasks than on the experimental task, is to stress the fact that the former included a continuous trial condition whereas the latter did not. EMT is a list of words, to be read in one minute, AVI is a series of texts, to be read within accuracy and speed limits. In contrast, in the oral reading task R isolated words were presented (discrete trial condition). Since the performance on the continuous trial oral reading tasks is a combination of speed and accuracy, reading disabled students can compensate low accuracy with high speed and vice versa. In the discrete-trial procedure, less opportunity for compensation strategies is allowed. Indeed, some authors suggest that discrete-trial procedures elicit the lower, automatic processes of reading more than the continuous-trial procedures (e.g. Bowers & Swanson, 1991; Stanovich, 1986; Wolf, 1986). If this is true, the discrepancy between EMT and AVI on the one hand and the experimental task R on the other, supports the hypothesis that dyslexics have problems particularly with the automatic processes in reading, a viewpoint that was expressed earlier by us (Van der Leij, 1990) and will be discussed in the next section.

When expert performance is described as smooth, automatic, (relatively) mentally effortless, and (relatively) unaffected by stress (Norman, 1982), our reading disabled students at the age of 12 were worse experts in single word processing than reading-age controls readers of about 8. Moreover, our longitudinal findings indicate that the performance of dyslexics was increasingly more affected by complicating task demands on fast visual analysis and repetition in working memory. In addition, phonological decoding was more influenced by phonological complexity at both assessment times. As was mentioned before, the relative weakness of dyslexics in the aforementioned processes is quite well documented, apart from the longitudinal evidence for a growing deficit which is revealed by our results. However, the question may be raised whether there are three independent deficits involved, or maybe two or even only one. In recent publications, a controversy between 'phonological deficit' and 'visual deficit' hypotheses is evident (see for example, Whyte, 1989). As is argued elsewhere (Yap & Van der Leij, submitted), we support the idea that 'dyslexia' is not only caused by a phonological deficit but also by a deficit in speed of processing (mainly indicated by visual tasks). Possibly, a general automatisation deficit is involved, independent of mode of stimulus or response, as is suggested by Nicolson & Fawcett (this volume). Although our experiment does not allow for any conclusions at this point, our results seem to be in favour of this hypothesis. Indeed, the effects of the three complicating task demands, relating to different stimulus and response conditions, were not very specific, but, on the contrary, quite general.

Interestingly, we also found that deficits in single word processing may be masked when task demands allow for compensating strategies, shown by the increasing difference in performance on the matching tasks (EMT and AVI) and the experimental tasks. The reading disabled students were able to progress in general reading level, but the progress depended increasingly less on single word processing. We regard both the possibilities of an automatisation deficit and of strategic compensation, as important perspectives for further research and practice.

ACKNOWLEDGEMENTS

The research project was supported by a grant from NWO (Nederlandse Organisatie voor Wetenschappelijk Onderzoek) and carried out by Dr. Harry Smeets. Requests for further information should be addressed to Dr. Aryan van der Leij, Department of Special Education, Vrije Universtiteit, Van der Boechorststraat 1, 1081 BT Amsterdam, The Netherlands.

REFERENCES

Baddeley, A. (1987). *Working memory*. Oxford: Clarendon Press.

Bouma, H., & Legein, C.P. (1980). Dyslexia: a specific recoding deficit? An analysis of response latencies for letter and words in dyslectics and in average readers. *Neuropsychologia, 18,* 285-298.

Bowers, P.G., & Swanson, L.B. (1991). Naming speed deficits in reading disability: multiple measures of single processes. *Journal of Experimental Psychology, 51,* 195-219.

Brus, B.Th., & Voeten, M.J.M. (1973). *EMT. Een-minuuttest, vorm A en B.* Nijmegen: Berkhout.

Nicolson, R., & Fawcett, A.J. (1990). Automaticity: A new framework for dyslexia research? *Cognition,* 30, 159-182.

Norman, D.A. (1982). *Learning and Memory.* San Francisco: Freeman.

Olson, R.K., Kliegl, R., Davidson, B.J. & Foltz, G. (1985). Individual and developmental differences in reading disability. In T.G. Waller (Ed.), *Reading Research: Advances in theory and practice.* Vol. 4. New York: Academic Press, 1-64.

Perfetti, C. A. (in press). The presentation problem in reading acquisition. To appear in P. Gough, L. Ehri, & R. Treiman (Eds.), *Reading acquisition.* Hillsdale, N.J.: Lawrence Erlbaum.

Snowling, M. (1987). *Dyslexia. A cognitive developmental perspective.* Oxford: Basil Blackwell.

Stanovich, K.E. (1986). Matthew effects in reading: Some consequences of individual differences in the acquisition of literacy. *Reading Research Quarterly,* 21, 267-283.

Treiman, R., & K. Hirsh-Pasek, K. (1985). Are there qualitative differences in reading behaviour between dyslexics and normal readers? *Memory & Cognition,* 13, 357-364.

Van den Berg, R.M., & Te Lintelo, H.G. (1977). *AVI-pakket.* Den Bosch: KPC.

Van der Leij, A. (1987). Netherlands, Special Education in the. In C.R. Reynolds & L. Mann (Eds.), *Encyclopedia of Special Education.* Volume 2. New York: Wiley & Sons.

Van der Leij, A. (1990). Can dyslexic children become automatic readers? In D. Hales, M. Hales, T. Miles, & A. Summerfield (Eds.), *Proceedings of the First International Conference of the British Dyslexia Association.* Reading: British Dyslexia Association, 103-107.

Van der Leij, A., & Smeets, H. (in press). The assessment of the autonomous lexicon: differences between severely reading-disabled students and reading-age controls. In L. Verhoeven & J. de Jong (Eds.), *The construct of language proficiency.* Amsterdam: John Benjamins (to be published in 1992).

Whyte, J. (1989). Progress in dyslexia research: An overview of Dyslexia: Current Research Issues. *The Irish Journal of Psychology,* 10, 465-478.

Wolf, M. (1986). Rapid alternative naming in the developmental dyslexias. *Brain and Language,* 27, 360-379.

Yap, R., & Van der Leij, A. (submitted). Word processing in dyslexics: a phonological and a speed deficit? (Submitted to the Journal of Reading and Writing).

V. EMOTIONAL CORRELATES OF DYSLEXIA

Facets of Dyslexia and its Remediation
S.F. Wright and R. Groner (Editors)

PERSONALITY CHARACTERISTICS OF ADULT DYSLEXICS

A.J. Richardson* and J.F.Stein
*Department of Experimental Psychology and
University Laboratory of Physiology, Oxford

INTRODUCTION

Although developmental dyslexia is defined in terms of an unexpected difficulty with written language skills, it is often accompanied by many other unusual characteristics involving motor function, sensory perception, and attention. It is now widely accepted that there is a neurological basis for this syndrome; and if so, it would be extraordinary if this was not reflected in some way in the personality profile of dyslexics. The purpose of this study was to examine the way in which adult dyslexics themselves describe their experiences and characteristics, and to see whether these could be quantified using personality measures.

Dyslexics often report organisational problems which extend beyond the practical difficulties they may face in keeping written records; some claim to lack an internal concept of time; they may be prone to drift or daydream, or find themselves absorbed by whatever is happening in the present moment. Many describe heightened sensory acuity, hearing slight sounds or observing visual details which others may fail to notice. In company they often find it difficult to attend to one conversation at a time, and many describe more general problems in attending selectively to one aspect of their environment. Other reported perceptions are more unusual; a doctor of physics who still finds reading and writing difficult says that while listening to music she has to close her eyes to fully appreciate it, but can then `see` the sounds in her head. Another dyslexic, very musically gifted, reported that he perceived the sound of each instrument in his orchestra `in a different colour`. Such comments may simply reflect a vivid descriptive imagination, but they do suggest that individuals may differ in the way they interpret sensory information.

Auditory-linguistic problems in dyslexia have been well documented (Bradley & Bryant, 1983; Snowling, 1981). There may be subtle difficulties in the perception and segmentation of speech sounds; and similar problems with spoken language are often evident. An excessive tendency to `Spoonerisms` is one of the most characteristic complaints (one student wanted a ticket for the `Cross Flannel Cherry`), together with word-finding difficulties. However, dyslexic difficulties are not confined to the language domain, and frequently include more general problems with judgements of temporal order, motor co-ordination and sequencing (Kinsbourne, Rufo, Palmer & Berliner, 1991). Many report visual disturbances, such as letters and words appearing to move on the page; and one student referred to us had found while watching films that for him any moving objects appeared blurred, though his friends reported that they could see these clearly.

Analogous perhaps to their perceptual style, dyslexics often show a preference for `lateral` thinking, finding it difficult to structure their ideas in linear form. Many describe their cognitive style as divergent and holistic rather than convergent and sequential, and report that they tend to think in images rather than words. This may confer advantages; such individuals often show originality in problem solving or an unusual creativity. Albert Einstein (who did not talk until he was four, nor learn to read until he was nine) had the following to say on his own mental processes:

" The words of the language, as they are written or spoken, do not seem to play any role in my mechanism of thought. The psychical entities which seem to serve as elements in thought are certain signs and more or less clear images which can be voluntarily reproduced and combined.... .this combinatory play seems to be the essential feature in productive thought - before there is any connection with logical construction in words or other kinds of signs which can be communicated to others. The above mentioned elements are, in my case, of visual and some of muscular type. Conventional words or other signs have to be sought for laboriously only in a secondary stage, when the mentioned associative play is sufficiently established and can be reproduced at will."

(Einstein, 1945)

Superior visuo-spatial, mechanical or mathematical skills are perhaps the most widely cited advantages associated with the dyslexic profile (Geschwind & Galaburda, 1985), or exceptional artistic talents, as with Auguste Rodin, whose father had complained "the boy is uneducable!" Isaac Newton, the inventor Thomas Edison, statesman Woodrow Wilson and writer Hans Christian Andersen feature amongst many notable individuals who also suffered specific learning difficulties (Thompson, 1969).

Outstanding creative achievement clearly requires certain traits in addition to the capacity for original thought, however. Motivation, initiative, persistence and a firm belief in one`s own capacities might all be considered crucial. By contrast, research into personality factors in dyslexia often highlights problems of emotional adjustment and self-esteem (Bell, Lewis & Anderson, 1972). These are usually seen as understandable consequences of the experience of scholastic failure from an early age; but it is suggested here that there may be some aspects of personality which are fundamentally related to the constitutional factors that give rise to specific reading difficulties. If so, the first question to address is how best to measure those personality characteristics which might relate to the unusual perceptual and cognitive styles associated with the dyslexic profile.

PERSONALITY SCALES

There are few instruments for the assessment of personality which are backed by any attempt to relate the traits they measure to underlying biological factors. The scales developed by Eysenck (Eysenck & Eysenck, 1975) and those devised by Claridge (Claridge & Broks, 1984) are an exception, having been extensively researched in this respect. Furthermore, an inspection of the item content of one of Claridge's scales - the STA - revealed many items bearing a resemblance to the kinds of experiences that dyslexics themselves report. Examples include:

Do you ever become oversensitive to light or noise?

Do you ever suddenly feel distracted by distant sounds that you are not normally aware of?

Have you ever thought you heard people talking only to discover that it was in fact some nondescript noise?

Do you ever feel that your speech is difficult to understand because the words are all mixed up and don't make sense?

Does it often happen that almost every thought immediately and automatically suggests an enormous number of ideas?

With regard to a possible biological basis for personality traits, it has been found that high scores on the STA are associated with unusual hemispheric specialisation (Broks, 1984; Rawlings & Claridge, 1984; Broks, Claridge, Matheson & Hargreaves, 1984), which is often suggested to be a feature associated with dyslexia. It was felt that a well-studied inventory which included this scale might be an appropriate instrument with which to investigate the personality characteristics of adult dyslexics; and therefore the Eysenck and Claridge scales were the ones selected for this study. A brief description of the traits measured by each scale is given in Table 1.

On the view that individual differences in personality are based on differences in nervous system functioning, Eysenck sought to measure these using three independent variables - E, N and P. E measures "Extraversion-Introversion", and scores on this dimension are attributed in Eysenck's theory to variation between individuals in general cortical arousability (as mediated by the reticular activating system). N, "Neuroticism-Stability", is said to reflect emotional arousability (as mediated by limbic system activity). P, "Psychoticism" purports to measure a general predisposition to mental disorder in the normal population, though it may be argued that it primarily taps anti-social traits. As their names suggest, Eysenck's scales were formulated on the view that traits which in the extreme may predispose to various clinical disorders are in fact continuously distributed in the normal population.

TABLE 1

Personality Scales

LIE SCALE L	primarily used as an indicator of dissimulation; high scores are regarded as reducing the validity of scores on most other scales. However, it may be interpreted as measuring personality traits involving conformity to social ideals.
EXTRAVERSION E	high scorers are generally outgoing and optimistic, showing traits of sociability, activity, impulsivity and sensation-seeking. Low scorers are quiet, shy, withdrawn, and generally seek to avoid high levels of external stimulation.
NEUROTICISM N	measures emotionality - the sensitivity or lability of feelings. High scorers are emotionally reactive, prone to anxiety and low self-esteem. Low scorers are emotionally more stable, self-confident and less sensitive or prone to worry.
PSYCHOTICISM P	measures traits of social non-conformity or individuality, impulsivity or risk-tasking and 'tough-mindedness'. High scorers are generally loners, often insensitive to the feelings of others and tend to pursue their own interests.
SCHIZOTYPAL PERSONALITY STA	high scorers report unusual perceptual experiences, an openness to superstition or other magical beliefs, and a tendency towards 'cognitive disorganisation' , i.e. divergent thinking, unusual attention-al strategies etc. Other items concern feelings of isolation or mistrust of others and a sensitivity to criticism.
BORDERLINE PERSONALITY STB	taps emotional instability and conflict. High scorers are prone to mood swings, ambivalent feelings about self and others, and intense stormy relationships. They may also report a general sense of futility, and addictive or self-destructive behaviours. Low scores indicate a more secure sense of self-identity, a more even temperament and a less dramatic view of life

Claridge introduced two new scales, the STA and STB, which are generally used in conjunction with the Eysenck inventory. They correlate with each other, with Eysenck's

N, and to a lesser extent with P; but each can be said to be measuring a distinct aspect of personality. The STA, "Schizotypal Personality", largely taps unusual aspects of perceptual and cognitive function; while the STB "Borderline Personality" scale measures unstable emotional attitudes and behaviour. Like the Eysenck scales, the STA and STB were based on the view that there are normal personality traits which in very extreme form may resemble clinical symptoms. Items for these scales were derived from the DSM III criteria for Schizotypal Personality Disorder and Borderline Personality Disorder respectively (APA 1980), syndromes which are themselves regarded as non-psychotic forms of schizophrenic and manic-depressive disorder. To construct the STA and STB these criteria were cast in milder form; and when used in the general population both scales yield a normal distribution of scores, with items endorsed on average by one in three adults (Claridge & Hewitt, 1987).

It was predicted that dyslexics as a group would score more highly on the STA scale, the content of which reflects many of the unusual perceptual and cognitive experiences that they report. Since dyslexic difficulties in themselves might be likely to lead to some increase in anxiety, frustration or lowered self-esteem, it was expected that they might also show elevated scores on N (Neuroticism) and STB (Borderline Personality). However, it should be noted that the adults in this study were as a whole relatively successful in their educational and occupational achievements. No differences between groups were predicted for the other personality scales.

SUBJECTS AND METHODS

52 adult dyslexics (27 male, 25 female) took part, with a mean age of 23.7 years (range 17-51). All had been previously diagnosed by educational psychologists on the basis of a significant discrepancy between general ability and written language skills; however, the majority had pursued educational courses up to or beyond the age of 18, and all could read sufficiently well to take part in a questionnaire study. Many were recruited from schools and colleges around Oxford and from local support groups; others were found via psychologists to whom they had been referred for assessment, and a few were known to the authors personally. Adult controls were chosen on the basis of having no history of a need for remediation in reading or spelling, and were matched with the dyslexics for age, sex, educational level and/or occupation. In some cases more than one control was found, yielding 77 control subjects (43 male, 34 female), with a mean age of 24.1 years (range 17-55).

All subjects completed the scales from the Eysenck Personality Questionnaire - L, E, N and P, together with the two Claridge scales - STA and STB. Subjects were required to indicate Yes/No responses to a total of 145 items (L 21, E 21, N 23, P 25, STA 37,

STB 18), and were allowed to complete the questionnaire in their own time. For each scale, T-test comparisons of mean scores for the dyslexic and control groups were computed, for all subjects and also for each sex.

RESULTS

Figure 1 shows the mean scores and standard errors for dyslexic and control groups on each of the six scales, and detailed results are given in Table 2.

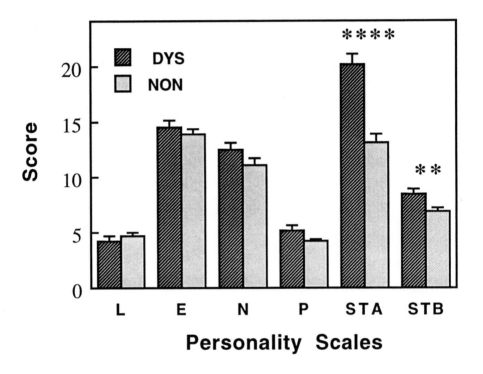

FIG. 1. Dyslexics v controls - all subjects

As predicted, the mean score for the dyslexic group on the STA `schizotypal` scale was very much higher than that of controls (p < 0.0001). This was found for both sexes (males p < 0.0001, females p < 0.003). A much smaller difference was found for the STB `borderline` scale (p < 0.03), though this was entirely attributable to males (p < 0.008 for males only). There was a trend towards higher P scale scores for the dyslexics, again due to males only, but this failed to reach significance (p < 0.07 for the whole group; p < 0.06 for males only). No differences were found between the two groups for Extraversion, Neuroticism or the Lie-Scale; though for males only, dyslexics scored more

TABLE 2
Mean Scores and Standard Errors

	ALL SUBJECTS			MALES			FEMALES		
	DYS n = 52	NON n = 77		DYS n = 27	NON n = 43		DYS n = 25	NON n = 34	
L	4.27 (0.385)	4.66 (0.370)	---	3.96 (0.429)	4.28 (0.494)	---	4.60 (0.658)	5.15 (0.554)	---
E	14.56 (0.496)	13.84 (0.538)	---	14.81 (0.686)	13.02 (0.749)	---	14.28 (0.734)	14.88 (0.740)	---
N	12.44 (0.709)	11.04 (0.630)	---	12.48 (1.019)	9.65 (0.853)	p < 0.04	12.40 (1.003)	12.79 (0.857)	---
P	5.10 (0.460)	4.14 (0.291)	---	6.22 (0.680)	4.72 (0.443)	---	3.88 (0.524)	3.41 (0.311)	---
STA	20.08 (0.887)	13.14 (0.795)	p < 0.0001	21.15 (1.108)	12.67 (1.197)	p < 0.0001	18.92 (1.390)	13.74 (0.986)	p < 0.003
STB	8.37 (0.543)	6.79 (0.451)	p < 0.03	9.41 (0.738)	6.60 (0.655)	p < 0.008	7.24 (0.751)	7.03 (0.607)	---

highly than controls on Neuroticism (p < 0.04). Slight sex differences are usually found on three of these scales: N and STA usually yield slightly higher scores for females, while P scale scores (though very low in general) are reliably higher in males. As shown in Table 2, the control group showed these patterns, but within the dyslexic group only the P scale showed the expected sex difference, and to a greater degree. Male dyslexics scored as highly as females on Neuroticism, and on the STA male dyslexics` scores exceeded those of females, contrary to the usual pattern.

Dyslexics` STA scores were so much in excess of controls` that an analysis of the items was conducted to illuminate this finding. Factor analysis of items from the STA generally produces three main factors (Hewitt & Claridge, 1989). The first involves unusual perceptual experiences and `cognitive disorganisation`; the second has been termed `magical ideation`, and includes items concerning a strong sense of intuition, or belief in paranormal experiences such as telepathy, `deja vu` or precognition. The third, `social anxiety` relates to feelings of isolation or mistrust of others. It was found that dyslexics differed most from controls in their endorsement of items relating to the first two factors - perceptual and cognitive anomalies, and `magical ideation`. In addition to items cited earlier (all of which dyslexics endorsed much more frequently than controls), examples of these are given below with % endorsement rates (Dyslexic / Control):

Do everyday things sometimes seem unusually large or small? (56 / 14)

Does your own voice ever seem distant, faraway? (63 / 21)

Does your sense of smell sometimes become unusually strong? (46 / 21)

Are you sometimes sure that other people can tell what you are thinking? (52 / 28)

Do things sometimes seem as if they weren`t real? (42 / 18)

Dyslexics differed least from controls on items relating to the third factor - `social anxiety`, though there was a slight tendency for dyslexic males to endorse these items more frequently. Examples of these include:

Do you feel that you cannot get `close` to other people? (28 / 28)

Do you feel at times that people are talking about you? (60 / 50)

Are you very hurt by criticism (46 / 56)

It is perhaps notable that the latter was the only one of the 37 items which dyslexics endorsed *less* frequently than controls.

Thus dyslexics` elevated scores on the STA primarily reflected their increased endorsement of items relating to unusual perceptual and cognitive experiences, rather than being due to the `social anxiety` component of this scale.

DISCUSSION

In this study it was found that adult dyslexics scored more highly than controls on a personality scale designed to measure schizotypal characteristics in the normal population - Claridge's STA. They did not differ from controls on Eysenck's measures of Extraversion (E), Neuroticism (N), Psychoticism (P) or the Lie-Scale (L). A small difference was found for Claridge's STB - a scale measuring borderline (manic-depressive) characteristics in normals; this was due to higher scores for dyslexic males, who also showed slightly higher scores on Neuroticism. Emotional problems, as indicated by raised scores on N and STB, were associated with dyslexia only in males in this sample, and to a much lesser extent than was schizotypy. An analysis of responses to individual items from the STA confirmed that the higher scores of dyslexics on this scale were not due to to its `affective` component (`social anxiety`), but primarily to items concerning unusual perceptual experiences,`cognitive disorganisation` and `magical ideation`. It would therefore seem that the STA scale succeeds in capturing and quantifying some aspects of dyslexics` perceptual and cognitive style.

One problem with this study is that the sample of adult dyslexics used may not be representative, since they were as a whole relatively successful. It would clearly be inappropriate to use questionnaire assessment for very severe dyslexics. However, severe adult dyslexics are unlikely to differ qualitatively in terms of their underlying neurological predisposition from those who have succeeded with remediation or were more mildly affected, as found by Kinsbourne et al (1991). By contrast, it is very likely that more severe dyslexics might show more emotional problems than the current sample. Further studies would be needed to find whether scores on N or STB might relate to severity of reading and spelling difficulties, since standardised assessments were not available for this group; but the problem of distinguishing primary personality characteristics from secondary consequences of dyslexic difficulties would remain.

In this respect the sex differences found in this study are difficult to interpret. Male dyslexics scored more highly than all other subgroups on the STB scale, meaning that they reported more frustration, mood swings and emotional conflict than others. They also had higher scores than is usual for males on Eysenck's N, which taps worry, anxiety and low self-esteem. Though the absence of similar findings for female dyslexics may suggest that there are perhaps greater pressures felt by males regarding their level of academic achievement, the data are simply descriptive and other interpretations are possible.

Similar caution is warranted in interpreting the main finding of a very significant association of dyslexia with schizotypal characteristics. This was found for both sexes, though it was more pronounced in males. It might be suggested that dyslexics` elevated scores on the STA simply reflect secondary stress associated with their written language problems, or that these problems might have affected their interpretation of the questionnaire items. These are distinct possibilities; however, they seem less likely in the absence of similar effects on their scores on other scales. Furthermore, analysis of items suggested that the difference between dyslexic and control groups largely stemmed from STA items relating to unusual aspects of perception and cognition, rather than to its `emotional` content. It therefore seems pertinent to consider what is already known from psychological and physiological studies about individuals who score highly on this scale.

In studies of hemisphere function, high scorers on the STA have shown unusual performance asymmetries in divided visual field tasks using nonsense syllables or letters, and local versus global processing of letters (Broks, 1984; Rawlings & Claridge, 1984). These seemed largely due to an unusual LVF / right hemisphere performance, and the effects were more consistently found for males. In listening comprehension, high STA scorers showed poorer recall when material was presented binaurally rather than monaurally (Broks et al 1984). Thus unusual hemispheric specialisation or interaction may be a feature of normal schizotypal individuals.

Cognitive studies suggest that high STA subjects may show weak inhibitory selection of information at the early, preconscious stage of processing (Beech & Claridge, 1987). Experiments demonstrating semantic activation without conscious identification (SAWCI) - such as enhanced target detection following subliminal presentation of a priming stimulus - show that this effect correlates very strongly with scores on the STA (Evans & Claridge, 1991). These findings suggest that a central feature of the schizotypal nervous system may be an unusual `openness` to environmental stimuli. On visual tasks, increased susceptibility to the effects of backward masking are consistently found in normal schizotypal subjects (Steronko & Woods, 1978; Merritt, Balogh & Leventhal, 1986), suggestive of abnormalities in the early stage of visual information processing. It is further claimed that these reflect abnormal functioning of the transient visual system (Merritt & Balogh, 1989).

As discussed elsewhere in this volume, dyslexics show subtle visual problems indicative of abnormal transient system functioning. Lovegrove, Martin & Slaghuis, 1986) review the case for a visual deficit in dyslexia. Their findings are in accord with those of Stein, Riddell & Fowler (1987) who find that two thirds of dyslexic children show unstable binocular control and poor visual direction sense. They are also consistent with reports of unusual interactions between central and peripheral vision in dyslexia (Geiger & Lettvin, 1987), the `crowding` phenomenon (Atkinson, 1991) where visibility of a target is affected by the presence of adjacent distractors, and the deficiencies in parallel visual search reported by Ruddock (1991). A recent study by Livingstone, Rosen, Drislane & Galaburda (1991) further supports the view that the transient system may be abnormal in dyslexics; and the authors go on to speculate that if similar `fast` and `slow` processing channels operate in other sensory modalities, these may be similarly affected.

Unusual hemispheric specialisation in dyslexia has been suggested since the time of Orton (1937), and neuroanatomical studies provide support for this hypothesis (Geschwind & Galaburda, 1985). Though neuropsychological test profiles often suggest relative deficiencies on tasks inferred to involve the left hemisphere (Gordon, 1988), other evidence implicates right hemisphere functions (Stein, Riddell & Fowler, 1989). The neurological abnormalities found by Galaburda, Sherman, Rosen, Aboitiz & Geschwind (1986) commonly involve both hemispheres; and evidence from studies mapping brain electrical activity (Duffy, Denckla, Bartels & Sandini, 1980) and regional cerebral blood flow (Naylor, Wood & Flowers, 1990) also suggest that differences between dyslexics and controls are often bilateral. However, whether one or both hemispheres are primarily involved, the evidence suggests that unusual cerebral lateralisation of function may be a core feature of the dyslexic syndrome.

In view of the similarities found in the experimental literature, it seems possible that the association found in this study between schizotypal personality characteristics and dyslexia may reflect some common features of central nervous system organisation and functioning. In particular, a range of independent studies suggest that both dyslexics and normal schizotypal individuals may exhibit unusual hemispheric specialisation and anomalous functioning of the transient visual system. The present study involved only self-report measures of personality traits, but the very strong results suggest that experimental investigation of the differences found might provide useful insights into perceptual and cognitive function in dyslexia.

ACKNOWLEDGEMENTS

This research was part of a project funded by the Wellcome Trust

REFERENCES

American Psychiatric Association (1980). *Diagnostic and Statistical Manual of Mental Disorders*, 3rd edition. Washington, A.P.A.

Atkinson, J. (1991). Review of human visual development: Crowding and dyslexia. In J.F.Stein (Ed) *Vision and Visual Dyslexia*: Vol 13 of Vision and Visual Dysfunction. New York: Macmillan, 44-57.

Bell, D.B., Lewis, F.D. & Anderson, R.F. (1972). Some personality and motivational factors in reading retardation. *Journal of Educational Research*, 65, 5, 229-233.

Beech, A.R. & Claridge, G.S. (1987). Individual differences in negative priming: relations with schizotypal personality traits. *British Journal of Psychology*, 78, 349-356.

Broks, P. (1984). Schizotypy and hemisphere function - II. Performance asymmetry on a verbal divided visual field task. *Personality and Individual Differences*, 5, 649-656.

Broks, P., Claridge, G.S., Matheson, J. & Hargreaves, J. (1984). Schizotypy and hemisphere function -IV. Story comprehension under binaural and monaural listening conditions. *Personality and Individual Differences*, 5, 665-670.

Bradley, L. & Bryant, P.E. (1983). Categorising sounds and learning to read - a causal connection. *Nature*, 271,746-747.

Claridge, G. & Broks, P. (1984). Schizotypy and hemisphere function - I. Theoretical considerations and the measurement of schizotypy. *Personality and Individual Differences*, 5, 633-648.

Claridge, G. & Hewitt, J.K. (1987). A biometrical study of schizotypy in a normal population. *Personality and Individual Differences*, 8, 303-12.

Duffy, F.H., Denckla, M.B., Bartels, P.H. & Sandini, B. (1980). Dyslexia: Regional differences in brain electrical activity by topographical mapping. *Annals of Neurology*, 7, 412-420.

Einstein, A. (1945). Quoted by Hadamard, in *The Psychology of Invention in the Mathematical Field*. Princeton University Press.

Evans, J. & Claridge, G.S. (1991). Poster presentation, ISSID, Oxford, June 1991.

Eysenck, H.J. & Eysenck, S.B.G. (1975). *Manual of the Eysenck Personality Questionnaire*. Hodder & Stoughton, London.

Galaburda, A.M., Sherman, G.F., Rosen, G.D., Aboitiz, F. & Geschwind, N. (1985). Developmental dyslexia: four consecutive cases with cortical anomalies. *Annals of Neurology*, 18, 222-223

Geiger, G. & Lettvin, J.Y. (1987). Peripheral vision in persons with dyslexia. *New England Journal of Medicine*, 316, 1238-1243

Geschwind, N. & Galaburda, A.M. (1987). *Cerebral Lateralisation: Biological mechanisms, associations and pathology*. MIT, Cambridge, Mass.

Gordon, H.W. (1988). The effect of `right brain/left brain` cognitive profiles on school achievement. In D. Molfese & S.J.Segalowitz (Eds.) *Developmental Implications of Brain Lateralisation*. New York: Guilford Press.

Kinsbourne, M., Rufo, D.T., Palmer, R.L. & Berliner, A.K. (1991). Neuro-psychological deficits in adults with dyslexia. *Developmental Medicine and Child Neurology,* 33, 763-775

Livingstone, M.S., Rosen, G.D., Drislane, F.W. & Galaburda, A.M. (1991). Physiological and anatomical evidence for a magnocellular defect in developmental dyslexia. *Proceedings of the National Academy of Sciences*, 88, 7943-7947

Lovegrove, W., Martin, F. & Slaghuis, W. (1986). A theoretical and experimental case for a visual deficit in specific reading disability. *Cognitive Neuropsychology*, 3, 225-267.

Merritt, R.D., Balogh, D.W. & Leventhal, D.B. (1986). The use of a metacontrast and paracontrast procedure to assess the visual information processing of schizotypics. *Journal of Abnormal Psychology*, 94, 74-80

Merritt, R.D. & Balogh, D.W. (1989). Backward masking spatial frequency effects among hypothetically schizotypal individuals. *Schizophrenia Bulletin*, 15, 4, 573-583

Naylor, C.E., Wood, F.B. & Flowers, D.L. (1990). Physiological correlates of reading disability. In G. Th. Pavlidis (Ed.), Perspectives on Dyslexia, Vol 1. New York: John Wiley.

Orton, S. T. (1937). *Reading, writing and speech problems in children*. New York: Norton.

Rawlings, D. & Claridge, G.S. (1984). Schizotypy and hemisphere function III - Performance asymmetries on tasks of letter recognition and local/global processing. *Personality and Individual Differences*, 5, 657-633

Ruddock, K.H., (1991). Visual search in dyslexia. In J.F.Stein (Ed.) *Vision and Visual Dyslexia*: Vol 13 of Vision and Visual Dysfunction. New York: Macmillan, 58-83.

Snowling, M.J. (1981) Phonemic deficits in developmental dyslexia. *Psychological Research*, 43, 219-234.

Stein, J.F., Riddell, P.M. & Fowler, M.S. (1987). Fine binocular control in dyslexic children. *Eye*, 1, 433-438

Stein, J.F., Riddell, P.M. & Fowler, M.S. (1989). Disordered right hemisphere function in developmental dyslexia. In C. von Euler, I. Lundberg and G. Lennerstrand (Eds.) *Brain and Reading*; Wenner Gren International Symposium Series Vol 54. London: Macmillan.

Steronko, R.J. & Woods, D.J. (1978). Impairment in early stages of visual information processing in non-psychotic schizotypic individuals. *Journal of Abnormal Psychology*, 87, 481-490.

Thompson, L.J. (1969). Language disabilities in men of eminence. *Bulletin of the Orton Society*, XIX, 113-120

Facets of Dyslexia and its Remediation
S.F. Wright and R. Groner (Editors)
© 1993 Elsevier Science Publishers B.V. All rights reserved.

THE EMOTIONAL EFFECTS OF DYSLEXIA

J. Edwards

Dyslexia Institute, Staines, UK.

INTRODUCTION

Dyslexia is recognisable as a measurable discrepancy between cognitive ability and literacy level. It can be clearly seen as a substantial difference between reading/spelling and IQ, as measured by traditional tests. Perhaps the clearest, simplest and most effective definition of dyslexia or specific learning difficulty is that given by the World Federation of Neurology in 1969, quoted by Stirling (1978). Dyslexics are categorized as those who "despite conventional classroom experience fail to attain the language skills of reading, writing and spelling commensurate with their intellectual abilities." It is this inability of dyslexic pupils to take advantage of normal classroom instruction which can lead to teacher frustration regarding failure to achieve pupil potential, and consequent negative labelling of the children as disruptive, lazy or stupid. Alternatively, in the event of a teacher becoming aware of the quality of intellect locked up within an illiterate child, and finding it impossible to get commensurate results, it may be assumed that the withholding of achievement is a semi-deliberate act. Tension and misunderstanding may consequently build up on both sides.

The emotional effects of dyslexia are important to understand, given that the emotional stability of the dyslexic pupil is a pre-requisite which underlies all teaching and research, and one which has never been conclusively documented. There is little consensus of evidence which leads us to a thorough knowledge of how these specific pupils react to particular teaching methods, remediation styles and school régimes. How can we hope to teach them successfully or explore their cognitive profiles if we have not taken time first to study their typical experiences, and reactions to them? What sort of emotional make-up underlies, or is produced by, the condition of dyslexia? We are all involved in charting the aetiology and manifestation of this condition, often in minute detail; so what effect is observable on the personality?

Hampshire, (1981) in an articulate adult account of a dyslexic childhood, describes a gradual build-up of damaging humiliation, class-room misery and frustration in trying to learn to spell and cope with memory problems. Hampshire experienced intense anxiety and recalls "just knowing that I was not mentally retarded or lazy, or backward, or emotionally disturbed....would have made all the difference" (p. 55). Attempts to cope with severe feelings of inadequacy are listed, like desperate over-helpfulness, giving extravagant gifts, pilfering and attempting to barter friendship for homework-copying. "This need to give, for fear of otherwise not being tolerated, has remained with me all my life... I assumed that no-one could like someone as "stupid" as me... Nothing has changed: the doubts of childhood remain".

Miles & Miles (1983) give a comprehensive over-view of the emotional climate surrounding dyslexic children and their families. Extreme reactions from children are registered, varying from withdrawal to dogged determination or anti-social behaviour. Attention is drawn to the cumulative deprivation of information involved in literacy failure, and the lack of stimulation which multiplies hidden general and educational problems and can exasperate sufferers. Miles & Miles corroborate the view that continuing problems with spelling and literacy failure can compound and accumulate in everyday situations, and their accounts lend support to the view that Hampshire's reaction is typical. Scars from treatment in primary and secondary school have also been found to negatively influence tutor-student relationships at college, due to a back-log of extreme sensitivity (Miles & Gilroy, 1986).

Tomatis (1978) presents some composite case-study descriptions, based on many years of observing dyslexic students and their families. He emphasizes "how crushing the anxiety caused by all the various possible perception distortions can be." (p. 157). Tomatis is perhaps unique in postulating that emotional reactions can be of primary importance, and he is convinced that dyslexia has a deep origin in early trauma. He takes an audiometric approach with underlying psycho-therapeutic principles. As a psycho-physiologist, he believes mixed laterality can represent a specific listening block.

If long-term follow-up studies are considered, the majority seem to suggest that dyslexics do not achieve qualifications or careers commensurate with their intellectual skill or potential (Carter, 1964; Preston & Yarrington, 1967; Rackham, 1972; Rawson, 1968; Silver & Hagin, 1964). There is also evidence from Saunders & Barker (1972) and from Yule & Rutter (1976) of long-term damage to individuals relating to sensitivity, concealments, matrimonial friction and resistance to help; and of a range of adult embarrassments concerning jobs, friendships, social interaction, clubs, committees and even sports.

Several studies have found less negative outcomes (e.g. Abbott & Frank, 1975; Edgington, 1975; Gottfredson, Finucci & Childs, 1983, 1984; Hinton & Knights, 1971; Robinson & Smith, 1962) and no evidence for an increased incidence of emotional and behavioural difficulties in dyslexic adults. Given this discrepancy in results, it would be useful to establish whether critical cognitive variables, such as locus of control, self-concept, and self-efficacy interact to determine overall adult adjustment.

Williams & Miles (1985) investigated aspects of the dyslexic personality and its coping strategies in relation to Rorschach responses. It seems that the educational and social pressures on a dyslexic child frequently make him lack confidence and so limit the adventurous quality of his responses. Knasel (1982) has found that on similarity responses to "triads" (for example, "myself, my mother, my favourite teacher") that dyslexics display fewer constructs. This supports the theory of a personality-based cause for this self-limiting effect. There are a variety of possible explanations for the differences revealed. Dyslexic children are often reported by parents and teachers to lack maturity; though Halpern (1953) and Francis-Williams (1968) both agree that even during the earliest years at school the number of responses increases for normal children.

Another interpretation is that dyslexics are less in the habit of matching things accurately. This is supported by the work of Stirling (1978) on inaccuracies of spoken language and confusions of similar words in dyslexics. She has suggested that in word-definition work verbal rambling shows not just lack of confidence about what is known, but insecurity about what is not known, a tendency to go "over the top" and to extremes. Williams & Miles interpret these findings as a difficulty in giving the socially appropriate response to a situation. The link between extremity of behaviour tendencies, either self-limiting or over-elaborative, organisational problems, and severity of dyslexia, needs further investigation. The concept of extremity of behaviour tendencies fits in with Knasel's (1982) identification of the dyslexic's "minimax" strategy to guard against exposure to failure and ridicule.

It is interesting to note that in fields where dyslexics have been known to excel, (like the scientific, art and design, the mechanical or practical), complexity can be dealt with comparatively easily. It is, however, a more detached, non-verbal, spatial kind of complexity which is more clear-cut than the demands of satisfying a person expecting a fluent response to an academic question.

A number of researchers have also reported low self-esteem in dyslexic children (eg. Critchley, 1970; Rosenthal, 1973) and evidence of compensatory manoeuvres. These include efforts to excel in non-literate skills like sport; or because of loss of self-esteem, becoming the classroom clown. Rosenthal found that behaviour was further influenced by the parents' degree of understanding that dyslexia is a specific abnormality. He also found that friendship-patterns revealed insecurity or low self-image as dyslexics, and their families, chose younger children or those with obvious difficulties (though of other types to associate with).

Stott (1978), though not writing specifically about dyslexics, shows how teaching styles and children's individual learning strategies can clash in the classroom and cause permanent damage to ordinary pupils in terms of their perceived image, potential, and behaviour-patterns. This is relevant because of the adverse classroom response many dyslexics document. Farnham-Diggory (1978) also records the effect which an adverse learning situation for children with learning difficulties can have on both pupils and teachers. She gives examples of pupils treated with mental cruelty by teachers who could not cope with continuous chronic failure, and accounts of clinical psychological problems directly induced by school failure.

Parental reactions to a child with literacy problems have also been examined (Owen, 1971). Owen found that parents perceived the learning-disabled children as having a range of problematic tendencies concerning global perseverance, verbal and motor skills, self-control and organisation. Relative to siblings, they were also thought lacking in general coping ability and anxiety control. Parents were negative about school experience even when a special program existed. For educationally handicapped children only did mothers tend to withhold affection if the child was irresponsible or disorganised. Fathers, however, withdrew affection if that child was apathetic or worried, lacked concentration, or showed lack of control. This did not apply to siblings. Academic pressure from fathers was related to their views of the learning-disabled child's perseverance and ability to carry responsibility.

In summary, it can be seen that the majority of these authorities, from their different disciplines and perspectives, converge in projecting dyslexic students as seriously affected both at home and school by academic failure. The original purpose of the study reported in this chapter was to examine factors which might have contributed towards good morale of a group of dyslexic boys and discover which ones had been most significant and helpful. I was concerned with variables like IQ, degree of dyslexia, behaviour patterns, family background, sibling and peer relationships, personality type, tendency to truant or persevere, teacher/pupil relationships, and early identification or help. However, the bitterness and trauma associated with some of their experiences led to a change in the line of enquiry and an emphasis on:

(a) what being dyslexic actually feels like,

(b) how it affects one emotionally,

(c) what kind of interaction with the educational system the dyslexic child has,

(d) what are the most important elements of support,

(e) presenting recommendations from teachers and parents for the future treatment of
those in the same difficult minority position.

METHODS

Subjects

All the eight boys I chose had either volunteered to stay on at a Special school for Dyslexics for an extra year or were prefects in responsible positions, two achieving Head Boy status. I felt that this proved they had overcome their aversion to schools to a large extent. Fairly tough "survivors" were deliberately selected as it was felt that it might be damaging to a very withdrawn or nervous pupil to dwell upon his experiences. These boys were at the end of their school careers, but still attending, so that school life was fresh in their minds and strongly felt. This seemed a suitable time at which to look back and review their school experiences.

Data was collected by means of:

- Free interview in which the subject chose the mode of recording responses, which could be tape-recorded, typed or hand-written.
- Structured interview with questionnaire.
- Parental interview with questionnaire in which the subject chose mode of response, which could be either written, telephoned, or by personal interview.
- Triangulation and comparison with school records/reports.
- Participant observation.

RESULTS

Of the 8 boys, all except one had been pushed to extremes of misery during primary school, and he had temporarily emigrated. All had been teased, humiliated and insulted, by staff, children, or both. There were five examples of extreme violence from teachers, three of which occurred under the age of nine (see Table 1).

TABLE 1

Negative Education Experiences Recorded for the Sample based on Free and Structured Interviews as well as Parental Interview and Records.

	John	Trevor	Gareth	Clark	George	Mark	William	Oliver	Total
A. Violence from teachers	√	-	√	√	√	√	-	-	5
B. Unfair treatment	√	√	√	√	-	-	√	-	5
C. Inadequate help/ neglect	√	√	√	√	√	√	√	√	8
D. Humiliation	√	√	√	√	√	√	-	√	7
E. Teasing/ Persecution	√	√	√	√	√	√	√	√	8

Of those who had not been physically attacked by teachers who could not cope, there was evidence of truancy, total demoralisation, psycho-somatic pains and isolation, and the risk of permanent physical damage being done while inducing artificial sickness (see Table 2). The boys registered a great deal of intensity, pain or embarrassment in recalling the worst incidents which showed in voice, gesture, eyes, increased colour, feeling hot, drawing odd doodles, or coffee consumption.

TABLE 2

Damaging Associative Reactions to the Negative Educational Experiences
Reported in Table 1

	John	Trevor	Gareth	Clark	George	Mark	William	Oliver	Total
1. Truancy/ School Refusal	√	-	√	√	√	√	√	√	7
2. Psycho-somatic pain	-	√	-	√	√	-	-	-	3
3. Isolation / Alienation	√	√	-	√	√	√	-	-	5
4. Lack of confidence	√	√	√	√	√	√	√	√.	8
5. Lack of communication	√	√	-	-	√	√	√	-	5
6. Self-doubt / Denigration of intelligence	√	√	√	√	√	√	√	√	8
7. Competitive-ness disorders	√	√	√	√	√	√	-	√	7
8. Sensitivity to criticism	√	√	√	√	√	√	√	√	8
9. Behaviour problems	√	√	√	√	√	√	√	√	8

These results raise the question of what can be done to stop teacher violence, child humiliation, and persecution by peers? It is the author's opinion that the responsibility must rest with the individual teacher, the education system and the government. Outlined below are suggestions, at both the national and school level, for creating a far greater awareness of dyslexia and its concomitant problems.

A. National/Political Issues
Ongoing publicity about dyslexia is essential.

A big influx of public money into education to reduce general class size is urgently needed.

Public money must be made available to provide more specialist help within schools.

More special day schools or integrated units for dyslexics.

An open learning situation for dyslexic adults who were not adequately helped in school should be established.

B. The Educational System

Early Identification is important and should involve:

Screening at age 7 alongside the normal medical checks.

Thorough pre-school checks of co-ordination and motor skills for those families "at risk" for dyslexia due to hereditary factors.

A specialist nursery school "head-start".

Upon identification, specialist treatment of an intensive nature should follow immediately (some families in this study had to wait six years).

Dissemination of knowledge about dyslexia to the individual and the family concerned can be critical. This should be the responsibility of a specific agency with government funding.

More information to student teachers and more provision of in-service training is essential.

More assistance and awareness in colleges of further education and universities would be helpful for dyslexic individuals.

More awareness of the pastoral needs of dyslexic individuals is desperately needed at all levels of the educational system.

C. The Student's View

The boys themselves emphasised the need for consistent genuine help - teachers who showed they cared for, listened to and respected the individual. They strongly registered the need for specialist help and contact with other dyslexics, but suffered social alienation through boarding school life. The majority advised that schools should capitalize on their strengths and build up their career prospects rather than continuously alerting them to their weaknesses. Finally an overwhelming resentment of being treated as "thick" was expressed.

CONCLUSION

The students in this study had all had years of intensive specialist help, yet their reported experiences make it clear that teachers also need to bear in mind the importance of individual reactions to failure. It is clear from this study that the the type of educational failure experienced by these boys not only scars the personality but can have damaging side behavioural side effects.

REFERENCES

Abbot, R.C. & Frank, B.E. (1979). A Follow-up of Learning - Disabled Children in a Private Special School. *Academic Therapy*, 10, 291-298.

Carter, B. (1964). *A Descriptive Analysis of the Adult Adjustment of Persons once Identified as Disabled Readers*. Ed. Dissertation, Indiana.

Critchley, M. (1970). *The Dyslexic Child*. Springfield; Charles C. Thomas.

Edgington, R.E. (1975). SLD children: a ten year follow up. *Academic Therapy*, 11, 53-64.

Farnham-Diggory, S. (1978). *Learning Disabilities* London: Fontana/Open Books.

Francis-Williams, J. (1968). *Rorschach With Children*. Oxford: Pergamon Press.

Gottfredson, L.S., Finucci, J.M. & Childs, B. (1983). *The Adult Occupational Success of Dyslexic Boys: A Large-Scale, Long-Term Follow-up*. Reprint No. 334. John Hopkins University.

Gottfredson, L.S., Finucci , J.M. & Childs, B. (1984). Explaining the Adult Careers of Dyslexic Boys: Variations in Critical Skills for High Level Jobs.*Journal of Vocational Behaviour*, 24, 355-373.

Halpern, F. (1953). *A Clinical Approach to Children's Rorschachs*. New York: Grune & Stratton.

Hampshire, S. (1990). *Every Letter Counts*. London: Bantam Press.

Hinton, C.G., & Knights, R.M. (1971). Children with Learning Problems: Academic History, Academic Prediction, and Adjustment Three Years after Assessment. *Exceptional Children*. 37, 513-519.

Knasel, E.G. (1982). *Towards a Science of Human Action*. Ph.D. Thesis, University of Wales.

Miles, T.R. (1983). *Dyslexia: The Pattern of Difficulties*. St. Albans: Granada.

Miles, T.R. & Gilroy, E. (1986). *Dyslexia at College*. London: Methuen.

Owen, F.W., Adams, P.A., Forrest, T., Stolz, L.M. & Fisher, S. (1971). Learning Disorders in Children: Sibling Studies. *Monographs of the Society for Research in Child Development*, 1971, 36, No.4.

Preston, R.C., & Yarrington, D.J. (1967). Status of Fifty Retarded Readers Eight Years after Reading Clinic Diagnosis. *Journal of Reading*, 11, 122-129.

Rackham, K. (1972). A Follow-up of Pupils at One Word-Blind Centre. *ICAA Word-Blind Bulletin*, 4, 71-79.

Rawson, M.B. (1968). *Developmental Language Disability: Adult Accomplishments of Dyslexic Boys*. Baltimore, M.D : The John Hopkins Press.

Robinson & Smith (1962). Reading Clinic Clients Ten Years After. *Elementary School Journal,* 63, 134-140.

Rosenthal, J. (1973). Self esteem in dyslexic children. *Academic Therapy*, 9, 27-39.

Saunders, W.A. & Barker, M.G. (1972). Dyslexia as a Cause of Psychiatric Disorders in Adults. *British Medical Journal*, 4, 759-761.

Silver, A. & Hagin, R.K. (1964). Specific Reading Disability. Follow-up Studies. *American Journal of Orthopsychiatry*, 34, 95-102.

Stirling, E. (1978). *Naming and Verbal Fluency in Dyslexic Boys*. M.Ed Thesis, University of Wales.

Stott, D.H. (1978). *Helping Children with Learning Difficulties*. Ward Lock Educational.

Tomatis, A.A. (1978). Education & Dyslexia. Association Internationale d'Audio-Psycho-Phonologie. Switzerland.

Williams, A.L. & Miles, T.R. (1985). Rorschach responses of dyslexic children. *Annals of Dyslexia*, 35, 51-66.

Yule, W. & Rutter, M. (1976). Epidemiology and Social Implications of Specific Reading Retardation. In R.M. Knights and D.J. Bakker, (Eds). *The Neuropsychology of Learning Disorders*. Baltimore: University Park Press.

VI. DEFINITION AND
EARLY DIAGNOSIS OF DYSLEXIA

Facets of Dyslexia and its Remediation
S.F. Wright and R. Groner (Editors)
© 1993 Elsevier Science Publishers B.V. All rights reserved.

DYSLEXIA: ISSUES OF DEFINITION AND SUBTYPING

S.F. Wright and R. Groner

Psychology Institute, University of Bern,

Switzerland

INTRODUCTION

An individual may experience difficulties learning to read for a wide variety of reasons. Vellutino (1979) for example, has distinguished between "extrinsic" and "intrinsic" causes of reading failure. Possible extrinsic causes for reading difficulty include social background, lack of opportunity, frequent changes of school and bad teaching. Intrinsic causes on the other hand comprise emotional and behavioural problems such as hyperactivity, difficulties in attention, low intelligence, brain damage and sensory deficits. Developmental dyslexia is typically defined as a severe and persistent difficulty with the written form of language in the absence of any apparent intrinsic or extrinsic cause.

This notion of unexpected reading failure is implicit in the majority of definitions of dyslexia. For example, the World Federation of Neurology defines dyslexia as a "disorder manifested by difficulty in learning to read despite conventional instruction, adequate intelligence and social-cultural opportunity" (Cited in Critchley, 1970, p.11). Later definitions have tended to become more specific listing, in addition to reading, other characteristics associated with dyslexia, such as "erratic spelling", and "a lack of facility in manipulating written as opposed to spoken words" (Critchely & Critchley, 1978) but still include statements excluding from definition children whose reading difficulties could be the result of extrinsic or intrinsic causes. Thus a child is not typically labelled as dyslexic if sensory, neurological, emotional or educational factors could be deemed responsible for the reading disability, but may be considered instead reading backward.

Some scepticism exists about the validity of this distinction between dyslexia and general reading backwardness and certainly there has been little research which has systematically explored the extent to which reading disability associated with low IQ, sociocultural inopportunity, or emotional disturbance is different from reading disability in the absence of these factors. A number of researchers have found scant evidence in support of differences between those defined as dyslexic and those as reading backward (e.g. Jorm, Share, Maclean & Matthews, 1986; Seidenberg, Bruck, Fornarolo & Backman, 1985; Siegel, 1988; Taylor, Satz & Friel, 1979). Although it cannot be denied that the absence of a clear difference between the two groups, dyslexic and reading backward, presents a challenge to the traditional notion of dyslexia as easily disassociated from other reading disorders, the central point of interest nevertheless remains - why children who are in other ways intelligent do not learn to read as easily or as effectively as their peers.

Identification of underachievement in reading, and a diagnosis of dyslexia, presupposes that there exists a reliable method of measuring poor reading. Although it can be argued that a diagnosis of dyslexia is rarely based simply on the degree to which an individual is underachieving in reading, reliable determination of the degree of underachievement associated with dyslexia is essential if there is to be any compatibility between groups defined as dyslexic. Existing methods and definitions can be grouped into at least four broad categories.

GRADE/AGE DISCREPANCY DEFINITIONS

According to this class of definitions, a child is labelled as dyslexic if his or her reading score is significantly below that which would be expected on the basis of grade or chronological age. There is, however, no agreement on the number of years a child must be below age or grade to be considered dyslexic and there is variation both across studies and across age groups. If chronological age is used as a measure of ability, it is common for a discrepancy of one or more years between ability and reading achievement to be considered significant (e.g. Byrne & Shea, 1979; Holligan & Johnston, 1988; Johnston, Rugg & Scott, 1987).

This group of methods have, however, been severely criticised (Bennett, 1982; Reynolds, 1981; Shepard, 1980, 1983). Used in isolation, both age and grade discrepancies have the tendency to identify slow learners rather than children whose performance is significantly different from their ability. While such children may have some problems with reading, and while some are undoubtedly dyslexic, a number are likely to be functioning academically at a level consistent with their intellectual ability. Consequently no severe discrepancy exists between their expected and obtained achievement levels and one could argue that the definition of dyslexia is inappropriate.

A further difficulty relates to the use of a fixed discrepancy to select subjects as dyslexic regardless of their age. The increase in dispersion of achievement scores with age means that a child of nine years who is reading one year below age may have problems as severe as an eleven year old reading two years below age level (Anastasi, 1982). If a fixed discrepancy is used to select subjects in a study, in which children from different age groups are included, the severity of the disability will vary as a function of age and the older children will tend to have less severe problems. On the Wide Range Achievement Test, for example, Gaddes (1976) found that if a cut-off of two years below grade level was employed, one percent of six year olds was identified as dyslexic, two percent of seven year olds and 25% of 19 year olds. Gaddes argues that "using a fixed academic retardation lag at all school ages to defined educationally impaired children is illogical and scientifically untenable" (p.15).

A related problem with the use of age as an indication of expected reading performance is that age has a decreasing correlation with reading. For example in a sample of 510 school children, Wright (1991) found that the correlation between reading and age was .25 at age eight and .10 at age thirteen which suggests that the age becomes an increasingly less accurate predictor. In view of the higher correlations found between intelligence and reading, which have been reported to fall between 0.3 and 0.7 (Stanovich, Cunningham and Feeman, 1984), the use of IQ to predict the expected reading score, instead of age would seem to be more appropriate measure of potential ability, particularly in older samples.

IQ: DEFINITIONS BASED ON STANDARD SCORE FORMULAS

According to this group of methods, an individual is classified as dyslexic if the reading standard score is significantly below the IQ standard score. The size of the discrepancy taken as severe varies between studies and a number of different approaches have been adopted: children have been classified as dyslexic if the discrepancy between IQ and reading falls in the bottom 10% of those whose IQ score is higher than the reading score (e.g. Erickson, 1975), in the presence of a half a standard deviation discrepancy between IQ and achievement scores (e.g. Jorgenson et al., 1985, 1987), or a one and a half standard deviation difference between the two scores (e.g. Nussbaum & Bigler, 1986).

One serious drawback to this group of methods is that the statistical comparison of two correlated test scores is associated with regression to the mean (Shepard, 1980). The result of the imperfect correlation is that, given a score on IQ, the optimal linear prediction of the score on reading will tend to be closer to the mean of all the other reading scores than IQ was to the mean of all other IQ scores. Thus, if a direct comparison is made between measures the greater the likelihood is for extremely intelligent children to be identified as underachievers. Rispens & van Yperen (1990) found that if IQ and reading were compared directly in this manner, a number of the supposed underachievers with high IQ actually had reading scores above the mean thus they argue "causing a serious distortion of the idea of dyslexia " (p.38).

IQ: DEFINITIONS BASED ON PREDICTION USING LINEAR REGRESSION

It has been argued that where possible the most appropriate method of defining a dyslexic group is through the use of a regression equation (Evans, 1990; Rutter & Yule, 1975;

Shepard, 1980; Yule, 1967). By accounting for the correlation between reading score and IQ, the regression approach determines an expected reading score avoiding the problems associated with regression to the mean and provides the best way of calculating a child's reading age based on his IQ and chronological age. However, once an expected reading score has been calculated, no universal criteria exists for determining whether the actual reading score is significantly different from this expected reading score. Yule and his colleagues initially suggested a cut-off of two standard errors whereby children whose actual reading score was more than two standard errors below their predicted reading score were identified as dyslexic (Yule, 1973; Yule et al., 1974; Rutter & Yule, 1975; Rutter et al., 1976). However, this two standard error cut-off has not been universally adopted. Other studies have found that it identifies too small a group and prefer to employ a less stringent cut-off of one and a half standard errors from prediction (Horn & O'Donnell, 1984; Jorm et al., 1986; Silva et al., 1985).

In spite of the argued methodological superiority of regression, this kind of analysis is not often possible as a relatively large sample size is required. Regression equations computed for small samples are associated with a large standard error and one would need at least 200 cases to make this a valid approach. Furthermore because a regression equation only applies to the population from which it was derived the possibilities of using the same regression equation on different populations are severely limited unless the new population is highly similar in terms of age, IQ, and social and economic characteristics.

IQ: DEFINITIONS BASED ON MULTIVARIATE PREDICTION.

One limitation of the regression approach is that is effectively prohibits the simultaneous consideration of a number of IQ-achievement discrepancies. An expected achievement score for a given individual is based on his or her achievement on a single test. Thus if an expected achievement score is to be calculated on a number of tests, e.g. reading, spelling and comprehension measures, unless these measures are combined to form a single score (e.g Jorm et al., 1986; Scarborough, 1984), several regression equations would have to be calculated. Neither solution is ideal. Combining measures results in a loss of information about differential performance on the individual tests and the use of separate regression equations will increase the error associated with the equation.

An alternative approach which enables a number of ability achievement discrepancies to be considered simultaneously is cluster analysis. It is the least used of all the methods for defining a dyslexic group. This is in part because clustering methods require large data sets and in part because cluster analysis is by no means universally accepted as a useful statistical tool. Nevertheless such methods have been widely used for identifying subgroups of dyslexic individuals (cf. Rourke, 1985) and some studies have employed clustering methods for initial identification of the dyslexic group.

The first study to use cluster analysis to identify a dyslexic group was the Florida Longitudinal Study (Fletcher & Satz, 1985; Satz & Morris, 1981). The sample comprised learning disabled boys and their matched controls (n=236). For each child three discrepancy scores were created based on the difference between grade equivalent score and reading score, between grade equivalent score and spelling score and between grade equivalent score and arithmetic score. These discrepancy scores were then entered into a cluster analysis. Two dyslexic groups were identified, representing 39% of the sample, which performed uniformly poorly on all three tests: reading, spelling and arithmetic. Van de Vlugt & Satz (1985) in a replication of the earlier study, using a Dutch translation of the original tests, identified five low achieving groups. While one of these showed a specific arithmetic disability, the remaining four showed general delays in achievement across all three subtests. These four groups represented 29% of the sample.

Results from a number of other studies also suggest that clustering methods provide a valid empirical methodology of subject selection. Spreen & Haaf (1986) identified nine clusters in a group of adult dyslexics and normal controls. Ninety eight percent of the controls fell into the three highest performing clusters. The remaining clusters represented different degrees of impairment on the achievement and neurocognitive tests employed. Korhonen (1988), using neuropsychological and cognitive test results as the classification criteria, divided a group of mildly learning disabled and controls (n=168) into six subgroups. With the exception of one subgroup, he found a relatively small overlap between learning disabled and controls within subgroups. Morris et al. (1986) were also able to divide a combined sample of 200 disabled and normal readers into three largely reading disabled groups and two largely normally reading groups.

One limitation of the studies listed above is that the cluster analysis has been performed on preselected samples. For example in the case of the Florida Longitudinal Study (Satz & Morris, 1981) the sample was representative of neither black nor female children. Furthermore, all the children had previously been defined as learning disabled or controls. In addition, the inclusion in the various studies of differing numbers of predefined dyslexics and controls means that no figures for prevalence exist when dyslexia is defined in this way. In contrast, Newman, Wright & Fields (1991), used cluster analysis in a largely unselected school sample in an attempt to derive one or more dyslexic groups.

Newman et al., (1991) examined the reading and spelling abilities of 462 children in relation to their intellectual abilities using cluster analysis. For each child six discrepancy scores were created based on the difference between the standard scores on two measures of ability (Verbal and Performance IQ) and three measures of achievement (reading single words, reading prose and spelling). The analysis identified five groups, one of which consisted of a number of children (11%) with large discrepancies between their intellectual and reading and spelling ability, in the direction of poorer reading and spelling. In addition to large discrepancies on all the diagnostic measures, poor reading and spelling and a high percentage of males in this group (92%), a further analysis indicated that this group also had a higher family incidence of dyslexia, more remedial help, performed more poorly on the memory tests than the comparison groups and were less able to distinguish between left and right. This study suggests that a dyslexic group can be identified without the application of a pre-determined cut-off beyond which a discrepancy score is considered significantly large for inclusion in the dyslexic group.

Unfortunately, such a technique as cluster analysis of multiple discrepancy scores does not translate easily into educational practice. While it supports the value of discrepancy scores in identifying dyslexics, it requires a large sample on which to perform the cluster analysis and multiple measures of cognitive ability for each participant. Further criticisms pertain to the instability of cluster analysis. Different methods may generate strikingly different solutions (Blashfield, 1986). Cluster analysis in not a single technique and the fact that it consists of a number of different methods means that not only must the researcher choose between different methods, he or she must make a number of subjective decisions, in particular, concerning the number of clusters or groups present. In addition, the nature of the clustering algorithm is such that it is possible for perfectly meaningless groups to be derived, thus it becomes essential to demonstrate that the resulting groups have some external validity. Arguably what is needed is a more detailed examination of the relationship between diagnosing a dyslexic group on the basis of a number of ability-achievement discrepancy scores and identifying an individual on the basis of a single discrepancy score.

EFFECTS OF DEFINITION

In view of the fact that the identification of groups of dyslexics, either within the clinical, educational or research setting, is not carried out using standard measures or agreed methods, it becomes important to determine the influence of definition not only on the number of children identified as dyslexic but also on the kind of children identified. Indeed this task has recently been regarded as a "prerequisite for an adequate understanding of the cognitive and biological correlates of reading disabilities" (Fletcher et al., 1989, p.338). Certainly, if the results from different studies are to be directly comparable we must arrive at a better understanding of the impact of definition on the resulting dyslexic group.

The influence of different definitions on the kind of children identified has seldom been directly investigated, in spite of an awareness of the problems concerning definition and of the definitional confusion that exists in the field. Many researchers prefer not to define dyslexia in quantitative terms and rely instead on clinic or referred populations. Dyslexic subjects have been defined as those who display "persistent and chronic difficulties with reading and language skills " (McKay & Neale, 1988, p.217) or those who meet definition "according to common criteria" (Stanley et al., 1982). Selection criteria such as these are impossible to replicate and may result in different samples being identified as dyslexic. Furthermore, conflicting results from different studies become difficult to evaluate when the selection criteria are vaguely stated. Even when procedures for identification are relatively clearly stated, detailed specification of the tests used to define 'reading achievement' may be omitted (e.g. Baker et al., 1984) or vary appreciably from study to study.

One recent study which investigated the concurrence of different definitional approaches was carried out by Rispens & van Yperen (1990). They examined the extent to which five different methods, belonging to two different categories (age-grade based discrepancies and ability-achievement discrepancies), identified different numbers and different kinds of children. In common with a number of other studies the authors found that the number of children identified as dyslexic depended on which measurement method was employed (Cone & Wilson, 1981; Epps et al., 1983; Forness et al., 1983; Fields et al., 1989). In the normal school sample this number ranged from between 4 and 18.3%. The type of reading task was also found to be a relevant factor: applying a word recognition task identified a lower percentage of children compared with a comprehension test. Concerning the concurrence of the models, Rispens and van Yperen distinguished between 'core discrepants', i.e. those children identified as dyslexic by all methods and 'core non discrepants', those children identified consistently as non-dyslexic. In the clinic sample, they found that about 70% of the children and more than 10% of the school sample changed category at least once. Differences were found between the different groups on IQ and reading ability.

The extent to which the theoretical confusion surrounding dyslexia can be attributed to the state of definitional confusion which exists in the field is unclear. On the one hand it is tempting to argue that the number and kind of children identified as dyslexic by different definitions could account for the lack of agreement concerning the etiology and prognosis of the disorder. There is certainly evidence to suggest that differences exist between clinic and school identified samples of dyslexics (Shaywitz, Shaywitz, Fletcher & Escobar, 1990). On the other hand, the small number of studies which have compared groups identified by different definitions find few significant differences, a state of affairs which lead Fletcher et al., (1989) to conclude that "there is little evidence suggesting any specificity of reading disability according to definition" (p.334). This is supported by Wright (1991) who found that variations in the selection criteria used to define dyslexic groups at two ages had little impact on the performance of the dyslexic groups relative to the controls on a cognitive test battery. Regardless of how the dyslexic groups were defined, the same associated difficulties relative to controls were found.

Definition has, however, been found to influence the relative numbers of male and females identified as dyslexic. Traditionally, consistently more boys than girls are identified as dyslexic: ratios range from 3:1 to 15:1 in the United States (Finucci & Childs, 1981) and from 2:1 to 5:1 in the United Kingdom (Critchley, 1970). However, in a recent paper Shaywitz et al., (1990) argue that reports of increased prevalence of dyslexia in boys may simply reflect referral biases, i.e. that there are different rates of referral for the two sexes. In support of this argument, they found that while the prevalence of dyslexia in a referred population was two to three times more common in boys than girls, this ratio was markedly lower in a research identified population, 1.2:1 when a regression equation was used to identify the dyslexic group. Share, Silva, & Adler (1987) have also argued that current definitions of dyslexia may lead to a "de-emphasis" of dyslexia in girls if their performance is always considered relative to that of the boys, as no account is taken of the finding that reading expectancies are higher for girls than boys (Finucci & Childs, 1981; Vogel, 1990). Thus, a girl may be underachieving relative to peers of her own sex but not relative to members of the opposite sex. The authors suggest that separate regression equations should be computed for boys and girls or sex should be included as an independent variable in addition to IQ and or age. The former approach has already been tried by Scarborough (1984). She computed separate regression equations for males and females and virtually eliminated the traditional sex ratios, identifying eight males and seven females using a two standard error cut-off.

The effect of definition extends beyond the selection of the dyslexic group. Not only is there enormous variability in the diagnostic criteria for dyslexia, similar variability exists in the criteria for defining a control group. Thus dyslexics have been compared to backward readers, unmatched, age-matched, IQ-matched and reading-age matched controls. Recent research has been simultaneously strengthened and complicated by a more explicit awareness of the importance of the control group in pinpointing causal factors in dyslexia. Strengthened - as the questions that can be validly answered are determined by the choice of a control group. If a group of dyslexics perform significantly more poorly on a given task than a group of younger reading-age matched children, it is highly likely that the variable being measured does not have a causal role to play in reading development, given that the two groups do not differ significantly on reading level. Complicated - as evidenced by the debate concerning the interpretation of positive and negative findings when different control groups are employed (e.g. Backman, Mamen & Ferguson, 1984; Bryant & Goswami, 1986; Goswami & Bryant, 1989; Pavlidis, 1990; Vellutino & Scanlon, 1989).

SUBTYPES OF DYSLEXIA

A related problem of definition comes from the growing realisation that dyslexia, even when it is narrowly and clearly defined, may not represent a single disorder but may consist instead of a number of subtypes, each possibly having different etiological correlates. This may explain the range of deficits on a large number of cognitive tasks which have been found when dyslexics have been compared to controls. Deficits, which as Stanovich (1986) has argued "are indirectly threatening to the assumption of specificity because they suggest widespread cognitive deficits rather than a highly localised dysfunction that selectively impairs reading" (p. 233). Whilst consensus amongst the majority of researchers might exist concerning the variability in the dyslexic population, the extent to which this can be neatly partitioned into discrete subtypes remains undetermined. Indeed, Seymour (1990) concluded that it is doubtful whether the heterogeneity in the dyslexic population can be "reduced to a categorical scheme of subtypes" (p.58).

Attempts to establish valid subtypes of dyslexia have been undertaken from two different perspectives. The first has been to divide groups of individuals into subtypes based on clinical observation of performance on reading and/or neuropsychological measures. The second has been to classify individuals using statistical methods of analysis such as Q-factor and cluster analysis (see Hooper & Willis, 1989 for a review of clinical and statistical approaches). There is little agreement on the number of subtypes, and this is independent of whether the subtypes are derived statistically or clinically. In a review of subtyping studies, estimates ranged from between two and six subtypes (Hooper & Willis, 1989). The most commonly reported are an auditory linguistic subtype and a visual perceptual subtype. Dyslexics who fall into the former tend to be characterised by difficulties in the discrimination of speech sounds, in 'sound blending' and in naming visual stimuli. Those who fall into the visual-perceptual subtype are characterised by problems in visual perception and visual discrimination. In almost all studies the language subtype is reported more frequently than the visual subtype. In addition to these two broad subtypes a 'mixed' subtype is often described comprising children with both auditory and visual-perceptual difficulties.

A number of researchers have argued that the repeated appearance of a 'mixed' subtype suggests that the variability in the dyslexic population cannot be neatly partitioned into discrete subgroups and they have proposed an alternative view of the heterogeneity in the dyslexic population (Ellis, 1985; Stanovich, 1988). Instead of searching for discrete subtypes which together account for the syndrome of dyslexia, these theorists argue that dyslexic children differ only in degree from normal readers. Kinsbourne (1986) has called this the 'continuity model of dyslexia' and contrasts it with the discontinuity model in which a finite number of subtypes is sought. Ellis (1985) argues that the most useful analogy for dyslexia is obesity. He argues:

"For people of a given age and height there will be an uninterrupted continuum from painfully thin to inordinately fat. It is entirely arbitrary where we draw the line between "normal" and "obese", but that does not prevent obesity being a real and worrying condition, nor does it prevent research into causes and cures of obesity being both valuable and necessary" (p.172).

The evidence for a statistical separation of dyslexia from the normal distribution of reading ability initially reported by Yule, Rutter, Berger & Thompson (1974) and Rutter & Yule (1975) has been largely refuted. Rutter & Yule (1975) argued that there was a 'hump' at the bottom end of the reading distribution, with a larger number of children reading poorly (4%) than would be expected from a normal distribution (2.28%). This provided evidence for a discrete group of dyslexics readers distinct from poor readers. More recent evidence, however, has suggested that this 'hump' was a statistical artefact due to ceiling effects on the tests employed (Rodgers, 1983; Share et al., 1987; Silva et al., 1985; van der Wissel & Zegers, 1985). Rodgers (1983) found no hump in a sample of over eight thousand ten year olds.

The absence of a "hump", however, as Thomson (1990) stresses, does not "argue against an etiologically distinct group of children with learning difficulties" (p.10) and certainly there is ample evidence of qualitative differences between children with dyslexia and normal readers of similar reading skill. Rather it argues that as selection of a group essentially involves the application of an arbitrary cut-off at a point along a distribution of discrepancies between IQ and reading scores. Where this cut-off is made will affect both the number and the severity of those identified as dyslexic. Similarly, the continuity theory of dyslexia does not exclude the possibility that it is possible to partition the variance in the dyslexic population into discrete subtypes. What it implies is that the resulting subtypes will not necessarily be homogeneous as they represent an arbitrary partitioning on a continuum. For example a reading related skill such as phonological coding ability is distributed on a continuum from very good to very poor, with some dyslexics showing very poor decoding skills and some superior decoding skills. Between these two extremes, Ellis (1985) suggests, "there will be a complete and unbroken graduation of intermediate dyslexics" (p.192). Thus it is argued that dyslexics will not form clear ability/disability clusters and that those that do form are just one possible division along a continuum. This viewpoint is supported by a recent study by Volger, Baker, Decker, Defries & Huizinga (1989) who performed a cluster analysis on factor scores derived from a battery of psychometric tests. They found little evidence for distinct subgroups and suggested that clustering routines were likely to be defining clusters "which were due to minor, possibly random deviations from a contiguous spread of points in the psychometric measurement space" (p.174).

CONCLUSION

Despite almost a century of research on dyslexia and ample descriptive definitions, there is as yet little consensus concerning a reliable operational definition. Existing definitions are based on individual diagnosis, symptomatology or on the extent to which an individual is reading significantly below the level expected on the basis of his or her IQ and or age. These differences in approach to definition have the effect that the number and kind of children identified as dyslexic can vary according to the technique by which they were defined.

It could be argued that variation in selection criteria are unaviodable, given the extensive range of diagnostic measures employed and the wide range of uses to which definitions are put in education and research. Within this framework, it is possible that the search for a universal definition is overly optimistic, particularly as it seems that dyslexia is, to a greater or lesser extent, comprised of a number of subtypes, each possibly having different deficits. Nevertheless, it remains an important challenge for both existing and future researchers in dyslexia is to find a means of reliably separating those whose reading difficulties can be attributed primarily to a specific deficit or deficits - the dyslexics - from the general pool of those with poor reading - the reading backward. The advantage of a more concrete definition would be twofold - not only would it provide a realiable means of identification and diagnosis, it would justify the distinction made between dyslexia, or specific reading difficult and general reading backwardness.

REFERENCES

Anastasi, A. (1982). *Psychological Testing* (Fifth Edition). New York: Macmillan.

Backman, J., Mamen, M. & Ferguson, M. (1984). Reading level design : Conceptual and Methodological Issues in Reading Research. *Psychological Bulletin,* 96, 560-568.

Bennett, R. E. (1982). The use of Grade and Age Equivalent scores in Educational Assessment. *Diagnostique*, 7, 139-146.

Blashfield, R. K. (1976). Mixture model tests of cluster analysis: Accuracy of four agglomerative hierarchical methods. *Psychological Bulletin*, 83, 377-388.

Blashfield, R. K. (1980). Prepositions regarding the use of cluster analysis in clinical research. *Journal of Consulting and Clinical Psychology*, 48, 456-459.

Bryant, P. & Goswani, U. (1986). Strengths and weaknesses of the reading level design: A comment on Backman, Mamen and Ferguson. *Psychological Bulletin*,, 100, 101-103.

Byrne, B. & Shea, P. (1979). Semantic and phonetic memory codes in beginning readers. *Memory and Cognition*, 7, 333-338.

Cone, T.E. & Wilson, L.R. (1980). Quantifying a severe discrepancy: a critical analysis. *Learning Disability Quarterly*, 17, 446-454.

Critchley, M. (1970). *The Dyslexic Child*. London: Heinemann Medical Books.

Critchley, M. & Critchley, E. (1978). *Dyslexia Defined*. London: Heinemann Medical Books.

Ellis, A.W. (1985). The cognitive neuropsychology of developmental and acquired dyslexia: a critical survey. *Cognitive Neuropsychology*, 2, 149-205.

Epps, S., Ysseldyke, J. E. & Algozzine, B. (1983). Impact of different definitions of learning disabilities on the number of students identified. *Journal of Psychoeducational Assessment*, 1, 341-352.

Erickson, M. T. (1975). The z Score discrepancy method for identifying reading disabled children. *Journal of Learning Disabilities*, 8, 308-312.

Evans, L.D. (1990). A conceptual overview of the discrepancy model for evaluating severe discrepancy between IQ and achievement scores. *Journal of Learning Disabilities*, 23, 406-413.

Fields, H., Wright, S. F. & Newman, S. (1989). Techniques for identifying a group with spelling difficulties in the absence of reading difficulties. *The Irish Journal of Psychology*, 10, 657-663.

Fletcher, J. M. (1985). External Validation of Learning Disability Typologies. In B. P. Rourke, *Neuropsychology of Learning Disabilities: Essentials of Subtype Analysis*. London: Guildford Press.

Fletcher, J. M., Espy, K. A., Francis, D. J., Davidson, K. C., Rourke, B. P. & Shaywitz, S.E. (1989). Comparisons of cut-off and regression based definitions of reading disabilities. *Journal of Learning Disabilities*, 22, 334-338.

Fletcher, J. M. & Satz, P. (1985). Cluster analysis and the search for learning disability subtypes. In B. P. Rourke, *Neuropsychology of Learning Disabilities : Essentials of Subtype Analysis*. London: Guildford Press.

Forness, S.R., Sinclair, E. & Guthrie, D. (1983) Learning disability discrepancy formulas: Their use in actual practice. *Learning Disability Quarterly*, 6, 107-115.

Gaddes, W. H. (1976). Prevalence estimates and the need for a definition of learning disabilities. In R. Knight & D.J. Bakker (Eds.). *The Neuropsychology of Learning Disorders*. Baltimore: University Park Press.

Goswami, U. & Bryant, P. (1989). The interpretation of studies using the reading level design. *Journal of Reading Behaviour*, 21, 413-424.

Holligan, C. & Johnston, R. S. (1988). The use of phonological information by good and poor readers in memory and reading tasks. *Memory and Cognition*, 16, 522-532.

Hooper, S. & Willis, W. (1989). *Learning Disability Subtyping*. Springer Verlag.

Horn, W. F. & O'Donnell, J. P. (1984). Early identification of learning disabilities: a comparison of two methods. *Journal of Educational Psychology*, 76, 1106-1118.

Johnston, R. S., Rugg, M. D. & Scott, T. (1987). The influence of phonology on good and poor readers when reading for meaning. *Journal of Memory and Language*, 26, 57-68.

Johnston, R. S., Rugg, M. D. & Scott, T. (1987). Phonological similarity effects, memory span and developmental reading disorders : the nature of the relationship. *British Journal of Psychology*, 78, 205-211.

Jorgenson, C. B., Jorgenson, D. E. & Davis, W. P. (1987). A cluster analysis of learning disabled children. *International Journal of Neuroscience*, 35(July), 59-63.

Jorgenson, C. B., Stone, R. B. & Opella. (1985). The univariate and multivariate relationships among environmental, sociocultural, biological and developmental variables and the identification of learning disabled children. *International Journal of Neuroscience,* 26, 283-288.

Jorm, A. F., Share, D. L., Maclean, R. & Mathews, R. (1986). Cognitive factors at school entry predictive of specific retardation and general reading backwardness: a research note. *Journal of Child Psychology and Psychiatry*, 27, 45-54.

Jorm, A. F., Share, D. L., Matthews, R. & Maclean, R. (1986). Behaviour problems in specific reading retarded and general reading backward children : A longitudinal study. *Journal of Child Psychology and Psychiatry*, 27, 33-43.

Korhonen, T. (1988). *Learning Disabilities in Children : An Empirical Subgrouping and Follow -up*.

Lundberg, I., Olofsson, A. & Wall, S. (1980). Reading and spelling skills in the first school years predicted from phonemic awareness skills in kindergarten. *Scandinavian Journal of Psychology*, 21, 159-175.

McKay, M. F. & Neale, M. D. (1988). Patterns of performance on the British Ability Scales for a group of children with severe reading difficulties. *British Journal of Educational Psychology*, 58, 217-222.

Morris, R., Blashfield, R. & Satz, P. (1986). Developmental classification of reading-disabled children. *Journal of Clinical and Experimental Neuropsychology*, 8, 371-392.

Newman, S., Wright, S.F. & Fields, H. (1991). Identification of a group of children with dyslexia by means of IQ-achievement discrepancies. *British Journal of Educational Psychology*, 61, 139-154.

Nussbaum, N. L. & Bigler, E. D. (1986). Neuropsychological and behavioural profiles of empirically derived subtypes of learning disabled children. *International Journal of Clinical Neuropsychology*, 8, 82-89.

Pavlidis, G. T. (1990). Conceptualisation, symptomatology and diagnostic criteria for dyslexia. In G. T. Pavlidis, *Perspectives on Dyslexia Volume 2: Cognition, Language and Treatment.* Chichester: John Wiley & Sons.

Reynolds, C. R. (1981). A note on determining significant discrepancies among category scores on Bannatynes regrouping of WISC-R subtypes. *Journal of Learning Disabilities*, 14, 468-469.

Reynolds, C. R. (1984/5). Critical measurement issues in learning disabilities. *Journal of Special Education*, 18, 451-476.

Rispens, J. & van Yperen, T. A. (1990). The identification of specific reading disorders: measuring a severe discrepancy. In G. T. Pavlidis, *Perspectives on Dyslexia Vol 2: Cognition, Language and Treatment..* Chichester: John Wiley & Sons.

Rodgers, B. (1983). The identification and prevalence of specific reading retardation. *British Journal of Educational Psychology*, 53, 369-373.

Rourke, B.P. (1985). Overview of Learning Disability subtypes. In B. P. Rourke, *Neuro-psychology of Learning Disabilities: Essentials of Subtype Analysis.* New York: The Guildford Press.

Rutter, M., Cox, A., Tupling, C., Berger, M. & Yule, W. (1975). Attainment and adjustment in 2 geographical areas. 1. The prevalence of psychiatric disorder. *British Journal of Psychiatry*, 126, 493-509.

Rutter, M., Tizard, J., Yule, W., Graham, P. & Whitmore, K. (1976). Research report: Isle of Wight Studies 1964-1974. *Psychological Medicine*, 6, 313-332.

Rutter, M. & Yule, W. (1973). Specific reading retardation. In L. Mann and D.Sabatino, *The First Review of Special Education.* Philadelphia: Burntwoood Farms.

Rutter, M. & Yule, W. (1975). The concept of specific reading retardation. *Journal of Child Psychology and Psychiatry*, 16, 181-197.

Satz, P., Fletcher, J., Clark, W. & Morris, R. (1981). Lag, deficit, rate, and delay constructs in Specific Learning Disabilities: a re-examination. In A. Ansara, N. Geschwind, A. Galaburda, M. Albert & N. Gartrell (Eds.), *Sex Differences in Dyslexia*. Maryland: The Orton Dyslexia Society 129-150.

Satz, P. & Morris, R. (1981). Learning disability subtypes : a review. In F.J. Pirozzolo & M. C. Wittrock. (Eds.), *Neuropsychological and Cognitive Processes in Reading*. New York: Academic Press. 109-141.

Satz, P. & Morris, R. (1981). *The Search for Subtype Classification in Learning Disabled Children*. New York.

Scarborough, H. S. (1984). Continuity between childhood dyslexia and reading. *British Journal of Psychology*, 75, 329-348.

Seidenberg, M. S., Bruck, M., Fornarolo, G. & Backman, J. (1985). Word recognition processes of poor and disabled readers. Do they necessarily differ? *Applied Psycholinguistics*, 6, 161-180.

Share, D. L., McGee, R., McKenzie, D., Williams, S. & Silva, P. A. (1987). Further evidence relating to the distinction between specific reading retardation and general reading backwardness. *British Journal of Developmental Psychology*, 5, 35-44.

Share, D. L. & Silva, P. A. (1986). The stability and classification of specific reading retardation, A longitudinal study from 7 - 11. *British Journal of Educational Psychology*, 56, 32-39.

Share, D. L. & Silva, P. A. (1987). Language deficits and specific reading retardation : cause or effect? *British Journal Disorders of Communication*, 22, 219-226.

Share, D. L., Silva, P. A. & Adler, C. J. (1987). Factors associated with reading plus spelling retardation and specific spelling retardation. *Developmental Medicine*, 29, 72-84.

Shaywitz, B. A., Shaywitz, M. D., Fletcher, J. M. & Escobar, M. D. (1990). Prevalence of reading disability in boys and girls: results of the Connecticut Longitudinal Study. *Journal of the American Medical Association*, 264, 998-1002.

Shepard, L. (1980). An evaluation of the regression discrepancy method for identifying children with learning disabilities. *Journal of Special Education*, 14, 79-91.

Shepard, L. (1983). The role of measurement in educational policy , lessons from the identification of learning disability. *Educational Measurement: Issues and Practice*, 2, 4-8.

Siegel, L. S. (1988). Evidence that IQ scores are irrelevant to the definition and analysis of reading disbility. *Canadian Journal of Psychology*, 42, 201-215.

Siegel, L. S. & Ryan, E. (1988). Working memory in subtypes of learning disabled children (Abstract). *Journal of Clinical and Experimental Neuropsychology,* 10, 55.

Silva, P. A., McGee, R. & Williams, S. (1985). Some characteristics of 9 year old boys with general reading backwardness of specific reading retardation. *Journal of Child Psychology and Psychiatry*, 266, 727-739.

Spreen, O. & Haaf, R. G. (1986). Empirically derived learning disability subtypes : A replication attempt and longitudinal patterns over 15 years. *Journal of Learning Disabilities,* 19, 170-180.

Stanley, G., Smith, G. & Powys, A. (1982). Selecting intelligence tests for studies of dyslexic children. *Psychological Reports,* 50, 787-792.

Stanovich, K. E., Cunningham, A. E. & Cramer, B. B. (1984). Assessing phonological awareness in kindergarten children: Issues of task comparability. *Journal of Experimental Child Psychology,* 38, 175-190.

Stanovich, K. E., Cunningham, A. E. & Feeman, D. J. (1984). Intelligence, cognitive skills and early reading progress. *Reading Research Quarterly,* 14, 278-303.

Taylor, H. G., Satz, P. & Friel, J. (1979). Developmental dyslexia in relation to other child-hood reading disorders: Significance and clinical utility. *Reading Research Quarterly*, 15, .

van der Vlugt, H. & Satz, P. (1985). Subgroups and subtypes of learning disabled and normal children: A cross cultural replication. In B.P. Rourke., *Neuropsychology of Learning Disabilities: Essentials of Subtype Analysis.* New York: The Guildford Press.

van der Wissel, A. & Zegers, F.C. (1985). Reading retardation revisited. *British Journal of Developmental Psychology,* 3, 3-9.

Vellutino, F. R. (1979). *Dyslexia: Theory and Research.* Cambridge: MIT Press.

Vellutino, F. R. & Scanlon, D. M. (1989). Some prerequisites for interpreting results from reading level matched designs. *Journal of Reading Behaviour,* 21, 361-385.

Volgel, S.A. (1990). Gender differences in intelligence, language, visual-motor abilities and academic achievement in students with learning disabilities. *Journal of Learning Disabilities*, 23, 44-52.

Volger, G.P., Baker, L.A., Decker, S.N., Defries, J.C. & Huizinga, D.H. (1989). Cluster analytic classification of reading disability subtypes. *Reading and Writing: An interdisciplinary Journal,* 2, 163-177.

Wright, S.F. (1991). *The Identification of Reading and Spelling Problems in Schoolchildren: a Longitudinal Study.* Unpublished doctoral thesis. University College and Middlesex School of Medicine, London.

Yule, W. (1967). Predicting ages on Neale's Analysis of Reading Ability. *British Journal of Educational Psychology*, 37, 252-255.

Yule, W. (1973). Differential prognosis of reading backwardness and specific reading retardation. *British Journal of Educational Psychology*, 43, 244-248.

Yule, W., Rutter, M., Berger, M. & Thompson, J. (1974). Over and under achievement in reading distribution in the general population. *British Journal of Educational Psychology*, 44, 1-12.

PATTERNS OF DEVELOPMENT IN GOOD AND POOR READERS AGE 6 - 11.

J. Whyte,

Clinical Speech and Language Studies,

Trinity College, Dublin.

INTRODUCTION

So much has been written about aspects of reading development and disorders and has been made available in useful publications, such as those edited by Bakker & van der Vlugt (1989), Dumont & Nakkem (1989), von Euler, Lundberg and Lennerstrand, (1989), Whyte (1989), Snowling & Thomson (1991) among others, that one hesitates before adding to the mountain. But we have not yet solved the problem of why some children have so much difficulty in learning to deal with written language in spite of the myriad of hypotheses, approaches, models and research studies.

For reasons, usually of economic expediency, and perhaps because results are sometimes needed fast, many of the studies reported until recently were once-off, or at best cross-sectional studies. Longitudinal studies are acknowledged to yield more reliable and more valid findings, and are probably the key to the unravelling of the enigma of reading difficulties, but they actually take more time and more resources. Data on progress in reading from longitudinal studies has been reported, among others, by Lundberg, 1988; Bradley & Bryant, 1985; Ellis, 1987; Scarborough, 1990; Tallal & Katz, 1989; Butler, Marsh, Sheppard & Sheppard, 1985; Juel, 1988; Jorm, Share, Maclean & Matthews, 1986; Share & Silva, 1986; Mackay, Neale & Thompson, 1985; Cox, 1987; Miles, 1991; Wright, 1991. Some of these have focussed on children thought to be at risk for reading failure because of familial or other factors, others have employed unselected samples or cohort samples and identified children with reading difficulties among them. But we are only at the beginning of what longitudinal studies can offer and anything they can offer should be useful.

This paper arises from a longitudinal study of the school progress of a cohort sample of 89 boys attending the same elementary school in Belfast, Northern Ireland (Whyte, 1980, 1983, 1989). The study commenced in 1975 and ended in 1991. The measures administered annually from age 4 to 6 were the Wechsler Intelligence Scale for Children (WISC; Wechsler, 1971), Burt Word Reading Test (BURT), English Picture Vocabulary Test (EVPT; Brimer & Dunn, 1972), Boehm Test of Basic Concepts (Boehm, 1969), Draw-a-Man (DAM; Goodenough & Harris, 1963) and the verbal expression subtest of the Illinois Test of Psycholinguistics (ITPA; Kirk & McCarthy, 1968). From age 7, in addition to the first three named above, all the children completed each year the group reading, spelling and intelligence tests developed by Young (1973). Records were also available on age and school attendance as well as of some family background characteristics such as employment status of parents, occupational level of parents, and number of siblings.

While the measures may seem to to-day's researchers to be somewhat coarse-grained and to lack the refinement of those currently being adopted in investigations of reading development and disorders from a phonological or a visual processing standpoint using a 'bottom up' approach, the fact that practically complete data sets for these variables were available on 66 normally distributed children for annual measures from age 5 to age 11 provides a unique opportunity to investigate questions of wider import.

Previous studies such as those by Share & Silva (1986) Jorm et al (1986), Ellis & Large (1987) have followed their subjects for shorter periods or have measurements recorded at intervals of more than a year. At this stage of children's development so many changes are happening that annual measurements would seem essential to chronicle the shifting relationships between cognitive and other variables, to pick up patterns of variation and their underlying constituents, and to identify critical points where progress became stable, increased, or declined. Cross-sectional studies and studies using clinical samples have contributed greatly to our knowledge of the processes involved in reading and spelling but some basic issues, from the 'top-down' point of view, are as yet unresolved and some could, perhaps, be addressed in part at least by longitudinal studies such as the present one.

The issues to be investigated in the present paper include: (i) prognosis for progress - whether there are differential growth rates for readers of differing status; (ii) the existence and identification of variables which discriminate between early good readers, slow starters, perpetually poor readers, and readers who decline relative to what might be expected; (iii) the relationship between reading and the development of other skills useful in the school context - is there a chicken-and-egg situation? (iii) the usefulness of specialised assessment and treatment ? What happens if there is none?

Prognosis for Progress

The 'normal' rate of progress for children's reading ability has in the main been determined from norm-groups based on cross-sectional surveys. But we do not know whether this 'normal' rate of progress should be equal in every year even for 'normally developing' individuals longitudinally. It has been acknowledged that individual children will not necessarily conform to the norms, yet regression analysis, commonly used to determine whether or not a child is achieving up to his or her level, is based on an acceptance of these norms. (Newman, Wright & Fields, 1991; Cox, 1987, Dobbins & Tafa, 1991).

It is possible that in real life, on a longitudinal basis, children stop and start and make erratic progress; sometimes this is due to factors outside the control of the school, sometimes not. It might therefore be considered unrealistic to expect similar progress in every year and at every stage of their reading development. The present study charts a pattern of progress in reading for a normally distributed, but socio-economically and culturally deprived group and for subgroups within that group. In this way a baseline is provided to which the progress of children with and without problems could be compared.

Variables Related to Early and Later Reading

For the vast majority of poor readers at school, performance in reading is predictable from their general cognitive abilities - which possibly, after a certain point, have been influenced by their lack of reading (see Stanovich, 1986 for a discussion of the "Matthew effect"). The question of whether there exists a subgroup of poor readers ('dyslexics'/'underachievers') who can be distinguished from the majority ('low achievers", Cox, 1987), 'garden-variety' poor readers (Gough & Tunmer, 1986) is as yet not resolved satisfactorily. The dyslexic child, has been characterised as one who has an 'unexpected' disability in the domain of reading, one not predicted by his or her general intellectual functioning and socio-cultural opportunities. Stanovich (1991) refers to much of the salient literature on the question of the definition of dyslexia by discrepancy criteria and is severely critical of this approach.

The question of stability of progress is also relevant here. The identification of underachievers and the stability of their classification over time has concerned many researchers (Share & Silva, 1986), and recently the question of the stability of this classification across measures has received attention. Dobbins & Tafa (1991) concluded that there was qualified support in their study which employed a great number of comparisons over a wide range of measures for the stability of identification of underachieving readers. They accepted the conclusions of previous research in the area (Yule, 1973; Rutter & Yule, 1975; Dobbins, 1984; Jorm et al 1986; Share & Silva, 1986). But other investigations, cited by Stanovich (1991) have found it difficult to differentiate discrepancy-defined dyslexic readers from low achievers (Fredman & Stevenson, 1988; Siegel, 1988, 1989).

Apart from the difficulties of operationalising the assessment of 'intelligence' as something that can be separated completely from the results of achievement, one of the problems with studies of dyslexia, and indeed reading difficulties in general, has been that until comparatively recently children were only assessed when the problem was manifested. Their IQ was then tested. The criteria for classification as dyslexic were in fact confounded.

It is acknowledged that reading and verbal ability are closely related and that lack of reading could influence the development of verbal ability. However, when children are not tested until they have actually been failing at reading for some time we do not know whether verbal ability had been within the normal range before reading instruction began. Neither can we tell whether, if it had been within the normal range, its development might have been influenced adversely by the lack of reading development. Lawson & Inglis (1984) claimed that the amount of deficit on any WISC-R subtest shown by learning disabled children, as identified by a learning disability index derived from the WISC-R, was closely proportional to the degree of its verbal content. Some longitudinal studies are currently addressing this issue by commencing with children at the preschool stage or earlier (e.g. Scarborough, 1990) and monitoring their progress.

While non-verbal ability has been treated by some researchers (reviewed by Stanovich, 1991) as independent of reading, it is possible that in the early stages it actually *is* important for reading and that it too has a reciprocal relationship with reading as time goes by. Mackay et al., (1985) distinguished between a group of continually failing readers and a subgroup of children making good or average progress in reading from age 4.5 - 10.5 with a measure of sequential ability at age 6 derived from the WISC. But this measure did not discriminate significantly between the continually failing group and a group of 'slow starters' whose initial level of reading was low but who subsequently improved their performance. The present study includes readers in all these groups and should contribute to the discussion.

Provision of Assessment and Treatment

A further relevant question which can be answered by the present data relates to the issue of provision and the politics of learning disability. What happens if children are just left alone and not assessed and remediated with an approach that is rigorous and systematic?

This study began for some subjects when they were aged 4 before reading instruction had been initiated and it followed them through to age 20-21. The present paper focuses on the subjects during the period from their second year of reading instruction to age 11. The researcher came from outside the school system and provided feed-back annually only in the most general terms in speaking to individual teachers about individual children, providing no written records. No action was taken as a result of his/her annual tests that would not otherwise have been taken.

The school was in an area of extreme social deprivation and had a well-developed system of remedial teaching by withdrawal. All children were considered to benefit from being withdrawn from their class, in small groups. The groups were formed by the class teachers on the basis of reading ability and all groups had two or three periods per week of special attention from one of the remedial teachers at ages 7, 8 and 9. All levels of reading ability enjoyed this extra attention. It was not systematic in terms of programme or evaluation, nor was it particularly regular, being subject to disruption for the usual variety of reasons in a big school. This condition was similar for all the children and therefore has not been taken into account in the analysis.

Description of Subjects

Table I gives the total number of subjects in the study. They comprised three cohort groups who had been with the same class teacher in three successive intakes. The groups did not necessarily stay together after the first year, but were usually dispersed across four or five parallel mixed-ability classes of 25-30 children. All classes had regular changes of teacher, usually after years 1, 3 and 5, and children did not usually change from the class group in which they were placed in year 2 at age 6. No significant differences were found between the cohort groups in the variables of interest here and they have been amalgamated into a single group for analysis in this paper. The mean age, Verbal and Performance IQ (WISC-V, WISC-P) , English Picture Vocabulary Test (EPVT), Draw-a-Man (DAM) , Illinois Test of Psycholinguistics (ITPA) and Burt raw scores at the end of their second year in school - Primary 2 - are shown in Table I

TABLE I

Subject Characteristics (N=89)

	Age	WISC-V	WISC-P	EPVT	DAM	ITPA	BURT
Means	6.5	101.7	103.4	99.8	22.0	27.1	16.6
Standard Deviations	0.34	12.1	12.3	12.8	8.2	4.7	9.9

It may be noted that while the IQ and EPVT scores are well within normal limits, at this stage of schooling, the mean reading score of the total group was lower than expected by the norms (a raw score of 27 at age 6.4 ; norms are not provided for scores lower than 27). As stated above, the school was in an area of multiple socio-economic deprivation and home and school life was being further disrupted during the period of this study by street disturbances and clashes between military and para-military groups, so this may reflect some effects of the environment on school readiness and school performance.

Ranking of Subjects by Reading Scores at age 6.5 and age 11.3

Subjects were selected from the above group for the part of the study reported in this paper on the basis of their being present for assessment again on the Burt , EPVT, and group measures of reading and spelling at age 11.3, at the end of their final year in Primary school.

Grouping of subjects

Share and Silva (1986) identified underachievers by regressing Burt reading scores on performance IQ and designating as the specific reading retardation group those children whose residual reading scores fell in the bottom 5 per cent of the distribution. They then examined the overlap between the group at age 7, 9, and 11. Newman et al (1991) used cluster analysis based on the size and pattern of the discrepancies between Verbal and Performance IQ scores and three measures of reading related skills. In the present study it was of concern to examine separately the possible contributions of Verbal and Performance IQ without confounding them with reading scores. Therefore a more low-tech method was employed: the subjects were ranked according to their score on the Burt word reading test at age 6. 5 and divided into five groups of approximately equal size as follows:

Ranked Group 1: Burt Score 0 - 6:   N = 15 (originally 16 but 1 left)
Ranked Group 2: Burt Score 7 - 14:   N = 14 (originally 15, but 1 left)
Ranked Group 3: Burt Score 15 - 20: N = 8   (originally 15, but 7 left)
Ranked Group 4: Burt Score 21- 25:   N = 13   (originally 16, but 3 left)
Ranked Group 5: Burt Score > 26 :   N = 18 (originally 20)

Subjects were then ranked on their Burt scores at age 11.3 and grouped as follows:

Ranked Group 1:   Burt Score = 0 - 50   (approx RA < 8.3)   N = 18
Ranked Group 2:   Burt Score = 51 - 60   (approx RA 8.4 - 9.2) N = 11
Ranked Group 3:   Burt Score= 61 - 70   (approx RA 9.3 - 10.2) N = 8
Ranked Group 4:   Burt Score = 71 - 80   (approx RA   10.3 - 11.1) N = 7
Ranked Group 5:   Burt Score = > 81   (approx RA > 11.2)   N = 24

Data for each subject were inspected to determine whether subjects had remained in their original ranked groups. It was found that there had been considerable change from all original groups. Given the possibility of error (referred to also by Dobbins & Tafa, 1991), it was decided that subjects who had remained in their own ranked group, or moved just one single group up or down, would be classified as stable in their categories. Thus a subject who had been in Group 1 at age 6.5 and was still in Group 1 or was in Group 2 at age 11.3 was classified as **Lo**, as was a subject who started in Group 2 and moved down to Group 1 or up to Group 3. Similarly a subject who had started in Group 5 and was still in Group 5 or was in Group 4 at age 11.3 was categorised as **Hi**, as was a subject who started in Group 4 and moved to Group 3 or Group 5.

Subjects who started in Group 1 or 2 and moved more than one rank were classified as Risers and subjects who started in Group 4 or 5 and moved down more than one rank were classified as **Decliners.** Subjects in Group 3 at age 6.5 were omitted from the next part of the analysis, since they were of average performance and the majority remained in that category. This left four groups.

Lo: N = 20
Risers: N = 8
Hi: N = 23
Decliners: N = 8.

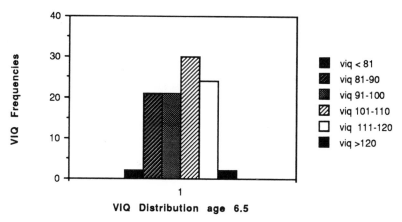

FIG. 1. WISC Verbal IQ Frequencies

The IQ scores of these subgroups were extracted from those of the total group to determine if they constituted a normally distributed sample. The frequencies are shown in Figs. 1 and 2 and indicate a normally distributed group overall, with if anything a bias towards the higher side of average for Verbal IQ on the WISC.

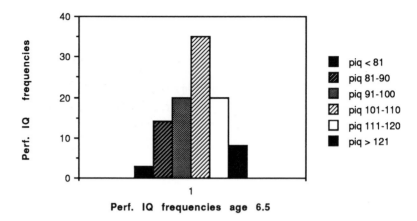

FIG. 2. WISC Performance IQ Frequencies

It may be noted that while the distribution of Performance IQ scores appears to conform to the normal curve, there are slightly more subjects with lower scores on Verbal IQ than on the Performance IQ. This difference was not significant however. For both Verbal and Performance IQ there were in fact more subjects with scores above than below 100. The sample, although from a socially deprived background, was achieving IQ scores within normal limits at age 6.5.

How many subjects changed ranked group for reading between age 6.5 and 11.3 ?

Thirty-nine percent of those in Ranks 1 and 2 at age 6.5 made a substantial move of more than one ranked group upwards, while 32% of those in the two highest groups at age 6.5 made a substantial move downwards. Of those in Ranked groups 1 and 2 at age 6.5, 61% remained in those groups; while those in the two highest groups at age 6.5 were more likely to remain high achievers since 68% of them remained in those groups. (The kind of movement involved may be seen in Figs. 3 and 4).

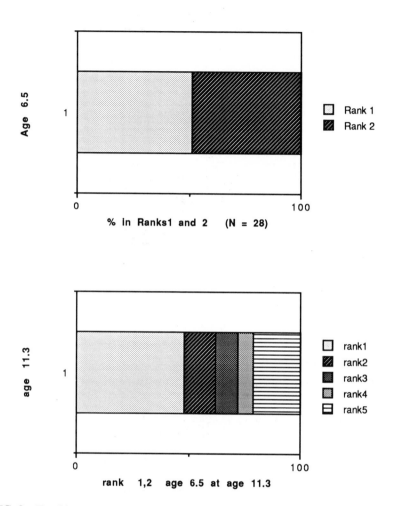

FIG. 3.   Total low scorers in Ranks 1 and 2 at age 6.5 and outcomes at age 11.3 in percentages

How can we account for these moves?

We looked first to see if there were any differences between the groups at age 6.5 on the variables measured which might account for the differences in reading score at that time (Table 2).

Already at this stage there were significant differences on the Burt between the two groups (F (3,34) = 88.02, p<.001), but their Verbal and Performance IQ's as measured by the WISC were not significantly different. On WISC Vocabulary the Risers group scored significantly higher than the other groups, but this had obviously not made any difference to their reading at this stage. Their Verbal IQ was higher than that of the Lo group and higher even than that of the high scoring groups and their performance IQ was close to that of the Hi group. The Lo group were still well within the normal range for IQ however and reading was their only low score. In fact they had hardly started reading at all.

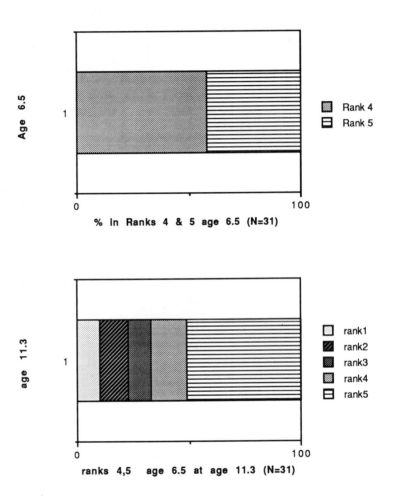

FIG. 4. Total in ranks 4 & 5 at age 6.5 and outcome at age 11.3 as percentages

Two other variables turned out to be relevant: school attendance record and age. Attendance at school is important and perhaps especially so when these early skills are being taught. It is interesting to note here that the Hi group, whose reading scores remained consistently high throughout their years at elementary school had in fact the best attendance record during this year. This had been true also for them during the previous year, at age 5.5 with significant differences at that stage between their record and that of the Lo and Riser groups. Attendance for the Hi group at age 6.5 was significantly correlated with reading score (r = .64; p < .001), and stepwise regression showed that it was the only one of the variables entered (Vocabulary, VIQ, PIQ and attendance) to contribute significantly to the variance for that subgroup.

The groups also differed significantly in age (F (3,35) = 3.99, p<.05). Given that all the children had started school on the same day in the September following their fourth birthday and that they had all received the same formal reading instruction for comparable lengths of time, the differences between the groups cannot be attributed to the amount of teaching and may reflect instead a lack of readiness for reading in the low scoring groups.

TABLE 2

Scores at Age 6.5 years (SD's in parenthesis)

	N	Age	Att.	EPVT	BURT	WISC VIQ	WISC PIQ	WISC Info.	WISC Arith.
Lo	20	74.6 (4.4)	162 (21.1)	99.3 (8.9)	6.1 (4.1)	99.7 (12.6)	98.4 (11.1)	7.9 (2.6)	9.9 (4.0)
Risers	8	72.6 (4.2)	163 (20.4)	111.4 (8.9)	6.5 (2.8)	106.5 (6.9)	105.8 (9.3)	10.4 (1.6)	13.1 (2.6)
Hi	23	76.8 (2.5)	174.0 (8.9)	99.8 (11.9)	26.1 (4.2)	101.9 (13.0)	107.1 (14.3)	9.6 (2.7)	10.3 (3.6)
Decliners	8	78.0 (1.5)	167.9 (30.0)	95.5 (15.6)	27.6 (3.5)	95.6 (13.8)	107.0 (12.4)	7.4 (2.7)	9.6 (2.1)

Correlations between reading and other variables

It would seem that at this stage, some factors not taken into account by the overall WISC scores must be important for reading. The groups were not significantly different on WISC Verbal or Performance scores, where all were performing within normal limits for their age, however, they differed in the correlations between reading and IQ and they differed significantly on reading, for which certain other skills must have been immature, or perhaps malfunctioning in the younger children.

Reading was significantly correlated with Performance IQ (r = .63 ; p < .01) for the Lo group and with Verbal IQ (r = .62; p <.05) for the Decliners Group, but these variables were not significantly correlated at this stage for either of the groups who were to become high-ranking readers at age 11.3. Receptive vocabulary is clearly not enough - the Risers group had the highest mean vocabulary score on the English Picture

Vocabulary Test and yet their reading score was like that of the low group, whose vocabulary score was significantly lower.

Differences between the Lo group and the Risers

Vocabulary was correlated significantly with reading for the Lo group, (r =.62; p< .01) but not for any of the others. Perhaps it is important at the very beginning stages of reading if other abilities are developed to a certain level. It is possible that for the Lo group, vocabulary development was in line with other abilities brought into play in the early stages of reading. For the Risers, Vocabulary level was way ahead and not significantly correlated with reading. Perhaps because of their immaturity or of problems in other areas not measured here, they did not make the same progress in reading as might have been expected of children with those language skills.

The Risers had some other abilities which differentiated them from the Lo group. They had significantly higher scores on the Information and Arithmetic subtests of the WISC and their scores on these subtests were also higher though not significantly so, than those of the Hi and Decliner groups. There were no differences on the other subtests between the groups .

Differences between the Hi Group and the Decliners.

At this stage there were no significant differences between the Hi group and the Decliners, although the Decliners had the lowest scores of all the groups on the EPVT and Verbal IQ. They were also lowest on Information, Arithmetic and Similarities. Perhaps this should be taken as a sign of being at risk, whereas high scores on these subtests even with low reading could be seen as an indication that progress is likely.

Progress at reading from age 6.5 to age 11.3

The chronological age difference between the groups remained steady throughout the period of the study. The differences between the groups in terms of reading age however changed substantially as may be seen from Fig. 5 which shows reading scores for the groups in terms of reading ages set against the mean chronological age for the whole group.

While age differences may explain in part the slow start of the two lower-scoring groups at age 6.5 they obviously did not explain everything since one of the low scoring groups, the Risers, eventually became as good as one of the original high-scoring groups. Since the age differences between the groups remained stable, it is necessary to look at other evidence which might indicate the reasons for the differential progress of these groups. This will be done for each year of school progress in turn.

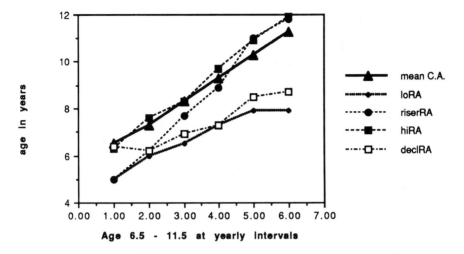

FIG. 5. Reading age and chronological age

Rates of Progress in Reading 6.5 - 7.3

Table 3 shows that by the mean cohort age of 7.3, the Hi group were reading at or above the age level with a mean score on the Burt of 42.7, giving a reading age of 7.6. They had made an average gain of 15 months during the 12 months after reaching the baseline on the test. The Lo and Risers groups had both almost reached the test baseline of 6.4 with scores of 23 and 25 and were reading about 12 months below chronological age, having gained the equivalent of about 9 months each. The Decliners had started to decline, having made no gains at all and remaining at a reading age of approximately 6.2 as measured by the Burt Word Reading Test.

TABLE 3

Results of Group tests at age 7.3 (SD's in parenthesis)

	Reading	Spelling	GRP IQ	WISC VIQ	BURT
Lo	87.5	85.0	97.0	102.8	23.3
	(8.9)	(3.7)	(8.8)	(10.5)	(20.3)
Risers	101.7	92.7	103.8	111.2	25.0
	(12.1)	(1.2)	(3.9)	(5.3)	(8.5)
Hi	106.8	102.2	105.8	103.4	42.7
	(8.5)	(13.6)	(6.8)	(11.4)	(9.9)

At this point, the Hi group scored significantly higher than all the other groups on the Burt reading test. Their attendance and that of the Decliners was higher than that of the Lo and Riser groups, but not significantly so and attendance was not significantly correlated with reading score for any group. There were no significant differences between the Hi group and the other groups on Vocabulary, Verbal or Performance IQ. Their EPVT score was slightly lower than that of the Lo group (97.8; 98.95) and their Verbal IQ was similar (103.4; 102.9) as was their Performance IQ (101.8; 101.4). There were significant differences however between the Hi and the Lo groups on the three group tests of reading, spelling and non-verbal intelligence taken by all subjects for the first time at this stage.

Differences between Lo Group and Risers

It might be assumed that these tests had been influenced by the degree of oral decoding word-reading ability of the Hi subjects, but the Risers were also significantly higher than the Lo group (p < .05 for reading and p< .01 for spelling and intelligence) on all of these tests, although they had scored similarly to them on the oral word reading test. It appeared that the Risers, although poor at orally decoding single words were able to use other skills such as those required in the written group tests.

It is perhaps important to briefly describe the group tests. The reading test had two parts: in Part I subjects chose one word from four to describe a pictured object. The words were similar in length and demanded quite fine discrimination over 16 items. In part 2, subjects chose one word out of four to complete a sentence. Again the words had features in common, but comprehension was required as well as some degree of decoding. These results would indicate that children can use other skills to read, even when their oral decoding is not apparently adequate. The IQ test had four subtests, vocabulary, differences, analogies and pairs. Children marked one choice having heard four responses.

Conclusions at age 7.3.

If measured by the oral word decoding test and the WISC alone, both the Lo and Riser groups would possibly have been designated dyslexic by discrepancy criteria at this stage; but the Risers were already showing that they had other skills as shown by their performance on the group tests of reading and IQ, perhaps foreshadowed by their superior performance on some subtests of the WISC and on the EPVT the previous year. Interestingly, their spelling was below their reading on the group test at this stage. It was better than that of the Lo readers, but not as close to the Hi Group as was their group reading score.

Progress from 7.3 - 8.3 and Correlations with the Burt

Significant differences were found between groups for the group reading ($F(3,42) = 10.43$, $p<.01$), spelling ($F(3,41) = 12.19$, $p<.001$) and IQ tests ($F(3,54) = 7.01$, $p<.01$) (for means and standard deviations see Table 4).

TABLE 4

Group Tests at age 8.3

	Reading	Spelling	Group IQ
Lo	87.1	85.4	91.5
	(2.6)	(3.5)	(13.9)
Risers	102.4	97.2	100.9
	(8.6)	(7.8)	(5.0)
Hi	102.1	110.8	103.7
	(7.6)	(13.9)	(7.5)
Decliners	98.1	87.4	88.0
	(6.1)	(8.4)	(12.9)

Between the ages of 7.3 and 8.3, the Hi Group made a mean gain of about 9 months, considerably less than during the previous phase. It meant however that their mean reading and chronological ages were similar, 8.3 years, at the end of this period. Perhaps they were entering a different phase of reading acquisition and decoding skills were not being developed at the same rate as before. Their scores on the Burt were significantly correlated with verbal IQ and with the group reading and spelling test scores ($p< .05$) for the first time.

One of the questions of interest at this point was, however whether the Lo and Riser groups would make as swift progress as the Hi Group had made at the equivalent stage of reading acquisition, having attained the baseline on the Burt. The results showed that the Risers did indeed make swift progress, gaining an average of 17 months during this period with a reading age of 7.7 - still behind the mean cohort chronological age of 8.3 (and 8.0 for their own group), but showing that once the baseline had been reached in their case, progress appeared to be assured.

The Lo group did not make equivalent progress. They averaged only 5 months progress between the ages of 7.3 and 8.3. Perhaps it was because their reading age the previous year had been 6.2 approx, not quite the baseline of 6.4 which may constitute a jumping-off point. Their Burt scores were significantly related to the group reading, spelling and intelligence scores and to their WISC Verbal IQ ($p< .01$ in all cases).

The Decliners made 8 months progress having started from a similar mean reading score to that of the Lo and Riser groups at the beginning of the year. Their scores on the Burt were significantly correlated with the group reading and spelling tests (r=.98, p<.001) but not with their WISC scores. The differential rates of progress during each chronological year may be clearly seen in Fig. 6.

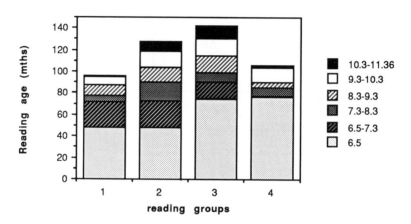

FIG. 6.   Mean gain scores in reading by Lo, Risers, Hi and Decliners age 6.5 - 11.3

Differences between the groups at age 8.3

Once again, there were no significant differences between the groups on Vocabulary (EPVT), attendance, or Performance IQ. The Decliners Group was significantly lower than the others on Verbal IQ and significantly lower than the Hi group on the group reading spelling and IQ tests. We might expect the Lo group to be like the decliners, since their reading scores were quite close, but the Lo group was very similar to the Hi group on vocabulary and Verbal IQ (92.3 / 93.2;   103.2 / 108.7) and the decliners had the lowest scores of all the groups on vocabulary (88.6) and Verbal IQ (92.5). There were obviously different mechanisms at work for each of these groups. The Decliners' IQ score had started to decline at this stage, but not that of the Lo group.

It is likely that at this stage, for children making normal progress in reading, reading contributes to Verbal IQ. These children had initially made normal progress but then for reasons which are not clear, ceased to do so, with the result that their Verbal IQ scores declined. The Lo group, on the other hand, had been making slow progress in reading and perhaps had not yet reached the stage where reading was important for their Verbal IQ, so their Verbal IQ remained constant.

Conclusions at age 8.3

While the Risers Group were scoring eight months behind the Hi group on the Burt and if measured on the Burt and WISC alone, might have been categorised as dyslexic, results from the group reading tests showed that they could be classified as falling within the normal range for reading, as they were at or close to the scores attained by the Hi group on reading and intelligence. They were still somewhat behind them in spelling, but were functioning within normal limits with results indicating a delay in this area which took longer to compensate for than the reading delay.

It was at this stage that substantial differences became apparent between the groups on all variables. Scores on the Burt were related to Verbal IQ scores for the two stable groups - the Lo and Hi Groups, but not for those whose reading rank changed significantly over the period. This could mean that the groups who changed were not integrating subskills to the same extent or in the same way as the other groups; the Lo group perhaps had perhaps slightly less well developed skills overall - but this was critical - so that although they were integrated, they were not in fact promoting their reading to the same extent as they did for the Hi Group.

Progress from 8.3 to 9.3

Significant differences were found between the groups on the Burt (F(3,53) =8.6, p<.001), the WISC Verbal IQ (F(3,34) = 7.34, p<.001), and the 3 group tests: Reading (F(3,53) = 14.1, p<.001), Spelling (F(3,53) = 14.1, p<.001) and IQ (F(3,53) = 3.93, p<.05).

The Hi group made another leap forward at this point and recorded an average gain of 16 months, ending the year with a mean reading age of 9.7 when the mean cohort chronological age was 9.3. Their performance on the word reading test was significantly correlated with that on the group reading (r = .64) and spelling tests (r = .71; p< .001 for both) and with the vocabulary sub-test of the WISC (r=.45; p<.05).

The Risers also made substantial gains - an average of 14 months on the Burt, giving them a mean reading age of 8.9 at mean cohort chronological age of 9.3. Burt reading was significantly correlated with the group reading test (r=.77, p<.05) but not with the group spelling test. On both of these group tests they were achieving standardised scores well within the normal range, not registering as poor performers. Their spelling had improved relative to that of the Hi group and the gap between them had lessened, but spelling for them was not significantly correlated with the Burt unlike all the other groups.

The Lo group made a mean gain of 10 months between the chronological ages of 8.3 and 9.3 - from reading age 6.5 to 7.3. This was a greater increase than in the previous year and it happened after they had achieved just over the baseline reading age. It was at this point in reading development that other groups had made substantial gains as well. Their low reading age, 2 years below chronological age, was mirrored in their case however, unlike that of the Risers, by low standard scores on the group tests and they scored significantly lower than the Risers or the Hi group on all these tests, including the intelligence test, for the first time, and also the EPVT.

The Decliners made a gain of approximately 5 months on the Burt test giving them a mean reading age of 7.3, like the Lo Group, at age 9.3. They were again significantly different , as they had been at age 8.3 , from the Hi Group with whom they had started out on equal terms, on WISC Verbal IQ, on four of the subtests - information, arithmetic, similarities and vocabulary and on the three group tests.

Associated variables

The vocabulary test differentiated significantly here for the first time in that Decliners and the Lo Group were both significantly lower than the Risers. The Hi group were still the best attenders, but not significantly so, and the Decliners were significantly lower than the Hi Group and the Risers on Verbal IQ. The group tests discriminated between the poor and better readers more effectively than did the WISC. The Risers and Hi Group had significantly higher scores on the group reading, spelling and intelligence. Their standard scores were all above 100, whereas those of the Lo Group and the Decliners were below 90.

Conclusions at age 9.3

It is tempting to see this age, i.e. around 9 years, as the age in which vocabulary growth is greatly influenced by reading. The Risers, whose reading was approximately 15 months higher, scored significantly higher than the Lo group on the EPVT. However, the scores on the Burt and on the EPVT were not significantly correlated for either the Risers nor the Lo group. In addition, we must ask why the EPVT scores for the more stable Hi group were not significantly higher than those of the low reading groups. Their vocabulary scores were higher but not significantly different from those of the Lo group at this stage although their reading age was over two years higher, and their scores at earlier stages had been very similar to those of the other groups. If it were reading level alone that boosted vocabulary, they should surely have had the highest vocabulary scores from the outset. Figure 7 shows the vocabulary scores for all groups.

Looking at the Lo Group and the Decliners however, it might be possible to conclude that poor reading *hinders* the development of vocabulary, although good reading may not necessarily boost it. Their vocabulary scores were similar (85.0) at this stage. Although all groups had actually logged a decline in vocabulary scores during this period, that of the Lo group was the greatest. Perhaps this was because their reading was so poor for so long and this would also explain the lower scores on the WISC verbal and performance scores (though these were still around the mean standard score). They could have missed out on the development of skills through reading which would normally happen before age 9.3 , if reading is at the normal level. When reading is two years behind as it was in this case, it resulted in low functioning on other tests in related areas. In support of this it may be seen that the gap between them and the Risers and Hi group on WISC Verbal and Performance IQ was widening slowly, but was not significant at this point (Fig. 8).

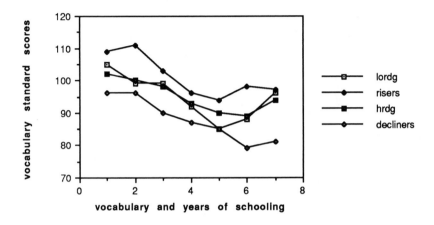

FIG. 7. Vocabulary scores over the period for all groups.

Progress from age 9.3 to 10.3

The Hi Group continued to make good progress, and recorded a gain of 15 months during this 12 month period. For them the only significant correlations with the Burt were scores on the group tests (p< .01 for reading, spelling and intelligence). The Risers however, really took off at this time, making an average gain of 27 months from their previous year's base of 8.9. This gave them a reading age of 11, when the average cohort age was 10.3 and their own average age was 10.00. The Burt was significantly correlated for them with Information and Vocabulary on the WISC (p<.05) and with the group reading and spelling tests (p<.01).

The Lo Group made an average of 8 months gain, less than during the previous 12 months, giving them a mean reading age of 7.11 at age 10.3. Their depressed scores on the Burt were mirrored by low scores on the EPVT and on the group tests. If assessed by the Burt and group tests alone at this stage they would probably have been classified as 'garden-variety' - poor readers (Gough and Tunmer,1986). Yet their cognitive abilities had been within the normal range as measured by the WISC at an earlier time. On the WISC they had dropped just below the mean by age 10.3. But this was not significant as their Verbal and Performance IQ's were about 99. They were, however, significantly lower than the Risers on Information and than the three other groups on Arithmetic. It is interesting that it was in these areas that the Risers had been particularly strong at age 6.5. There were significant correlations for the Lo group between their scores on the Burt and the Information, Arithmetic, Similarities Vocabulary and Object Assembly subtests of the WISC (p<.05).

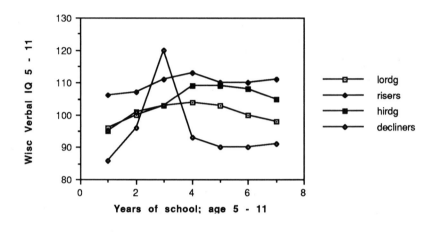

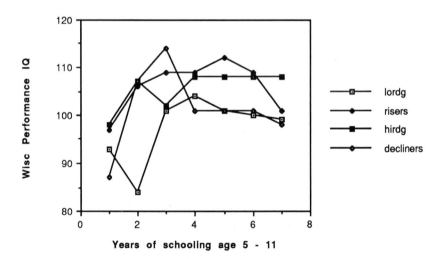

FIG. 8. Changes in Verbal and Performance IQ over time

The Decliners made an average gain of 14 months, almost as good as the Hi Group, and more than the Lo Group who had been at the same level at the end of the previous year. This brought them up to a reading age of 8.5 at age 10.3. Again, their WISC scores were well within the normal range (Verbal = 109; Performance = 101) and their group test scores were below 90, close to those of the Lo Group.

Progress from age 10.3 - 11.3

The Hi Group this year made the progress expected by the norms for the first time - about 12 months - ending their primary school years with a reading age of 11.10 and a mean cohort chronological age of 11.3. The Burt was correlated again with the group test scores (p< .05 for group reading and spelling and p<.01 for group intelligence), but not with vocabulary or WISC.

The Risers made a gain of 8 months on the Burt, giving them a reading age of 11.8 - possibly a ceiling effect was in operation both for this group and for the Hi Group. At this stage the Burt was correlated significantly for them with EPVT, Performance IQ and the Comprehension, Similarities, Picture Completion, Block Design and Object Assembly subtests of the WISC, but not with the group tests.

The Lo group made slower progress, ending up with a mean reading age of 7.9 at age 11.3. Their WISC scores were still in the high 90's, though they were significantly lower than the Risers on Verbal IQ (97.6/110.8; p<.05). They would still not qualify for classification as 'garden-variety' poor readers. Their group reading, spelling and IQ scores were in the mid to low 80's.

The Decliners also made very little progress - about 2 months on average. Perhaps they had given up trying. Their Burt scores were significantly correlated with spelling and they were significantly lower than the Hi and Riser groups on the group tests of spelling, reading and IQ. Their EPVT score was significantly lower than that of any other group.

Overall Progress

The results show that in this total group progress in reading was by no means steady, or equal for everyone. When the group was subdivided according to reading outcome at age 11 compared to reading level at age 6.5, it was found that there were no significant differences between the groups on verbal or performance IQ at the age of 6.5. Each group therefore must have had a share of those with lower IQ scores as well as those with higher IQ scores.

Those who had made a good start (the Hi Group) averaged a gain of 12.8 months per year on the Burt reading test (s.d. = 3.1) and were therefore at or about the expected gain rate according to the norms over the whole period. But from year to year the mean gain for the group varied from 9 months to 16 months and the mean gain of individuals inevitably covered a wider range.

The rate of improvement and the steadiness did not appear to be related to whether or not the group had made a fast start to reading, however, as the Risers had made a very slow start, and were already a year behind at age 7. Their average increase when they got going was 15.8 months (s.d. = 6.7) per year and they certainly made up for lost time. But here again the mean annual gain varied from 7 to 24 months. The Lo group which had also made a slow start, continued to move much more slowly than the other groups and had an average annual gain of 8.2 (s.d. = 6.5) per year with a range of 0 to 17 months. The Decliners' gain scores were lower than those of any of the others a mean gain of 5.8 months (s.d. 5.0) per year with a range of 0 to 11.

Years of Greatest Score Increases

All groups except the Decliners made their greatest gains in the year from age 6 to 7 and all groups except the Lo group also made high gains in the year when they were aged 9 to 10. This could be related to stages of reading acquisition as different stages of the code are cracked (see Frith, 1985; Ehri, 1991). The Lo group did not get to the second stage of rapid progress within the time period analysed in this paper.

Contribution of Family Background

This was a cohort sample from an economically deprived area in which 50% of heads of household were unemployed and the average family size was 5 children. Were employment status of father and number of children in the family different for the groups with different reading outcomes?

TABLE 5

Family Variables by Reading Outcome Group

	Father Employed	Mother Employed	5 or more children
Lo Group	38%	44%	80%
Risers	20%	20%	40%
Hi Group	73%	80%	69%
Decliners	77%	62%	62%

The two groups who had a slow start were also the most disadvantaged in one generally accepted sense in that more of them had unemployed parents. It would suggest that pre-school stimulation may have been affected by this aspect of family conditions. Cox (1987) concluded that a paucity of literacy experience in the pre-school years, as could happen in a home with low income, and a limited oral vocabulary resulting in a relatively poor interest in and concentration upon pre-reading and reading tasks in school, would be salient features in long-term reading retardation. In the present study, the Risers who had the highest rate of parental unemployment and therefore might be expected to have even more limited means of gaining literacy experiences also had very low reading scores at age 6.5. But their vocabulary was higher than that of the Lo group at age 6.5 suggesting a better quality of linguistic interaction in the preschool stage, and it may be significant that they came from smaller families than the Lo group.

The Lo group and the Risers also differed significantly at age 6.5 on the Arithmetic and Information subtests of the WISC. These tests may also be more predictive than other measures - the Decliners' scores were similar to those of the Lo group.

IMPLICATIONS

These results indicate that reading progress can vary widely between individuals of similar IQ on starting reading instruction. The rates of progress will also vary widely not only between those who make a slow start and those who make a good start, but within these groups from year to year.

Expectations of gain based on individual IQ and reading scores could be misleading, depending on the stage when the initial measures were made. If made at an early stage, IQ, individually measured, did not appear in this study to have much bearing on reading outcome since those who failed to make progress had very similar levels of IQ, at age 6.5, to those who shot ahead. If made at a later stage - say 9 years or more, the chances are that some IQ measures will have been affected by the failure to read, and true potential will either have been blocked, or will not be measurable. Performance IQ appeared to be important at the very early stages for those who made a good start and remained good readers, but was not enough in itself as shown by the Riser group.

The relationship of receptive vocabulary to reading score appeared also to vary over the years and to vary with level of reading skill. It is generally agreed that this relationship is not a simple one (Francis, 1974). It should not be taken for granted that the two should necessarily be highly correlated at all stages.

In addition to cognitive variables this study found that environmental variables could be important at the very early stages of schooling. Those who made a faster start to reading were less disadvantaged economically. But it was encouraging to note that the Risers, who made a poor start and were in fact very disadvantaged socio-economically, eventually made excellent progress. Their high receptive vocabulary, together with their strong performance on Performance IQ and the DAM suggest that, perhaps parsimoniously, part of their problem may simply have been lack of exposure to print. They were also differentiated from the other early low-achieving group by their scores on the Arithmetic and Information subtests of the WISC.

In this study we have seen what happens without specialised remediation and what some of the implications are for a group of socio-economically disadvantaged boys of average intelligence going through an elementary school system. One large group (the Lo Group) made very poor progress in reading. Although at age 6.5 they appeared to be functioning within the normal range of intelligence, their later IQ scores were substantially depressed. Another group started well, but for reasons which this study could not elucidate, their progress was not in line with their initial promise. It could be that some of the children classified as dyslexic in other school systems belong to one of these two groups. Although by a certain point their reading scores were similar, the routes by which they had arrived at that level of reading were very different. The aetiology of their problem was probably therefore dissimilar. The group who started off as poor readers and eventually caught up with their chronological age peers might also have been classified as 'dyslexic' in other systems, yet their initial difficulties must have had different roots from those of the Lo or Decliner group and their progress was made without specialised assessment and teaching.

An important variable not investigated here was the influence of individual teachers, stability of teaching and the application of teaching methods and materials on the outcomes for the different groups. More precise assessment of subskill differences between groups would also have been useful as more fine-grained investigative techniques would give detail that could be useful for remediation and provide an understanding of aetiology. The use of longitudinal data relating outcomes to initial measures will be essential to this end.

REFERENCES

Bakker, D.J. & van der Vlugt, H. (1989). *Learning Disabilities Vol I. Neurological Correlates and Treatment.* Amsterdam: Swets & Zeitlinger.

Boehm, A.E. (1969). *Boehm Test of Basic Concepts.* New York: Psychological Corporation.

Bradley, L. & Bryant, P.E. (1985). *Rhyme and Reason in Reading and Spelling.* Ann Arbor: University of Michigan Press.

Brimer, M.A. & Dunn, L.M. (1962). *English Picture Vocabulary Test.* Slough: NFER.

Burt, C. (1974). *Burt Word Reading Test.* London: Hodder & Stoughton.

Butler, S.R., Marsh, H.W., Sheppard, M.J. & Sheppard, J.L. (1985). Seven-year longitudinal study of the early prediction of reading achievement. *Journal of Educational Psychology,* 77, 349-361.

Cox, T. (1987). Slow starters versus longterm backward readers.*British Journal of Educational Psychology,* 57,1,73-86.

Dobbins, D.A. (1984). *The prevalence and characteristics of children with specific learning difficulties.* Final Report to the Department of Education and Science. Vol 1. London: HMSO.

Dobbins, D.A. & Tafa, E. (1991). The stability of identification of underachieving readers over different measures of intelligence and reading. *British Journal of Educational Psychology,* 61, 2, 155-163.

Dumont, J.J. & Nakken, H. (1989). *Learning Disabilities Vol II . Cognitive, Social and Remedial Aspects.* Lisse, Netherlands: Swets Publishing Service.

Ehri, L. (1991). Development of the ability to read words. In R. Barr, M. Kamil, P. Mosenthaland & P. Pearson (Eds.) *Handbook of Reading Research,* Vol.2. New York: Longman.

Ellis, N.C. & Large, B. (1987). The early stages of reading: a longitudinal study. *Applied Cognitive Psychology,* 2, 47 - 76.

Francis, H. (1974). Social background, speech and learning to read. *British Journal of Educational Psychology,* 44, 290-299.

Fredman, G. & Stevenson, J. (1988). Reading processes in specific reading retarded and reading backward 13 year olds. *British Journal of Developmental Psychology* 6, 97 - 108.

Frith U. (1985). Beneath the surface of developmental dyslexia. In K.E. Patterson, J.C. Marshall & M. Coltheart (Eds.) *Surface Dyslexia.* London: Erlbaum.

Goodenough, F.L. & Harris, D.B. (1963). *Draw-a-Man Test.* Harcourt Brace and World.

Gough, P.B. & Tunmer, W.E. (1986). Decoding, reading and reading disability. *Remedial and Special Education,* 7, 6-10.

Jorm, A.F., Share, D.L., Maclean, R. M. & Matthews, R. (1986). Cognitive factors at school entry predictive of specific reading retardation and general reading backwardness: a research note. *Journal of Child Psychology and Psychiatry,* 27, 45 - 54.

Juel, C. (1988). Learning to read and write: a longitudinal study of 54 children from first through fourth grades. *Journal of Educational Psychology,* 80, 437 - 447.

Kirk, S.A. & McCarthy, J.J. (1968). *Illinois Test of Psycholinguistic Abilities.* Urbana: University of Illinois Press.

Lawson, J.S. & Inglis, J. (1985). Learning disabilities and intelligence test results: a model based on a principal components analysis of the WISC-R. *British Journal of Psychology,* 76, 35-48.

Lundberg, I., Frost, J. & Peterson, O.P. (1988). Long-term effects of a preschool training program in phonetical awareness. *Reading Research Quarterly,* 28, 263-284.

Mackay, M.F., Neale, M.D. & Thompson, G.B. (1985). The predictive ability of Bannantyne's WISC categories for later reading attainment. *British Journal of Educational Psychology,* 55, 280-287.

Miles, T.R. (1991). On determining the Prevalence of Dyslexia. In M. Snowling and M. Thomson (Eds.) *Dyslexia: Integrating theory and practice.* London: Whurr.

Newman, S., Wright, S., & Fields, H. (1991). Identification of a group of children with dyslexia by means of IQ-achievement discrepancies. *British Journal of Educational Psychology,* 61, 2, 139-154.

Rutter, M. & Yule, W. (1975). The concept of specific reading retardation. *Journal of Child Psychology and Psychiatry,* 16, 181-197.

Scarborough, H.S. & Dobrich, W. (1990). Development of children with early language delay. *Journal of Speech and Hearing Research,* 33, 70 - 83.

Share, D.L. & Silva, P.A. (1986). The stability and classification of specific reading retardation: a longitudinal study from age 7 to age 11. *British Journal of Educational Psychology,* 56,1, 32 - 39.

Siegel, L.S. (1988). Evidence that IQ scores are irrelevant to the definition and analysis of learning disabilities. *Canadian Journal of Psychology,* 42, 210-215.

Siegel, L.S. (1989). IQ is irrelevant to the definition of learning disabilities. *Journal of Learning Disabilities,* 22, 469-478.

Snowling, M. & Thomson, M. (1991). *Dyslexia: Integrating Theory and Practice.* London: Whurr.

Stanovich, K. (1986). Matthew effects in reading: some consequences of individual differences in the acquisition of literacy. *Reading Research Quarterly,* 21, 360-407.

Stanovich, K. (1991). The theoretical and practical consequences of discrepancy definitions of dyslexia. In M. Snowling and M. Thomson (Eds.) *Dyslexia: integrating theory and practice.* London: Whurr.

Tallal, P. & Katz, W. (1989). Neuropsychological and Neuroanatomical Studies of Developmental language/reading disorders: recent advances. In C. von Euler, I. Lundberg & G. Lennerstrand (Eds.) *Brain and Reading.* Wenner-Gren International Symposium Series. London: MacMillan.

von Euler, C., Lundberg, I. & Lennerstrand, G. (1989). *Brain and Reading.* Wenner-Gren International Symposium Series. London: MacMillan.

Wechsler, D. (1971). *Wechsler Intelligence Scale for Children.* Windsor: NFER.

Whyte, J. (1989). *Dyslexia: Current Research Issues* (Ed.) *Irish Journal of Psychology Special Issue.* Vol 10, 4.

Whyte, J. (1980). Can we relieve the stress? A language stimulation programme in a reception class. In J. Harbison and J. Harbison (Eds.) *A Society under Stress*. London: Open Books.

Whyte, J. (1983). Educational enrichment with deprived children: the long-term consequences. In J. Harbison (Ed.) *Children of the Troubles*. Belfast: Stranmillis College.

Whyte, J. (1989). The long-term implications of a pre-school language intervention programme. In J. Harbison (Ed.) *Growing up in Northern Ireland*. Belfast: Stranmillis College.

Wright, S.F. (1991). The identification of reading and spelling problems in school children: a longitudinal study. Unpublished doctoral thesis. University College, London.

Young, D. (1973). *Group Reading Test*. London: Hodder and Stoughton

Young, D. (1973). *Spar Reading and Spelling Tests*. London: Hodder and Stoughton

Young, D. (1973). *Non-readers Intelligence Test* . London: Hodder and Stoughton

Young, D. (1973). *Oral Verbal Intelligence Test*. London: Hodder and Stoughton

Facets of Dyslexia and its Remediation
S.F. Wright and R. Groner (Editors)

DEVELOPMENT OF THE DEST TEST FOR THE EARLY SCREENING FOR DYSLEXIA.

A. J. Fawcett, S. Pickering, and R. I. Nicolson

Department of Psychology, University of Sheffield, UK

INTRODUCTION

It has been for many years a dream of dyslexia researchers to devise a screening test which allowed one to pick out those children likely to have problems *before* they start learning (and failing) to read (eg. de Hirsch, Jansky & Langford, 1966; Bender, 1970). Such early information should allow the prophylactic support of reading acquisition in the 'at risk' children, thus allowing them to learn to read at the same rate as their peers, and avoiding the vicious circle of failure, negative expectations, possible disruptive behaviour, reading avoidance, and continuing failure which besets people suffering early set-backs (Holt, 1970; Stanovich, 1986). There seems little doubt that the approach is fundamentally sound, in that studies which have identified 'at risk' children early have been able to help them to learn to read, and subsequent problems have rarely developed (Lundberg, Olofsson & Wall, 1980; Bradley, 1988; Clay, 1991). Perhaps most telling in this regard is the analysis by Strag (1972) which showed a massive increase in reading failure in children as a function of the age at which dyslexia was diagnosed. Strag (p 52) noted;

"When the diagnosis of dyslexia was made in the first two grades of school nearly 82% of the students could be brought up to their normal classroom work, while only 46% of the dyslexic problems identified in the third grade were remediated and only 10-15% of those observed in grades five to seven could be helped when the diagnosis of learning problems was made at those grade levels."

Unfortunately, however, researchers wishing to develop an early screening test for dyslexia face several formidable problems. The major problem is the lack of a definition of dyslexia except in terms of reading outcomes, which means that it is impossible to diagnose dyslexia definitively before a child fails to learn to read. Consequently, it is necessary to undertake longitudinal studies, testing a cohort of children at the age of say 5, 6 and 7 years, identifying which of them are dyslexic at the end of of the study, and then using statistical methods to decide which of the tests used were, in retrospect, the best predictors at the age of 5 years. It is fair to say that the early studies were by no means conclusive, and even now there is a disappointing failure to account for a high proportion of the variance in the reading data.

We believe, however, that it now timely to undertake a further effort to construct a 'Dyslexia Early Screening Test' (DEST), principally for two reasons. First, new theoretical developments in dyslexia research have suggested that there may well be indicators of dyslexia which are quite independent of reading, and are identifiable using carefully designed tests of information processing. Furthermore, we believe that such characteristics, though capable of improvement over the long term, are much less susceptible to short-term training effects which account for much of the variance in reading skill. Sensitive tests of information processing require computer presentation and scoring. The second factor which supports our belief that the time is ripe for the DEST is the advent of the new generation of microcomputers such as the Apple Macintosh — cheap, portable, very powerful, and capable of multimedia input and output, using visual and auditory modalities. In early research (Nicolson, note 1) we have shown that it is possible to provide a faithful replication of laboratory tests of information processing, with the computer capable of automatically presenting the stimuli, administering the test, and scoring and interpreting the results for a battery of computer-based tests. Furthermore, not only are the results objective and reliable, as one would expect, with carefully designed programs they are valid, in the sense that they mirror the results obtained by traditional methods.

The organisation of the remainder of the chapter is as follows. First we shall present a brief overview of the recent studies of prediction of dyslexia, leading to an analysis of the desirable requirements for the DEST. Next we shall briefly assess the theoretical developments which encourage us to believe that there are robust early indicators of dyslexia. Finally we shall present an overview of the design of a research study we have just embarked upon which has the objective of constructing a suitable battery of tests.

APPROACHES TO PREDICTING READING PROBLEMS AND DYSLEXIA

Much of the early work on prediction of dyslexia was carried out by researchers interested primarily in the development of language and reading. More recently valuable contributions have been made by experimental psychologists and by cognitive psychologists. Although there has been some useful cross-fertilisation of ideas, it appears that, as with many areas of science, this cross-fertilisation has been disappointingly incomplete. The consensus of several decades of work on cognitive psychology was that understanding of 'big' problems can come only from a multi-disciplinary approach to the problem, attacking it from a range of perspectives (Newell, 1973; Allport, 1975). There is no doubt that dyslexia represents a diffuse and complex problem, with many perspectives having much of worth to contribute. In this section we present a brief overview of several strands which we believe can be brought together.

(i) Studies using Reading as the Outcome Measure

Many predictive studies have been reported (eg. Badian, 1982, 1986,1990; Bradley & Bryant, 1978, 1985; Butler, Marsh, Sheppard & Sheppard, 1982, 1985; Lundberg, Olofsson & Wall, 1980; Satz, Taylor, Friel & Fletcher, 1978; Stevenson & Newman, 1986; and Wolf, 1991) some investigating 'pure' dyslexia and others investigating the general issue of 'poor readers'. A useful review of the early literature may be found in Horn and Packard (1985) who conducted a meta-analysis of 58 studies. More recent reviews have been provided by Satz & Fletcher (1988), and, as introductions to their studies, by Badian (eg. 1990), by Jansky (eg. 1989) and by Scarborough (eg. 1991).

Badian's series of studies provide prototypical and good examples of the approach. In early studies (Badian, 1982), Badian had derived the Holbrook Screening Battery which incorporated a variety of tests found to be useful in the prediction of general reading problems. The HSB included 14 tests in four sections (verbal, preacademic, fine motor and gross motor skill). Badian (1990) reported a study in which the HSB was administered at 4 years, with reading performance measured at grades 3 and 8. Background factors (test behaviour, speech delay, handedness, birth history, birth order, socio-economic status and family history of learning disabilities [FHLD]) were also tested, and also proved important in reading attainment, with the group identified as positive on background factors having an average attainment 2.5 years below that of the 'no background factors' group. The HSB tests accounted for 33% of the variance in reading at grade 8 for the boys (36% for the girls), whereas HSB plus background factors accounted for 35% and 43% respectively.

The results for predicting individual children were not particularly encouraging, in that only 18 (out of 38) poor readers at grade 8 had been predicted as 'high risk' (out of 56) via the HSB. On the other hand, of the 337 children in the 'low risk' HSB group only 20 became poor readers. These results are reasonably representative of similar studies, with relatively good prediction of good readers, but relatively poor prediction of poor readers.

Discriminant function analysis was then undertaken to identify those sub-tests which with hindsight gave the best predictions. Taking the HSB and Background Factors together, for boys the most predictive tests were: Colour naming, FHLD, gross motor skills and Information. For girls, the most predictive tests were Colour naming, FHLD, Speech Delay, Similarities, Information (Weschler, 1963), and Test Behaviour. Naturally, use of these tests significantly improved 'predictive' performance on the given sample to 29 (out of 38) of the poor readers who were 'high risk' (out of 70). Of the 316 in the low risk group only 9 became poor readers.

A further study (Badian, McAnulty, Duffy & Als, 1990), in which Badian and her colleagues applied a similar approach to predicting dyslexia, provides a fascinating contrast. In this study a considerably larger battery of tests was administered at age 6 (a total of 35 tests, including 7 suites of 'nonverbal neuropsychological tests', 9 suites of 'verbal neuropsychological tests', 6 pre-academic tests and a full electrophysiological mapping analysis). Outcome measures were reading scores at age 8 and 10. Interestingly, however, girls were not included, and neither were boys with both verbal and performance IQ below 90 or with neurological or emotional problems or serious sensory impairment. This exclusionary procedure concentrated the study on those children most likely to be dyslexic.

Dyslexic children were identified in grade 4 using the criterion of a reading standard score at least 1.5 standard deviations below the mean. This led to identification of 8 dyslexic children (who were, of course, the 8 worst readers) out of a total of 163. A stepwise regression was performed on the 35 tasks as predictive variables and reading as the dependent variable. This led to identification of the 'top 6' independent predictors of reading performance. Of these, three variables were the best differentiators of the dyslexic from the normal readers, namely Sounds (speaking the sound associated with each letter), Rapid Automatised Naming of numbers (the time taken to complete naming each digit in turn on a chart with 50 digits), and Finger Localisation (Benton, 1959, in which the child has to indicate on a diagram of two hands which finger or fingers were touched by the examiner under various conditions). A 'predictive mean' was then derived, based on normalised scores for the three tests. Appropriate setting of the cutoff for predictive mean performance led to excellent performance, with identification of all 8 dyslexic children as 'high risk' (out of 11) and correct identification of 160 (out of 163) non-dyslexic children as 'low risk'.

The contrast between these two studies illustrates several important issues in test design. First, one must expect a high difference in apparent success rate between truly predictive tests and retrospective tests, in that in the latter the 'predictors' are guaranteed to do a pretty good job because they are attempting to fit the self-same data from which they were derived. As Badian and her colleagues noted, the value of their 'predictive mean' can only be determined when it is used in a further, predictive study. Second, the choice of population to sample can have a significant effect. The Badian et al study effectively excluded non-dyslexic 'backward readers' from their sample by use of the IQ-based selection criterion. Consequently, there is no way of telling whether their 'predictive mean' would also serve to discriminate the dyslexic children from the non-dyslexic 'backward readers'. Other studies suggest that it probably would not. For instance, a similar study by Horn and O'Donnell (1985) which explicitly contrasted 'low-achieving' and 'learning disabled' (LD) children found that " ... *many of the variables that significantly discriminated low achievers from higher achievers did not significantly discriminate LD children from non-LD children*" (p1114). In particular " *.. finger localisation was associated only with low reading achievement, not with reading* *disabilities"*. A third issue raised by the Badian experiments also relates to the selection of participants. In the Badian et al study, only 8 dyslexic children (out of a total of 163 participants) were found, that is 5%, roughly in line with the usual population analysis for dyslexia (eg. Badian, 1984; Jorm et al, 1986). Consequently, 95% of the total effort in the study is devoted to basically 'uninteresting' children from the point of view of dyslexia. Not only is this a depressing picture, but it means that the generality of the findings is doubtful, based on a sample of only 8 dyslexic children. Scarborough (1990, 1991) has adopted a more pragmatic approach in attempting to round up a reasonable sample of dyslexic children by recruiting younger siblings where there is known familial incidence of dyslexia, though it must be said that this may well lead to the need for very much more careful statistical analyses.

In summary, although there has been some progress since the mid-1980s, subsequent reading-based studies give little reason to dispute Bryant's gloomy analysis of the prospects for predictive tests (1985, p45) *"There is no point in anyone trying to prevent children ever becoming dyslexic without having a precise way of detecting which child among a group of three or four year olds is likely to do much worse at learning to read than would be expected from his or her general intelligence.... The problem is however that no such measure exists."*

(ii) Intervention Studies

One problem with purely correlational studies is that a correlation between some aspect of early performance and subsequent reading problems does not demonstrate that the reading problems are *caused* by the initial performance deficit. The causal link may be via some third aspect which underlies the two, or maybe some complex interaction between several processes. In an incisive analysis of the situation Bradley & Bryant (1983) argued that the only way true causality could be demonstrated was via a training programme — if factor X correlates with subsequent reading problems **and** training on factor X prevents subsequent reading problems **then** it is reasonable to assume that factor X has a causal role in the development of the reading problems. They applied this approach to phonological factors, in particular, phonemic segmentation and rhyme detection skills, and showed that phonological skill did indeed fulfil the role of Factor X, in that early phonological deficits did correlate with later reading difficulties, and that early phonological training led to much improved subsequent reading.

This and related studies provided strong support for the 'Phonological Deficit' theory for the origins of dyslexia (see Stanovich, 1988, for a powerful espousal of the approach). It should be noted, however, that the strength of the support is somewhat reduced by three factors. First, there is little doubt that phonological skills do underlie the acquisition of reading, and so it is hardly surprising that training therein benefits subsequent acquisition of reading, a point discussed in subsequent publications (eg. Bryant, Mclean & Bradley, 1990). Second, while the results do support the causal role of phonological skills, they are of course silent on whether the phonological deficit is itself attributable to some deeper underlying deficit, which might perhaps be uncovered via some skill unrelated to reading. Third, it should be noted that the research design makes no distinction between dyslexic children and ordinary backward readers. Since the training study appeared to be equally effective for all the participants, it seems less likely that it is uncovering a deficit specific to dyslexia.

(iii) Studies of Dyslexic Children

An alternative approach to the reading-based studies is to work with older children, in particular those already diagnosed as dyslexic, and to identify factors which discriminate them from non-dyslexic children matched on various criteria. Methodological problems also arise with this approach, with the main issue being that of the appropriate control group.[1]

In an early attempt to devise a screening test for dyslexia which was based on positive indicators rather than exclusionary criteria, Miles (1983) developed the Bangor Dyslexia Test. This test is intended as a quick screening device for use by doctors, teachers, speech therapists etc. to identify whether a child's difficulties are or are not typically dyslexic, and has been used with children from 7 to 18 years old. Basically the test looks for areas of 'incongruity' which have been found to be typical of dyslexia, and if sufficient 'positive indicators' are present, this makes it likely, according to the author, that other manifestations will be found. Tests used are: left-right confusions; repetition of polysyllabic words; subtraction; tables; saying the months forwards; saying the months backwards; forwards and reverse digit span; b-d confusion; and familial incidence.

Studies of skill development suggest that different skills may be critical in different stages of development (see, for instance, Gathercole & Baddeley, 1991). Consequently, it would be extremely valuable if some proven indicators of dyslexia could be found that were less dependent upon the developmental sequence of acquisition of reading and of taught skills such as tables and reading, and which might tap the underlying cause of the difficulties. An early finding was in Rapid Automatised Naming (Denckla & Rudel, 1976) and, as discussed above, RAN deficits of colours and numbers appear to be useful predictors of dyslexia (see also Wolf, 1991), thus providing some support for the

[1] Naturally if one matches on chronological age (CA), then by definition the dyslexic children will be impaired on reading-related measures, and thus one must exclude reading-related measures from comparative analyses. An alternative approach (see Bryant and Goswami, 1986, for an analysis of its strengths and weaknesses) is to use younger children of the same reading age (RA), that is, the RA-match design. With an RA-match design, it is quite legitimate to analyse the various components of the reading process, although the overall power of the analysis will be diminished because RA is held equal overall. In either CA or RA designs, it is important to note that verbal or performance IQ is usually matched also, and so this will artifactually reduce any IQ-related contributions.

approach. Much research has concentrated upon phonological difficulties, with the most promising for diagnostic tests being early deficits in the ability to detect rhyme (Bradley & Bryant, 1983), and to repeat nonsense words (Snowling, Goulandris, Bowlby & Howell, 1986). Interestingly, the latter 'nonword repetition' task is now thought to be more a test of memory span than of phonological skill (Gathercole & Baddeley, 1990), and hence it may be used as a more natural alternative to the standard digit span tasks.

Of considerable potential significance is our recent research (Nicolson & Fawcett, 1990, this volume) which has identified problems in a range of skills unrelated to reading or to phonological processing. Probably the most promising findings for diagnostic purposes here are the dual task deficits in balance, the speed of information processing, and the rate of articulation, all of which appear capable of successfully differentiating our groups of around 25 dyslexic adolescents from groups of non-dyslexic children matched for age and intelligence. Indeed, a longitudinal analysis (Nicolson, Fawcett & Baddeley, note 2) has suggested that it is likely that the dyslexic children's deficits in such tasks are probably most easily identified early on, for instance, in the pre-school period, and actually become less marked as they grow older. Of course, in the absence of data on pre-school performance on these tasks there is no firm foundation for such assertions, but the approach nonetheless seems likely to prove fruitful as a further approach to the prediction of dyslexia, especially as new educational technology has become available.

RECENT DEVELOPMENTS IN EDUCATIONAL TECHNOLOGY

Information processing speed is best tested using a microcomputer and so has been beyond the scope of traditional pencil, paper and stop-watch testing methods. Furthermore, evaluations of early computer-based implementations of existing tests demonstrated that though the results were reliable and objective, they were not valid in that they differed significantly and uninterpretably from the pencil and paper results (Beaumont, 1985). However, the recent development of the new generation of microcomputers such as the Apple Macintosh with multimedia capability allows the storage of digitised sound, together with very natural interaction via touch screen or mouse. A recent demonstration of the potential of these techniques was the COTAL test (Van der Leij & Reitsma, 1990) which provides a sensitive test of reading automaticity. In an explicit comparison of multimedia with traditional approaches to administering a memory span experiment, Nicolson (1991) concluded that the multimedia test was not only reliable and objective but it was also valid, administering the test in essentially identical fashion to the traditional method, and it was also significantly easier to administer. This technological breakthrough provides the opportunity for constructing a new generation of psychometric tests, more sensitive than the traditional tests and more easily administered, thus de-skilling the administration requirement, and enabling large scale, low cost screening for dyslexia (and other problems).

REQUIREMENTS ANALYSIS FOR THE DEST

The above sections complete our review of those factors to be considered in designing the DEST, and formed the background against which we attempted to design our DEST study. Before starting on a longitudinal project of this nature, it is important to be clear about factors such as what and who we intended to test, why we were testing them, how we were going to do it, and, of course, what the long-term objective of the research would be. These are by no means trivial design issues, and we discuss the decisions taken below.

Issue 1. The Objective of the DEST
This issue was easily resolved for us. We wanted to design a screening test, administered to pre-school children, which would identify individual children as 'at risk' for dyslexia or not. It should also be able to discriminate between high risk for dyslexia and high risk for 'backward reading'. In short, the objective is to develop a screening test which satisfies four criteria:
- it yields a direct index of degree of dyslexia
- it may be used to diagnose 'at risk' children before they fail to learn to read.
- it is objective, reliable and valid
- it is easily administered

Issue 2. Mode of Use
Following most other researchers (eg. Butler, 1985, Satz & Fletcher, 1988) we would argue that ideally the prediction process should be two stage, with first an initial screening, which should be over-inclusive if anything, followed up if necessary by further more detailed individual testing. As Satz et al argue, *'True screening is rapid and cost effective and does not require professional interpretation'*. Note that this implies that the screening test must be quick, and easy to administer to large numbers of children. This rules out intensive studies such as the brain electrophysiology studies undertaken by Badian, McAnulty, Duffy & Als (1990). Fortunately, the advent of the multimedia testing programs allows rapid, accurate, and objective testing to be undertaken by only semi-trained experimenters, providing an ideal environment for individual screening.

Interestingly, the need for individual testing offers an exciting opportunity for providing individualised diagnostic testing in that, following the initial screening, it might be possible to construct a diagnostic computer program which automatically identified the likely diagnostic category, and administered further tests as appropriate selected to further investigate any underlying problems. Of course, this capability is as yet a long way off. The first priority is to devise a screening test!

Issue 3. Populations of children to be sampled.

In view of the need to end up with a reasonable pool of say 30 dyslexic children, and the need to attempt to dissociate dyslexia from backward learning, it would be necessary to have a very large initial sample of children (say 1000) if no sampling bias were to be included. For our initial study with a wide range of tests this seemed prohibitively time-consuming, and so we felt it best to adopt the method used by Scarborough (e.g.1991), and to incorporate a set of children thought to be at high risk for dyslexia via familial factors (ie. pre-school children with siblings who had already been identified as dyslexic). Clearly such an approach has attendant hazards in terms of statistical sampling bias, but it seems preferable in the first instance to avoid generalising from an over-small pool of dyslexic children. In order to avoid the critique of failure to discriminate between dyslexic children and backward readers, it is necessary to find a pool of children at risk for reading backwardness, again through familial incidence, and perhaps via information from health visitor records and the like.

Issue 4. Types of test

Since the objective of this initial study was to administer a wide range of tests, with the intention of selecting those which proved the most 'predictive' (in retrospect, once the children involved had tried to learn to read), we felt it important to administer tests drawn from each of the various approaches discussed above, and thus ended up with four different classes of test:

- General tests
- Tests identified in the reading-based prediction literature
- Tests identified in the dyslexia-based literature
- Theoretically based tests of information processing and of learning.

The specific tests involved are detailed in a subsequent section. The COMB test (Combined Operations Multimedia Battery, Nicolson 1991) is the suite of theoretically based computer administered tests presented on the Apple Mac using Hypercard.

TABLE 1

Major Categories of Tests

Test	Reference	Method
Psychometric Tests B.P.V.S.	Dunn, Dunn and Whetton (1982)	Paper
British Ability Scales (10 sub-scales)	Elliott, Murray and Pearson (1978)	Paper
Predictive Tests		
Left-Right confusions	Bangor Dyslexia Test	Paper
Finger Localisation	Benton (1959)	Paper
Phonological/Memory Skills		
Non-Word Repetition	COMB[2]	Mac
Phonological Discrimination	COMB[3]	Mac
Prepositions		Verbal
Rhyming Task	COMB	Mac
Word Segmentation		Verbal
Motor Skill and Automaticity		
Blindfold Balance		Motor
Articulation Rate	COMB	Mac
Beads task		Beads
Draw - a - Person	Koppitz (1968)	Pencil
Pegboard Task	Annett (1970)	Board
Information Processing Speed		
Visual Search		Paper
Simple Reaction	COMB	Mac
Choice Reaction	COMB	Mac
Lexical Access	COMB	Mac
Naming	COMB	Mac
Background Factors		
FHLD		Questionnaire
Handedness		Scissors etc.
SES		Interview
Test Behaviour		Rating
Tests of Learning		
Paired Associates	COMB	Mac
Choice Reaction Learning	COMB	BBC micro

THE DESIGN OF THE STUDY

The design is based on the following plan:

(i)   Prepare a large battery of objective tests, suitable for 4 to 7 year old children, and based on the above four general types of test.

---

[2] Based on the technique used by Gathercole and Baddeley, 1990.

[3] A modification of the technique used by Bishop (1990)

(ii)   Select a panel of subjects, aged just under five years, including a large cross-section of 'normal' children, plus as many children 'at risk' for dyslexia as we can find, plus 'at risk' children with backward reading.

(iii)   Administer the battery of tests to the panel, then again after one year, and again after two years.

(iv)   Using the standard tests for dyslexia, determine those children who would be diagnosed as dyslexic and those who would be diagnosed as 'backward readers' at 7 years.

(v)   Using stepwise regression and discriminant function techniques, determine which items of the battery were the best 'predictors' of dyslexia and those which were the best 'predictors' of non-dyslexic backward reading.

(vi)   Construct a small, easily administered battery which includes the most predictive tests.

(vii)   Distribute the tests widely, undertake genuinely predictive studies, and refine the norms over the course of the next few years.

Of course the tests outlined in Table 1 are merely one selection from the range of possible tests. They have been selected following both theoretical and pragmatic criteria. If any researchers feel that we should include some others we would be delighted to hear from them.

CONCLUSIONS

We believe that there is now a better opportunity than ever to construct a predictive test for dyslexia, owing to the recent development not only of new theoretical tests for dyslexia which are less dependent on the reading process, but also because of the availability of much more powerful, yet affordable, multimedia testing batteries. We believe that it is only through international cooperation and sharing of ideas that a solution to the problems of diagnosing, predicting, remedying and understanding dyslexia will be found. We offer this analysis of the possibilities for developing the DEST in a constructive spirit, offering the analysis as a starting point for cognate researchers to suggest modifications and introduce improvements.

ACKNOWLEDGMENTS

The research discussed here was funded by a grant from the Leverhulme Trust to the third author and by a research studentship from the British Science and Engineering Council to the second author. We are particularly grateful to Professor Tim Miles for several long discussions on the basis of the predictive tests.

Note 1: Nicolson, R. I. (submitted). Hypermedia: An enabling technology for empirical psychology. Currently available as Report LRG 6/90, Dept. of Psychology, University of Sheffield.

Note 2: Nicolson, R.I, Fawcett, A.J. & Baddeley, A.D. (submitted). Working Memory and Dyslexia. Currently available as Report LRG 3/91, Department of Psychology, University of Sheffield.

REFERENCES

Allport, D.A. (1975). The state of cognitive psychology. *Quarterly Journal of Experimental Psychology*, 27, 141-152.

Annett, M. (1985). *Left, Right, Hand and Brain: the Right shift theory*. London: Lawrence Erlbaum.

Badian, N.A. (1982). The prediction of good and poor reading before kindergarten entry: A four year follow up. *Journal of Special Education*, 16, 308-318.

Badian, N.A. (1984). Reading disability in an epidemiological context: Incidence and environmental correlates. Journal of Learning Disabilities, 17, 129-136.

Badian, N.A. (1986). Improving the prediction of reading for the individual child: A four year follow-up. *Journal of Learning Disabilities*, 19, 262-269.

Badian, N.A., McAnulty, G.B., Duffy, F.H. and Als, H. (1990). Prediction of dyslexia in kindergarten boys. *Annals of dyslexia*, 40, 152-169.

Beaumont, J.G. (1985). The effect of microcomputer presentation and response medium on digit span performance. *International Journal of Man-Machine Studies, 22*, 11-18.

Bender, L. (1970). Use of the Visual-Motor Gestalt test in the diagnosis of learning disabilities. *Journal of Special Education, 4*, 29-39.

Benton, A. (1959). *Right-left discrimination and finger localisation: Development and Pathology*. New York: Hoeber-Harper.

Bradley, L. & Bryant, P.E. (1978). Difficulties in auditory organisation as a possible cause of reading backwardness. *Nature*, 271, 746-747.

Bradley, L. & Bryant, P.E. (1983). Categorising sounds and learning to read: A causal connection. *Nature*, 301, 419-421.

Bradley, L. & Bryant, P.E. (1985). *Rhyme and reason in reading and spelling*. International Academy for research in learning disabilities series. Michigan: University of Michigan Press.

Bradley, L. (1988). Making connections in learning to read and to spell. *Applied Cognitive Psychology*, 2, 3-18.

Bryant, P.E. (1985). The question of prevention. In M.J.Snowling, (Ed). *Children's writ-ten language difficulties.* Dorset: NFER Nelson

Bryant P., Mclean, M. & Bradley, L. (1990). Rhyme, language and children's reading. *Applied Psycholinguistics,* 11, 237-252.

Butler, S.R., Marsh, H.W., Sheppard, M.J. & Sheppard, J.L. (1982). Early prediction of reading achievement with the Sheppard School Entry Screening Test.: A four year longitudinal study. *Journal of Educational Psychology,* 74, 280-290.

Butler, S.R., Marsh, H.W., Sheppard, M.J. & Sheppard, J.L. (1985). Seven year longitudinal study of the early prediction of reading achievement. *Journal of Educational Psychology,* 77, 349-361.

Clay, M. (1991). Reading recovery: a guilt edged investment. Paper presented to 'Meeting the Challenge' 2nd International Conference of the BDA, Oxford, UK.

de Hirsch, K., Jansky, J.J., & Langford, W.S. (1966). *Predicting reading failure.* New York: Harper and Row.

Denckla, M.B. & Rudel, R.G. (1976). Rapid 'Automatized' naming (R.A.N.). Dyslexia differentiated from other learning disabilities. *Neuropsychologia, 14,* 471-479.

Dunn, L. M., Dunn, L. M. & Whetton, C. (1982). *British Picture Vocabulary Scale.* Dorset, NFER-Nelson.

Elliott, C.D., Murray, D.J., & Pearson, L.S. (1978). *The British Ability Scales.* Windsor: NFER-Nelson.

Gathercole, S.E. & Baddeley, A.D. (1990). Phonological memory deficits in language disordered children: Is there a causal connection? *Journal of Memory and Language,* 29, 336-360.

Holt, J. (1970). *How children fail.* Middlesex: Penguin.

Horn, W.F. & Packard, T. (1985). Early identification of Learning Problems: A meta-analysis. *Journal of Educational Psychology,* 77, 597-607.

Jansky, J.J., Hoffman, M.J. , Layton, J. & Sugar, F. (1989). Prediction: A six year follow-up. *Annals of Dyslexia,* 39, 227-246.

Jorm, A.F., Share, D.L., Maclean, R. & Matthews, D. (1986). Cognitive factors at school entry predictive of specific reading retardation and general reading backwardness: A research note. *Journal of Child Psychology and Psychiatry and Allied Disciplines,* 27, 45-54.

Koppitz, E.M. (1968). *Psychological Evaluation of Children's Human Figure Drawings.* New York: Grune and Stratton.

Lundberg, I., Olofsson A., & Wall, S. (1980). Reading and spelling skills in the first school years predicted from phonemic awareness skills in kindergarten. *Scandinavian Journal of Psychology,* 21, 159-173.

Miles, T.R. (1982). *The Bangor Dyslexia Test*. Cambridge: Learning Development Aids.

Miles, T.R. (1983). *Dyslexia: The Pattern of Difficulties*. London: Granada.

Newell, A. (1973). You can't play 20 questions with nature and win. In W.G. Chase (Ed.) *Visual Information Processing*. New York: Academic

Nicolson, R.I. & Fawcett, A.J. (1990). Automaticity: a new framework for dyslexia research? *Cognition*, 30, 159-182.

Satz, P. & Fletcher, J M. (1988). Early identification of learning disabled children: An old problem revisited. *Journal of Consulting and Clinical Psychology*, 56, 6, 824-829.

Satz, P., Taylor, H., Friel, J. & Fletcher, J. M. (1978). Some developmental and predictive precursors of reading disabilities: A six year follow-up. In A.L. Benton & D. Pearl (Eds.). *Dyslexia: An appraisal of current knowledge* (pp. 457-501). New York: Oxford University Press.

Scarborough, H. (1990). Very early language deficits in dyslexic children. *Child Development*, 61, 1728-1743.

Scarborough, H. (1991). Antecedents to reading disability: Preschool language development and literacy experiences of children from dyslexic families. *Reading and writing* (in press).

Snowling, M.J., Goulandris, N., Bowlby, M. & Howell, P. (1986). Segmentation and speech perception in relation to reading skill: a developmental analysis. *Journal of Experimental Child Psychology*, 41, 487-507.

Stanovich K.E. (1986). Matthew effects in reading Some consequences of individual differences in the acquisition of literacy. *Reading Research Quarterly*, 21, 360-407.

Stanovich, K.E. (1988). The right and wrong places to look for the cognitive locus of reading disability. *Annals of Dyslexia*, 38, 154-177.

Stevenson, H.W. & Newman, R.S. (1986). Long-term prediction of achievement and attitudes in mathematics and reading. *Child Development*, 57, 646-659.

Strag, G. (1972). Comparative behavioural ratings of parents with severe mentally retarded, special learning disability, and normal children. *Journal of Learning Disabilities*, 5, 52-56.

Van der Leij, A. & Reitsma, P. (in press). The development of a computer assisted training program of reading disabled children. In L. de Leeuw, R.J. Simons, and B.J.M. Pieters (Eds.). *Computer and instruction*, Lisse: Swets and Zeitlinger.

Wechsler, D. (1963). *Wechsler Preschool and Primary Scale of Intelligence*. New York: Psychological Corporation.

Wolf, M. (1991). Naming speed and reading: The contribution of the cognitive neurosciences. *Reading Research Quarterly*, 26, 123-141.

VII. REMEDIATION:

PRINCIPLES AND TECHNIQUES

Facets of Dyslexia and its Remediation
S.F. Wright and R. Groner (Editors)
499

DYSLEXIA THERAPY: IN SEARCH FOR A RATIONALE

F. Gaillard
Lausanne University, Institute of Psychology,
Switzerland

RESEARCH, EDUCATION AND THERAPY: THREE DIFFERENT CONTEXTS

Whereas single case research studies are necessary in order to provide the scientific community with foolproof evidence of the different conditions and processes which may determine child illiteracy, the actual education and therapy of that child has to be carried out within the social context.

Firstly, education at school is primarily concerned with overall equal opportunity, a fact that militates against too strong an involvement of the normal teacher in any particular child's case. Furthermore, teachers neither have the time, nor the range of methodological alternatives, to adapt their teaching to specific problems encountered by a child in difficulty. Without a different pedagogical strategy, we know that repeating normal teaching methods over and over again with a dyslexic child only leads to saturation and phobic reactions against school and against the collusion of adults who want the child to do what he is simply unable to do.

Special education is then required to establish new training methods and, above all, to preserve the desire to learn in general, whatever the difficulties with reading might be.

Secondly, therapy often starts by reconciling certain anxieties about the dyslexic child: the child himself, his parents and teachers come to a crisis point where everyone admits the need for special attention and is ready to accept an alternative teaching method.

With this point in mind, we might ask ourselves where the differences lie between researcher, teacher and therapist working together. While, it is clear that the researcher must find a way of reconciling those involved to the idea of accepting the stigma which is attached to the 'patient' visiting the examination clinic, he or she, nonetheless, starts a process of isolating the different variables at play, disentangling the cognitive processes in order to categorize the related problems. His/her mode of thinking is 'partition', i.e. to come to some model showing default or dysfunction in single units.

In complete contrast, both educational and therapeutic perspectives are holistic in essence: their goal is to unify, or, if necessary, reunify the image of the child instead of symbolically slicing the problem up, like histopathologists are able to do. Furthermore, the familial and school environment is taken into consideration with a view to assessing its importance in relational terms: the 'problem' in question touches every one in this social environment and, contrary to what the researcher observes, appears more important at the level of emotional repercussions than the realistic and objective dysfunction. Figure 1 depicts the complex net of influences in which the child is caught when designated as a patient.

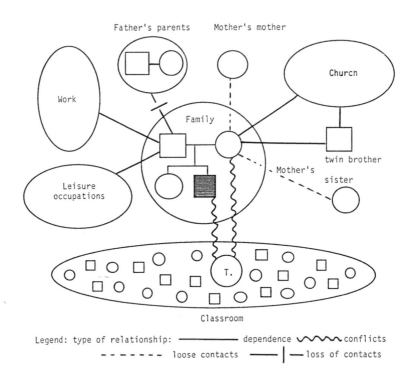

FIG. 1. Ecomap of case ■

This is perhaps why most therapists first adopt a general view of the problem before looking at particular aspects. Historical reasons can also be found for this approach. Indeed psychoanalytic, existentialist and behavioural therapies, - to name but a few - addressed the suffering individual long before dyslexia was even recognized as a specific learning disability. On the other hand, it is impossible to deny that dyslexic children are under stress, depressed and neurotic when confronted with stigmatizing tasks. They are tense, hypertonic, sometimes hyperkinetic, and awkward since they are for the most part acting hurriedly and unreflectively.

Psychotherapy for relational and emotional disorders works in some way, as do relaxation and counteracting therapies through play, art, music, movement and dance, and group activities.

The parent-child relationship itself is sometimes more difficult to treat, for the learning problem has disturbed the family homeostasis, i.e. the internal rules governing interaction and balance in role distribution. One major question for parents is to find out ' to what extent they can go too far' or how to measure their effort in solving the child's problem: should they mobilize all the therapeutic efforts and methods available on the planet ? At what point should they stop their search for the ideal remedy? To what extent can they afford to confront the teacher already in charge of the child at school? To what extent can they hurt the child him or herself with their often exaggerated concerns and interventionism ?

It is sometimes strange for researchers to observe the difficulties parents may have in understanding the subtle interplay of the different factors to be taken into consideration. They may hear about predisposition in the child (what does this mean for them ?) about differentiation and brain specialization during foetal life, about metabolic influences and relations to other vulnerabilities, about physical and psychic development and relations to learning experience, about educational influence and pressure and finally about previous specialized interventions. They very often hunt for the unique causal factor, a form of behaviour which is definitely an illusion. It may be difficult to make them understand what "concomitant" means, referring to the reciprocal influence between two variables with little chance of disentangling the forces at work and of knowing which factor is the stronger. It seems as if the human being naturally creates causal links between two concomitant facts. This tendency is probably more innate than dyslexia. It is a kind of fantasy that is particularly difficult to destroy.

Holistic therapy begins with the steady acceptance of these unknowns. In fact, the child and his parents need to make a much greater effort to accept the unknown than to search for certainties. Only then do we have some chance of liberating the child from his or her imprisonment in a whirl of relational misunderstanding (s/he can very easily be stereotyped as retarded, lazy, blocked and confrontational) which can lead to progressively greater eccentricity and social isolation.

The argument presented in this chapter is that to learn such a complex and integrative task as reading demands, on the part of the child, a rather rich choice of strategies. Therapy, therefore, needs just as rich a choice of calibrated procedures in order to be able to help each dyslexic child. We are not sufficiently advanced in our knowledge of the wide variety of reading difficulties to be able to respond to each case's needs. As single-case studies in this area are only just beginning to be published, with well-illustrated comparative data, only a few therapies are known which can serve as remedies for specific disabilities. It is for this reason that it is probably wise, at the present stage of scientific knowledge, to give some credence to holistic therapies which are of obvious help to the child.

It is clear that much remains to be done once a sufficient number of careful examinations have been carried out on individuals suffering from reading problems. Certainly the day will come when specific training will be able to respond to specific needs. It is even conceivable that, in the near future, global psychotherapy will not lose its importance, for generalized support and an atmosphere of openness and empathy will set up the best conditions for specific therapies to work.

BEYOND THE VERBAL-VISUAL DILEMMA

At the present time, there are a variety of pragmatic remediation techniques available. Some are based on language development, from oral to written, some not. Some pinpoint the relevant eye-movements in reading, which are considered by the author to be a typical example of the concomitance cited above. Moreover, it is known that visual problems which are not of optical but of orthoptical origin, such as disparity arising from poor vergence control during reading, can also lead to reading difficulties.

Dyslexia is an instrumental deficit located at the crossroads between language and visual perception, both routes likely to be travelled from top to bottom and vice versa.

Beyond the visual perception system, the entire realm of visual memory for symbols is involved. Visual symbolization is sometimes the target of therapy since difficulties very often arise at that level in dyslexics: some of these children do not manage to learn the graphemes easily and lag behind with grapho-phonemic identification.

Are confusions phonemic or visual in this case ? We know that they can be of visual origin since we sometimes observe later on that the same children have a poor visual lexicon, rendering them incapable of directly recognizing even short words. Deciphering words by the grapho-phonemic route appears to them to be the only way of approaching words. Then they very often get stuck in the middle of words and struggle with phoneme identification and combination. This is called surface dyslexia, a syndrome well illustrated in some brain injury patients suffering from acquired reading difficulties. With learning-disabled children also, this form of dyslexia has been isolated as a specific sub-type (Coltheart, Masterson, Byng, Prior & Riddoch, 1983).

Probably more frequent is developmental phonological dyslexia, where the child avoids the use of the grapho-phonemic route (does not read non-words) and relies upon his/her visual memory to confirm his/her comprehension. After having illustrated this form of acquired dyslexia in brain-damaged patients, a few researchers have published illustrative cases of developmental phonological dyslexia (Temple & Marshall, 1983; Snowling, Stackhouse & Rack, 1986).

A third sub-type of dyslexia mainly shows body-scheme difficulties, which translate themselves into spatial confusion and disorientation (Benton & Kemble, 1960). Finger agnosia, left-right confusion and constructive apraxia can be observed besides writing and reading problems (Galifret-Granjon, 1959). These children may show no sign whatsoever of oral language disturbance nor of visual perception difficulties.

SEQUENCING

A fourth group of children reveals itself to be particularly poor in sequencing tasks. Mentally to put words or images in the right order can be a specific problem of working memory and we know that short term memory problems are often cited in the dyslexia literature (Jorm, 1983; Torgesen, 1985).

Neuropsychological testing is used to examine sequencing in different modalities. However, it is only recently that commercialized large-scale testing has proposed the introduction of a multimodal examination of this function. For example, the K-ABC battery for children is one such scale. Three tests are involved: the traditional digit span test (direct and inversed order), a sequence of hand movements, and a sequence of words given orally by the examiner and to be translated into sequential designation (by pointing) from a series of line-drawings (Kaufman & Kaufman, 1983).

TABLE 1
Summary of the K-ABC Individual Test Record

Name: John   Parents: middle class
Age: 15 years 5 months   school: 9th Grade

Subtests (Mean=10, s.d.=3)		**Scales quotients** (Mean=100, s.d.=15)	
Hand movements	9	Sequential	83
Gestalt closure	13		
Number recall	7	Simultaneous	120
Triangles	15		
Word order	6	Mental Processing	
Matrix analogies	11	Composite	105
Spatial memory	11		
Photo series	14	Non verbal	101

Table 1 shows the results obtained from the K-ABC battery by a 15 year old dyslexic boy whom we followed from the age of 9. The reading age of the child is below 9; the child managed to gain some costly knowledge in reading (though after some time and at some cost to himself), while developing fully normal abilities in domains other than written language. The interesting thing is that the sequential scale of the K-ABC shows the most dramatic residual symptom, while the verbal and the visual scales show above average performances. Sequencing appears to be relatively lower than reading on the age-related scale of development since it does not go beyond the performance of a six year old.

Any clinical psychologist would immediately look at the ability to reproduce rhythms in this case (Galifret-Granjon, Stambak & Santucci, 1953). This task does not rely on auditory or visual symbols, but rather expresses the skill of repeating a pure temporal sequence. Our adolescent made 10 mistakes on 21 items, a score which places him in the normal reference group for six year olds (Stambak, 1964).

The concentration of difficulties on the task of sequencing argues in this case for a specific dysfunction of which dyslexia only represents a subsequent symptom.

Also interesting is the total independence of this function from traditional social learning. Even if oral language, nursery rhymes, songs and skilled movements cannot develop without the participation of some sequencing abilities, we must admit that sequencing is not taught in itself, neither in the family nor in kindergarten. The case presented here is interesting, therefore, since it illustrates an incapacity which has little to do with social negotiation. We should keep such an example in mind when confronted with hypotheses concerning the responsibility of the social environment in the dysharmonic development of the child.

One might ask whether there is any anatomo-functional model of the central nervous system which would support the relative independence of the sequencing syndrome. In fact, we already have such a model valid for short-term memory, indicating implication of the Papez' hippocampo-mammillo-thalamic circuit.

Furthermore, the K-ABC battery, based as it is on neuropsychological principles, should lead to a closer look at Luria's pioneering work on frontal lobe functions. The go/no-go capability with its interference of action and inhibition, the separation of past and on-going tasks, the discrimination between internal impulse and external stimulus represent precisely the major challenge for the sequencing tasks. It could be that a sub-group of learning-disabled children suffers specifically from difficulties related to the weakness of functions depending on the most recently organized structures of the brain, both in evolutionary and developmental terms. These functions, recognized as representing typically the work of the frontal lobes, are responsible for regulating the continuous flow of action and perception, that of past and present events, eventually that of internal and external information.

A RATIONALE FOR SPECIFIC THERAPIES

Having reviewed the classical non-reading difficulties encountered in dyslexic children, a rationale for a range of therapies all aimed at helping the dyslexic child can be proposed (Fig. 2).

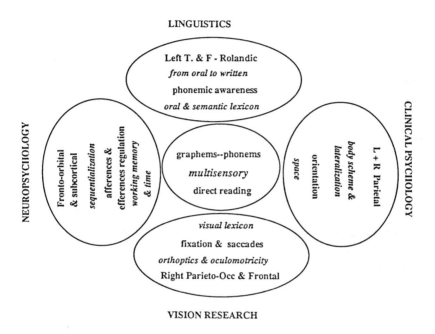

FIG. 2. A Rationale for Therapies in Dyslexic Children

As a result of the fast development in linguistics, speech therapy developed from the 1960's on. Many children benefit from therapy based on the progressive transformation of oral into written language. Phonological awareness is both a powerful predictor of reading ability (Bradley & Bryant, 1983) and a useful tool for remediation (Naidoo, 1981). The question arises how do we know precisely what kind of child will benefit the most from such therapy ?

Some authors, like Bakker (1990), suggest that the speed of reading and the quality of mistakes made by the child might prove to be good indicators: the child who reads relatively quickly, and who relies on visual cues and the overall appearance of a word to confirm some oral expectation, might encounter difficulties with grapho-phonemic conversion, which would handicap the reading of new words. It might be the case that this child is diagnosed relatively late, because of his alertness and apparent ease with texts. Nevertheless he puts too much distance between what he would like to read and what is actually written, so that the teacher might well ask herself why the difficulty is so severe with unfamiliar text. Referred to the speech therapist, the child might cause some surprise by his complete inability to read simple short regular words without any sign of aggression against the world of written language.

This is just an example, and it should not be believed that every case is as clear as this, especially after years of suffering at school due to educational misunderstanding.

Problems of a very different kind are encountered with the child who has difficulties of a visual or visuo-spatial origin. He/she might be diagnosed much earlier, for reading and writing are still a closed book whilst other children have already achieved syllabation and the reading of short regular words.

As a caricature of the visual deficit, we may find children who do not perceive the sequence of the letters correctly. This must not be confused with visual acuity problems. Indeed, we know that myopia in children tends to make for normal or even good readers (Grosvenor, 1977). But misperception can occur through vergence defects (Cornelissen, this volume). In this case, disparity between the images provided by the two eyes disturbs the skilful pick-up of information. Visual devices can be of some help, such as the use of contrast enhancement and letter enlargement.

Ocular motricity can be responsible for some exploration awkwardness and it is true that reading implies strategic gazes which must be accurately placed on the string of letters and words in order to free the reader from the necessity of backward control. There are a few problems with the saccades: images from the peripheral field must be cut out in favour of images perceived by the fovea. Some dyslexics suffer from a lack of this suppression, resulting in image rivalry. These dyslexic subjects have been found to read better in the peripheral field than do non-dyslexics.

Recently, the dynamics between the visual sustained and transient channels have been put forward as a possible explanation for such a deficit. Normally, the saccade from one target to the other, which is controlled by the transient system, suppresses the information taken from the first target. The result is the avoidance of superimposition of images belonging to two distant targets and having nothing to do with each other. It has been found that some dyslexic children suffer from a lack of saccadic suppression, i.e. from a transient-channel deficit. Such difficulties are found before reading instruction and failure, so that this is by no means a functional and behavioural consequence of the frustration at not being able to read.

In these cases, reading vertically or through a window can help the child to overcome his difficulties. Moreover, as has been shown, enhancing visual contrast of the material to be read can improve reading without much alteration of the reading conditions: wearing tinted glasses could be sufficient.

Spatial problems in reading, i.e. orientation difficulties, may be just as severe as phonological problems. Indeed twisted symbols were among the first symptoms of development dyslexia described by Hinshelwood a century ago. Reference to mis-orientation of drawings or praxic copies are often cited as accompanying this kind of dyslexia. Confusion about orientation with one's own body is observed together with reversals in orientation judgment.

Since the dyslexic child was seen to make mistakes concerning letter form and orientation and since he/she can hardly distinguish between two mirror images, researchers immediately thought that the two sides of the body are involved in this symbolic confusion, and that this might have something to do with the two sides of the brain. Research on dyslexic children has very often reported left-right confusion (eg. Wechsler & Hagin, 1964), as well as weak and inconsistent laterality (Harris, 1957).

To reconcile the brain with the task of orienting symbols might start with full lateralization regarding gestures and postures (writing, for example).

Psychomotor therapy aims precisely at training the child to move with a clear mission assigned to each side of the body, foot or arm, and to solve the coordination problem between balance maintenance and addressed gestures.

Another set of therapies concentrates on the coordination between two independent gestures, required, for instance, in writing with one hand while holding the paper with the other. Kinesiology proposes a therapy of this kind, seeking harmony in coordinated gestures, mobilizing for example the left foot and the right arm in definite movements. Kinesiology claims to have strong therapeutic effects on dyslexics' difficulties.

The fourth leaf of our clover (Fig.3) is sequencing and working memory. Reading is achieved through close top-down control, i.e. by the permanent role of intention and search for meaning. However the reader experiences two separate realities, the one residing in his understanding system and the other, external, on the sheet of paper. The inner reality is cognitively represented by working memory which always saves and enriches integrative information. The outer reality is to be scanned by the oculomotor system. This can result in a temporal overlap between what has to be done and what one does in the present moment..

One example can be taken from the traditional frontal tests used in clinical neuropsychology. The Weigl-sorting-test and its later improved versions (Berg, 1948) show that when one card is concealed beneath another, the visual stimulus dominates and the child cannot put out of his mind what he sees, and therefore finds great difficulty in discovering a series confirmed by preceding choices. More precisely, if he has made a mistake which might confirm the series, he cannot look on it as a mistake: it becomes the beginning of the series in his mind.

The same effect can occur in a probability task where we ask the child to predict which colour he is going to see. If one tries to teach the child a regular series (such as red-red-black, for example), the child takes his errors for granted since he confuses what he says as a prediction with what he sees as the reality. In this case, the series can never be completed. It is to be noted that this frontal task is particularly perverse because the observer induces the confusion by asking the child to say predictions out loud. The series could otherwise be more easily recognized if one were merely to ask the child simply to look on in silence.

Again, regulation between action and memories of actions is poor. In imitation of rhythms, it appears as if the engaged action of starting the first and then the second stroke erases the symbolic image which would have led to correct repetition. When we show the child a pictorial representation of the rhythm heard, he can easily recognize it: his answer does not belong to the series itself as would be the case when reproducing the series.

How can we remedy such difficulties ? This is hard to say, for 'frontal therapists' do not exist. However, one good way of insisting on sequencing in written language would be to use a type-writer or a computer. Here the child develops the skill of focusing a great deal of attention on the immediate effect of each of his actions. Furthermore, he adopts his own pace and is sanctioned only by her or himself. Also there are correction procedures which s/he engages without any sense of shame or inadequacy with respect to the demand.

Another step-by-step procedure was found to be very useful with the most severe dyslexics who did not learn to read at all by the age of 9. When such a child still struggles with the string of letters, recognizing no whole word except his own name, a phonological sequencing task without letters can greatly help. Carol Rose, from the Hôpital Henri Rousselle in Paris, reported about the method of "words in colour" developed in France by C. Gattegno (Rose, 1991). It appears to be an efficient way of helping the child to discover and train himself in the basic rules of phonological fusion and to become aware of the relative value of each independent sound. Therapists reported a great deal of success using this method with non-reading preadolescents and even adolescents. Hunter-Carsch (see this volume) confirms the effect of the coloured words method, used namely in Lancashire with subjects aged 16-18.

Thus one cannot say that we do not know what to do in the case of sequencing difficulties. However, such therapies really are basic and are introduced as a form of brake mechanism in the disordered cognitive behaviour of the child. We certainly have to explore the development of these techniques more thoroughly. The use of computers can be extended given that now almost every classroom is equipped with one and every specialist has access to a machine. It can be a powerful tool since it can be adapted to every stage in the progress of written language. With highly skilled users, it again recapitulates the fundamental abilities of bringing together the mentally heard phoneme, the manual action of typing and the visual control of the trace.

CONCLUSIONS

Many therapeutic strategies have been developed to help dyslexic readers. Some of them are the result of long-lasting and careful research into childrens' difficulties, some are simply developed as an empirical solution to the problem of a specific case. It has to be admitted that none can be considered as the panacea nor are any absolutely indispensable in a series of remedial measures. We think that many of them are not adapted to a specific child, a fact which hinders any exaggerated generalization and any use by non-experts.

Here we come back to research and diagnosis and it should not be surprising to discover that our synopsis of remedial procedures leads to the enhancing of the bonds linking practice and research.

One question arising at this point, is whether there exists a single specialist capable of performing a comprehensive diagnosis. Or, on the contrary, whether we have to place the problem in the hands of an interacting team of specialists, performing vision and visual examination, language observation, psychomotor and sequencing testing. And that is to say nothing about the necessary accompaniment of the child and the significant persons surrounding him in order to prevent any misunderstanding regarding the procedure and to avoid destructive build-up of stress and concern and social relational problems. Some might also say that chemical medication could be of some help during both diagnosis and therapeutic procedures, a point of view which seems legitimate on the part of the physician.

In the present chapter, it was not our intention to compile a list of all the therapeutic methods available nor to highlight any specific technique used in remediating dyslexia problems. Instead the focus was on the richness in experience that has been accumulated with different sub-groups or single cases of poor readers. At the same time, we wanted to stress the absence of any real rationale that might help one pick one's way through the labyrinth of different proposals.

Research in dyslexia clearly leads to subtyping the different forms that the child's difficulties may take. Some writers on therapy start by considering two opposing patterns of dyslexia, dysphonological and dyseidetic dyslexia, but from there on, they rapidly have to admit the existence of at least a third form which could be a mixed type. The same thing happens with the opposition of verbal and visual functions which leads to the identification of difficulties that demand linguistic and visual therapeutic strategies, respectively.

Some researchers divide dyslexics into three subgroups when the body scheme syndrome is added to the verbal and visual forms of the handicap. For example, Mattis, French & Rapin (1975) stress the importance of articulation problems and graphomotor dys-coordination for defining a third subgroup of dyslexic children and adolescents.

In this paper it is suggested that a fourth subgroup be taken into consideration represented by "frontal" dysfunction, encompassing working memory deficits and weakness in continuous regulation of actions. It seems to us that evidence is now sufficient to use the tools available to diagnose more precisely this last form of dyslexia. The K-ABC battery was mentioned which contains a sequencing scale and connected this with the long cited difficulties encountered in reproducing rhythms.

On the therapeutic side, some tools are ready and available for use, with the presence in every classroom and in every specialist's office of the PC computer. Literature on its use in dyslexia remediation is now available. More specific tools can also be used in the most severe cases of this kind, like Gattegno's coloured words device.

We desperately need to develop remediation specialists. Speech therapists are on the market, thanks to the needs of orally impaired children. When treating written language, even some logopedists step back or do not feel themselves concerned, for example when arithmetic is involved and no longer just the alphabet and spelling.

Remediation for arithmetic problems is very insufficiently developed, although it must be said say that some therapists do an excellent job in this specific area.

Psychomotor therapy has developed with a variety of useful specializations like kinesiology. No one such therapist should work on his/her own, a situation which seems unfortunately to be the case.

Orthoptics needs to develop in the direction of a dynamic assessment and treatment of visual exploration.

To whom should we turn to when faced with the case of a child with sequencing difficulties ? There is no specialist in this matter, although all of the above mentioned specialists are aware of this kind of difficulty and of the kind of assistance the child would need.

Our hope is that we will carry this reflexion to the point where we will not only propose a rationale for diagnosis but also for remediation. We are sure that interdisciplinary teams will in the near future develop more complete models of remediation and perhaps propose new ways of treating dyslexic disturbances. They certainly will lead to new professions in remediation and to new interesting work for tomorrow's therapists.

REFERENCES

Bakker, D.J. (1990). *Neuropsychological Treatment of Dyslexia*. New-York: Oxford University Press.

Benton, A.L. & Kemble, J.D. (1960). Right-left orientation and reading disability. *Psychiatry and Neurology*, 139, 49-60.

Berg, E.A. (1948). A simple objective technique for measuring flexibility in thinking. *Journal of General Psychology*, 39, 15-22.

Bradley, L. & Bryant, P.E. (1983). Categorising sounds and learning to read: a causal connection. *Nature*, 301, 419-421.

Coltheart, M., Masterson, J., Byng, S., Prior, M. & Riddoch, J. (1983). Surface dyslexia. *Quarterly Journal of Experimental Psychology*, 35A, 469-496.

Gattegno, C. (1966). *La Lecture en Couleurs*. New York: Educational Solutions Inc.

Galifret-Granjon, N. (1959). L'élaboration de rapports spatiaux et la dominance latérale chez les enfants dyslexiques, dysorthographiques. *Bulletin de la Société A. Binet*, 452, 1-35.

Galifret-Granjon, N., Stambak, M. & Santucci, H. (1953). Des débilités et des dyslexies. *Les Cahiers de l' Enfance Inadaptée*, 17.

Grosvenor, T. (1977). Are visual anomalies related to reading ability ? *Journal American Optometry Association* , Apr., 48, 4, 510-517.

Harris, A.J. (1957). Lateral dominance, directional confusion and reading disability. *Journal of Psychology*, 44, 283-294.

Jorm, A.F. (1983). Specific Reading Retardation and Working Memory: A Review. *British Journal of Psychology*, 74, 311-342.

Kaufman A.S. & Kaufman, N.L. (1983). *K.ABC. Kaufman Assessment Battery for Children*. Circles Pines, MN: American Guidance Service.

Mattis, S., French, J.H. & Rapin, I. (1975). Dyslexia in Children and Young Adults: Three Independent Neuropsychological Syndromes. *Developmental Medicine and Child Neurology*, 17, 150-163.

Naidoo, S. (1981). Teaching Methods and their rationale. In G. Th. Pavlidis & T.R. Miles (Eds). *Dyslexia Research and its Applications to Education*. New York: Wiley.

Rose, C. (1991). What about letters and know-how? The rich harvest from the remediation of a group of non-reader normal children. *Paper presented at 18th. Rodin Remediation Scientific Conference, Bern*.

Snowling, M., Stackhouse, J. & Rack, J. (1986). Phonological Dyslexia & Dysgraphia - developmental analysis. *Cognitive Neuropsychology*, 3, 3, 309-339.

Stambak, M. (1964). Trois épreuves de rythme. In R. Zazzo, R., M. Stamback, H. Santucci, M-G. Pecheux and N. Galifret-Gransons (Eds.). *Manuel pour l'examen psychologique de l'enfant.*. Neuchâtel: Delachaux et Niestlé.

Temple, C.M. & Marshal J.C. (1983). A case study of developmental phonological dyslexia. *British Journal of Psychology*, 74, 517- 533.

Torgesen, J.K. (1985). Memory Processes in Reading Disabled Children. *Journal of Learning Disabilities*, 18, 6, 350- 357.

Wechsler, D. & Hagin, R.A. (1964). The problem of axial rotation in reading disability. *Perceptual and Motor Skills*, 19, 319-326.

Facets of Dyslexia and its Remediation
S.F. Wright and R. Groner (Editors)
© 1993 Elsevier Science Publishers B.V. All rights reserved.

REASON, RHYTHM, RELAXATION AND THE NEW LITERACY: IMPLICATIONS FOR CURRICULUM DIFFERENTIATION TO MEET THE SPECIAL EDUCATIONAL NEEDS OF PUPILS WITH SPECIFIC LEARNING DIFFICULTIES

C. M. Hunter-Carsch

University of Leicester School of Education,

Leicester, U.K.

A paper of this title was originally presented by invitation at 'Meeting the Challenge', the second international conference of the British Dyslexia Association and European Dyslexia Association in Oxford in April 1991. The original paper has been developed further for the 18th conference of the Rodin Remediation Academy in Berne and subsequently updated prior to publication. The title of the Berne conference, 'Reading and Reading Disorders: Interdisciplinary Perspectives' prompts the following introduction to an educator's perspective.

Children with specific learning difficulties (SpLD) appear to respond positively to highly structured teacher-directed approaches to learning in the classroom. These approaches involve intensive individual attention from specialist teachers. The paper addresses the issue of how best to meet these children's special educational needs within the mainstream classroom in a context which currently includes the seemingly conflicting trend towards the more child- directed approaches in what is described by Willinsky (1990) as 'the new literacy'.

It is hoped that although many other European countries already have a national curriculum, reference to the new national curriculum in England and Wales and the development of curriculum and assessment policy in Scotland may be of interest at a time of increasingly closer links within Europe. Furthermore, the writer's views are informed by experience of teaching and teacher education on both sides of the Atlantic (in Scotland, England, Canada and the United States of America).

The Problem

The problem addressed by this paper involves the urgent need for special educators and mainstream educators to share professional knowledge and skills in order to maximise opportunities for informed individualised teaching to take place within the classroom context. The introduction of the national curriculum in England and Wales makes it mandatory to provide equal access to the curriculum for all children and thus the implications for curriculum differentiation become matters of professional responsibility for all teachers.

The first question to be raised is thus, what are the crucial elements of professional knowledge and skills that should be communicated and where relevant shared between generalist primary teachers (GP) and special educational needs specialist teachers (SENS) of children with SpLD?

To be acceptable to class teachers, a framework for curriculum differentiation should concern not only the special educational needs of children with SpLD but should be sufficiently comprehensive to address the problems presented by all pupils with basic literacy difficulties.

A framework aimed to provide sufficient breadth to encompass the range of factors affecting all such pupils has been outlined elsewhere (Hunter-Carsch, 1990). The next question to be addressed is to what extent can the proposed framework reconcile two apparently contradictory approaches to teaching literacy, firstly those approaches which have more recently been described collectively as 'the new literacy' (Willinsky, 1990) and secondly, what might be collectively termed 'the structured approach' including, for example the methods outlined by Cruickshank (1963) and proponents of so-called * 'multisensory methods' ? (* e.g. Orton, 1937; Fernald, 1943; Gillingham & Stillman, 1956; Bannatyne, 1966; Slingerland, 1977; The Mills Learning Centre, 1970; Critchley, 1970; Moseley, 1971, Newton & Thomson, 1974; Hickey, 1977, Hornsby & Miles, 1979).

The method of addressing these questions does not, at this point, involve empirical studies as such, but seeks to explore wider issues which affect both the theoretical bases and practical implications of current approaches to teaching basic literacy, teaching children with reading difficulties and children with SpLD. The discussion is influenced by the writer's first hand experience of teaching both as a generalist and as a specialist. The discussion is organised under four subtitles which concern the following matters:

I Requirements of the National Curriculum

This section concerns the requirements of the curriculum and teachers' changing roles and responsibilities as well as the theoretical underpinning of curriculum guidelines with reference to 'English' (in England and Wales) 'English Language' (in Scotland), 'language', 'literacy' and 'reading'. It recommends a form of grouping children to permit focused observation and highly individualised teaching.

II Specific Learning Difficulties

This section explores the writer's observations of SpLD. These are based on wide experience of mainstream classes and special education including work in a multidisciplinery assessment centre. A hierarchy of difficulties is postulated. The problems described earlier (Hunter-Carsch, 1990a.) as characterised by difficulties in 'motivation, meaning and imagery' are explained as relating firstly to problems which affect memory which in turn affects and is affected by the child's way of trying to make sense of experience, i.e. 'meaning', which in turn affects motivation.

III A Framework for Curriculum Differentiation.

This section briefly reviews work by King (1989) and Spillman (1990) and concentrates on developing and discussing the writer's relevant model with reference particularly to a series of 'microstrategies' for adapting or 'differentiating' the curriculum to meet the special needs of pupils with SpLD.

IV A Context for Both 'The New Literacy' and 'The Structured Approach'.

This section introduces an approach to developing children's writing which is illustrative of 'the new literacy' approaches in that it shifts the locus of control for learning increasingly into the hands of the learner. The approach, 'systematic drafting and redrafting' (Binns, 1978) has been found to be helpful not only for pupils with SpLD but was designed for all children and appears to be equally effective for (a) children with literacy difficulties and (b) children who are 'normal' and 'able' writers. The approach is seen to be of assistance also in that it endorses the tradition of promoting 'the writing route to reading' while providing a context for structuring the development of reading, writing, handwriting and spelling, and, of course, listening and speaking. For the child working within this context, specialist support which is tailored to meet his or her specific difficulties can be the more easily mediated.

The overriding factors of reason, rhythm and relaxation are then related to the framework for curriculum differentiation. The concluding discussion returns to the two initial questions and implications for teacher education.

I Requirements of the National Curriculum: English in the National Curriculum (England and Wales), English Language (Scotland): Implications for Classroom Organisation, Management and Control of Learning.

The 1988 Act is based on the belief that all children should have equal access to the curriculum and brings that aim closer to becoming a reality by enshrining it in law. The practical challenge for the class teacher charged with this legal responsibility, is to orchestrate the range of adaptations which are required to be made in order to mediate learning effectively to all children.

This requires the class teacher to be able to differentiate between children whose reading difficulties relate to the following:

i) *developmental gaps* (missed out the relevant learning at school but simply need to be taught, i.e. to 'fill up the gaps')

ii) *mild learning difficulty* (generalised 'slower pace' of learning but, with sufficient experience and practice, can learn)

iii) *'crisis' problems* resulting in arrest of reading progress or regression in ability to read (requiring immediate referral for specialist advice / medical and psychological)

iv) *specific learning difficulties* (requiring intensive observation to discern optimal modes of learning and to plan individualised teaching in consultation with multi-disciplinery specialists)

The challenge is the greater as some children's reading difficulties relate to more than one of the above categories (Hunter-Carsch, 1985). Furthermore, it is important to consider children's reading difficulties within the wider picture of their literacy and learning problems.

To do this within the educational context in England and Wales in which the focus is on the even greater breadth of the concept of 'special educational needs' (S.E.N.) (D.E.S/Warnock Report 1978), can be daunting. The concept, in its magnitude, does not immediately make clear to teachers, at what point and how they should address the problem of delineating children's 'learning difficulties'.

There may be some lessons to be learned from contemporary educational developments in Scotland in which the focus is on 'learning difficulties'. In 1978, the HMI Report,' The Education of Pupils with Learning Difficulties in Primary and Secondary Schools in Scotland' delineated the four- part role of the new specialists which it suggested would be 'a far cry' from the role of the' remedial teacher' as it then existed. The resultant training of these specialist teachers was begun through the development of a Diploma in Professional Studies: Special Educational Needs (Non-recorded Pupils).

In the delineation of the four-part role there was, from the outset, the recognition of the need for trained specialists to support class teachers in working with 'non-recorded pupils' (non-statemented) as well as to provide appropriate support for those pupils who were designated as requiring special education support within the mainstream school (Joint Committee of Colleges of Education in Scotland) Advisory Committee Report, 'Ten Years On, 1987).

It is perhaps pertinent to note also, that it was extremely difficult for many former 'remedial teachers' who did not have the benefit of training, to learn how to cope with the four roles which include the following:

Role 1 : Acting as a consultant to other members of staff.

Role 2 : Personal tuition and support for pupils with specific and severe learning difficulties in the processes of communication and computation.

Role 3 : In co-operation with class and subject teachers to offer tutorial supportive help in their normal classes to pupils with learning difficulties in the later stages of these processes and in any other areas.

Role 4: Providing, arranging for, or contributing to special services within the school for pupils with temporary learning difficulties.

While there does not appear to be a government document delineating the roles of the S.E.N. specialist in England and Wales, the Guidelines produced in 1985 by the National Association for Remedial Education (N.A.R.E.) describe the following seven broad features:
1. An assessment role
2. A prescriptive role
3. A teaching / pastoral role
4. A supportive role
5. A liaison role
6. A management role
7. A staff development role

Both the documents note the ' cooperative teaching' or 'collaborative teaching' (see under Role 3 and 3. above).

The difficulties which the specialist teachers face include the fact that some preparation of non-specialist class teachers and subject specialists is required in order for them to make the necessary adjustments on their part to work in the new pattern of cooperative/ collaborative teaching and planning.

Although there have been dramatic developments in England and Wales during the decade between 1978 (Warnock Report) and 1988, which include the 1981 and the 1988 Education Acts, the problems facing the class teacher and S.E.N. 'specialist' (whether trained or not) remain substantially the same with reference to the following issues:

1. The communication of the specialist's professional knowledge and skills regarding the 'structuring ' of learning for individual pupils and strategies for the class as a whole

2. The recognition of the need for training and the provision of training in collaborative teaching and for quality time for planning together

3. How best to meet the S.E.N. of 'non-recorded' pupils as well as those who are already recognised and 'statemented' or in the process of being recorded as having 'special educational needs'. Additionally, at this time teachers in England and Wales are facing the implications of new school management patterns, new methods of assessment and recording and, increasingly, concern about subject specialism.

While arguably the most important task of the primary school remains that of equipping children with competence in basic literacy and the confidence to seek for themselves any information they may require through reading, writing and talking, there is no single agreed theoretical model of normal development of reading and writing.

Despite the publication of the Kingman Report (D.E.S. 1988) which dealt with a model of' what teachers should know about language', there remains an absence of any single agreed model of the language process or of relating language development to teaching methods, let alone to learning difficulties and more pressingly, to SpLD.

In addition to knowing what is meant by 'literacy and 'language', the challenge for teachers involves knowing how individual pupils learn and how best to support the learning of those with language and communication disorders as well as difficulties in the classroom context.

In relation to these wider issues there is a need to understand the theoretical bases as well as practical implications of ' structured approaches' within the teaching of reading (e.g. use of schemes) and 'open approaches' which appear to have little structure (e.g.'real books'/story approach/apprenticeship reading). Because of the centrality of language and communication across the curriculum, these methodological issues have an impact both in and beyond the subject specialism of English. The interrelationship between listening, speaking ,reading and writing was well recognised in the Bullock Report (1975) entitled, 'A Language for Life'. It was acknowledged in the curriculum papers which then formed the basis of the national curriculum. The emphasis has shifted to the more detailed specification of Attainment Targets for each aspect of English in the national curriculum in England and Wales. The Attainment Targets are specified as follows:

AT 1 : listening and speaking

AT 2 : reading

AT 3 : writing

AT 4 : spelling

AT 5 : handwriting and presentation.

The understandable and valuable retaining of the area of reading within one single attainment target becomes curious, however, when followed by a different logic applied to the specifying of separate areas within writing (AT3,4,5). Because the targets for writing have been separately stated, the implications for curriculum differentiation, for example,in handwriting (a skill area) as distinct from composing or editing within the process of writing for different purposes, are the more easily related to differences in methods and organisation.

While agreeing with the overriding issue of recognising reading to be a unitary process rather than separate bundle of skills, there may be some merit in specifying what aspects of reading are being practiced or developed at a given time. The Scottish curriculum guidelines for 'English Language 5-14' may be helpful in this respect as four purposes for reading are delineated as follows, then the Attainment Targets in Reading are presented under seven strands each of which is graded from level A(lowest) to E(most advanced):

Purposes of Reading

1. To obtain information and respond appropriately.

2. To be responsive to the feelings of others.

3. To reflect upon ideas, experiences and opinions.

4. To gain satisfaction and aesthetic pleasure.

Strands

1. Close reading

2. Reading aloud

3. Reading for information

4. Reading for enjoyment

5. Finding and handling information

6. Awareness of genre

7. Knowledge about language

In order to feel confident about their complex responsibilities and the demanding role they must play , teachers need to have clarity about the wider aims of the curriculum for

English as well as relating them to methods which will assist them to provide 'balance, continuity, coherence and progression' in the curriculum. Necessarily, distinctions must be made in their own planning between methods which are appropriate for achieving progress with reference to different aspects of reading development and reading for different purposes.

The importance of making these distinctions becomes the more clearly recognised when they are related to the matters addressed in the notion of 'the new literacy'. As distinct from the 'old' literacy in which the control of the learning materials and methods was substantially in the hands of the teacher, the 'new' literacy seeks to shift the locus of control progressively towards the hands of the learner (or more precisely, in the head of the learner). Thus, matters of choice, independence, relationships and management of resources and time are affected.

For the teacher, implication for curriculum differentiation include not only recognising that in any interaction in the classroom the actual 'locus of control' as well as the perceived locus of control constitutes a crucial factor in pupils' learning. The matters concern the finer points of definition of 'control within the classroom', 'control of the classroom' and 'control within the child'. The way in which the teacher perceives his or her role is likely to have an impact on communication with the individual pupil as well as affecting interaction in the classroom as a whole. For children with specific learning difficulties (SpLD) which frequently affect and are affected by the processes of language and communication, these issues are even more critical.

Approaches which have proved to be particularly helpful for such children may be broadly described as 'structured approaches'. The dilemma for the class teacher is thus how best to employ what seem like opposite approaches (one affording choice of materials and placing the child as the 'controller' and director of aspects of learning related to reading and writing; the other, more directedly presenting the child with 'controlled' interactions and preselected resources.

Clearly, to implement the national curriculum effectively it is necessary to increase the amount of individualised teaching and to employ a range of organisational strategies which include working with groups and with the whole class where relevant (See Alexander, Rose & Whitehead, DES 1992).

Inevitably this requires teachers to train their classes to work in such a way as to permit the extent of individualisation which is required. It is in order maintain this aim as a priority, and to seek to establish the prerequisite 'routines' in terms of ways of organising the classroom with the potential for considerable flexibility, that the following suggestions are made as part of a framework for curriculum differentiation.

Classroom management becomes the easier if, instead of having to deal with a whole class, for some of the time, after routines have become established, the class can be divided into three groups and work on three distinct kinds of tasks, only one of which requires teacher-attention. To do this in practice appears requires not three 'set activities' but only two. One of these involves an activity which requires practice or review in which if children require help, they are taught how to work with a friend or to signal their need for assistance by recording a message on a tape recorder after trying to get help from others, and then to proceed with an alternative task which has been preselected by the teacher.

The second group who are not with the teacher are freed to relax in ways that include, where relevant, resting, playing or choosing activities which do not interfere with any one else's work, thus should be controlled in volume and possibly restricted to 'relaxing' alone rather than in a group. If this 'gap' in the flow of directed active learning is built in routinely as a part of classroom timetables, there is usually a corresponding decrease in interruptions during work times with the teacher and practice/review work times and also a greater availability of children's energy and attention for the two other types of activity.

The groups rotate. During their intensive time with the teacher there are opportunities for individualisation which can be systematically lengthened as children realise how enjoyable and important these times are for their learning. The success of the organisational pattern depends not only on the enjoyment of work produced in the teacher's group time, but on teacher recognition of good work during practice time, and for 'self management' in 'relaxation time'.

Of course, this group rotation for intensive teacher attention (including individual tuition) is not the only organisational pattern for teaching and learning. It serves only as an illustration of one way of setting up a pattern within which it becomes possible to increase the input to 'conferencing', an approach developed particularly with reference to children's writing (Graves, 1983; Barrs, Ellis, Hester & Thomas, 1989; Czerniewska, 1989). 'The National Writing Project', 1989 ; Binns 1980, 1990; Hunter-Carsch, Binns & Sobey, 1991). The kinds of problems which can be more particularly addressed during individualised discussions within conferences include those described below under 'Specific Learning Difficulties'.

During special tuition times it is necessary both to work from developing the children's strengths and interests (including extending their interests) and to attempt to find ways of coping with intractable learning problems.

II Specific Leaning Difficulties

In England and Wales the term 'specific learning difficulties' was adopted in the Education Act (1981) following its use in the Warnock Report (1978) to differentiate a subgroup of children from others with special educational needs. The subgroup consists of those who experience subtle and persistent learning difficulties yet are of around average intelligence or above average, and whose problems cannot be attributed primarily to poor health, lack of formal schooling, physical, sensory, emotional or social factors.

'Specific Learning Difficulties / Dyslexia' is the subject of an enquiry carried out by the Division of Educational and Child Psychology of the British Psychological Society (D.E.C.P.1984) and followed by a wide-ranging national survey carried out together with the Association of Educational Psychologists. The Report of the survey (Pumfrey and Reason 1991) includes detailed discussion of theory, research, policy and practices with regard to SpLD. The writer's views are included in the submission made on behalf of the United Kingdom Reading Association to the enquiry. The UKRA submission was discussed and further developed by an international study group. The following brief descriptive notes are based on fuller discussion elsewhere (Hunter-Carsch, 1991).

Children with SpLD characteristically show difficulties in one or more of the following areas:
  i) handwriting (motor-coordination / clumsiness)
  ii) speech and language (verbal communication and 'body language'/ social awareness)
  iii) reading and spelling (decoding and encoding)

In all three groups there appear to be impediments which affect 'fluency' in 'making connections' (see also Leong's 1987 analysis of perceptual-cognitive patterns of disabled reader) suggesting that there were information processing problems concerning simultaneous synthesis at a perceptual level and successive synthesis. Leong's theoretical postulate was derived from Luria's two basic forms of integrative activity; simultaneous (primarily spatial) and successive (primarily temporally organised) syntheses at the perceptual, memory and intellectual levels). Leong shares with Luria (1966), an interest in the qualitative aspects of functioning, not only the quantitative findings, i.e., how a function suffers not only what functions suffer.

The learning problems may be further categorised as including difficulties with the following, particularly as they affect each other in terms of the integration and generalisation of learning:

(a) memory (attention / discrimination / sustained attention / recall - revisualising, reauditorising, motor-memory / retrieval/ word-finding / association)

(b) the attribution of meaning (imagery ; imagination / analysis / synthesis / phonemic- segmentation / blending / sequencing / pacing / spacing / directional factors / integrating / transferring / generalising)

(c) motivation (energy level / anxiety / frustration)

To understand the children's problems it is useful not only to observe and analyse their behaviour in terms of their reactions to learning situations, but also to analyse the learning tasks with which they are faced in the classroom.

Firstly the child has to select that feature or set of features to which the teacher intends that they should give their attention. Usually this requires the child to be able to ignore stimuli which are for the purpose of the moment, irrelevant. It is then necessary to fix and sustain attention on the required object or idea for the required amount of time not only to internalise the perceptions and to have recognised (re-cognised) or 'registered' (labelled / conceptualised) the message but so that it can be retrieved later (thus moving the message from short term /working memory into longer term memory). Furthermore, the child must shift attention when appropriate.

This does not appear to be a linear process and impediments can occur within the dynamics of interaction of more than one factor. Also, it is necessary to take into account : (a) within-child variables which include intra psychic factors as distinct from extraneous factors which include: (b) within-curriculum variables, and (c) 'between' variables, e.g. the child's interaction with the teacher and other adults and with the peer group (Hunter, 1982).

While recognising that for practical purposes the subgroup of children with SpLD is currently differentiated from others with special educational needs on the basis of having a general intelligence level of around or above average, it was not their level of intelligence as such which differentiated the children in the pioneering studies by Strauss, Werner, Lehtinen & Kephart (1942) and Cruickshank (1963) but their homogeneity in terms of positive response to 'structured teaching methods'. This is perhaps a point which might be considered along with Stanovich's (1991) argument that the 'intelligence factor' is perhaps less relevant than the reading difficulty if one is seeking to differentiate on the basis of specificity of learning difficulty.

However, clustering the children with SpLD for purposes of economy to try to teach them in groups rather than individually has proved to be a complex matter requiring very detailed and frequently checked test results on profiles covering their functioning on a range of learning tasks and processes. For example, Frostig & Maslow 's (1983) basic profile included Kirk's Illinois Test of Psycholinguistic Abilities, the Weschler Intelligence Scale for Children and the Frostig Test of Visual Perception. Furthermore, in teaching groups of children of different ages but with similar difficulties, the writer found that the teaching of those under eight years of age could effectively address their 'weaknesses' but for those over eight, 'compensating via their strengths' was the preferred emphasis (1973 University of Strathclyde film, Remediation in Reading).

It is important to be clear about exactly what kinds of learning are considered to be most efficiently delivered in groups. The writer's experience suggests that group work for certain 'drills for skills learning' would constitute both an efficient and effective way of delivering aspects of the national curriculum for English for 'normal learners as well as for children with SpLD. ' Dictation' is one such 'drill' (see e.g. Hornsby & Shear, 1979). It should be taught, however, in a way that is enjoyable, constructive and informative, helping the teacher to discern children's individual ways of learning, not simply to mark words as 'correct' or 'wrong'. The art of devising games for word-building and word-analysis should be developed and many of the good ideas' which formed an integral part of structured teaching using older reading schemes can be rescued and updated so, that children have sufficient opportunities to practice, review and consolidate learning not only to skate over the surface of new experiences in a race to 'cover the topics in the curriculum'.

However, there is an overriding consideration in the case of many children with SpLD; their need to recognise the reasoning behind what they are being asked to do. Once explained, they are usually the more willing as learners. It is thus a matter of harnessing necessary energy from motivation as well as providing ideally, non-distractable circumstances for the presentation of carefully graded steps towards achieving adequate practice for skills to become learned to the level of 'automaticity ' required for their fluent use.

This kind of carefully prepared and teacher-directed approach may be described as 'the 'structured approach '. It can include the use of 'multi-sensory teaching methods', a descriptor which is liable to misinterpretation as it is not the simultaneous presentation of many different sensory stimuli which facilitates learning as much as the selective sequential presentation of information via different modes with a view to discerning which are the preferred and relatively more effective modes of presentation and 'tailoring' the teaching to meet the special needs of the individuals.

III   A Framework for Curriculum Differentiation

To be sufficiently comprehensive, the framework for curriculum differentiation must take into account the range of variables discussed above. The following section includes a brief report of the views of King and Spillman and further development of the writer's model originally presented in 1990.

King (1989) produced a 'differentiation menu' which included the following 23 factors to be differentiated in order to meet pupils' special educational needs (i.e. not solely SpLD): 'aims / content / learning contexts / breadth / depth / pace / language / materials / teacher-pupil;pupil-teacher interaction / forms of pupil recording / forms of pupil grouping / teaching styles / forms of classroom organisation / uses of support teacher / forms of assessment / kinds of marking / resources for self study / reinforcement / feedback / levels of motor skills / levels of confidence in oral work / levels of maturity of response.'

Looking at the problem of 'decoding differentiation' more than a year later, Spillman (1991) discusses differentiation on the basis of classroom organisation and by the ways in which tasks are (a) introduced, (b) carried out, and (c) the kinds of outcomes which are required. The value of paired work and joining together of pairs into groups is illustrated. A checklist of 'good practice' is provided and the final words of advice indicate that 'the key to the differentiated curriculum is the flexible use by teachers of a wide range of activities and lesson organisations'.

There are many constructive practical suggestions made by King & by Spillman. Both also address theoretical issues. While an extensive combined list can be made of all the factors they note, there remains the difficulty of priorities and how to translate into practice, ways of bridging the seeming gaps between 'the new literacy' with its 'open' methods and 'the structured approach' ' with its 'routines' and seeming inflexibilities.

The writer's attempt to devise a model for curriculum differentiation which would address such issues (Hunter-Carsch, 1990) seeks to illustrate the fact that teachers need to operate at more than one level at any one time. It is here perhaps, that the relative emphases of generalist and specialist most require to be shared.

The model concerns the relationship between curriculum issues about WHAT has to be learned (content) and HOW it is to be learned (processes) and the impact on these factors which is exerted by the teacher's employment of SUPPORT STRATEGIES which concern motivation and attitude. The balance of the whole picture is 'held in place' by MACROSTRATEGIES and MICROSTRATEGIES which can be employed by the class teacher (GP) or a collaborative teaching pair or team of teachers (including SENS).

The teacher(s) could use the framework for checking their understanding of their own and each other's expectations and plans. The three 'macrostrategies' (wider concepts and stages in the process of meeting S.E.N. for individuals) were described as including firstly, a series of steps towards meeting special educational needs. The second involved an understanding and ability to make operational in the classroom a wider appreciation of the concept of 'literacy' and relatedly the concepts of 'language' and 'reading' in ways which connect them in the child's mind with human interest , curiosity, the will to communicate, to learn and to know.

The third involves a dynamic system for recording (making notes on) the interaction between child and teacher within a 'conferencing' situation (one-to-one tuition within the class). The eleven' microstrategies' (factors to be adapted and controlled in terms of both the extent and the nature of learning problems) are listed below and briefly discussed :

Microstrategies for Adapting Teaching and Learning:

(i) choice   (ii) challenge (iii) space (iv) time (v) materials (vi) relationships (vii) methods , also ; (a) whole-part learning (b) sequencing (c) speed, and (d) accuracy.

Support for the child with SpLD should be mediated in such a way as to provide empathy conveyed through understanding of the nature of the difficulties, encouragement which is unobtrusive, practical strategies and routines which facilitate learning, and ways of communicating which mobilise and sustain the child's interest, imagination and ability to think. The following comments on each of the 'microstrategies' seek to illustrate ways in which that variable can be adjusted to mediate learning more successfully for the child with SpLD affecting that particular aspect of his or her functioning.

(i) **choice** : Some children with SpLD find it difficult to make choices. When faced with having to make a choice, for example between several activities, they become anxious and /or over-excited and appear to be unable to 'reason' or consider the alternatives. It is helpful to remove choice about work tasks for such children, then, gradually, to introduce limited choice, and eventually to open up the possibilities to include a range of tasks (choices). There need be no sense of punishment in the absence of choice at the initial phase. Rather, there is a routine suggesting by the very way in which materials are laid out, that there is an order. (e.g. This is what you can do to start... making the model.) Explanations and negotiations can be avoided and reassurance provided by the structuring of the activity itself.

For other children, there may be insufficient choice for them to feel able to adapt the work in ways that they can handle. The increments in extending choice for them can be faster.

(ii) **challenge**: This is related to choice, but involves also, the amount of work which the child can be expected to complete in a given time, and the level of difficulty of the work.The tasks should not be so difficult that they reduce the child's resilience and will to tackle them with interest and enthusiasm. Nor should they be so easy that there is virtually no challenge. Getting the level exactly matched is likely to involve considerable observation and getting to know the child over some time, particularly as children with SpLD often are erratic in their functioning even on the same kind of tasks. There is obvious value in starting at a level which is predictably within the child's capabilities and through gradually escalating the challenge, to build increasing confidence as well as the relevant knowledge or skills.

The management of both choice and challenge concerns the gradual decrease in teacher-directedness of the work and the corresponding increase of child-directedness.

It is often helpful to consult with the child regarding the increases in the length, complexity and nature of the challenge, once there is enough confidence built up on the basis of successful learning experiences.The child will need to see the evidence of the successful learning. This can be provided by a simple record keeping system which involves the child in completing a part if not all of the record as each piece of work is completed. The aim is to build a visible/tangible note of successful learning. Later, it is hoped that there will be generalisation of learning sufficient to reduce the need to see, touch, check the 'evidence' that the child can learn.

(iii) space: For some children, the size of the classroom (even a normal classroom not an open plan space) can seem to be a vast space that is over stimulating. The child may dash about or else show the opposite kind of reaction and stand as if fixed and unable to organise himself comfortably within the space. The child's reactions to the space in a gymnasium may provide even clearer evidence of difficulty in learning within that context, whether of a panic induced racing about or a rigid stillness.

Adjustments of space can include creating a series of small working carrels (study-booths) in the classroom, either by use of freestanding heavy cardboard screening or by placing a desk/table against a wall and flanked by two screens /backs of library shelves or cupboards. the tiny 'offices' which are created then become work places for scheduled 'office time' which the children can all enjoy for specified amounts of time, booked in advance. There should be no sense of punishment in being 'sent to the office'. This approach works in the opposite manner. The children can be awarded extra office time to work quietly on their own, undisturbed, as a reward for specially good work as well as by choice as a routine part of classwork.

The structuring of the space and placing of furniture can help to control the amount and direction of movement in the classroom. It can slow down active children, avoid creating enticing corridors or bottlenecks and relate clearly to the kind of work to be done in the space(s).

It is helpful for some children if the spaces in the classroom are clearly delineated,for example, having work areas split up by room dividers such as long (possibly moveable) shelf units.The establishments of routines about what is done in which area, can also be helpful in the building of an external order which appears to affect the internal ordering of experiences.

Open shelf units are to be avoided where possible in favour of closed doors. The open shelves invite the distractible child to explore them visually if not also by touch and movement. Where possible the surroundings should be functional but not fussy (e.g. avoid mobiles and too many dazzling, busy illustrations). The aim is to avoid clutter and distractions from the current task.

(iv) **time**: Time may be a very difficult concept for some children if they have no awareness of, for example, what three minutes feels like. Their seeming anxiety in some cases, and 'head in the cloudness' in others, may be related to limited sense of the passage of time. It is helpful to provide a visual picture of 'time passing' e.g. by use of attractive coloured sand-timers. Once the children have watched the sand run through a three minute egg timer, for example, and thus been shown the time for the task that is set for them to do in the office, they are often both relieved that they can perceive that the task will be 'finishable' (both because it has been designed to be within their capabilities and because it does not require more than three minute, the time which can be experienced/paced by using the timer).

It can easily be seen how an individualised timetable (or simply a list of two or three tasks) can be drawn up for the child in a way which indicates (by illustrations) how many 'eggs' are allocated for each task. It may be necessary to start with a single task of less than three minutes duration, in order to reinforce success and the chance for the child to 'rest' while watching until the sand has completely run through the timer.

The idea of pacing themselves becomes the easier in such circumstances and the sense of accomplishment is the greater if a record of successful timed tasks is maintained. (Perhaps as few as two three minute tasks per day; one in the morning and one in the afternoon, can bring about a sense of self worth and turn a child's attitude from one of self doubt to one of willingness to try and hopefulness that there can be a better future.)

After successful work in the office with timed individualised tasks, there is often a rapid transfer of success to working on a timed basis in other places in the classroom, perhaps with the use of the timer at first then gradually to use of the classroom wall clock and / or digital or other watches (and self direction).

(v) **materials**: The structured approach seeks to assist the child to concentrate only on those features in the environment that are relevant to the particular aspect of learning that is the concern of the moment. Once the immediate surroundings are 'controlled' by making them as distraction free as possible (e.g. using an 'office'), it is easier to heighten interest in materials which are attractively presented. The resources the child uses are thus the more memorable if they are tailored to acknowledge and extend the child's interests. They might include exercise books covered in the child's favourite colour, a specially colour coded folder with the child's timetable and attractively prepared tasks for the short work session in the office. Technology can play a special role in that there are commercially produced programmes as well as word processing facilities which may be employed for developing learning.

Materials might include published schemes, kits, or teacher-made resources (or materials made by children). The special nature of the materials is closely linked to the methods of using them. A tape-recorded programme made by the teacher and with invitations to the child to participate in various 'game-like' tasks, can provide the basis for a move from teacher-directed work to more independent work if the child follows the taped instructions which can extend work to include gradually more challenging tasks. The use of headphones can individualise the work and avoid the sounds becoming a distraction to others. The child can be taught how to use the tape recorder to record messages for the teacher or others and in this way manage to continue to work independently, but retain the means of requesting assistance .

(vi) **relationships**: The structuring of relationships between teacher and pupil does not imply anything ominous or cold. It involves respecting the child's need for calm and detachment from emotional demands so that all the necessary effort can be directed into the learning which is intended. If the child is already aware of having learning difficulty, it is not always a kindness to draw attention to the child's efforts by being effusive. A 'low-key', more matter-of-fact tone and routine is frequently felt to be less intrusive. A quieter voice, slower pronunciation and shorter rather than longer interactions may provide the basis for a relationship which gradually can accept more energetic praise and extended interactions.

A helpful discussion of structuring relationships is provided in Hewitt's (1972) hierarchy of educational tasks which is based on Maslow's hierarchy of basic needs.

(vii) **methods**: This is perhaps the most complex variable as it can be regarded as an overriding approach, thus subsuming the other six variables, or, as intended here, as the way in which the materials are used within the carefully structured context. 'Essentially reading is a matter of getting it together' commented Goodman (1975) who added that the children who have reading difficulties are those who experience difficulty in 'getting it together'. For them a structured approach such as that of Stott (1964) which was based on a distillation of 'good practice' of experienced class teachers, may not be sufficiently differentiated to support the learning of individuals who require reinforcement of each tiny step mediated through their preferred modality or combination of modalities and orchestrated with reference to pacing and spacing uniquely designed to meet their needs. It is thus a professionally sophisticated and demanding challenge to discern how to interrelate the components from each of the variables noted above and to provide an individually adapted version of, for example, 'multisensory methods'. This subject requires a separate and full discussion.

The additional microstrategies listed on page 14 constitute further variables any or all of which may need to be adapted in order to facilitate the child's learning. Briefly, the child's difficulties may be eased if attention is drawn appropriately to (a) the relationship of the part of learning which is the focus of the moment to the whole of the relevant topic; (b) the sequencing, (c) speed and (d) accuracy of the teacher's modelling, any or all of which may need to be adapted individually or in combination to assist the child.

IV   A Context for both 'NEW LITERACY' and STRUCTURED LEARNING.

Provision of the above constellation of support strategies is easier when working on a one-to-one basis than in a class or even group teaching situation. A teaching method which permits extensive individualisation and potential for group and whole class interaction, is exemplified by the approach known as 'systematic drafting and redrafting' pioneered by Binns since the early seventies. It constitutes a means of developing writing that has proved to be flexible enough to permit every child to progress at their own pace within the classroom.

The main idea involves freeing the writer to concentrate on the emerging meaning, unimpeded by problems of spelling, handwriting or presentation . It is also proving to be valid in using word processor programmes .

The approach is described elsewhere (Hunter-Carsch, Binns & Sobey, 1991b; Binns 1980, 1989, 1990). For the purpose of this paper, it should perhaps be noted also, however, that the approach highlights the value of using a dynamic visual focus provided by the overhead projector which acts as a means of 'welding together' the class's attention while they read aloud for meaning, and are encouraged to listen as they read . They become aware of the mechanical aspects of phoneme-grapheme matching, characteristic letter sequences and irregular words which are enjoyed incidentally in the course of choral and individual readings. The illustrations provided by Binns (1989) demonstrate also the way in which the technique lends itself to helping children grasp the concept of 'stance', one which is vital to their development of 'self-concept' and to aspects of self- control, not only control of their writing. It can thus be recognised as a technique which represents the shift of the locus of control of learning to the learner.

Individual tuition via ' conferences' can build further on the basis of the evidence of the child's understanding of the writing process and of how to maintain clarity of purpose and cope with their particular learning difficulties. This 'writing road to reading' can be considered as a basis for extending skill, competence and confidence in both encoding and decoding.

Matters of frequency of individual tuition and length of tuition sessions must be considered on the basis of a range of factors, not least the fairness of allocating time to all of the class in a regular rota. The approach has been found to assist with the development of self-reliance as well as collaborative learning. The teacher remains responsible for the orchestration of all of the differentiated work which permits the child's initiatives to be recognised through the priority given to the content of the writing.

Reason

The above discussion with reference to use of the draft/ redraft technique can be regarded as illustrative of employing reasoning, which, it is suggested is one of the three overriding components governing the success of the proposed wider framework for curriculum differentiation.

With reference to this component of the framework, through developing writing the child learns how to match his or her own intended meaning against each successive effort made to encode the message, whether it is in a story or other form.   The child's frustrations and anxieties about having literacy difficulties are both 'contained' and resolved within the progressively clearer sketches which the child makes through drawings or words.

Rhythm

The second major component is that of 'the Rhythm of Education' (Whitehead, 1932). This principle involves the ideal cycle of educational experiences which Whitehead describes as beginning with 'the age of 'Romance' (the exciting introduction), followed by the age of 'Precision' (when detailed study and analysis is required) and finally, the age of 'Generalisation' (being able to relate the earlier experiences to appropriate contexts , to apply the ideas and to 'transfer' the learning).

Critics of 'the structured approach' might consider it to start and finish within the second stage in the cycle and thus to constitute what Whitehead referred to as 'dead ideas'. For the ideas to come to life they should be connected to the flow from 'romance' through 'precision' to 'generalisation'.

The good teacher can facilitate this cycle of experience for the learners by becoming the choreographer as well as composer and conductor of the interaction with the child and all of the children in the class.

Relaxation

The third component, that of the relevance of relaxation is bound up with the importance of pacing and spacing of work which requires concentration and effort in order to learn (remember). Many children, not only those with SpLD, may be tense and anxious and thus unable to apply themselves easily to the learning tasks. If the previously suggested rota of three groups is established with the timetabled cycle of tasks one of which involves relaxing or resting, the other work times are more likely to be productive. A cycle of activities which is punctuated regularly by rest periods which can be recorded by the children themselves. It may also be necessary to include training in relaxation. (The teacher's group then becomes the one that is learning (being taught) how to relax. The other two groups work on practice or review tasks; see also Sauter, 1991 for information on relaxation training.)

Training in relaxation can be related to breathing exercises, to listening and 'image-making' (visualising or auditorising) and to various patterns of movements, particularly movement to music. Some children are unaccustomed to experiencing silence. It may be necessary to employ a gentle background music firstly then gradually work on introducing longer periods of silence punctuated by listening to a series of musical sounds, different patterns of sounds and rhythms. The aim is to increase the periods of time in which individuals can feel in control of themselves, make and stay with their own choice of restful 'work' or play materials.

It may be possible to design the classroom in such a way as to include resting places which are available all of the time. If space permits this, and if the routines can accommodate such differentiation, these places can be labelled and used by one or two of the children throughout the working time. They can request to take a rest time when they think it will improve their work. It is essential to maintain records to note which children use these areas and how frequently. The children themselves can keep simple record charts noting their regular rests as well as the length of time they request for additional relaxing times.

The above discussion has broadly covered a range of components within the framework for curriculum differentiation . Many of the strategies can be applied, when appropriate, within the context of individualised tuition times which are envisaged as part of the teacher's group time in the rotation of group activities. It may be necessary, however, for some children to have more individual tutorial support than can be provided solely by the class teacher. The support of a trained specialist teacher of children with SpLD may be required in addition to training for both teachers (GP and SENS) to develop fully their roles through cooperative teaching which employs both 'open' and 'structured' methods.

CONCLUSION

The problem of how to differentiate the curriculum to meet the special educational needs of children with SpLD in the normal classroom has been discussed in terms of the proposed desirability of sharing a conceptual framework within which a range of factors are taken into account by both the class teacher and the specialist teacher. It is suggested that the factors are not discrete but relate to each other in a dynamic manner. They include the priority of reason, the 'rhythm of education' and the relevance of relaxation for creating conditions to facilitate learning.

For the child with SpLD who requires individualised teaching, there is a need for a practical approach to literacy learning which permits tuition to be mediated within a structured context and employing structured teaching methods. If this can be provided within the ethos of an open approach to learning in which the locus of control is increasing felt by the child to be within himself or herself, there is the greater likelihood that the child will be able to learn successfully and go on learning . The necessary 'learning of how to learn' can be facilitated through 'systematic drafting and redrafting', a means of developing reading and spelling 'incidentally' through writing about experiences and subjects which matter to the child.

The material that constitutes the content of the writing then renders the written communications the more easily memorable, meaningful and envisagable by directing the imagination rather than by 'capturing' it. Thus, while sharing and enjoying the flexibility of the 'new literacy' along with members of the peer- group, the child can be offered highly structured teaching within a safe, clearly delimited 'office 'space, during specified times and with the shared professional support which derives from both specialist and generalist teacher working cooperatively together in the classroom. It is suggested that in this way, the child becomes more in 'control' of his or her learning and can recognise and get ' around ' if not ' over' the hurdles which are presented by his or her SpLD.

To return to the original questions, firstly with regard to the crucial elements of professional knowledge and skills which should be communicated and perhaps shared between GP teachers and SENS teachers,the above discussion may be considered as providing support for the following recommendations: that teacher education for both GP and SENS teachers should include:

1. Information about individual differences in learning which are illustrated with reference to children with SpLD. The illustrations should include not only reading / literacy difficulties, but speech and language and also motor problem as well as other learning difficulties.

2. Information and some practice in differentiating between reading difficulties which are associated with different causal factors including (a) 'gaps' in teaching (b) mild learning difficulty (c) 'crises' and (d) SpLD , as well as appreciating multiple causal factors in any one case.

3. Information and training in a range of teaching methods which include systematic drafting and redrafting, multisensory methods and an understanding of the principles of 'the structured approach'.

4. Training in collaborative teaching and working with parents. This should include planning together and evaluating the effectiveness of employing various strategies for curriculum differentiation and particularly, for monitoring and recording children's progress and problems and for home-school liaison.

5. Classroom design to include 'offices' and space for relaxing.

The above list is neither detailed nor comprehensive. It represents a 'handful' of priorities. Additionally there is the need for advanced levels of training which provide opportunities for research and teaching to progress together and for multidisciplinery parameters to be maintained throughout.

With reference to the second question concerning the extent to which the proposed framework could reconcile the apparently contradictory approaches ,' the new literacy' and 'the structured approach', it is concluded that with a balanced and positive attitude and with the relevant knowledge and skills which could be fostered through training, GP and SENS teachers can work in such a way as to avoid the unfortunate outcomes of poor teaching of either approach.

Poor teaching of 'the new literacy' puts the child's learning in jeopardy in cases of SpLD by requiring the child to handle too much 'freedom' without the means of bringing the range of impinging variables under control at any one moment in a way that is required for successful learning of those features to which the teacher intends the child to respond. The provision of 'offices' within the classroom can make it possible for fairly distractible children to concentrate more easily on thinking and creating their own stories. The offices can be just large enough for paired work including special 'structured ' teaching.

Likewise, poor teaching of the structured approach renders the child vulnerable to what Meek (1991) warns as the dangers of converting an enjoyable and 'natural' shared experience of, for example, rhyming, into a form of 'therapy'.

Good teaching, particularly in a collaborative teaching context, can promote professional confidence and competence on the part of the teachers and for the children, can bring about a shared experience that permits (a) an appealing, harmonious, rhythmical and relevant sustaining of the sense of 'romance' in the engagement with learning while also providing (b) the relative detachment, pleasure in skill mastery and ease of progress that characterise the carefully 'scaffolded' route towards higher levels of complex learning which can be facilitated by appropriate attention to ' precision '. By examination of the impact of each of the two approaches and their effect on the children's perceptions of literacy as a result of having both, we may be better able to understand how 'generalisation' of learning may be facilitated as well as monitored and assessed in the classroom context.

It is perhaps in this way also, that we may approach a more precise definition of what it is to be 'a good teacher' (Warnock, 1991) and particularly, what that means for children with subtle and complex learning difficulties. To meet their special educational needs in basic literacy it has been suggested that the teacher will require to bring together aspects of both 'the new literacy' and 'the structured approach' in ways which take into account the factors discussed within the proposed framework for curriculum differentiation.

ACKNOWLEDGEMENTS

Thanks are due to Professor Maurice Galton, Director of the School of Education for support to attend the conference, to Dr. Roger Merry for encouragement to write on the subject of 'differentiation', to Dr. Harry Chasty for the invitation to speak at the conference on 'Meeting the Challenge' and to UKRA colleagues who contributed to relevant discussions. Also thanks are due to all the children and teachers who have influenced the development of the framework.

REFERENCES

Alexander, R., Rose, J. & Whitehead, C. (1992). *Curriculum Organisation and Classroom Practice in Primary Schools: A Discussion Paper*, D.E.S.

Bannatyne, A. (1967). The Colour Phonics System. In J. Money (Ed.) *The Disabled Reader*. Baltimore: Johns Hopkins U. Press.

Barrs, M., Ellis, S., Hester, H. & Thomas, A. (1988). *The Primary Language Record*. London: ILEA/ C.L.P.E.

Binns, R. (1978) in M. Hunter, unpublished Interim Research Report on *Drafting and Redrafting*. Edinburgh: Scottish Council for Research in Education .

Binns, R. (1980). A Technique for Developing Written Language. In M.M Clark *Reading and Writing for the Child with Difficulties..* Education Review Occasional publications No.8, University of Birmingham.

Binns, R. (1989). Recreation through Writing. In M. Hunter-Carsch. *The Art of Reading*. Oxford: Blackwell.

Binns, R. (1990). Monitoring the Development of Writing. In M. Hunter-Carsch. *Primary English in the National Curriculum..* Oxford: Blackwell.

Bradley, L. & Bryant, P.E. (1983). Categorising Sounds and Learning to Read: A Causal Connection. *Nature*, 301,419-21.

Critchley, M. (1970). *The Dyslexic Child*. London: Heinneman.

Cruickshank, W. (1963). *A Teaching Method for the Brain Injured Hyperactive Child*. Syracuse: Syracuse University Press.

Czerniewska, P. (1989). The National Writing Project. In M. Hunter-Carsch. *The Art of Reading*. Oxford: Blackwell

Division of Educational and Child Psychologists (1984). *Dyslexia/Specific Learning Difficulties*. British Psychological Society.

Department of Education and Science (1975). *A Language for Life*. The Bullock Report. H.M.S.O.

Department of Education and Science (1978). *Children with Special Educational Needs*. The Warnock Report H.M.S.O.

Department of Education and Science (1988). *Report of the Inquiry into the Teaching of English Language*. The Kingman Report H.M.S.O.

Department of Education and Science & the Welsh Office (1989). *English in the National Curriculum* H.M.S.O.

de Hirsch, K. Jansky, J. & Langford, W. (1966). *Predicting Reading Failure*. New York: Harper & Row.

Donaldson, M. (1989) *Sense and Sensibility: Some thoughts on the Teaching of Literacy.* , Occasional Paper No.3 Reading and Language Centre, University of Reading.

Fernald, G. (1943). *Remedial Techniques in the Basic School Subjects.* New York: McGraw Hill.

Frostig, M. & Maslow, D. (1983). *Learning Problems in the Classroom..* New York: Grune & Stratton.

Gillingham, A. & Stillman, B. (1940). Remedial Training for Children with Specific Disabilities in Reading, Spelling and Penmanship. New York: Sackett & Wilhelms.

Goodman, K. (1975). *How do you read?* Interview on BBC Horizon Film.

Graves, D. (1983). *Writing: Children and Teachers at Work..* London: Heinneman.

Hewett, F. (1966). A hierarchy of task relationships. In D. Hammill and N. Bartel (Eds.) *Educational Perspectives on Learning Disabilities.* New York: Grune & Stratton.

Hickey, K. (1991). *A Language Teaching Course for Teachers and Learners.* Staines: The British Dyslexia Institute.

Hunter, M. (1980). Becoming a Better Teacher of Children with Learning Difficulties, in M.M Clark. Reading and Writing for the Child with Difficulties. *Education Review,* University of Birmingham.

Hunter, M. (1982). Reading and Learning difficulties: relationships and responsibilities. In A. Hendry. *Reading: the Key Issues.* London: Heinemann.

Hunter, M. (1985). Reading (main unit) and Asssessing and Recording Reading. In R. Dawson.*Teacher Information Pack: Special Educational Needs.* London: Macmillan

Hunter-Carsch, M. (1990). Learning strategies for pupils with literacy difficulties: Motivation, meaning and imagery. In P.D. Pumfrey & C.D. Elliot (Eds). *Children's Difficulties in Reading Spelling and Writing.* London: The Falmer Press.

Hunter-Carsch, M. (1991a). Celebrating and defending Oracy as well as Literacy in Inservice Education in Specific Learning Difficulties. In C. Harrison, (Ed). *Celebrating Literacy : Defending Literacy.* Oxford: Blackwell.

Hunter-Carsch, M., Binns, R. & Sobey, J. (1991b). Creating Young Writers. *Language and Learning* , 5, February 1991 pp 9-12.

Hornsby, B. & Miles, T.R. (1980). The effects of a dyslexia centred teaching programme British Journal of Educational Psychology, 50 (3) 236-42.

Joint Committee of Colleges of Education in Scotland. Advisory Committee Report. *Ten Years On.* Edinburgh: Moray House College of Education.

King, V. (1989). Support Teaching: Differentiation Menu. *Special Children*, Vol.33 Oct.89 Practice Papers 11.

Leong, C.K. (1987). *Children with Specific Reading Disabilities*. Amsterdam: Swets and Zeitlinger.

Luria, A.R. (1966). *Human Brain and Psychological Processes*. New York: Harper and Row.

Meek, M. (1989). What do we know about reading that helps us to teach? *Language and Learning*, 1, p 3-8.

Mills Learning Test (1970). The Mills Centre, Fort Lauderdale Fla. USA.

Moseley, D. (1971). *English Colour Code Programmed Reading Course*. NSMHC, London.

National Association for Remedial Education (NARE) (1985). Guidelines 6. *Teaching Roles for Special Educational Needs*. Stafford: NARE Publications

Newton, M. & Thomson, M. (1979). *Readings in Dyslexia*. Wisbech:London Dyslexia Association.

Orton, S. (1937). *Reading, Writing and Speech Problems in Children.*. New York: Norton.

Pumfrey, P. & Reason, R. (1991). *Specific Learning Difficulties (Dyslexia) Challenges and Responses*. Windsor:NFER-Nelson.

Sauter, J. (1991). Emotional stress, visual impact and reading disorders. Paper presented at the 18th Rodin Remediation Scientific Conference, University of Bern, August 1991.

Scottish Education Department (1978). *The Education of Pupils with Learning Diffixulties in Primary and Secondary School in Scotland*. H.M.I. Report.

Scottish Education Department (1990). *English Language* 5-14.

Slingerland, B. (1977). Public school programme for the Prevention of Specific Language Disabilities in Children. Modified and reprinted from 1966 *Educational Therapy*, 1, 389-424.

Spillman, J. (1991). Decoding Differentiation. *Special Children*, Jan 1991, 7-10.

Stanovitch, K. E. (1991). The Theoretical and Practical Consequences of Discrepancy Definitions of Dyslexia. In M. Snowling & M. Thomson (Eds). *Dyslexia: Integrating Theory and Practice*. London: Whurr Publications.

Stott, D.H. (1962). *Programmed Reading Kit*. Glasgow: Holmes & McDougall.

Strauss, A., Werner, H., Lehtinen, L. & Kephart, N. (1947). *The Psychopathology and Education of the Brain Injured Child*, Vols 1 and 2. New York: Grune.

Whitehead, A. N. (1932). *The Aims of Education* . London: Williams and Norgate.

Warnock, M. (1991). Good Teaching. In P. Snelders (Ed). *Papers of the Celebration Conference of the Philosophy of Education Society of Great Britain.* London: The Froebel Institute.

Willinsky, J. (1990). *The New Literacy.* London: Routledge.

Facets of Dyslexia and its Remediation
S.F. Wright and R. Groner (Editors)
© 1993 Elsevier Science Publishers B.V. All rights reserved.   541

SYSTEMATIC PHONOLOGY: THE CRITICAL ELEMENT IN TEACHING READING AND LANGUAGE TO DYSLEXICS

Analysis of Project Read, a systematic phonology, multi-sensory reading/language arts program piloted in regular education classrooms of three Louisiana school districts during the 1990-1991 academic year.

J. Greene

Basics Plus Education Centres

New Orleans, USA

INTRODUCTION

A critical element in the design, implementation and evaluation of efficient programs for teaching dyslexics to read, write, and spell is the systematic presentation of phonological elements. Dyslexics demonstrate difficulty in processing receptive and expressive written language. Tutors, teachers and therapists often make the fatal error of using traditional methods and materials and then presenting linguistic concepts more slowly, attempting "mastery learning" of randomly presented phonological and morphological elements. Because dyslexics evidence lack of phonological awareness and difficulty in establishing visual memory for the sequence of letters within words, they require a systematic and sequential presentation of phonological elements. Moreover, the presentation of non-phonetic words must be systematic as well. Sight words presented randomly in large numbers for automatic memory cause frustration and result in little or no measurable gain.

Basal reading texts provide for scope and sequence of more general reading skills; however, in the introduction of phonological elements, most are random, rather than systematic. Further, currently popular "whole language" programs are by definition unstructured, controlling neither scope nor sequence, permitting random and unlimited "natural language" introduction of non-phonetic words. The result for the dyslexic is increased frustration and decreased ability to process written language. Unless materials introduce phonology systematically, even multi-sensory methods fail.

Establishing Appropriate Sequences for Systematic Phonology

To analyze phonological elements of a language and develop a logical chronology by which these elements may be taught to build words, one must assess the elements' complexity. In most languages, consonants are more stable than vowels in their phoneme-grapheme correspondences, so it is possible to introduce as many as four to five consonant phonemes with a single vowel phoneme, and then construct the possible single-syllable words of the language.

Logical progression results in the cumulative addition of vowel phonemes and the mastery of each prior to the introduction of additional structures. Further linguistically-based introductions progress (here: English examples) from short vowels to long, to r-controlled, to the schwa, to diphthongs. They progress from single consonants to blends, to clusters, to digraphs, and eventually to phonological/morphological elements with less linguistic stability and/or frequency.

This approach identifies phonological elements and accumulates possible words from each new phonological element. The result is significant measurable gain for dyslexics in decoding and encoding. The element which is critical throughout is that the phonological presentation be systematic and cumulative.

Determination of an appropriate sequence of phonological elements (in any language) is essential. The ability to analyze a language and design an appropriate sequence of its phonological elements is vital for therapists who work with dyslexics. Critical evaluation of selected materials is equally important.

Lack of phonemic awareness among dyslexics has been demonstrated repeatedly (Catts, 1986; Liberman, Mann, Shankweiler & Werfelman, 1982; Liberman 1989; Lundberg, 1988; Rapala & Brady, 1990; Shankweiler, 1989; Treiman & Baron, 1981; Vellutino, 1987; Wagner & Torgesen, 1987). Findings indicate that therapists should select materials based on linguistic analysis and systematic (sequential and cumulative) phonological representation.

The Orton-Gillingham method was the first widely employed method shown to be successful in teaching dyslexics to read, write, and spell. It has been modified several times for use in classroom and clinical settings. This approach provides for systematic, cumulative introduction of phonological elements combined with multi-sensory methodology. Orton-based methods which have produced excellent results with dyslexics in diverse settings include: Alpha to Omega; Alphabetic Phonics; Project Read; and Slingerland. Of these, a systematic phonology program which has produced significant measurable improvement for poor readers in regular education classrooms across the United States is Project Read. Now implemented in more than fifty school districts across the country, Project Read has produced remarkable statistical evidence to support the use of systematic phonology.

A PROJECT READ STUDY

A pilot study was was implemented in regular education classrooms of seven parish school districts of Region III, State of Louisiana, during the 1990-91 academic year, to investigate the effectiveness of a systematic phonology program for students at risk for severe reading/language disability.

Selection of **Project Read** from among available systematic phonology programs was based on the following documented information:

Project Read was developed 23 years ago in the public schools of Bloomington, Minnesota. Its objectives were: (a) to reduce costs of expensive pull-out programs for at-risk students; (b) to provide more efficient education of students at risk for reading/language problems; and (c) to train classroom teachers to work effectively with at-risk students. Although lower socioeconomic families now comprise the district, Bloomington's current average reading scores (including special education subjects) consistently rank 90th percentile.

- Project Read is an alternative reading/language arts program initially designed for use in the *regular education* classroom with at-risk students (generally, students whose standardized scores measure performance below the 25th percentile).

- Project Read provides a highly structured, sequential phonology program for students in grade one through the junior high level.

- Project Read is a *total* reading/language arts program which has three interwoven components: (1) Systematic Phonology; (2) Written Expression; and (3) Comprehension.

- Project Read provides teacher manuals and lesson plans for each grade level.

- Project Read provides for ongoing in-servicing and demonstration teaching through its demonstration teacher/facilitator model. Once demonstration teachers/facilitators are proficient, the school system conducts its own in-servicing and teacher training.

- Project Read has been replicated in over 50 school districts and in over 20 states, in special education and in Chapter I programs as well as in regular education. Administrators country wide consistently praise its concurrent cost-effectiveness and academic effectiveness.

Review of Data from Project Read Replication Sites
A review of Project Read data from schools across the nation included the following:
  1. Project Read has reduced referrals to special education by 75%.

2. Students in Project Read have consistently outperformed students in traditional programs with *double* their annual gain scores, based on comparison of standardized test scores.

3. The documented cost-effectiveness of Project Read in each replication site has consistently shown it to save school systems thousands of dollars annually. These data are available for review from representatives at the replication sites.

4. Data related to the implementation of Project Read in school systems/schools were reviewed from the following eight school districts and one private school:

Bloomington, Minnesota	Lansing, Michigan
Irvine, California	Hemlock, Michigan
Bay City, Michigan	Greenwich, Connecticut
Tampa, Florida	Lincoln, Massachusetts
Portland, Oregon	The Carroll School for Dyslexic subjects

METHODS

Subjects

The population for this study consisted of: (1) traditional classroom placement in grade one, two, or three in a Louisiana Region III public elementary school; and (2) reading scores ranked at or below the 25th percentile on standardized reading tests. From a population of more than 400 participating students, 112 matched pairs of subjects were selected for a total N of 224. Data available from three of the participating districts which piloted Project Read are reported here.

Pre-testing

Subjects' pre-test scores were based on the California Achievement Test (CAT) Reading composite, Form E (1986) which includes visual recognition, sound recognition, word analysis, vocabulary and comprehension.

Participation Criteria

Subjects included students in first, second, and third grade traditional classroom settings whose CAT reading percentiles ranked at or below the 25th percentile, with standard scores below 90. Subjects did not include special education students. Some of the subjects participated in supplementary Chapter One programs[1]. Their scores were also

---

[1] Chapter One is a U.S national program which provides reading lab assistance to children within school settings.

analyzed separately, as were scores of subjects who did not participate in Chapter One programs.

Experimental Group

From the Project Read groups, 112 subjects were selected for participation in the experimental (Project Read) group.

Control Group

A control group of 112 subjects was selected for inclusion in this study. Subjects were matched to those in the Experimental group on the following six criteria: (1) Parish/district of subject, which determines its students' reading programs; (2) Gender; (3) Ethnicity; (4) Grade placement in school (grade one, two, or three); (5) Whether or not the subject participated in the Chapter One supplementary reading program; and (6) Pre-test results from standardized (California Achievement Test) reading scores. Control group subjects continued in their traditional programs, with no change in instructional method or materials.

Research Design:

Matched pairs were subjected to pre-test post-test comparison.

Pre-test/Post-test Measurement

Measurement was based on comparison of pre-test post-test percentile ranks resulting from reading subtests of the CAT.

Analysis of Data

Percentile ranks cannot be arithmetically treated for statistical analysis, so CAT percentile ranks were converted to standard scores (mean=100; standard deviation=15) for statistical analysis. Differential scores were determined by measuring the difference between pre- and post-testing standard scores.

Gain scores for subgroups were also measured. The following variables were measured: 1) individual parish/district; 2) gender; 3) ethnicity; 4) grade in school (1, 2, or 3); and 5) whether subjects participated in Chapter One.

Limitations of the Study

Most districts were unable to fund the demonstration teacher model used in the original Project Read. Most teachers were inadequately supervised. There was no assurance that subjects received the minimum of 30 minutes of Project Read instruction daily. Furthermore, the majority of Chapter One teachers in Project Read replication sites are trained in Project Read to ensure that methodology is consistent between classroom and Chapter One. Since only one participating parish/district (Ascension Parish) trained its Chapter One teachers, subjects received two different (and sometimes conflicting) methods. Finally, most districts did not begin implementation of Project Read until two to three months into the school year. Most Project Read researchers report progress over three or more years.

RESULTS

The results are given below in terms of the annual gain score (in standard score points) made by each of the groups.

1. Inclusive of the three districts, which group made overall greater gain: the Experimental Project Read groups or the control groups?

Annual Gain Score by Control Groups:	-1.50
Annual Gain Score by Project Read Groups:	11.20
Difference between Project Read and Control Groups: Project Read by	+12.70

2. Within each of the three parish/districts, which group made the greater gain: Experimental Project Read groups or the Control groups?

Ascension Parish

Annual Gain Score by Control Groups:	-1.56
Annual Gain Score by Project Read Groups:	14.10
Difference between Project Read and Control Groups: Project Read by	15.66

St. Mary Parish

Annual Gain Score by Control Groups:	-8.40
Annual Gain Score by Project Read Groups:	14.66
Difference between Project Read and Control Groups: Project Read by	23.06

St. John Parish

Annual Gain Score by Control Groups:	6.62
Annual Gain Score by Project Read Groups:	-1.08
Difference between Project Read and Control Groups: Control Group by	7.70

3. Between male and female subjects in Project Read, who made the greatest overall gain?

Annual Gain Score by males in Project Read:	10.50
Annual Gain Score by females in Project Read:	12.06

4. Between black and white subjects in Project Read, who made the greatest overall gain?

Annual Gain Score by blacks in Project Read:	12.96
Annual Gain Score by whites in Project Read:	8.26
Difference between black and white subjects: black by	4.70

5. Among the Grade 1, 2, and 3 subgroups in Project Read and control groups, which made the greatest gain?

Annual Gain Score 1st graders in Project Read:	13.52
Annual Gain Score 1st graders in control groups:	-0.78

Annual Gain Score 2nd graders in Project Read:	4.72
Annual Gain Score 2nd graders in control groups:	1.00

Annual Gain Score 3rd graders in Project Read:	8.00
Annual Gain Score 3rd graders in control groups:	-0.40

6. Among the following groups:

> A) subjects in Project Read who participated in Chapter One;
> B) subjects in Project Read who did not participate in Chapter One;
> C) subjects in control groups who participated in Chapter One; and
> D) subjects in control groups who did not participate in Chapter 1

which group made the greatest gain?

Gain Score for Project Read subjects who also participated in Chapter 1	+13.12
Gain Score for Project Read subjects who did not participate Chapter 1	10.58
Gain Score for Control Group subjects who also participated in Chapter 1	-10.00
Gain Score for Control Group subjects who did not participate in Chapter 1	-1.10

CONCLUSIONS

1. In three school districts which piloted Project Read, a systematic phonology program, students at risk for reading/language disability who participated in Project Read gained an annual mean of 12.7 standard score points compared to a mean loss of 1.50 standard score points by matched-pair subjects in control groups. This result complies with findings of school districts across the nation who have gathered data on the effectiveness of Project Read.

2. No significant difference existed between male and female or Black and White subjects.

3. First grade students had the highest overall mean gain. This result complies with research which illustrates that the earlier intervention takes place, the more successful it is (Catts & Kahmi, 1986; de Hirsch et al., 1966; Lundberg, 1980; Masland & Masland, 1988; Sawyer & Butler, 1991). Even third grade students gained a mean of 8.40 standard score points over their controls.

4. Among Project Read subjects, there was a 2-point standard score improvement for those who also participated in Chapter One programs. Even those Project Read subjects who did not participate in Chapter One made a mean annual gain of 10.58 standard score points. Control group subjects who also participated in Chapter One lost 10 standard score points. Control group subjects who did not participate in Chapter One lost 1.1 standard score points.

These findings support the use of a systematic, cumulative phonology, language-based program such as Project Read in the *traditional classroom setting,* taught by a trained, *regular education teacher.* Project Read was significantly more effective than costly pull-out programs currently in use in Louisiana public schools.

REFERENCES

Catts, H.W. (1986). Speech production/phonological deficits in reading disordered children. *Journal of Learning Disabilities*, 19, 504-508.

Catts, H.W. & Kahmi, A.G. (1986). Intervention for reading disabilities. *Journal of Childhood Communicative Disorders*, 11, p.67-79.

de Hirsch, K. & Jansky, J. & Langford, W. (1966). *Predicting Reading Failure.* New York: Harper & Row.

Liberman, I.Y. (1989). Phonology and beginning reading revisited. In C. von Euler (Ed) Wenner-Gren International Symposium Series: *Brain and Reading.* Volume 54. Hampshire, England: Macmillan.

Liberman, I.Y., Mann, V., Shankweiler, D. & Werfelman, M. (1982). Children's memory for recurring linguistic and non-linguistic material in relation to reading ability. *Cortex*, 18, 367-375.

Lundberg, I., Frost, J. & Peterson, O.P. (1988). Effects of an extensive program for stimulating phonological awareness in preschool children. *Reading Research Quarterly*, 23, 263-284.

Lundberg, I., Olofsson, A. & Wall, S. (1980). Reading and Spelling skills in the first school year predicted from phonemic awareness skills in kindergarten. *Scandinavian Journal of Psychology*, 21, 159-173.

Masland, R.L. & Masland, M.W. (1988). Preschool Prevention of Reading Failure. Parkton, MD: New York Press.

Rapala, M.M. & Brady, S. (1990). Reading ability and short-term memory: the role of phonological processing. *Reading and Writing: an Interdisciplinary Journal*, 2, 1-25.

Sawyer, D. & Butler, K. (1991). Early language intervention: a deterrent to reading disability. *Annals of Dyslexia*, 41, 55-79.

Shankweiler, D. (1989). How problems of comprehension are related to difficulties in coding. In D. Shankweiler & I.Y. Liberman (Eds). *Phonology and Reading Disability: Solving the Reading Puzzle.* Ann Arbor: University of Michigan Press.

Treiman, R.A. & Baron, J. (1981). Segmental analysis ability: development and relation to reading ability. In G.E. MacKinnon & R.G. Walker (Eds). *Reading Research: Advances in Theory and Practice 3.* New York: Academic Press.

Vellutino, F.R. & Scanlon, D. (1987). Phonological coding and phonological awareness and reading ability: evidence from a longitudinal and experimental study. *Merrill-Palmer Quarterly*, 33, 332-363.

Wagner, R.K. & Torgesen, J.K. (1987). The nature of phonological processing and its causal role in the acquisition of reading skills. *Psychological Bulletin,* 101, 192-212.

Facets of Dyslexia and its Remediation
S.F. Wright and R. Groner (Editors)
551

COMPUTER-BASED SPELLING REMEDIATION FOR DYSLEXIC CHILDREN USING THE SELFSPELL ENVIRONMENT

R. I. Nicolson and A. J. Fawcett
Department of Psychology
University of Sheffield, UK

INTRODUCTION: Computers and Dyslexia

Dyslexic children typically have greater problems in learning to spell than in learning to read. Furthermore, spelling problems appear more resistant to remediation than reading problems (Thomson, 1991). It is perhaps somewhat surprising, in view of the theoretical importance of spelling for evaluating existing frameworks for dyslexic deficits, that relatively little emphasis has been placed on comparative studies of remedial methods. This would be understandable if it could be assumed that any improvement in reading would naturally lead to a corresponding increment in spelling. However, rather to the contrary, development in reading and spelling, Frith (1985) argues, involves a reciprocal interaction, in which the attainment of the alphabetic stage in spelling must precede its acquisition in reading. The alphabetic stage is that in which the correspondences between letters and their sounds are exploited both for decoding printed words (reading) and for analysing spoken words into their written equivalents (writing). Frith (1985) suggests that the alphabetic stage represents a bottleneck for dyslexic children in both reading and spelling, preventing their smooth transition to the mature orthographic stage.

The research reported here derives from a feeling of guilt, in that for some years we had been helped in our theoretical research by a dedicated panel of adolescent dyslexic children (see the chapter 'Towards the Origins of Dyslexia' in this volume), and we were keen to do something to help them with their problems. Discussions with the panel suggested that spelling remains a severe problem even for those who had successfully overcome the reading difficulties on which their original diagnosis was based.

This chapter presents an overview of the SelfSpell hypermedia environment for spelling support, which represents a new and promising approach to this previously intractable problem. Two evaluation studies will then be outlined, demonstrating that this new approach can be effective and fun. The first study involved a child-centred rule-based approach to learning spellings with adolescent dyslexic children, and the second study compared the efficacy of rule based and mastery approaches with younger dyslexic children who had made little or no progress with spelling.

THE SELFSPELL ENVIRONMENT

Our dream was to create a supportive adaptive environment for dyslexic children which allows them to consolidate their basic reading, spelling and arithmetic skills either at school or in their own home, at their own rate, with their own ideas and under their own control. The problem was that although Computer Aided Learning programs have the potential to be of great value for dyslexic children, in terms of sustaining attention and increasing motivation, the limitations of early computer programs in their ability to interact with the user, meant that they were particularly ill-suited to the needs of dyslexics. Even the simplest commands, for example, demand a degree of reading ability beyond the reach of the average dyslexic child, and all input demands the use of the keyboard. Therefore, the potential advantages of computer use for the dyslexic could not be realised. However, recent major advances in educational technology with the advent of the Apple Macintosh computer have unleashed the full potential of this approach, in the process revolutionising the pedagogical environment. The most exciting development here has been the availability of hypermedia, which allows the integration of text, graphics and digitised speech into a magical package. For dyslexic children, in particular, using icons eases communication and 'point and click' techniques using the mouse free them from the low-level processing demands of the complex keyboard. For the first time, successful interaction with the computer is more than a dream.

We designed the Selfspell program to capitalise on these recent technological advances, in particular to create an element of fun which is so often sadly lacking in learning for these children. Naturally enough, we wanted the program to be effective at remedying the spelling errors of the children, and we therefore built in immediate feedback to prevent the perseverance of earlier mistakes. It was essential that the child should be actively involved throughout in making decisions, and we were keen to provide a range of techniques to suit the needs of different learning styles.

In order to meet this specification, we developed the program within the Apple HyperCard environment, using a 1 Mbyte Macintosh Plus micro. The intention was that the child should be able to use the system unaided much of the time, but with initial support from a parent or teacher. Use of this SelfSpell prototype fell into four distinct phases: first a pencil and paper dictation. Next the child's passage was typed into the computer, and bug cards made for each error. Next the child went through the passage with help, identifying all the bugs, and for each one thinking up a rule to help them spell it right the next time. Finally, the child went through the passage fixing all the bugs, without human help but with support from the program which could either give a hint, a rule or 'speak' the correct spelling for a bug.

EVALUATION STUDY 1

Although pilot studies showed that dyslexic children interacted well with the program and derived considerable benefit from it, a more formal evaluation was necessary. The study is described in detail in Nicolson, Pickering & Fawcett (1991), and so an overall summary should suffice here.

Participants

The first evaluation study was undertaken using 15 of the worst spellers from our panel of 23 dyslexic children (mean chronological age 13.4, mean reading age 10.5, mean spelling age 8.6). None of them was familiar with use of the Apple Macintosh. It should be noted that all the children had been diagnosed as dyslexic several years previously (using the standard criterion of an 18 month or more deficit in reading age over chronological age; together with normal or above normal IQ and no primary emotional or neurological problems). By the time of the study several of the children were technically 'remediated' in terms of reading age (ie. their reading age was within 18 months of their chronological age). Nonetheless, even for the 'remediated' children the spelling performance was very poor. The standardised tests used were the Schonell test of single word reading and the Schonell test of regular spelling.

Experimental Method

For this first evaluation, a series of simple passages were constructed by the experimenters, matched to the reading ages of the participants. An appropriate passage was dictated to each child, and then typed into Hypercard by the experimenter and bug cards created for each error. Each child then completed three training sessions, identifying their bugs, and creating rules to help them remember the correct spelling. Finally, in order to evaluate the effectiveness of the computer program, the original dictation was repeated in pencil and paper format one month after session 3.

Results of the Evaluation

After just three sessions on SelfSpell, the children were able to identify 80% of the bugs; to fix 70% of them immediately and to recall 50% of their rules. The knowledge of the spellings transferred to a pencil and paper dictation identical to the pretest, with 70% of bugs spelled correctly. The improvement overall may be attributed to the targeted words, with over two-thirds of them spelled correctly, since the untargetted words did not improve overall. All these improvements in performance were highly statistically significant (t (14) = 5.58; p<.0001; and t (14) 10.34, p<.0001 for the passage overall and the bugs respectively).

Improved motivation and attitude were reported by 75% of the parents, and all the children and parents reported that it was fun to use SelfSpell! Open-ended comments from the comments were uniformly favourable. A selection includes:

She seems to be more aware of spelling mistakes and wants to put them right.

He has a new strategy for remembering correct spellings

He has been checking through his spellings more carefully.

He is more aware of spelling errors and seems to be enjoying reading more.

He seems to have grasped the spelling side and seems to have gained confidence.

N. has had lessons on a one-to-one basis, but has improved more from this method of teaching.

Of all the methods that have been used this is the most successful.

The computer talking makes it better and helps them remember what they have written.

TESTING RULE-BASED VS MASTERY BASED APPROACHES TO SPELLING REMEDIATION

Participants

The participants in the second study were 10 dyslexic children, aged between 10 and 12 years (mean age 11.4). As in the previous evaluation, none of the children was familiar with use of the Apple Macintosh.

Their age-normed performance in reading and spelling before the study reported here is shown in Table 1. It may be seen that the children in this study had made little or no progress in spelling, with a mean deficit between spelling and chronological age of 4.8 years.

Experimental Design

In order to compare the rule-based, undirected approach outlined above (Nicolson, Pickering & Fawcett, 1990) with a directive mastery learning approach, a mastery learning program (SpellMaster) was written which, following a technique developed by Nicolson (1979), presented each spelling to be learned in a sequence designed adaptively to assist the learning process. Each child learned 20 spellings, with 10 learned using a rule-based approach and a matched set of 10 learned using a mastery approach. The two training methods are described in detail below.

A pencil and paper pre-test on performance was administered by the parents before training began. Three training sessions, each of 20 minutes' duration, were run for each set of 10 words for each child. The order of presentation for the training methods was counterbalanced, so that half started with the Selfspell rules and half with the Spellmaster. The training took place over the Autumn half term holiday, to ensure that all children would be available to complete three training sessions within a week. One week after the last training session a pencil and paper post-test exactly equivalent to the pre-test was administered by the parents. One month after the training tests were completed, parents were asked to administer a paper and pencil delayed post-test, to compare the relative long term benefits of the training approaches.

TABLE 1

Reading and Spelling Ages for the Participants

Participant	IQ	Chron. Age	Reading Age	Spelling Age
CJ	119	11.6	8.9	6.0
CE	109	10.0	7.0	5.0
DP	109	11.1	7.3	6.0
IT	112	12.2	9.5	9.0
MC	114	11.5	8.3	9.0
MCr	109	12.7	9.8	6.0
RH	105	12.8	8.0	6.0
SA	128	11.6	9.6	8.0
TA	109	10.8	8.3	6.0
TF	109	10.0	6.9	5.0
Mean	**112.3**	**11.4**	**8.4**	**6.6**

The Pretest

We wanted to train the children in meaningful words, rather than the limited vocabulary geared to their spelling age. Each parent was therefore asked to identify 20 words with which their child had particular difficulty, dictate them to the child and help them to rank them in order of difficulty from 1 to 20. Two sets of words were derived for each child, matched for level of difficulty. In order to make conditions as realistic as possible, the words were not monitored in any way for regularity or objective ease of spelling. They ranged from complex words such as *dyslexic, persuade, humanities, umbrella and grateful*, to simpler confusable words such as *there*.

The Training

Selfspell Rules

A similar process was used to develop the Selfspell rules as in the previous study, the difference being that in this study an individual passage was designed for each child incorporating their problem words. This process is illustrated in Figs. 1a and 1b, where it may be seen that the child has difficulty with many spellings, in particular spelling 'might' as 'mintue' (Figs. 1a & 1b - see end of paper). Next the child went through the passage (helped by the experimenter) identifying all the bugs, and for each one thinking up a rule to help them spell it right the next time (in Fig. 1b, the child has typed in the rule 'green hero turtles', the title of a popular TV program, to help him remember the 'ght' at the end). In the final phase the child went through the passage unaided, fixing all the bugs, through judicious support from the program (Figs. 1c, 1d, 1e & 1f - see end of paper - the latter demonstrates the use of scoring to help with motivation).

TABLE 2

Results from Spelling Mastery: score = 42 out of 50

Word	Spellings on Session 2	Pre-test	Post-test	Delay
serious	shrece, sere, serious, serious, serious, serious	seros	serious	serious
certain	srten, sertain, certain, certain, certain, certain, certain	serton	certain	certain
comfortable	comfortable, comfortable, comfortable, comfortable	comfortuble	comfortable	comfortable
grateful	grateful, grateful, grateful, grateful	greateful	grateful	greatful
wearing	wearing, wearing, wearing, wearing	wharing	wearing	wearing
celebrate	slabrat, celebrate, celebrate, celebrate, celebrate	celabrate	celebrate	celebrate
absolutely	apslotly, abserlutly, absorlutely, absolutely, absolutely, absolutely, absolutely	abserlootly	absolutely	absolutely
receive	resevie, receive, receive, receive	reseve	receive	rese
through	through, through, through, through	throught	though	th
peace	peace, peace, peace, peace	peace	peace	peace

Spellmaster

In addition to the rule-based approach outlined above a more directive, mastery learning approach is also available under SelfSpell. The technique employed was that of 'overlearning' which is generally considered to be one of the most effective for dyslexic children (see e.g. Thomson, 1991). In the Spellmaster program, a list of words may be typed in (together with their homophone to ensure correct speech synthesis, and an appropriate context clue), and each word is then presented individually. Immediate feedback is used at all times, and the correct spelling is displayed immediately following any error. The spellings are introduced cumulatively, first with four spellings, then three more when the first four are learned, and so on. An example of the spellings made in the second session is given in Table 2.

It may be seen that the program requires the user to spell all the words correctly at least three times consecutively, and usually more in that even 'learned' spellings are re-tested from time to time. The children were encouraged to type in the letters themselves, but the experimenter took over when the child showed signs of tiring.

RESULTS

The results for each child are given in Table 3 and displayed graphically in Fig. 2. Although the children scored few marks on the pre-tests, both approaches led to excellent learning, as evidenced by the high scores on the post-test. The individual results are sufficiently striking to warrant presenting them in full. Note that all these tests were performed using the standard pencil and paper format.

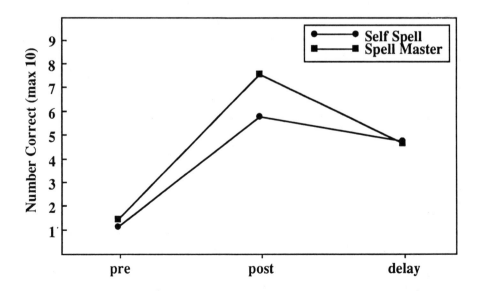

FIG. 2. Correct spellings for rule and mastery training at pre, post and delayed test.

An analysis of variance was performed on the data, with two factors, namely time of test (pre, post and delayed) and training method (mastery or Selfspell). The analysis indicated that time of test had a highly significant main effect whereas that of training method was not significant [$F(2,18) = 36.8$, $p<.0001$; $F(1,9) = 3.2$, NS respectively] and that the interaction was not significant [$F(2,18) = 3.0$, $p>.05$]. Both the post-test and delayed post-test performance were significantly better than the pretest ($p<.01$) for both Selfspell and Spellmaster training. The Spellmaster training was significantly better than the Selfspell rules at the post-test [$F(1,9) = 5.7$, $p<.05$], but there were no significant differences between the training conditions at pre-test or delayed post-test [$F(1,9) = 0.5$, NS and $F(1,9) = 0.1$, NS respectively].

TABLE 3

Individual Results for Pre, Post and Delayed Tests (scores out of 10)

Participant	Spelling Age	Rule-Based			Mastery		
		Pre	Post	Delay	Pre	Post	Delay
CE	5.0	0	7	4	0	4	4
TF	5.0	1	3	6	1	5	4
CJ	6.0	1	7	3	0	10	1
DP	6.0	0	3	3	0	4	4
MCr	6.0	1	3	6	5	8	5
RH	6.0	4	6	4	2	7	5
TA	6.0	1	6	5	1	9	7
SA	8.0	0	5	3	1	9	5
IT	9.0	3	7	6	3	10	4
MC	9.0	0	10	7	1	9	7
Mean		1.1	5.7	4.7	1.4	7.5	4.6

Questionnaire

As in the previous study, an informal questionnaire was administered which established that all the children had enjoyed both the Selfspell and Spellmaster program. Half the children preferred the Spellmaster program for its success while the other half preferred Selfspell because it was more fun. Interestingly this could be directly linked to the spelling age of the child. Typically, the lowest achievers preferred the Spellmaster, presumably because they generally mis-spelled even the most basic words. Open-ended comments were again uniformly favourable. A selection includes:

"Selfspell was good, it's got a story to it, like making rules because they're silly, like the computer speaking what you've put - the other one (Spellmaster) were only alright! Its better on computer than just on paper." (TF)

"I liked the Selfspell best. liked making rules, the computer saying words in funny way I'd spelled them. It was different to what I thought - thought I'd just have to read it over and over and then spell words after, but it was a lot better. Selfspell was more fun - the other one's quite good, but once you get to know what the words are it gets boring. Both helped a lot with the spellings".(MC)

"I liked the whole passage - it was good fun, I particularly liked the pronunciation, especially the sounds where it goes wrong. I liked the scoring at the bottom so you could tell how you were doing. It was quite hard to think of the words, and I'm not sure if they're spelled wrong or right, with words like friend need rule". (IT)

"More fun Selfspell, the way it spoke and said the words. Got a bit boring when you had to just write them in". (SA)

"Spellmaster was interesting, easy to do. When you repeated it over and over again it got better. (RH)

I liked the Spellmaster one best - it was easier, you didn't have to think as hard, didn't get bored with it.- but will use making rules elsewhere". (TA)

DISCUSSION

In summary, both the Selfspell and Spellmaster programs proved effective in remediating the spelling errors of a group of dyslexic children with low spelling ages. Performance on the delayed test showed some decline, but nevertheless was significantly better than at pre-test. The Spellmaster program led to significantly better performance on the immediate post-test but performance on the delayed test was equivalent for both training methods.

Let us consider the theoretical and practical questions which formed the focus of this investigation: which method would be most effective for the remediation of severe spelling impairments and what was the cause of the success of the approach? The first question was resolved clearly. Both programs proved extremely effective and produced lasting improvements for the majority of children. Such improvements are particularly striking in view of the typical resistance to remediation found with traditional pencil and paper training (see Thomson, 1991). Possibly it was the reinforcing nature of the computer presentation, non-judgemental, with immediate feedback and an element of fun which engendered the success. One might argue that any structured learning approach may prove to be of value within such a supportive environment.

In terms of the theoretical issues underlying the acquisition of spelling in the dyslexic child, Frith (1985) has argued that dyslexic children tend to get locked into the initial, 'logographic' stage for reading, in which performance is based on whole-word recognition, and are unable to make the transition to the phonologically mediated alphabetic stage. Owing to their problems in making connections between the graphemes and phonemes, unless new vocabulary was taught explicitly, they would have no mechanism for acquiring it. Unlike reading, spelling cannot be facilitated by the use of context, and therefore one might predict particularly severe problems. Spelling would break down in either the segmentation process or in phoneme/grapheme translation, preventing the child from spelling phonetically and generating the bizarre errors which characterise dyslexic performance (such as *mintue* for *might*), where the child is trying to reproduce the shape of the word, rather than its constituent sounds.

From the point of view of Frith's framework, therefore the good progress of the children is both encouraging and surprising. So it is worthwhile here to consider what learning strategy the rule based and mastery training techniques demand from the learner. The mastery training employs the traditional rote learning approach and encourages segmentation. The Selfspell program encouraged the speller to focus on both phoneme/grapheme translation and segmentation, emphasising the number of syllables (for example *al -to-get-her* for altogether) and typically using the initials of the missing letter for the clue (an example here might be *insects eat vegetables* for the *ieve* in *believe*.

Perhaps the design of our experiment, in which each child was taught to use both mastery and rule based approaches, led to the creation of a learning environment which encouraged segmentation coupled with the active learning and deeper processing engendered by the rule-based approach. It may well be that a combination of both these methods would be most effective, in particular in remedying the deficits of the younger group.

It appears therefore that with adaptive support, sufficient practice and motivation to succeed, even the most intractable spelling problems can be ameliorated. The fact that all the children in this study were able to use the alphabetic principle is particularly encouraging, especially in the light of the discouraging findings of earlier research on spelling remediation (e.g.Thomson 1991), and the theoretical rationale for expecting severe problems in spelling for dyslexic children. These results may be compared with those of Bradley (e.g. 1988) who demonstrated that children given early support in phonological processing showed much better subsequent acquisition of reading. Given early intervention, it seems likely that the initial problems can be substantially alleviated, leading to relatively normal acquisition of spelling skills.

In conclusion, we have argued that the creation of a hypermedia environment using open learning techniques proved both effective and enjoyable for all the participants regardless of their spelling age. Such environments appear to overcome the known drawbacks to computer-based techniques for dyslexic children. New developments in educational technology, added to the decreasing cost of powerful systems, make the creation of such environments a practical, affordable and very desirable target for the whole education system. The Selfspell environment provides an initial step in this direction, and multimedia programs for reading and arithmetic are now under development.

COMPUTER BASED SPELLING REMEDIATION

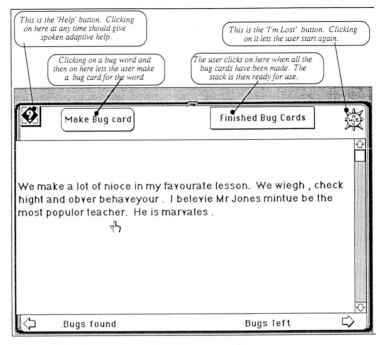

FIG. 1a. The Computer version of the transcript

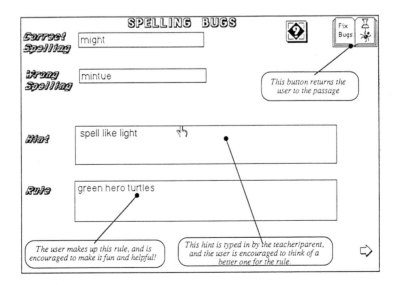

FIG. 1b. Making a Bug Card for 'might'

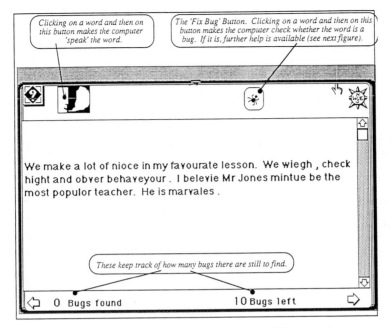

FIG. 1c. Childs's view of the support environment. Click on a bug

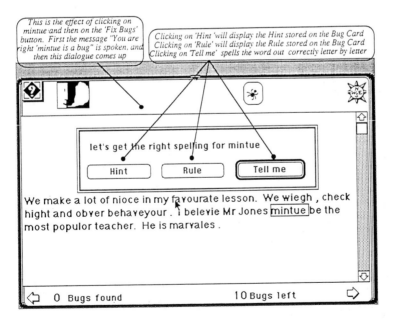

FIG. 1d. After selecting mintue then clicking on the 'bug' button

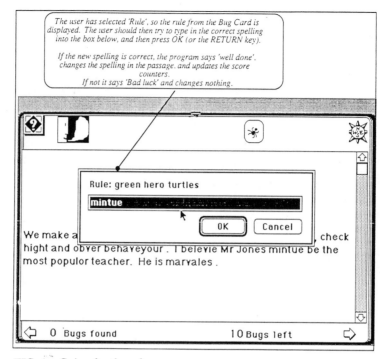

FIG. 1e. Going for the rule

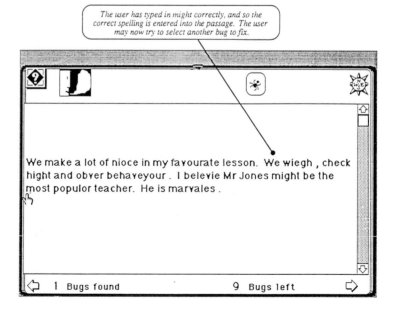

FIG. 1f. The Result of Entering 'might'

REFERENCES

Bradley, L. (1988). Making connections in learning to read and to spell. *Applied Cognitive Psychology*, 2, 3-18.

Frith, U. (1985). Beneath the surface of developmental dyslexia. In K.E. Patterson, J. C. Marshall and M. Coltheart (Eds). *Surface dyslexia.* London: Lawrence Erlbaum.

Nicolson, R.I. (1979). *Identification of stages in learning paired associates.* Unpublished Ph.D thesis, University of Cambridge.

Nicolson, R.I, Pickering, S. & Fawcett, A.J. (1991). Open Learning for Dyslexic Children using HyperCard™. *Computers and Education*, 16, 203-209.

Thomson, M.E. (1991). The teaching of spelling using techniques of simultaneous oral spelling and visual inspection. In M. Snowling and M. Thomson (Eds). *Dyslexia: Integrating theory and practice.* London, Whurr.

Facets of Dyslexia and its Remediation
S.F. Wright and R. Groner (Editors)

VISUAL AND LINGUISTIC DETERMINANTS OF READING FLUENCY IN DYSLEXICS: A CLASSROOM STUDY WITH TALKING COMPUTERS.

D. Moseley

School of Education,

University of Newcastle-upon-Tyne, UK.

INTRODUCTION AND THEORETICAL RATIONALE

This paper describes a study in which computers with a synthesised speech system were used as a means of providing intensive reading practice in the classroom. The visual presentation was in word, phrase or sentence units, appearing either in a central window or cumulatively as on a page. The quality of the speech used allowed the pupils to listen, look and understand while passages were read with appropriate phrasing and intonation. After this, the speech was used to provide feedback for the reader at the level of words, phrases, or sentences. Fluency was built over repeated readings by reducing the time available for reading each phrase or sentence.

The study involved two teachers and seven pupils aged between 11 and 14 years in a residential school for children with specific learning difficulties. The pupils had no known hearing or visual defects, but there was concern about their slow progress and two months before the intervention three of them had been referred to an optometrist with a special interest in reading problems and vision.

One pupil, LU, had good accommodation-convergence and stereopsis, but a correction had been prescribed for a moderate degree of long-sightedness in the right eye. The second, RK, also had good accommodation-convergence and stereopsis and was mildly long-sighted, especially in the left eye. He experienced discomfort when viewing an Amsler grid and was supplied with grey tinted lenses. The third pupil, ST, had an alternating right divergent squint and was long-sighted in the right eye. It is possible that she experienced some image confusion when reading. She seemed to fit the pattern described by Moseley (1988) who found that dyslexics who suppressed the right eye in a retinal rivalry paradigm read slowly and inaccurately and were also slow in a text-scanning task. ST reacted with distress when viewing the Amsler grid and was given grey-green tints. All three pupils, were provided with spectacles for reading, but they subsequently failed to use them, saying that they made no difference.

It is not clear which features of computer presentation (if any) are likely to help dyslexics. Gambrell, Bradley & McLaughlin, (1987) found that there was no advantage other than a motivational one for 8 year-olds in reading text on screen rather than from printed sheets and Cato, English & Trushell (1989) found that children generally found it more difficult to locate information on screen than on paper. However, computers can offer a wide choice of display formats, can reduce the effort of visual scanning by presenting text in windows, can help to build fluency by self-paced and by timed display and can efficiently provide speech feedback on demand. By combining these features and offering choice within a structured and progressive framework, it was hoped to sustain concentration through a number of readings of the same material, thereby encouraging readers to make use of all available cues in an active overlearning process.

In designing the 'Read and Speak' software used in the field trial, special attention was given to visual factors. Words and phrases can be presented in a central window which rarely exceeds the visual span of 21 characters established by Rayner (1983). When words, phrases and sentences are presented in page format, they appear one at a time, so that the reader cannot lose the place. The amount of text on the screen is limited, by a line-length of 35 characters, by a double-spacing option, and by having a maximum of six sentences on screen.

Evidence has accumulated in recent years (see Lovegrove in this volume) to suggest that dyslexics have a sluggish-acting transient visual system. This idea is consistent with the finding that they show long visual persistence and have long dark-interval thresholds (O'Neil & Stanley, 1976; Slaghuis & Lovegrove, 1985). Riding & Pugh (1987) found that children with long visual persistence read better in conditions of low illumination, which suggests that tinted lenses (which reduce contrast) may have some effect. Williams reports (see this volume) that blue overlays improved reading in dyslexics in 80% of cases, and attributes this to the idea that blue slows down the function of the sustained visual system, allowing it to act more in synchrony with a sluggish transient system. Certainly the blue cone mechanism is slower-acting than the other colour channels (Mollon & Polden, 1976). A blue background was adopted in the Read and Speak software because it was subjectively restful and because blue is apparently the colour most frequently chosen by dyslexics as an overlay or tint (Irlen Institute, personal communication). Brightness and contrast were set at low-to-moderate levels in view of the fact that two pupils had found that tints reduced visual discomfort produced by the Amsler grid.

Reitsma (1988) and Davidson, Coles, Noyes & Terrell (1991) are among those who have shown that beginning readers can benefit from computer-presented reading practice with speech feedback. Moseley (1969) with an extensive and flexible range of software for the Talking Typewriter and Fiedorowicz & Trites (1991) with a suite of programs called Autoskill CRS have reported successful results in controlled studies where similar technology was used with children and adults with reading problems. In both cases text was presented at different levels of difficulty and both visual presentation and auditory feedback was provided in letter, word, phrase and sentence units.

The basic features of the Read and Speak software have been described in a previous paper (Moseley, 1990). There are 45 different ways in which text may be presented (and even more if word-segmentation options are selected). Text can be sequentially displayed on the whole screen or in a single-line central window. Different sizes and colours are available, but the default is white text on a blue background. In the present study, the first presentation of text was accompanied by speech so that the overall meaning of the passage could be understood - a fairly rapid 'read-through'. The language used was topic-appropriate and vocabulary was in no way phonetically constrained. Word-segmentation into syllables or graphemes (Liberman, Shankweiler, Fisher & Carter, 1974) was an optional facility, which could be used to draw attention to the structure of particularly difficult words. However, this option was used only rarely. It was designed to demonstrate word-building visually and was not accompanied by segmented speech. It is worth noting here that in a recent study using similar techniques, van Daal & Rietsma (1990) found that segmented sound feedback was not significantly better than whole word spoken feedback in promoting learning.

A major emphasis in the present study was on fluency-building, using passages at readability levels one year in advance of instructional level as measured using an Informal Reading Inventory (IRI). After the first 'read-through' (Step 1) a further three steps were required, with choices within each step. Learner-paced phrase-presentation (with everything spoken or with optional speech) was used in Step 2. Here, phrases could be presented visually word-by-word or else as a visual unit - but in each case the speech followed as a fluent phrase. At Step 3 there was a choice between fluency-building with phrases and learner-paced reading sentence by sentence. Finally, at Step 4 the reader had to work towards paced oral reading of sentences at the adult rate of 120 words per minute, with no more than two accuracy errors.

As yet there is insufficient research evidence to justify using different kinds of teaching programme with different subtypes of dyslexia. Lyon & Flynn (1991) reviewed dyslexic subtype x treatment interaction studies and concluded that "attempts at predicting treatment options on the basis of subtype characteristics fall far short of the educational task confronting us." Different subtypes whether classified as accuracy disabled/rate disabled or as dysphonetic /dyseidetic did not respond very differently to treatments to which they had been randomly allocated. Broadly-based treatments emphasising a language-experience approach, reading in context, listening and reading comprehension, vocabulary development, syntactical elaboration and written composition were effective irrespective of subtype, as was the use of the Initial Teaching Alphabet. The main aim of the Read and Speak field trial was to provide a powerful impetus to learning in dyslexics by a flexible and broadly-based combination of features which could be adapted to individual learning needs.

The writer has previously authored an audiovisual package in which graphic, phonic, linguistic and contextual cues are provided for both reading and spelling (Moseley, 1972). By providing graphic cues (bold print and colour) for word and syllable segmentation and by simultaneous reading and 'phonic dictation' on tape, the task of establishing symbol-to-sound links was simplified. Lewis (1983) reported better results with the English Colour Code package than with Direct Instruction (Corrective Reading). The challenge in the present study was to improve on such results by taking advantage of the additional facilities provided through the use of a computer. The BBC B computer and the Computer Concepts Speech System were used, since this combination provides synthesised speech of acceptable quality at low cost.

In designing the study the following pedagogical and theoretical assumptions were made. Firstly, it was reasonable to assume that the pupils had adequate receptive language skills, since the mean full scale Wechsler IQ for the group was 100 (range 84-118). Secondly, it was assumed that by including options which reduced the need for eye-movements and by attempting to maximise visual comfort through the use of blue background and low-to-medium brightness and contrast, children with visual problems would be helped. Thirdly, it was assumed that both feedback and repetition are helpful in promoting mastery-learning. Text was presented and feedback given in meaningful linguistic units, so as to facilitate word recognition through the use of context. Repeated reading is thought to be helpful because with each reading less attention is taken up in decoding and more is available to support comprehension (Herman, 1985; Dowhower, 1987). Another important assumption (supported by the work of Breznitz, 1991) is that fast-pacing leads to fluent automatic reading in which attentional and working-memory resources are freed. This in turn leads to improved reading accuracy and comprehension. A fifth assumption was that there would be an interaction between improved word-recognition and phonological skills such that speech feedback would encourage children to examine sound/symbol correspondence, to become aware of recurrent patterns and to generalise from this in processing both aural and written forms of language.

The field trial was set up with the aim of demonstrating that rapid improvement which generalises and provides the basis for further progress is possible in severely dyslexic children. The intervention involved intensive reading practice of interesting and relatively challenging material, with speech serving both as a model for fluent reading and as a precise source of feedback controlled by the reader. It was hoped that children who had long and discouraging histories of reading problems would be convinced of their ability to read more effectively and that the sheer amount of repeated and paced reading experienced in a intervention totalling approximately 10 hours would help to establish more reliable automatic responses to common words and linguistic patterns. A further aim was to obtain feedback from the pupils in the hope of arriving at an optimal balance between predetermined structure and the active development of new strategies by the learner.

THE STUDY

The project involved six boys and one girl aged between 11 and 14 years. The deputy head identified all pupils in the school who had reading ages between 7 years 6 months and 8 years 11 months and for whom Schonell reading test scores were available for the previous year at the school. This range of scores was specified since it covers the stages when whole-word visual acquisition of sight vocabulary and letter-by-letter phonics gives way to an awareness of more complex phoneme-grapheme relationships and linguistic features involving clusters, digraphs, morphemic markers, prefixes and suffixes. Dyslexics typically experience severe problems in acquiring these skills, which play an important part in the development of both reading and spelling.

As the sample was small, and as there were considerable individual differences in rates of progress, the results will be presented, first in terms of mean scores and then, more informatively, by means of individual case vignettes.

The average Schonell Word Recognition score both at the start of the project and one year previously was found to be 8.0 years. The mean Schonell Spelling score one year prior to the Read and Speak intervention was 7 years 5 months and in every case spelling was weaker than word recognition. This suggests that the pupils were not familiar with common letter sequences. Despite conventional phonic-based teaching employing multisensory methods, little progress was being made in establishing connections between spelling patterns and sound patterns. A change of emphasis seemed appropriate.

Pre- and post-intervention assessment of oral reading accuracy, comprehension and speed was carried out by the writer using parallel forms of an informal reading inventory based on work by Brown (1986). The IRI consists of passages of between 50 and 200 words, graded in 6-month steps up to a 9-year readability level and then in 1-year steps. Comprehension questions are presented after oral reading, with the passage available for reference. In terms of oral reading accuracy, instructional level is defined at the 95% level of success, with hesitations of more than 4 seconds counting as an accuracy error. In terms of reading comprehension, instructional level is set at 75%. Pupils continue with progressively harder passages until they fail to reach instructional level in either accuracy or comprehension. Speed of reading was not used as a reason for discontinuing the reading assessment.

At pre-test the mean instructional level score was 7 years 6 months, which was somewhat below the Schonell results supplied by the school. Comprehension was in five out of seven cases satisfactory when accuracy was at or above 95%, and was at an average level of 73% for passages just failed in terms of accuracy. Oral reading speed for passages just passed in terms of accuracy was in all except one case below the norms. The average rate was 54 words per minute, which is appropriate at a 7-year reading level. This was not an unexpected result as children with specific reading difficulties are often slow readers (Lovett, Ransby & Barron, 1988). It appeared that of the seven pupils one (MJ) was a 'guesser' who read inaccurately with normal fluency, but with poor comprehension. The others read slowly, frequently encountering words which they tried to work out on the basis of phonetic clues or which they just looked at, hoping for inspiration. Often, the previous misreading of a function word had led them into a meaningless quagmire from which they could not extricate themselves. Very slow reading was in three cases associated with imperfect comprehension.

The pre-test instructional level was used to determine a starting point for the Read and Speak intervention, starting with passages one year above instructional level. Sixteen 'Main Idea' information passages were used, ranging from 77 to 222 words in length and at readability levels between 8 and 12 years on the Mugford index. The passages were used with the permission of the publisher (LDA). The passages were intended to be too difficult for the pupils to read without speech support. The idea was for the reader to work through each passage within the 4-Step structure described in the Introduction and Theoretical Rationale until he or she could read aloud accurately sentence by sentence at the rate of 120 words per minute. This is a normal speech rate and to reach it pupils had to more than double their initial speed.

The writer worked for 30 minutes each week with two or three pupils in a room with three computers. A further computer session of similar length was also organised each week by a teacher. In each session the passage was first presented and spoken by the computer at the 120 wpm rate. The teacher then asked the pupils to provide an account of the passage in their own words. This was generally well done, but if there were any gaps in understanding the teacher explained the concepts involved. It was felt that the assumption made that the pupils had adequate receptive language skills was upheld. After this, the pupils worked through Steps 2-4 and succeeded in reaching accuracy and fluency targets in 93% of all trials. (Each passage had to be read with no more than two errors and at a rate of 120 w.p.m.)

The average number of passages read over 10 weeks was 9 and the range was from 6 to 11. The average number of readings before reaching criterion on a passage was 7, with a range from 6 to 9. It is estimated that the average number of words read during the intervention was approximately 7000. While the first passage read was one year above instructional level, the last passage read was between 2 and 4 years above that level. It was predicted that the more words read during the intervention period, the greater the improvement in reading would be.

In terms of reading gains the project was successful (see Table 1). Whereas over the previous 10 months the group as a whole had made no progress in word recognition (Schonell test), an average increase of 4 months was recorded over the first two months of the project. On the informal reading inventory accuracy gains ranged between 0 and 18 months (average 7 months) over 11 weeks. In no case did an increase in accuracy take place without accompanying comprehension and on passages of equivalent difficulty mean comprehension scores increased from 73% at pre-test to 95% at post-test. At post-test no-one failed in comprehension items before reaching an accuracy ceiling and all met the comprehension criterion even on the first passage failed in terms of accuracy. The accompanying fluency gains were also substantial, since on passages of the same readability level, an average improvement of 30% was shown. This is equivalent to a 1-year increase in fluency. Only one pupil showed a decrease in fluency over the intervention period and he was initially reading relatively rapidly and relying on guesswork.

TABLE 1

Pre- and Post-Intervention Reading Means and (SD's) Compared by 1-tailed t test.

	Pre-test	Post-test	t	p
Schonell W.R.	7.97 (0.52)	8.34 (0.52)	4.77	<.01
IRI Accuracy	7.50 (0.58)	8.07 (1.06)	2.49	<.05
IRI w.p.m	54.29 (16.81)	70.43 (19.46)	2.87	<.05

Accuracy and fluency gains proved to be statistically significant when a paired t test was applied to pre- and post-test scores. As predicted, the gains were related to the amount of reading undertaken in the course of the Read and Speak lessons. The total number of passage readings was significantly associated with IRI gains in reading accuracy ($r=0.67$, $p<0.05$) and with gains in fluency ($r=0.67$, $p<0.05$), but not with gains in word recognition ($r=0.01$, n.s.). However, the number of passages successfully completed was a significant predictor of word recognition gains to be made in the twelve months after the the intervention ($r=0.78$, $p<0.05$). This suggests that the intervention succeeded in modifying reading strategies rather than improving phonic-based decoding. The possibility remains that the sheer amount of reading done did help to automatise recognition of 'sight' words. Moreover, the pupils who developed successful means of reaching the very demanding accuracy and fluency criteria may well have been encouraged by their success to develop better decoding skills during the following year.

Further light was shed on the nature of the reading gains by a miscue analysis carried out on the passages just failed on the informal inventory. A sequential non-overlapping five-category system was used:

(a) errors leading to syntactic distortion (SYN)
(b) morphological errors where correct syntax is maintained (MOR)
(c) errors with high frequency basic vocabulary words (BAS)
(d) errors with irregularly-spelt words (IRR)
(e) errors with regularly-spelt words (REG)

The percentages of errors falling into the above categories on pre-and post-test for the sample as a whole is given in Table 2.

TABLE 2

Results of Miscue Analysis

	SYN	MOR	BAS	IRR	REG	TOTAL
	%	%	%	%	%	n
Pre-test	19	13	19	27	22	74
Post-test	5	18	22	14	41	78

Inspection of the above figures shows a substantial reduction over the intervention period in the proportion of errors resulting in syntactic distortion, but even more clearly a big change in the type of word recognition error outside a core vocabulary. Errors in reading irregularly-spelt words were almost halved, while errors in reading regularly-spelt words were correspondingly almost doubled. However, these errors with 'regular' words tended to occur with vocabulary items that were more advanced and phonically complex than those encountered at pre-test (e.g. 'constellations', 'constructed', 'directions', 'eventually', 'location', 'estimate', 'interminable', 'observing', 'repress', 'stern', 'strengths', 'striped'). There was relatively little change in the morphological category (generally involving attention to word endings) and in the basic vocabulary category (words like 'as', 'are', 'go', 'his', 'it', 'of', 'one', 'saw', 'she', 'so', 'some', 'the', 'their', 'them', 'these' and 'to').

Follow-up testing one year after the pre-tests showed that the gains reported above had been consolidated, although the pupils (now only 6 in number) had not continued to use the Read and Speak system on a regular basis. Schonell word recognition scores had improved on average by a further six months, but on the IRI, accuracy scores had improved by a further 12 months, with comprehension remaining at a high level (94%) and fluency showing a further 19% improvement in relation to the baseline measures. However, while the mean gain in word recognition was statistically significant (t=3.57, p<0.01), the IRI accuracy gains had a large variance and did not reach significance.

During the project, records were kept of the text and speech options selected by each learner within the pre-determined 4-Step framework. For the read-through at Step 1, fast presentation of sentences was the most popular choice (53% of trials), but the next most popular choice was slow presentation in phrase units (19%). At Step 2 (learner-paced phrase presentation), speech was used for feedback on demand on 61% of trials and for continuous feedback phrase by phrase on 39% of trials. At Step 3 there was a slight overall preference for building fluency with phrases (56%) rather than for reading sentences at an unpaced rate. At Step 4 between 1 and 4 readings were required to reach criterion, the average being 2 readings.

At the end of the project the participants were asked to explain the strategies they had used and to comment on the system as a whole. Six of the seven felt that their reading had improved in some way, four spontaneously adding that they could deal better with hard words. Five mentioned the need to listen as well as look during the first speeded run-through. The speech option was used almost all the time to confirm reading accuracy, and it proved to be almost 100% intelligible. All learners said they preferred to achieve accuracy first with phrases and then with sentences before moving on to building fluency by selecting predetermined rates of sentence presentation. When difficulty was encountered, the most preferred strategy was to re-read and to try to memorise individual words. Three tried to split words up, but only one went further, saying that he would look back at the preceding words to try and work out the meaning and would try to think of similar words as an aid to memory. This particular boy made the greatest gains of all in fluency during the intervention period, although he was not positive about using the computer, since he found it visually fatiguing and complained of blurring. However the same problem arose during the post-testing (using printed sheets) when he had to read four passages in succession. He was not convinced that he had made any real progress and seemed to be relying on memorisation and guessing from context rather than on improved word recognition. He was an anxious lad, afraid of making mistakes, and at follow-up, while he had made further gains in fluency, he still found reading very stressful.

It has to be admitted that none of the pupils was enthusiastic about having to re-read the passages several times, and four of them were critical about reading from the screen rather than from a sheet of paper (which confirms the findings of Gambrell et al. - see

Introduction). It was noted that little use was made of the window-presentation option for phrase-reading, even though this would be expected to reduce visual fatigue. So far as the speech facility is concerned there were no adverse comments and it is clear from the answers given that the pupils were aware that they were coping successfully with relatively difficult material.

The results obtained at follow-up were encouraging in that five of the six pupils had moved forward in either fluency or accuracy and two had made substantial gains (2 years and 3 years six months respectively on the IRI accuracy measure). The best result was a clear case of a boy who had got 'hooked on reading' and had overcome his earlier problems. He was the 'guesser' referred to above who was initially adept at task avoidance but whose whole attitude changed when he achieved success. At the time he said that he preferred reading on the computer to reading from paper since he found it easier to understand bigger words. However, despite the progress made, the reading performance of five pupils was still, at follow-up, well below expectation for age and ability, with instructional levels between 3 and 8 years below chronological age.

Perhaps the amount of exposure to text was simply not enough for these pupils. Perhaps they needed to focus more on single words in order to establish a larger automatised lexicon before being expected to read whole passages. There had been no preparatory word-building study to improve decoding, and it is clear that both during the study and during the follow-up period, single word recognition improved at a slower rate than the ability to read passages where semantic and syntactic context was available. While sight-reading of irregularly-spelt words had improved, phonic decoding of harder regularly-spelt words was still a major obstacle. The assumption that these learners would become aware of recurrent patterns turned out to be ill-founded. Perhaps insufficient use had been made of the window presentation and word-segmentation options which could have been used to enhance awareness of visual, phonological and morphological word-structure. Separate word-building practice might have helped here.

It should be remembered that the Read and Speak intervention consisted of only 10 hours of instruction. Fiedorowicz (1986), using the Autoskill CRS computer program with speech feedback did not obtain unequivocal transfer of word-recognition gains to paragraph reading fluency and comprehension after 21.5 hours of instruction, but did so (Fiedorowicz & Trites, 1987) in a larger and well-controlled study involving 30 hours of instruction.

CASE VIGNETTES

1. **HT** (C.A. 13y 11m): WISC-R: Verbal IQ 87 Performance IQ 104 Full Scale IQ 94. No evidence of visual problems. Passages successfully completed: 10 Total trials: 58 Word-by-word phrases: 7% Phrases: 29% Sentences: 64%

	Pre-test	Post-test	Follow-up
Schonell W.R.	8.9	9.4	10.3
IRI Accuracy	8.5	10.0	12.0
IRI w.p.m	73.0	97.0	100.0

HT made excellent progress in accuracy and fluency and was aware of his general improvement. He often started with phrase presentation at a medium speed and generally adopted a visual approach. There were problems in reading harder regular words such as 'interminable' and 'disappeared' at post-test. He remained prone to morphological errors with word endings, e.g. 'early' for 'earlier' and 'ascent' for 'ascend'. He acknowledged trying to memorise hard words by looking at them 'in little bits'. There were some vocabulary gaps with more advanced passages, consistent with his verbal I.Q. level.

2. **KB** (C.A. 12y 3m). WISC-R: not known. No evidence of visual problems. Passages successfully completed: 4 Total trials: 55 Word-by-word phrases: 7% Phrases: 26% Sentences: 67%

	Pre-test	Post-test	Follow-up
Schonell W.R.	7.6	8.3	8.2
IRI Accuracy	7.5	7.5	7.5
IRI w.p.m	40.0	52.0	67.0

KB improved in fluency, but remained rather slow and did not gain in accuracy at all. However, he avoided syntactic and morphological errors at post-test and said he was understanding what he read better. Yet at follow-up, he was again making several syntactic errors. He said his strategies were to 'split up words' and to try to memorise phrases. He needed more repetitions than most at the final speeded stage with sentences and with two passages failed to meet the criteria. While he showed a gain in word recognition over the intervention period, there was little progress during the following year, except in fluency. He might have done better with single word practice designed to establish phoneme-grapheme correspondence.

3. **LU** (C.A. 12y 1m). WISC-R: Verbal IQ 124 Perf. IQ 114 Full Scale IQ 11. Moderately long-sighted in both eyes. No other visual problems. Passages successfully completed: 10 Total trials: 63 Word-by-word phrases: 8% Phrases: 33% Sentences: 59%.

LU made rapid progress in both accuracy and fluency. On the post-test he gave insufficient attention to short basic words like 'are', 'as', 'it', 'of', 'the', 'to' and to tense and plural markers. However, when using the program this occurred much less and he invariably required only one trial at the final stage of reading sentences fast to achieve success. He admitted to trying to memorise the words, but also referred to sounds, syllables and the final 'e' marker as cues. However, several errors at post-test were holistic visual approximations, such as 'scar' for 'saucer' and 'scattering' for 'skating'. **LU** clearly relied on good linguistic skills to predict, and often neglected visual detail (which may or may not be related to his long-sightedness). His fluency gain may have been facilitated by his continuing with the phrase options until he felt ready to succeed with sentences. He felt that the program had helped with longer words and that he had understood the passages well.

	Pre-test	Post-test	Follow-up
Schonell W.R.	7.7	8.0	changed
IRI Accuracy	7.0	8.0	school
IRI w.p.m	57.0	79.0	-

4. **MJ** (C.A. 11y 0m). WISC-R: Verbal IQ 72   Perf. IQ 81   Full Scale IQ 84
No evidence of visual problems. Passages successfully completed: 8 Total trials: 47
Word-by-word phrases: 4% Phrases: 34% Sentences: 62%

	Pre-test	Post-test	Follow-up
Schonell W.R.	8.2	8.4	9.2
IRI Accuracy	8.0	8.5	12.0
IRI w.p.m	76.0	63.0	112.0

Initial reading was not slow, but there were many signs of task avoidance and poor comprehension in the early stages of working with 'Read and Speak'. However, **MJ** became progressively more efficient, ending by reaching the accuracy and fluency criteria on the third reading of a passage. He preferred to start with a medium-paced reading of phrases and there is evidence at post-test that he had established an accuracy/speed trade-off, with improved accuracy but slower reading. With the more advanced passages, he had considerable vocabulary problems, as might be expected given his low verbal I.Q. At the end he said that he preferred reading on the computer because it was easier to understand and it helped him with bigger words. He said that he used the speech feedback to try to learn words that he had misread. The intervention certainly succeeded in establishing a positive attitude. He still had decoding problems at post-test, especially with longer regularly-spelt words, but went on to make remarkable all-round gains in the following year. At follow-up he clearly considered himself to be a reader and was an Enid Blyton fan.

5. **RK** (C.A. 12y 8m). WISC-R: Verbal IQ 97   Perf. IQ 101   Full Scale IQ 101. Mildly long-sighted, especially in the left eye. Visual discomfort when viewing Amsler grid. Passages successfully completed: 9 Total trials: 60 Word-by-word phrases: 3% Phrases: 40% Sentences: 57%

	Pre-test	Post-test	Follow-up
Schonell W.R.	8.3	8.4	9.0
IRI Accuracy	7.5	8.5	8.5
IRI w.p.m	60.0	89.0	91.0

Good progress was made during the intervention, especially in fluency. **RK** became almost too fast, since he was often insensitive to visual features. He said that he tried to memorise words by looking at the initial letter and this resulted in miscues like 'soldier' for 'sailors' and 'lodging' for 'location'. He used the phrase options a good deal, preferring fast presentation rates to the medium speed. He found the computer visually rather stressful. He made little or no further progress in accuracy and fluency during the follow-up period, but his Schonell word recognition score did improve.

6. **ST** (C.A. 14y 1m). WISC-R: Verbal IQ 102   Perf. IQ 102   Full Scale IQ 102. Alternating right divergent squint, with long sighted right eye. Viewing the Amsler grid was distressing and a teacher had noticed that she would sometimes rub her eyes during lessons. Passages successfully completed: 4 Total trials: 46 Word-by-word phrases: 4% Phrases: 26% Sentences: 70%

	Pre-test	Post-test	Follow-up
Schonell W.R.	7.5	7.8	8.3
IRI Accuracy	7.0	7.0	7.0
IRI w.p.m	33.0	43.0	35.0

ST was the slowest and most inaccurate in the group. She missed two sessions because she took a holiday and was very conscious of her poor reading, being rigid in her adherence to strategies which were manifestly not working. Her approach was to try to memorise whole sentences and she used the phrase options less than anyone else. On one occasion she re-read a passage 12 times and still failed to reach the accuracy criterion. Her basic sight vocabulary was very limited and many of her miscues showed that she was attending to little more than initial letters (e.g. 'at' for 'as', 'he' for 'his', 'soon' for 'so', 'went' for 'wind'). It is possible that her visual problems meant that she did not have stable images of words or found it hard to scan text from left to right. She made no progress in passage reading during the follow-up period, but continued to make modest gains in single word recognition which were apparently of little benefit to her.

7. **WK** (C.A. 12y 3m). WISC-R: Verbal IQ 96 Perf. IQ 105   Full Scale IQ 100. No evidence of visual problems, but he was tense when reading and sometimes complained of blurring, of his eyes hurting and of headaches. Passages successfully completed: 7 Total trials: 52. Word-by-word phrases: 4% Phrases: 40% Sentences: 56%

	Pre-test	Post-test	Follow-up
Schonell W.R.	7.6	8.1	8.5
IRI Accuracy	7.0	7.0	7.5
IRI w.p.m	41.0	70.0	79.0

This boy was a very slow reader, who hated making errors and clearly worried about his problems. He made little progress in accuracy, but improved in fluency more than anyone else, partly because he made fewer errors with basic sight vocabulary. Records of option choice and his own account show that he was highly flexible in trying different strategies. Overall, he used phrase presentation options more than anyone else, especially at the start. Like **MJ**, he became increasing efficient in reaching criterion as time went on, but at the end he did not think his reading had improved, except perhaps in understanding. At post-test he was able to read passages at a 9-year level with good comprehension, although far from accurately. He continued to make overall progress during the following year, albeit at a slower rate and without real confidence.

CONCLUSION

It is probably no coincidence that the three pupils where some visual problem had been identified (LU, RK and ST) appeared to be relatively insensitive to visual detail within words. Two of the three responded well to the intervention, but not ST, who doggedly avoided the least visually demanding display options. RK and ST made very little progress after the intervention, when the visually favourable features of text display were no longer available.

A successful outcome would seem to be associated with greater use of the phrase options and the two pupils who responded best overall preferred to begin by hearing the passage read in phrase units at medium speed.

It is felt that the Read and Speak system cuts down the frustrating delays which take place when longer and 'harder' words are encountered or when children mis-read words and lose the thread or are forced into syntactic blind alleys. It is likely that the gains obtained resulted from improved automaticity in the recognition of irregularly-spelt words, from the practice of reading in phrase units and from improved decoding strategies. While there was evidence that some new strategies had been acquired in the process of trying out different options, the overall impression was that inefficient approaches to decoding were hard to shift and that it is probably desirable to tighten the structure of the framework within which choices are offered. Without such a structure, it is less likely that phonological skills, automaticity and sensitivity to finer linguistic details will develop.

The results of this study confirm those of Fiedorowicz & Trites (1987) and those of Breznitz (1991) in showing that improved automaticity acquired by computer-aided instruction generalises to reading text off the computer. Moreover, long-lasting benefits were obtained, as also reported by Fiedorowicz & Trites. This may have happened because of the emphasis on word-recognition (as in the Autoskill program), because the speech feedback served as a model for pronunciation, stress, intonation and syntactic accuracy, because phrase presentation reduced visual fatigue and helped with semantic and syntactic processing, because of the sheer amount of reading and repetition or because of the use of paced reading. Breznitz showed that fast-pacing reduces the demands on working memory during reading (thereby improving comprehension). It also reduces distractability and increases vocalisation during reading. Breznitz also found that with normal beginning readers fewer substitutions of phoneme for grapheme occurred during fast-paced than during self-paced reading. At the same time, in their well-controlled study, van Daal & Reitsma (1990) showed that children learned to read visually segmented words more quickly when speech feedback was available on demand than when it was not available at all.

The combination of speech feedback, visual word- and phrase- building, the reduction of visual fatigue with the window technique, paced reading to build fluency and a high degree of reader-control over presentation options seems to be a powerful one. Now that substantial global benefits have been demonstrated with children with dyslexics, future work can usefully be directed at trying to understand which features are most important for which learners at which stages. Means must also be sought for helping children to link visual features within words with phonological and linguistic features. Here the most promising techniques would appear to be visual sequential word-building, highlighting, the use of onset-rime splits (Goswami, 1991), as well as syllabic and morphological splitting, with accompanying speech. The role of the teacher in this area should not be understated. It is likely that training in auditory discrimination, in clear articulation, in segmentation of the spoken word, in blending and in structural analysis are needed throughout the period of acquisition of decoding skills.

The teaching strategy employed in the present study differs in several important ways from the approach traditionally used with dyslexic children. With a structured multisensory phonic-based programme, the emphasis is on the laborious study of phoneme-grapheme correspondence. It would seem that there is value in attaching greater importance to the development of automaticity. The use of passages which are age-appropriate and not phonically controlled presents decoding problems, but solutions can be found if visual cues and speech feedback is used. If the pupil can become aware of an improvement in reading during a short intervention of this kind, the motivation is created for more focussed study, in which individual difficulties can be addressed.

REFERENCES

Breznitz, Z. (1991). The beneficial effect of accelerating reading rate on dyslexic readers' reading comprehension. In M. Snowling, and M. Thomson (Eds.), *Dyslexia: Integrating Theory and Practice*. London: Whurr, 236-243.

Brown, D.A. (1986). *Reading diagnosis and Remediation*. Englewood Cliffs, N.J., Prentice Hall.

Cato, V., English, F. & Trushell, J. (1989). Reading screens: mapping the labyrinth. *Reading*, 23, 168-178.

Davidson, J., Coles, D., Noyes, T. & Terrell, C. (1991). Books that talk. In Singleton, C. (Ed.) *Computers and Literacy Skills*. Hull: BDA Computer Resource Centre, 75-86.

Dowhower, S. (1987). Effects of repeated reading on second grade transitional readers' fluency and comprehension. *Reading Research Quarterly*, 22, 389-406.

Ellis, N. (1990). Reading, phonological skills and short-term memory: interactive tributaries of development. *Journal of Research in Reading*, 13, 107-122.

Fiedorowicz, C. (1986). Training of reading component subskills. *Annals of Dyslexia*, 36, 318-334.

Fiedorowicz, C. & Trites, R. (1987). *An Evaluation of the Effectiveness of Computer-Assisted Component Reading Subskills Training*. Toronto: Queens Printer for Toronto.

Fiedorowicz, C. & Trites, R. (1991). From theory to practice with subtypes of reading disabilities. In B.P. Rourke (Ed.), *Neuropsychological Validation of Learning Disability Subtypes*. New York: Guilford, 243-266.

Gambrell, L.B., Bradley, V.N. & McLaughlin, E.M. (1987). Young children's comprehension and recall of computer screen displayed text. *Journal of Research in Reading,* 10, 156-163.

Goswami, U. (1991). Recent work on reading and spelling development. In M. Snowling, and M. Thomson (Eds.), *Dyslexia: integrating theory and practice.* London: Whurr, 108-121.

Herman, P. (1985). The effects of repeated readings on read rate, speech pauses and word recognition accuracy. *Reading Research Quarterly,* 20, 553-564.

Lewis, A. (1983). An experimental evaluation of a Direct Instruction programme (Corrective Reading) with remedial readers in a comprehensive school. *Educational Psychology,* 2, 121-135.

Liberman, I.Y., Shankweiler, D., Fisher, M.F. & Carter, B. (1974). Explicit syllable and phoneme segmentation in the young child. *Journal of Experimental Child Psychology,* 18, 201-212.

Lovett, M.W., Ransby, M.J. & Barron, R.W. (1988). Treatment, subtype, and word type effects in dyslexic children's response to remediation. *Brain and Language,* 34, 328-349.

Lyon, G.R. & Flynn, J.M. (1991). Educational validation studies with subtypes of learning-disabled readers. In B.P. Rourke (Ed.), *Neuropsychological Validation of Learning Disability Subtypes.* New York: Guilford, 243-266.

Mollon, J.D. & Polden, P.G. (1976). Some properties of the blue cone mechanism of the eye. *Journal of Physiology,* 254, 1-2P (Abstract)

Moseley, D.V. (1969). The Talking Typewriter and remedial teaching in a secondary school. *Remedial Education,* 4, 196-202.

Moseley, D.V. (1972). The English Colour Code programmed reading course. In V. Southgate (Ed.), *Literacy at all Levels.* London, Ward Lock.

Moseley, D.V. (1988). Dominance, reading and spelling. *Bulletin d' Audiophonologie Annales Scientifiques de L' Université de Franche-Comte,* 4, 443-464.

Moseley, D.V. (1990). Lit-oracy: a technological breakthrough. In M. Hunter-Carsch, S. Beverton, and D. Dennis (Eds.), *Primary English in the National Curriculum.,* Oxford: Blackwell.

O'Neil, G. & Stanley, G. (1976). Visual processing of straight lines in dyslexic and normal children. *British Journal of Educational Psychology,* 46, 323-327.

Rayner, J. (1983). *Eye Movements in Reading: Perceptual and Linguistic Aspects.* New York, Academic Press.

Reitsma, P. (1988). Reading practice for beginners: effects of guided reading, reading while listening and independent reading with computer-based speech feedback. *Reading Research Quarterly,* 23, 219-235.

Riding, R.J. & Pugh, J.C. (1987). Dark-interval threshold, illumination level and children's reading performance. *Journal of Research in Reading*, 10, 21-28.

Slaghuis, W.L. & Lovegrove, W.J. (1985). Spatial-frequency-dependent visible persistence and specific reading disability. *Brain and Cognition*, 4, 219-240.

van Daal, V.H.P. & Reitsma, P. (1990). Effects of independent word practice with segmented and whole-word sound feedback in disabled readers. *Journal of Research in Reading*, 13, 133-148.

Facets of Dyslexia and its Remediation
S.F. Wright and R. Groner (Editors)
© 1993 Elsevier Science Publishers B.V. All rights reserved.   585

OPHTHALMOLOGIC ASPECTS OF DYSLEXIA: BINOCULAR FULL CORRECTION OF DYSLEXICS WITH PRISMATIC GLASSES

D. Pestalozzi,

Ophthalmologist, Olten, Switzerland

INTRODUCTION

Psychologists and other specialists who are concerned with dyslexics suspected for a long time that some cases of dyslexia might be connected with (bin)ocular problems, although ophthalmologic examinations generally did not confirm this. A special method for investigating binocular vision (The "Zeiss-Polatest") now makes it possible to test this hypothesis, advanced also by Stein (1989). It allows the researcher to obtain the critical data which could not be found by traditional methods.

In the following the term dyslexics will be used for subjects with dyslexic symptoms, i.e. with reading or writing problems. All cases reported were suffering from reading or writing problems, and almost all of them had been "labelled" as dyslexics by school-psychologists. A few did not meet criteria for a diagnosis of "dyslexia" although they had reading/writing problems.

BINOCULAR CORRECTION

At first some comment on terminology: *binocular* means with both eyes, i.e., the way the eyes are working together. *Heterophoria* is a hidden squint or a squint overcome or compensated by the subject and therefore invisible; the eyes are straight. *Prismatic glasses* are wedge-shaped and have the property to deviate the light. Thus, looking through, one can 'look around the corner'. They are used to make deviated eyes look in the same direction. The eyes therefore are able to fix an object on the centre of their retina although they actually are not looking in the same direction.

Normal and pathological binocular vision (Fig. 1) includes heterophoria. *Normal binocular vision* means that the eyes are able to compensate a position of rest which is different from the *ortho-position* (definition see "Begriffe der Physiologischen Optik DIN 5340, #278). If they can not, there is a pathological binocular vision: strabism (right side of the graph).

Ideally motor functioning (i.e. the balance of the eye-muscles) is undisturbed. Both eyes are looking without strain at the same object. Sensory conditions are then also ideal: the object is projected at the centre of both retinas. In presence of a compensated muscular imbalance a condition called heterophoria exists. If compensation is no longer possible, motor functioning is pathological (right side of the graph) and strabism occurs. If binocular vision persists, sensory functions become pathological as well: there is anomalous correspondence (for definition see DIN 5340 Begriffe der physiologischen Optik #221) A heterophoria at first may be overcome by muscle force. This condition is called *motor fully compensated heterophoria*, sensory functioning being still ideal. Both eyes fixate in the centre of the retina. If the imbalance is no longer motorically compensated, *diplopia* would result. To avoid this sensory compensation is required. Either there will be a central scotoma of the deviated eye which leads to lack of central binocular perception (as reported by Safra, 1991), or, if binocular vision persists, one or both eyes are no longer using the centre of the retina to look at a fixated object. Instead a paracentral spot assumes the sense of 'straight ahead'. In this way central binocular vision is maintained, although of a poorer quality. This condition is called *fixation disparity* (FD).

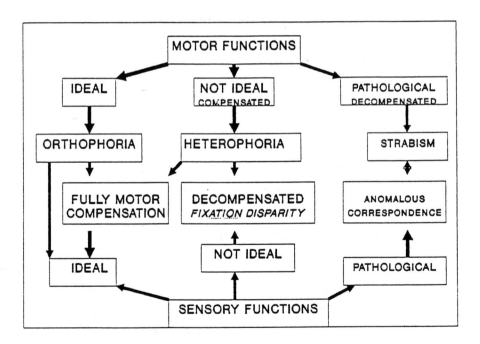

FIG. 1. Normal Binocular Vision (after Goersch, 1987, translated and adapted by the author)

Fixation Disparity (Figs. 2a & 2b) Goersch (1987) describes two different kinds of FD (type I and II).   Under natural conditions with FD-I the eyes (Fig. 2a) see only one distant dot, although, in the right eye, it is projected at a spot beside the very centre. In the graph the dot lies within the little ellipse surrounding the centre of the right retina, symbolizing Panum's area. Panums area is the region where a non centrally fixated object is still seen singly (and stereoscopically). With fixation disparity it is 'abused' to maintain binocular single vision resulting in poorer stereoscopic vision. (This can be measured with the stereo-charts of the Polatest). But if an object is projected outside Panums area, there is either diplopia or, more frequently, inhibitions (Central scotoma as mentioned by Safra, 1991) will develop preventing binocular vision and diplopia .

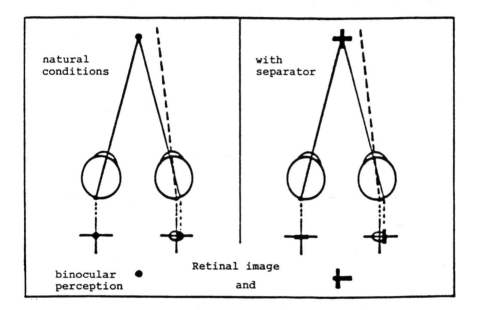

FIG. 2a. Fixation Disparity I (see text for explanation).

Those inhibitions can be overcome and binocular vision be restored by prismatic corrections using the Polatest as a measuring device (See also Fig. 4). However, using a separator, each eye sees a different part of the same object (in the graph: the horizontal and the vertical bar of a cross), the cross appears deviated. Conventional investigating methods for heterophoria work in this way. But in presence of a FD-II (Fig.2b) a cross is still seen in spite of the deviation of one eye, because the sense of 'straight ahead' has shifted from the very centre of the retina to a spot beside it, even for totally different objects.

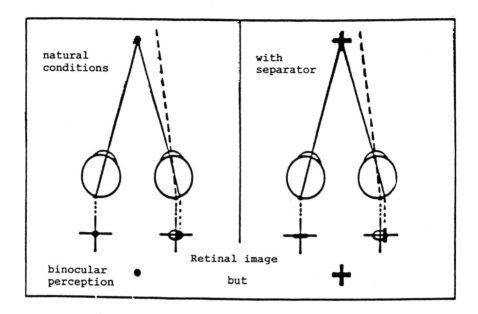

FIG. 2b. Fixation disparity II (see text for explanation).

To determine full binocular correction an examination method is needed which allows one to find and correct motor and sensory functioning simultaneously. Measurements of this kind belong to the field of optometry and are often performed by optometrists or in Switzerland by diploma eye opticians. It appears that only by full motor and sensory control can fixation disparities be overcome.

The above mentioned Zeiss-Polatest fulfils all requirements for analyzing in a detailed manner central binocular vision. It has been designed by Haase, a former teacher at the "Staatliche Fachschule für Augenoptik und Fotografie (SFOF)" in Berlin, in the 1950's and allows for "measurement by correction". The principle is to present each eye with different parts of the same object under almost natural conditions. The examination is performed in a bright room. Separation of pictures is done by polarization. Based on the literature on binocular problems and on thousands of observations, Haase (1980) derived his theory of fixation disparities and indicated methods for efficient Polatest-examination.

With heterophoria the different charts of the Polatest appear shifted to the examinee. The art of the examiner consists then in asking him/her how actually the charts are seen, and then in restoring their real shape by prismatic glasses. This task requires a clear knowledge of possible answers and how to react to them, as well as patience and time.

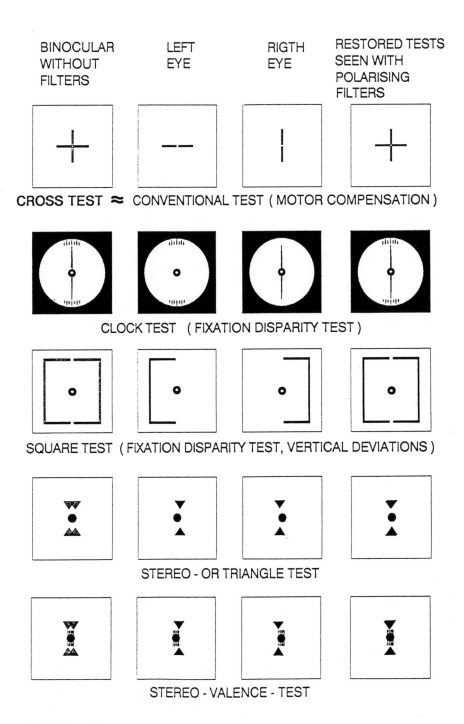

FIG. 3. The Zeiss-Polatest (by Haase)

The cross-test (to be displayed first) allows the detection of mainly motor compensations and corresponds to the conventional devices for examination of heterophoria.

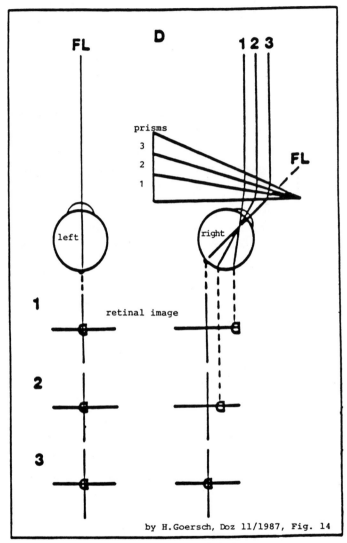

1 Restoring prism with conventional test
2 Restoring prism of fixation-disparity-charts
3 Full correction to achieve with stereo-charts

FIG. 4. Retinal images with Fixation Disparity: 1 represents the prism correction found by conventional methods; 2 that found by the FD-tests; and 3 the prism correction found by the stereo-charts of the Polatest. Reprinted with Permission from Goersch (1987).

The clock- and square charts are called "fixation-disparity-tests". They have a striking central ring as a fusion object (i.e. it must be seen simple with both eyes). However, in the presence of a fixation disparity the charts may appear shifted, although the ring remains single.- The stereo-charts allow a check to be made of not only the quality of depth perception but also the detection and correction of very old types of FD.

Figure 4 shows why binocular full correction cannot be found by conventional methods, and why only the Polatest with its three different types of charts allows for the determination of binocular full correction. The fixation line (FL) of the left eye is looking at a far distant "D", the right eye is deviated outside (designed exaggeratedly). Its FL is deviated by prismatic glasses of three different sizes to direct it's FL to the "D" as well: Prism 1 is the correction found by the cross-test or by conventional methods. Prism 2 corresponds to the value found by FD-tests. With both corrections the retinal image is still eccentric. While Prism 3 finally gives the value for binocular full correction which is found by the different stereo-charts of the Polatest.

Inhibitions of binocular vision as mentioned above may also hamper the reflex of convergence and lead therefore to a (pseudo-)exophoria (heterophoria directed outside) at near (Safra, 1991 and Pestalozzi, 1992), even in the presence of an opposite deviation at far. When binocular vision is restored, the pseudo-exophoria will disappear.

The Polatest-examination can be performed only with monocular best correcting glasses. Disturbed accommodation as found by Mühlendyck should be overcome before.

HETEROPHORIA AND DYSLEXIA

Figure 5 shows an interaction model by Lie (1987) "Interrelating visual anomalies, visual stress conditions and reading disorders" which demonstrates general relations between visual problems and dyslexia. Stein (1989) writes "we now have strong evidence to support our hypothesis that many dyslexic children experience unstable binocular control when attempting to read" (p 530). He considers the consequences of impairment of the visuomotor and visuospatial functions of the right hemisphere to be a cause of developmental dyslexia.

The author became aware of a link between binocular control and reading already in 1978 as children's mothers often reported an improvement, following binocular correction, of previous dyslexic symptoms. Following a lecture on that topic (1984) many dyslexic children were subsequently referred to us by psychologists and logopedists. In 1988 we reported on 175 dyslexic cases. (Pestalozzi, 1989).

If a heterophoria causes complaints we say it is decompensated. The complaints are summarized under the term of asthenopia. (See Table 1). Dyslexic symptoms include: confounding of characters b/d, m/n, o/a, etc., the omission of short words or endings,

FIG. 5. Visual deficiencies and reading problems: An interaction model, interrelating visual anomalies, visual stress conditions and reading disorders

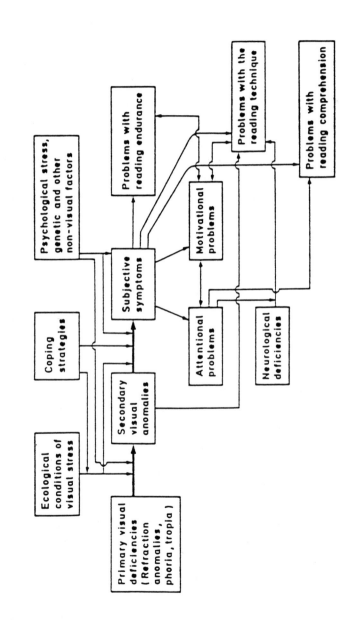

Ivar Lie

and an impossibility to read longer words. We also have seen some cases who were unable to read more than one line of a foreign language (e.g. English) without prismatic glasses, but who with binocular full correction (and possibly later a squint-operation) no longer had this difficulty. For the optometric trained ophthalmologist or optometrist, dyslexic symptoms constitute only a small proportion of the asthenopic troubles caused by heterophoria, as there are only few dyslexics among an immense number of heterophoric patients with asthenopic complaints.

TABLE 1

The Main Asthenopic Symptoms

Symptoms of Motor Decompensation	Symptoms of Sensory Decompensation
Headache (most frequently reported)	Blurred vision with change of focus
Migraine	Fixation problems
Stiffness and pain in the neck	Difficulties in keeping small objects fixated for long periods.
Concentration problems	**Dyslexic symptoms**
Disorders of behaviour (eg. restlessness)	
Occasional diplopia	
Glaring	

CASE STUDIES

The following data represent the experience of a practising ophthalmologist. During the last ten years 370 dyslexics were seen and 278 of them treated by the above mentioned method. Unfortunately in a fully operating practice, reasons of time and personnel mean that it was impossible to perform a scientifically perfect study. Especially since not all possibly interesting data could be gathered and evaluated, and no control group could be established. Of the 370 cases, 16% came only for a first check, 7% cannot yet be evaluated, and only 2% had no heterophoria. Most cases were of school-age, although a percentage were also adult (8%). With the exception of the adults, more boys were seen than girls, which is in contrast to "normal" asthenopics where females are predominant.

RESULTS

It is difficult to judge the influence of any treatment on dyslexia. However, two pilot studies performed in 1991 at the professional school for eye opticians (Optometrists) in

Olten (SHFA) showed some evidence that in the future the influence of binocular full correction on dyslexia might be electronically verified by registration of reading saccades before and after prismatic correction (Gilgen & Cajacob, 1991, Hüsser & Lengwiler, 1991).

The results of our study were based on the statements of parents, teachers, psychologists, logopedists and the children themselves. The symptoms reported to have improved are listed in Table 2. Listed are also some non-dyslexic symptoms, as we noted that often the general behaviour and mood would noticeably improve after prismatic correction (See Table 2).

We marked the result as *good* (60%) if at least two of these symptoms had improved. Cases in which writing or reading ability definitely improved were registered as *very good* (11%). If there was no relief of dyslexic symptoms, but at least of asthenopic complaints and/or increased visual acuity, sensory or behaviour improvement, results were marked as ± (17%). Finally we had 12% of failures with no positive effect of prisms at all.

TABLE 2

Suggestions for the Evaluation of the Result of Prism-Corrections.

1. Reading/Writing: Writing, speed, number of errors. Reading Speed, fluency. Relationship between number of errors and length of text. Does he/she read voluntarily?	**3. Social Criteria:** Average class-performance reached? Marks in language? Promotion to the next school grade?
2. Asthenopic complaints (Table 1) Improvements?	**4. Behaviour:** Stress-resistance? Concentration? General behaviour? Mood? Coordination?

The following cases illustrate *good* results with prismatic corrections:

A dyslexic girl (12) was unable to attend music lectures in the afternoon after school. Following correction this was easily possible. Her endurance improved and she became more cheerful and her dyslexic symptoms were relieved. A boy (12) was unable to sit quietly at school. He was noisy, disturbed classes and avoided reading whenever possible. After correction the opposite was noted: he sat still and quietly at school and started reading voluntarily. His dyslexia also improved.

The following cases illustrate *very good* results with prismatic corrections:

A boy at the end of the first school year was still unable to read. After prismatic full correction he learned it within a fortnight (Palm, 1991). Kathi (Cathleen), a 16 year old girl, was still almost unable to read. A check before correction in the consulting room showed she knew the letters but was hardly able to read any words correctly. With quite a low prismatic correction she could instantly read much better, when still sitting in the consulting room. The development of her writing is displayed in Fig. 6.

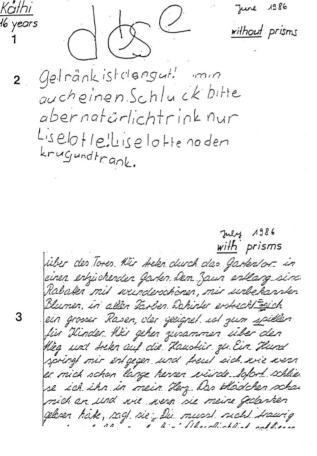

FIG. 6. Handwriting samples from a 16 year old girl before and after prisms

Her problems disappeared with full correction. On the way home she reported to her mother that she was able to read much better with the glasses, because the letters did no longer move around as they usually did. The phenomenon of moving words and letters in dyslexics has been described also by Stein (1989). However, such dramatic improvements are rather exceptional. Usually a period of adaptation is needed before such positive results can be achieved. A speech therapist argued that she had the impression that prisms would improve first the "internal order" of children and only later might there appear also positive effects on the dyslexic symptoms.

Mike (10) also had very bad writing although he wore well correcting spectacles. He needed a strong prismatic correction done by "press-on prisms" on his glasses. Although those prisms diminish visual acuity, his writing actually improved.

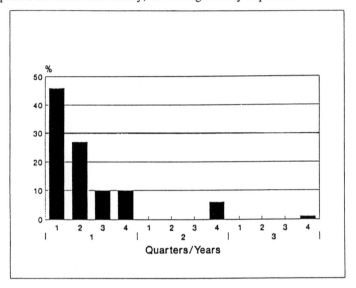

FIG. 7. Time taken for improvement to be seen with prism use.

Ninety percent of the improvements were seen within the first year, of which 46% in the first and 27% in the second quarter; a further 10% in each of the third and fourth quarters. The remaining 7% improvements occurred after two or three years (see Fig 7). Similar rapid improvements following optical full correction of visual anomalies have also been reported by Lie (1989). He found significant improvements in reading comprehension over an eight week period and some, although less marked, improvements in reading speed in the same time period.

Ophthalmologically we also noticed an improvement in visual acuity (VA) (in spite of best refractive correction). The percentage of eyes with VA ≥ 1.5 increased by 18%, while the number of eyes with VA < 1.5 decreased (see Fig. 8). Sensory development namely *Ideal Binocular Vision* (IBV) also increased under correction by 63%. Motor fully compensated (MFC) and Fixation disparity Type-I (FD-I) cases decreased by 5%

and FD-II cases by 46%. Even 12% of the pathological cases reached binocular single vision and even stereopsis (Fig. 9).

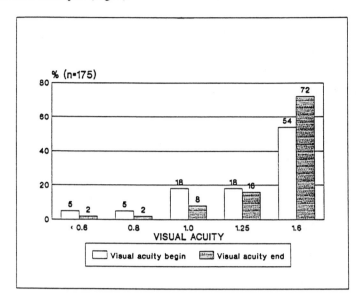

FIG. 8. Improvement of visual acuity with prims.

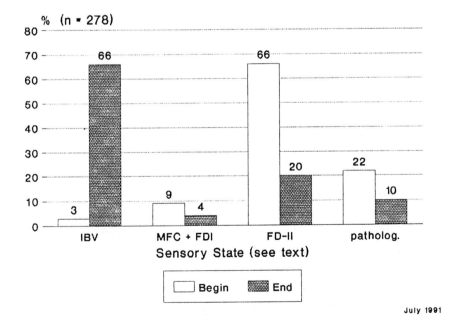

FIG. 9. Sensory state before and after therapy.

DISCUSSION

The examination by Polatest and prismatic correction of a heterophoria and - if necessary - a subsequent squint operation leads, in a high percentage of cases, to an improvement of the dyslexic and asthenopic symptoms or to the relief of other complaints or disturbed behaviour. However, a careful logopedic examination may still reveal some dyslexic symptoms, even in cases with apparently good results, as a small pilot study performed together with Schwarz indicated (Schwarz, 1991). According to Stein (1989) this may be due to abnormalities in the left hemisphere, whilst what we are doing may influence problems with the right hemisphere.

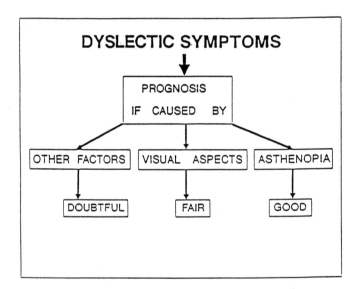

FIG. 10. Dyslexic symptoms

It should be stated we do not pretend to cure a dyslexia, and some of our good results may have been judged too optimistically, but rather to correct heterophorias and restore, if possible, ideal binocular vision, thereby improving the visual input. It remains the task of specialized teachers to improve - if this is possible - cerebral processing and "output" like reading or writing. However, we noticed that their work often became easier and more successful after binocular full correction; in a number of cases specific treatment could be brought to an end soon after binocular full correction was established.

We share the opinion of Sauter (1991) and other authors that compensation of heterophorias is a very energy-consuming process. We think that by correcting heterophorias energy is saved, enabling the patients to overcome their dyslexia better (cf.

the above mentioned girl who couldn't attend music lessons before correction). As Fig. 10, shows, our working hypothesis is that prognosis is best in cases with asthenopia alone (in the sense of binocular problems possibly correlated with abnormalities in the right hemisphere according to Stein, 1989) causing dyslexic problems. In cases with defects in both hemispheres prognosis might be still fair; but in cases due to other disabilities (i.e. in cases with exclusively linguistic problems) prognosis is doubtful. We suppose that our failures might belong to that group. - Our findings are consistent with those of Lie (1989). We feel that binocular full correction will have a good influence not only on dyslexic symptoms but on the patient and his or her whole personality.

In conclusion, in view of the improvement in dyslexic symptoms as a result of prismatic correction, we suggest that prospective studies on a large number of cases are carried out to establish the real value of binocular correction in the treatment of dyslexia. Dyslexics may profit from such interdisciplinary treatment by psychologists and optometrists.

REFERENCES

Gilgen, M. & Cajacob, K. (1991). *Computer-Lesetest 'Blitz-Test'. Schweiz.* Höhere Fachschule für Augenoptiker, Olten SHFA.

Goersch, H. (1987). *Die drei notwendigen Testarten zur vollständigen Heterophorie-bestimmung.* Deutsche Optikerzeitung Heidelberg Nr. 11.

Haase, H.J. (1980). *Binokulare Vollkorrektion.* Die Methodik und Theorie von H.J. Haase. Verlag Willy Schrickel.

Haase, H.J. (in press). Zur Fixationsdisparation.Verlag: Deutsche Optikerzeitung.

Hüsser, A. & Lengwiler, Ch. (1991). *Einfluss der prismatischen Vollkorrektion auf die Sakkaden beim Lesen.* SHFA.

Lie, I. (1987). *Ocularmotor factors in the aetiology of occupational cervicobrachial diseases.* European Journal of Applied Physiology, 56, 151-156.

Lie, I. (1989). Visual anomalies, visually related problems and reading difficulties. *Optometrie*, 4, 15-20.

Mühlendyck, H. (1991). *Reading disorders in the case of juvenile accomodation disturbance.* Paper presented at the 18th Scientific Rodin Remediation Conference, Bern, August 1991.

Palm, A. (1991). Lesen und Sehen, oder brauchen Legastheniker eine Brille? VBL-Bulletin 1.

Pestalozzi, D. (1991). Prismenverordnung bei Legasthenie. Lecture at the 1st Meeting of the International Association for Binocular Full Correction. (IABF). Schriftenreihe der IVBV No. 1, 16-22 and Neues Optikerjournal 1/89.

Pestalozzi, D. (1992). Pseudo-Nahexophorien. Neues Optikerjournal, 6.

Safra, D. (1991) Theoretische Betrachtungen zur orthoptischen Legastheniebehandlung. Poster presented at the 18th Rodin Remediation Conference, Bern, August 1991.

Sauter, J.M. (1991) Emotional stress, visual impact and reading disorders. Poster presented at the 18th Scientific Rodin Remediation Conference, Bern, August 1991.

Schwarz, C. (1992). Ist das Tragen von Prismenbrillen zur Behandlung von Legasthenikern eine sinnvolle Massnahme? *Klinische Monatsblätter für Augenheilkunde*, 200, 599-613. Stuttgart: Enke

Stein, J. (1989). Visuospatial perception and reading problems. *Irish Journal of Psychology*, 10, (4) 521-533.

Facets of Dyslexia and its Remediation
S.F. Wright and R. Groner (Editors)
© 1993 Elsevier Science Publishers B.V. All rights reserved.

THE ORTHOPTIC TREATMENT OF DYSLEXIA

D. Safra

St Gallen, Switzerland

DEFINITION OF DYSLEXIA

By defining dyslexia, as a difficulty in learning to read despite adequate intelligence and sufficient visual acuity, we distinguish dyslexic children from asthenopia cases. In contrast to dyslexics, who have difficulty right from the outset at school, asthenopics learn to read in exactly the same way as a normal child. They experience no difficulties until the third or fourth class or even later, when school reading material is normally in smaller, denser print and more complex. Common to both cases is badly controlled binocular vision requiring the same treatment, although the prognosis for asthenopics is usually more favourable. Both dyslexics and asthenopics may occur in the same family, and in marginal cases it is difficult to distinguish between them.

HYPOTHETICAL CAUSES OF DYSLEXIA

The nature and causes of dyslexia are still not clear, and opinions are divided into two main groups. The first group considers dyslexia to be the result of a disturbance in linguistic processing, while the second group maintains that dyslexia results from visuomotor abnormality. Both opinions are well-founded, but neither of them do justice to the complex overall picture. Based on observations of the general behaviour and reading habits of dyslexics together with a precise case study and the reports of pediatricians and psychologists referring cases to us, we have formulated a rather broader hypothesis as follows: dyslexia is part of an organic cerebral syndrome, either congenital or acquired in early infancy. This hinders translation of text into language or vice versa due to specific gnostic failures, while simultaneously impairing visual perception during reading due to unstable binocular vision control.

As a result of these gnostic problems, dyslexic children have difficulty in recognizing and remembering the form and significance of printed characters. In the same way as normal children when they first learn to read, dyslexics go on confusing ei with ie, for example, or letters which are distinguished by their sense of rotation or orientation such as a and e, b and d, b and p or p and q. Dyslexics also typically omit letters, above all at the beginning and end of words, guess the rest of a word from its initial letters, mingle up letters in the middle of a word, frequently lose the place in a text or jump over the next line. The latter reading errors can be interpreted as being due to faulty binocular vision control with alternating eye fixation, frequent changing from monocular to binocular vision, and alternating central scotoma due to mutual suppression. In the same way as dyslexia itself, we have found that the degree of binocular control instability varies from case to case. Nevertheless, given that not a single case of dyslexia with monocular vision is known as yet, this seems to be a conditio sine qua non.

FUNDAMENTALS OF ORTHOPTIC DYSLEXIA TREATMENT

Orthoptic dyslexia treatment as practised today has developed over many years. The first impulse for ocular orientated treatment came from Orton (1937). From his observations of dyslexic children, he concluded that dyslexia is the result of delayed cerebral lateralization comparable with stuttering, which he ascribed in some cases to competition between the two cerebral hemispheres for speech control which is normally the task of the left brain. Since then Orton's basic theory of competition between the cerebral hemispheres and between the eyes has been followed by all kinds of ocular dyslexia treatment, whereby the former can only be assumed, the latter can be measured in practice.

The lack of stable hand and eye dominance observed by Orton in dyslexics gave rise to occlusion treatment. By occluding one eye, the other can be induced to take over the lead, thus preventing competition between the eyes and improving reading ability. Successful results with this treatment are reported by Benton, McCann & Larsen (1966) Stein & Fowler (1985), Otto & von Frisching (1989), whereby each of these authors used different tests for determining eye dominance. Bishop (1989) critically examined the results of Stein & Fowler and their eye dominance test procedure according to Dunlop. Bishop found that according to the Dunlop test, occlusion treatment did not improve the reading ability of children with lack of eye dominance, and that there was no causal connection between positive test results and reading improvement under occlusion conditions. As far as the success of such treatment is concerned, these findings also contradict the reports of the other authors mentioned above. However, it can be argued that the other authors carried out occlusion more strictly and for longer time periods (up to two years of full occlusion nearly the whole day) than Stein & Fowler (6 months of occlusion, only during reading and writing).

Eye dominance means that preference is given to the visual information imparted by one eye while that from the other is suppressed if the information is not identical - for example regarding the location of an object, such as text characters during reading when both eyes are not precisely aligned on the object. Dominance thus prevents competition between the eyes. Undisturbed reading requires at least 3 sets of clear information: a) the location of the characters for saccades programming, b) image sharpness for accommodation adjustment, c) the contours o£ characters for their identification. Our own investigations have shown that the diverse dominances may not be confined to the same eye even in persons with normal reading ability (Safra & Otto, 1992). It thus appears that a single dominance test is inadequate for establishing or excluding a relationship between eye dominance and reading ability.

Other explanations for the success of occlusion treatment may be that with strict long-term occlusion of one eye, a strong dominance for all the information required for reading can indeed be developed in the free eye. A further explanation could be that monocular occlusion is also a common antisuppressive measure for breaking down interactive binocular inhibition and may result in stable visual correspondence.

Another way of avoiding informational competition between the eyes is by generating Orthophoria. Orthophoria means that both eyes precisely and simultaneously align their visual lines on the fixated object. Complete motoric balance of the two eyes and therefore spontaneous orthophoria is rare. In most cases (70 to 80 %) a well-functioning visually conditioned central control system corrects a deviating visual line, so that the object image falls on corresponding retinal points and thus allows fusion of the two images into a single one, or when failing to correct the deviation deletes the double image in order to prevent diplopia and confusion. Normally deviation of the visual line and deletion of the image concerns only the non- leading eye, while the leading eye keeps up fixation undisturbed. By using prisms, a kind of passive or so-called sensory orthophoria can be generated. Instead of corrective movements the prisms deflect the image towards their base on to corresponding points of the central retina. Otto and von Frisching (1989) combined occlusion with prism treatment since they found that acceptance of the latter was better following monocular occlusion, thus resulting in better stabilization of binocular vision. Pestalozzi (1991) reports considerable success using prisms alone, since this treatment allows correction of even the smallest fixation line deviation, whereby measurement was exclusively carried out by the Polatest.

Our dyslexia treatment method is based on the experience of Otto & von Frisching during their 6-year study. We do not carry out occlusion in all cases, and even then only for short periods of about 3 to 6 weeks, usually with weakly translucent occlusion foils covering the weaker monocular reading eye as an antisuppressive measure. We concentrate primarily on eliminating the intermittent, alternating central scotoma typical of dyslexics. This has been described amongst others by Rabetge & Kraus-Mackiw (1981) and Schuhmacher (1985), and tested objectively by Wenzel (1988) using a special VEP test. We agree with these authors that this sensory disturbance indicates competition due to inadequate central control of binocular vision as the main factor interfering with reading. Central scotoma (inhibitory suppression of the central retina) prevents stable fixation, precise accommodation (Safra & Otto, 1974) and convergence according to reading distance as well as precise following from one character to the next - the most important visual capabilities required for a child to learn reading and recognize characters. Our treatment is therefore aimed at preventing competition and central scotoma by generating stable binocular vision, thus improving the reading ability of dyslexics through better visual perception.

TREATMENT PROCEDURE

Interference between accommodation and convergence, or refractive asymmetry between the eyes, is first prevented by meticulous objective and subjective refraction testing at long and short range, with visual aid correction even of slight ametropia and anisometropia. Latent strabismus (heterophoria) is then compensated with adhesive prismatic foils to the extent of sensoric orthophoria. The low initial prismatic values are increased with the gradually increasing heterophoria angle normally encountered, to the point of complete relaxation. With prismatic values below 15 cm/m, the prisms are ground into the visual aid lenses, while in case of higher values a squint operation is undertaken. During prism fitting, measures are taken to eliminate central scotoma by exercises designed to stimulate the foveolae on the synoptophore, or by accommodation and convergence exercises at home with the accommodation- convergence bar specially designed for this purpose (Safra, 1990).

In order to determine the angle of heterophoria, we use the cover test, synoptophore, Polatest and Bagolini Test. Particularly suitable for diagnosing central scotoma are the Pola and Bagolini tests. The latter allows examination in all directions and at all distances, thus covering the entire reading range under more or less normal conditions. Improved reading ability was found to be an extremely reliable indication of central scotoma suspension.

We initially assess reading ability based on reading speed and number of errors. At each session the patient reads 3 text passages from a suitable school grade book (difficulty increases according to grade due to smaller and denser print and more complex words). Reading time is measured by stop-watch and the number of errors is recorded. Tests with normal children indicated a best time of about 30 seconds for faultless reading of 4 lines of suitably graded text. According to a special experimentally established grading system, we award 10 marks for the best reading performance and 1 for the poorest performance (4 lines in 90 or more seconds with 8 mistakes).

RESULTS OF TREATMENT

From January 1988 to June 1991 we treated 166 dyslexic children 9 to 14 years old, comprising 110 boys and 56 girls. After an average treatment period of 1.5 years, 116 (70 %) of them developed considerably improved reading ability (mark 7 to 8 or more), while 42 (25 %) achieved an excellent result (mark 9 - 10 for reading speed and precision, fluent reading with logical emphasis, voluntary reading). 73 subjects (45 %) improved their school grading by at least one mark (out of six in Switzerland). 50 children (30 %) hardly showed any improvement, partly due to lack of cooperation and above all inconsistent wearing of spectacles. The overall success achieved was also due in many cases to additional therapeutic reading instruction and above all the motivation of children and parents.

DISCUSSION OF OUR TREATMENT METHOD

Almost every treatment method has its drawbacks, for example the adhesive prism foils used in our treatment impair clear vision so that some children were not very happy about wearing their glasses. A further drawback of prism treatment is that it often leads to squint operations on children who previously showed no orthoptically unusual characteristics. Moreover a second operation frequently has to be carried out if the heterophoria angle was not completely corrected the first time, or if heterophoria with central scotoma appears again after initial success. On the other hand, after squint operations we often find stable binocular vision with better reading ability than after prism treatment alone. We can only explain this by the superiority of active sensomotoric orthophoria over the passive, purely sensoric orthophoria produced by prisms.

Another important aspect in our opinion is the positive influence this treatment seems to have on the psychological behaviour of many dyslexics. Parents and teachers repeatedly report that previously unruly, self-centred nuisances unable to concentrate on lessons turned into quiet, friendly and cooperative children. Such changes cannot be due to successful treatment alone, since they have been observed even before a noticeable improvement in reading ability. Benton and Pestalozzi (personal communication) report similar observations. This may lead back to Orton's theories, and in fact does not exclude the possibility that interruption of competition between the eyes may well have resultant effects on various cortical centres.

REFERENCES

Benton, C.D.Jr., McCann, J.W., Larsen, M. (1966). Dyslexia and Dominance. *Journal of Pediatric Ophthalmology* , 3, 53-57.

Bishop, D.V.M. (1989). Unfixed reference,monocular occlusion and developmental dyslexia - a critique. *British Journal of Ophthalmology*, 73,209-215.

Orton, S.T. (1937). *Reading, Writing, and Speech Problems in Children.*. Norton: New York.

Otto, J., von Frisching, D. (1989). *Ophtalmologisch-Orthoptische Therapie der Legasthenie.* Legasthenie, Ostschweizerische Orthoptik-Pleoptikschule St.Gallen.

Pestalozzi, D. (1991). Die Prismenverordnung bei Legasthenie. *Schweizer Optiker, 67,* 4-13.

Rabetge, G. & Kraus-Mackiw, E. (1981). Visuelle Storfaktoren bei der Legasthenie. *Tägliche Praxsis, 22,* 667-688.

Safra, D. & Otto, J. (1974). Ueber die Auswirkung der Brillenkorrektur bei hyperopen Schielern. *Klinische Monatsblatter der Augenheilkunde, 165,* 625-629.

Safra, D. (1990). Der Akkommodations-Konvergenz-Stab. *Klinische Monatsblatter fur Augenheilkunde, 196,* 101-102.

Safra, D. & Otto, J. (1992). *Asthenopische Heterophorie-beschwerdefreie Heterophorie.* Paper for the congress of the IVBV in Vienna, May, 15th-17th.

Schuhmacher, H. (1985). *Diagnostik und Therapie von visuellen Storfaktoren bei Kindern mit isolierter Lese-Rechtschreibeschwache.* Medizinische Dissertation, Heidelberg.

Stein, J. & Fowler, S. (1985). Effect of monocular occlusion on visuomotor perception and reading in dyslexic children. *The Lancet,* July 13, 69-73.

Wenzel, D. (1988). Visuelle und neurophysiologische Befunde bei Legasthenie. *Sozialpadiatrie,* lO.Jg., 3,197-200.

Facets of Dyslexia and its Remediation
S.F. Wright and R. Groner (Editors)
607

ILLITERACY IN ADULTS: RESULTS FROM A SURVEY STUDY OF A READING AND WRITING TUTORIAL PROGRAM FOR ADULTS

S.Burri & R. Hänni,

Psychology Institute, University of Bern, Switzerland

INTRODUCTION

Over the past few years it has become clear that the adult illiteracy problem still exists in Industrial Countries and is not only limited to the Third World. In spite of compulsory education and the obligation to go to school for at least nine years there are large parts of the adult population who have insufficient reading and writing skills. Although they have attended school, these adults did not acquire literacy skills. This phenomenon is called "functional illiteracy". Estimates of the number of illiterate adults in the population differ. This is probably due in part to the different definitions of "illiteracy" adopted. Some definitions deal mainly with the ability to read and to write while others stress intellectual integration into a society, or the difference between the formal and the effective educational level.

UNESCO estimated the percentage of adult illiterates in Industrial Countries all over the world to be 2.5% (Hadamache, 1984). Grissemann (1985) revealed that at the time they leave school about twenty to thirty thousand Swiss, between 3 and 4% of the population, will not have even reached the level of the 3rd educational year, after nine years of education. These alarming facts resulted in the formation of the association "Reading and Writing for Adults", founded in Switzerland in 1985. The aim of this association was to organize tutorial programs for adults with poor reading and writing skills all over Switzerland.

The Psychology Institute at the University in Bern was asked by the organizers to evaluate the effect of the tutorial programs offered to the population in the region of Bern. The following study covers two different programs with participants from both included. However, since the learning conditions in both programs were approximately the same they were not investigated separately. The data presented were collected during a year of participation and observation in the respective courses.

Problems were first discussed with the organizers and a number of teachers in order to determine the most important issues with respect to the organization of these tutorial programs. The first set of questions dealt with the participants' biographical data. There were three main parts:

(1) Are there common biographical factors which could have led to the development of functional illiteracy?

(2) To what extent are insufficient reading and writing skills affecting the participants' lives?

(3) Why did they decide to take part in a tutorial program and what consequences did this decision bring about?

The second set of questions concerned the benefit of the tutorial programs, i.e. whether they led to an improvement in the participants' reading and writing skills. The third set of questions dealt with the relationship between intelligence and the status of reading and writing skills at the beginning of the program and the individual possibilities to profit from the program.

As work is still in progress only selected parts of the study will be presented here.

METHOD

Subjects

Eight male and nine female adults who participated in one of two reading and writing tutorial programs were investigated in this study. They ranged in age from 24 to 65 years, with a mean age of 35.7 years. The mother tongue of 15 of the participants was Swiss German, one French, and one Portuguese. They had heard about the program through a variety of advertising strategies. All participants volunteered to take part in these programs to improve their reading and writing skills. At the beginning of the tutorial-period they were informed of the aim of the present study and asked to participate.

Biographical data

The demographic data were assessed by means of a half-standardized interview which concerned the participants' educational background, their socialization, their actual professional situation and their motivation to participate in a tutorial program.

Quantitative Assessment of Reading Skills

Research on adult literacy projects has been hampered by the fact that there are few standardized tests appropriate for adults. In this study the Zürcher Lesetest (ZLT, Zürich Reading Test, Grissemann, 1981) was used. It was developed in Switzerland and is suitable for both adults and children. The test was administered to measure the participants' entry reading skills as well as possible changes in their reading ability after a tutorial period of about eight months (pre-post measurement). All reading attempts were recorded on tape.

The ZLT was first analysed in the conventional way, ie. the time taken to read the different passages was measured and the errors which occurred during reading were recorded. Additional information was gained by the introduction of a further quantitative indicator, a quotient which represented the number of errors made in a given time. Some data from the ZLT were further analysed by means of an additional quantitative method. For this analysis only passage 3 (the dog story) of the ZLT was used. The tape recordings of this passage were examined with a speech-analyser which was developed by Hänni and Zaugg at the Psychological Institute in Berne (for an exact technical description of this analyser, see Zaugg, 1986). This enabled the number of vocalizations and pauses during the reading process to be determined. The length of vocalizations and pauses was also measured. This procedure resulted in the registration of the frequencies of vocalizations and pauses as well as their duration.

Goldman-Eisler (1968) considered speech pauses (p) as "Hesitation Phenomena" as signs of uncertainty or delay. There are of course differences between speech pauses during a conversation and speech pauses while reading a text, however, the impression formed by the delay seems to be very similar in both situations. The duration of vocalization (v) represents an aspect of the quantitative reading capacity of a person. It is no doubt further influenced by the syntactic characteristics of a text or the persons' reading style. Good readers show a long average duration of vocalization.

The quotient duration of vocalization per duration of pause (v/p) is further indicator of the organization of the text by the reader, e.g. a quotient greater than zero means that the pauses make up for less than half of the time needed to read a text. The number of events (ie., sum of pauses and vocalizations) illustrates the 'staccato' of the text, which means the frequency of changes between periods of silence and periods of voicing.

To enable a comparison with the results of normal readers to be made, the same text was read on tape by ten members of the Psychological Institute in Berne and subsequently analyzed in the same way.

Qualitative Analysis of Reading Achievement

In order to evaluate further reading achievement it was also necessary to consider the qualitative aspects of the reading process. An analyzing system was developed which allowed the identification of the different types of errors which occurred during reading. The system is is based on two main categories: (1) Distortion of the Word and (2) Grammatical Error, and two subsidary ones (3) Number of Corrections and (4) Number of Words Repeated. The four categories and their subcategories are shown below. Each error was placed in either of the two main categories (D,G).

(1) Distortion of the word (D)

    -1-2 letters misread

    -additional letters read

    -missing letters

    -interchanged letters

    -total deformation of the word (3 or more letters misread or unrecognizable word).

(2) Grammatical error (G)

    -error concerns time of verb

    -error concerns conjugation of verb

    -error concerns noun

    -error concerns article

    -error concerns pronoun

    -error concerns particle

(3) Number of Corrections (C)

(4) Number of Words Repeated (R).

Measurement of Intelligence

The participants' intellectual abilities were measured by means of the German version of the Reduced Wechsler Intelligence Scale (WIP, Dahl, 1986). The WIP consists of four subtests, the first two from the Verbal Scale and the second two from the Performance Scale: Information, Similarities, Picture Completion and Block Design.

RESULTS

Demographic Data

Twelve participants grew up with their parents. One participant was adopted and educated by a foster-father. Three lived in their early childhood with their parents but were later sent to a social institution. One participant spent her childhood in several institutions and with several different foster-parents. Most of the participants came from large families. The number of children in the families varied between two and 12, with a mean of five children per family.

Thirteen participants started their school period at a Primary School and three went to a Special School. One person did not attend school at all. For the majority of participants this first school was located in a rural area. More than half went to a "Gesamtschule" which means that there were at least two classes of different levels in the same classroom, taught by only one teacher. In general the participants attended school regularly. During their school days, nine subjects had to work in their spare time in order to contribute to the family income. Changes of school type and/or place were extremely frequent.

Eleven participants had no professional education, six learned a profession. At the time of the tutorial program 16 were employed and one person was out of work. The majority of the participants worked at places in which their reading and writing skills were hardly required.

Intelligence

The IQ ranged from 66 to 117. The median of the distribution was 94. The median was used as a criterion to divide the participants into two groups, a "high IQ-group" (IQ>94) and a "low IQ group" (IQ<94). Group means were 106.75 and 78.63, respectively. Both groups differed significantly on all subtests (p<0.01, t-test).

Quantitative Reading Fluency (ZLT, reading passage 3)

Temporal aspects of speech are relatively sensitive to the description of verbal activity (Jaffe & Feldstein, 1970; Siegman & Feldstein, 1979; Schlatter & Hänni, 1989). For this reason they seem to be appropriate to describe oral reading fluency. Duration and frequency of reading pauses and the ratio of the duration of pauses and vocalizations were used in order to quantify this aspect of reading skill.

TABLE 1

Mean Values of the Temporal Parameters (Reading Passage 3)

	P (sec)	V (sec)	V/P	NE
Control N=10	0.68	3.24	5.05	16.2
IQ>94 N=9	0.68 *	3.68 *	5.89 *	25.4 *
IQ<94 N=8	0.95	1.86	2.44	113

P = pause   V = vocalization   NE = number of events
* = p<0.05 (Mann-Whitney)

First assessment

Table 1 shows the average duration of pauses and vocalizations, the average number of events (number of vocalizations and pauses) and the average ratio of vocalizations and pauses in the two subgroups (IQ>94, IQ<94). Although they are of no importance to the statistical test used, the means are also presented to give an impression of the average time needed to read a text. As shown in Table 1 there are no differences between normal-readers and the high IQ group except for the number of events. The low IQ group, however, differs significantly from the control group and the high IQ group on all parameters (p< 0.05, Mann-Whitney).

Pre-post measurement

After about six to eight months the participants were asked to read the same text again. Unfortunately only ten persons could be included in this second assessment.

TABLE 2

Pre-Post Measurement: Mean Values of the Temporal Parameters (Reading Passage 3)

	P-pre	P-post	V-pre	V-post	V-pre/P-pre	V-post/P-post	NE-pre	NE-post
IQ>94 N=5	0.68	0.71	3.68	3.35	5.89	5.28	25.4	28.2
IQ<94 N=5	0.95 *	0.81	1.86	2.02 *	2.44	2.88 *	113.0	115.0 *

P = pause **V** = vocalization **NE** = number of events   * = p< 0.05 (Wilcoxon)

Table 2 displays the results of the pre-post measurement. It shows that there are differences between the two groups concerning the temporal parameters. No statistically significant changes can be observed for the high IQ group whereas for the low IQ group the differences are significant on all variables (p<0.05, Wilcoxon).

Conventional analysis of the ZLT

First measurement

For the conventional analysis only the results of the subtests word-reading 1, 2, 1+2 and reading test 3 (dog story) were included, since only these passages of the ZLT were read by all persons. The time represents the time in seconds needed to read a subtest. The error refers to the number of errors made and the quotient 'errors over time' describes the relationship between the number of errors and the time needed to read a test.

The two groups showed different results with regard to time and errors. The high IQ group needed less time to read a test and made fewer mistakes than the low IQ group, however, none of these differences were significant, with the exception of a significant difference of time in the case of passage 3 (see Table 3).

TABLE 3

First Measurement ZLT: Results of the Groups IQ>94 and IQ<94

ZLT	mean values		p (Mann-Whitney)
	IQ>94 N=9	IQ<94 N=8	
WL1 time	33.56	71.00	0.0603
WL1 errors	0.68	2.13	0.1928
WL1 errors / time	0.02	0.03	0.6475
WL2 time	27.78	73.50	0.1346
WL2 errors	0.78	2.25	0.2388
WL2 errors / time	0.03	0.02	0.7222
WL 1+2 time	61.33	144.50	0.0670
WL 1+2 errors	1.44	4.38	0.1663
WL 1+2 errors / time	0.03	0.03	0.8439
LA3 time	52.78	152.88	0.0385 *
LA3 errors	2.56	6.50	0.0803
LA3 errors / time	0.05	0.05	0.7355

WL = single word reading test   LA = reading passage
time: time in seconds needed to read a subtest
errors: number of errors
errors/time: relation between the number of errors and the time needed to read
          a subtest

Pre-post measurement

Table 4 displays the results of the pre-post measurement regarding the time, the number of errors and the relation of errors and time for the high IQ group.

No regular pattern of achievement can be seen. The results of word reading test 1 show a significant improvement of time, but also a higher amount of errors and therefore an impaired relation of errors and time. In word reading test 2 there is an improvement on all three variables. The summary of the two word reading tests revealed a significant improvement of time, but at the same time a higher number of errors and therefore an impaired relation of errors and time. In reading text 3 there is an improvement on all variables except time.

TABLE 4

Pre-Post Measurement ZLT: Results of the High IQ-Group (IQ>94), N=5

ZLT	mean values		p (Wilcoxon)	tendency
	pre	post		
WL1 time	34.20	29.40	0.0431 *	amelioration
WL1 errors	1.00	2.20	0.0588	deterioration
WL1 errors / time	0.03	0.08	0.0431 *	deterioration
WL2 time	30.20	27.60	0.2228	amelioration
WL2 errors	1.20	0.60	0.1797	amelioration
WL2 errors / time	0.04	0.02	0.2850	amelioration
WL 1+2 time	64.40	57.00	0.0431 *	amelioration
WL 1+2 errors	2.20	2.80	0.4076	deterioration
WL 1+2 errors / time	0.03	0.05	0.2249	deterioration
LA3 time	58.40	59.60	0.8927	deterioration
LA3 errors	4.20	2.80	0.1797	amelioration
LA3 errors / time	0.07	0.04	0.1441	amelioration

WL = single word reading subtest   LA = reading passage
time: time in seconds needed to read a subtest
errors: number of errors
errors/time: relation between the number of errors and the time needed to read
      a subtest
* = p< 0.05 (Wilcoxon)

Table 5 shows the corresponding results of the low IQ group. In this case a regular pattern of achievement emerges. Time is improved in all word reading tests and in reading passage 3 which implies that the members of this group read the passages the second time faster than they did the first time. On the other hand there were a larger number of errors which led to a poorer relation between errors and time.

TABLE 5

Pre-Post Measurement ZLT: Results of the Low IQ Group (IQ<94), N=5

ZLT	mean values		p (Wilcoxon)	tendency
	pre	post		
WL1 time	81.40	61.20	0.1041	amelioration
WL1 errors	1.60	3.60	0.1441	deterioration
WL1 errors / time	0.02	0.06	0.1441	deterioration
WL2 time	75.40	55.80	0.0431 *	amelioration
WL2 errors	2.20	3.80	0.0656	deterioration
WL2 errors / time	0.02	0.08	0.0431 *	deterioration
WL 1+2 time	156.80	117.00	0.0796	amelioration
WL 1+2 errors	3.80	7.40	0.0796	deterioration
WL 1+2 errors / time	0.02	0.07	0.0431 *	deterioration
LA3 time	184.40	139.40	0.0796	amelioration
LA3 errors	7.00	7.20	0.8927	deterioration
LA3 errors / time	0.05	0.06	0.5002	deterioration

WL = single word reading subtest   LA = reading passage
time: time in seconds needed to read a subtest
errors: number of errors
errors/time: relation between the number of errors and the time needed to read
        a subtest
* = p< 0.05 (Wilcoxon)

Qualitative analysis (ZLT, reading passage 3)

As mentioned above a qualitative analysis of the errors was carried out in order to examine reading performance further. Tables 6 and 7 show the results of this analysis regarding the first measurement and the pre-post measurement. The absolute values (noted in brackets) don't allow a comparison between the two groups. The percentage of the respective categories makes a comparison possible, although it must be considered that this measure refers to very small numbers.

DISCUSSION

With regard to the quantitative variables (reading fluency), the group of normal readers and the high IQ group differed only in the number of events. There were no differences concerning the speech-pauses, duration of vocalizations or the relation of duration of vocalizations and speech-pauses. These results indicate that this group of participants reads nearly as fluently as normal readers. The difference in the number of events could be due to repetitions of words and self-corrections. The normal readers hardly made any errors and read without correcting or repeating words whereas the high IQ group showed some corrections and repetitions, which might have led to the count of a higher number of events.

TABLE 6

Qualitative Analysis: First Measurement ZLT (Reading Passage 3)

	number of errors	percentage of corrected errors	*total number of errors 100%* distortions (D)	grammatical errors (G)	% corrected D	% corrected G	repeated words (means)
IQ>94 N=9	23	35% (8)	52% (12)	48% (11)	58% (7)	9% (1)	0.44 (4)
IQ<94 N=8	52	31% (16)	75% (39)	25% (13)	36% (14)	15% (2)	4.25 (34)

TABLE 7

Qualitative Analysis: Pre-Post Measurement ZLT (Reading Passage 3)

	number of errors	percentage of corrected errors	*total number of errors 100%* distortions (D)	grammatical errors (G)	% corrected D	% corrected G	repeated words (means)
IQ>94 pr	21	38% (8)	57% (12)	43% (9)	58% (7)	11% (1)	0.8 (4)
IQ>94 po	14	36% (5)	79% (11)	21% (3)	45% (5)	0% (0)	0.6 (3)
IQ<94 pr	35	29% (10)	69% (24)	31% (11)	33% (8)	18% (2)	3.8 (19)
IQ<94 po	36	5% (2)	58% (21)	42% (15)	9% (2)	0% (0)	3.4 (17)

pr = pre   po = post   IQ>94 N=5   IQ<94 N=5

The low IQ group on the other hand differed from both the group of normal readers and from the high IQ group on all quantitative variables. These participants read the text distinctly less fluently than the other two groups. Their vocalizations were shorter and they made longer pauses in speech while reading. The nature of their reading behaviour was similar to that of children who have spent between two and three years in school. This finding is in agreement with the results of a study by Fagan (1987) in which a group of low achieving adult Canadian illiterates was examined with regard to their reading behaviour. The study showed that the reading process of these adult illiterates was very similar to the reading behaviour of children who have spent about two years in school.

There were also differences between the groups on variables in the conventional and the qualitative analysis. The high IQ group made less distortions and grammatical errors. The distortions were corrected more often than the grammatical errors. There were hardly any repetitions of words. In comparison with the high IQ group the low IQ group made more errors. The distortions predominated and were corrected more often than the grammatical errors. Repetitions of words were very frequent. This may be due to a lesser degree of automaticity in decoding.

In the high IQ group a similar however less pronounced tendency could be observed. The participants of this group seemed to have fewer difficulties decoding single words, but instead the grammatical structure of sentences was sometimes misinterpreted. It is possible that they formed hypotheses about the grammatical structure of a sentence which subsequently proved to be inappropriate.

For a further evaluation of reading achievement the difficulty of the text has to be taken into account as well. The text dealing with the story of a little badger-dog who has lost his collar must be considered as an easy text: It consists of simple, concrete and short words, one exception being the word "gekollert" (rolled) which is rather unusual, and sentences of a relatively easy structure. There are only a few subordinate clauses and generally the sentences are fairly short. Thus the text does not impose great difficulties on the reader. In the context of the ZLT this text represents the most difficult part for second graders and the easiest part for pupils in their fourth year of school.

For the high IQ group this text was easy to read. In comparison with a sample of children who were in their second school year the participants results (errors /time) are in the rank of 76-100%. The results of the low IQ group showed a different picture. Some participants of this group made more mistakes and read the text slower than the weakest member of the sample of children. These findings show that there is indeed a difference between the two IQ groups at the beginning of the tutorial program. The participants are at different stages of the reading process and preconditions for learning to read are different.

Eight months later the results of the high IQ group's second reading of the dog-text hardly differed from their first reading with respect to the quantitative variables. The reading-fluency did not change. Given that at the beginning this group already read nearly as fluently as normal readers, this finding is not very surprising. The conventional and qualitative analysis of the ZLT show an improvement. The group made less errors and the number of errors per time also decreased. The relation of distortions and grammatical errors was different, however, there were still more distortions. The number of corrections stayed about the same but none of the grammatical errors were corrected.

The high IQ group was found to read the text in a more controlled way whereby the reading fluency remained the same and the number of errors decreased. Furthermore the text was read in a clearly understandable way. The results of the single word reading part of the ZLT confirmed this finding although there were still problems concerning the decoding which might partly be due to the missing context.

The comparison of the low IQ group's first and second reading of the text revealed significant differences on all quantitative variables of reading (duration of speech-pauses, duration of vocalization, ratio of duration of vocalization and pauses, number of events). From the first to the second measurement the reading fluency of this group improved. It approached the fluency of normal readers. The results of the conventional and qualitative analysis on the other hand limit the impact of this improvement. The members of this group made more errors during the second reading so the amount of errors as a function of time increased as well. The number of distorted words decreased, but these misreadings still predominated. Very few errors were corrected and of these it was only distortions and not grammatical errors which were corrected. The number of repeated words did not change.

The increased reading fluency therefore seemed to have a negative effect on the quality of reading (more errors and fewer corrections). For these reasons it can be concluded that reading fluency has improved, but the quality of the reading process cannot keep up with this quantitative improvement. Despite these qualitative deficits the results of this group allow for a certain optimism. The ability to read faster and more fluently was a source of motivation to the participants. They started to read in their spare time since reading was no longer as frustrating as it used to be. With the help of the tutorial program they became more self-assured and attempted to read the text in a more courageous way maybe overstressing speed as a crucial aspect in oral reading. A study of Smith (1990) points out that 'good readers' and 'poor readers' differ in their ideas about the important factors in reading. Good readers consider reading comprehension as most important whereas poor readers believe that successful decoding is the crucial point. Thus the differences in the way the two groups read the text could partly be explained by their different theories about reading. These divergent ideas may have led to two different reading strategies, a more comprehension oriented strategy and a strategy which stresses the importance of decoding.

Given the fact that the sample was split into two subgroups on the basis of intelligence and that these subgroups performed differently on several of the variables it seems viable to use IQ as an indicator for degree of success in teaching reading skills to adults. It is clear however that intelligence and reading skill interact. This means that low scores in an IQ-scale are not necessarily considered the reason for poor reading performance.

The participants' biographies have a lot of aspects in common. Regarding the demographic data no differences between the IQ groups were found. Most of the participants came from families with a relatively high number of children. Participants felt that their parents did not have time to look after the achievement of their children in school and could not help them with their homework. Difficulties at school led to problems with parents and teachers and therefore time at school especially reading and writing were experienced in a negative way. A large number of participants had to change schools very often and had to adapt to many different learning situations. In particular, the change to a Special school was described as a very traumatic experience. The participants' attitudes towards reading and writing are still influenced by these experiences. Many of them failed in the first two years of school and never caught up in later periods. Today most of the participants work in jobs where reading and writing has little importance which results in the little knowledge they have acquired is fading. These findings are supported by Oswald & Müller (1982) in a study on functional illiterates in Germany. The authors reported comparable family situations and the need for children to contribute to the family income etc. Experiences at school and actual professional background were also reported to be very similar.

From our findings and those of earlier studies we can conclude that a) that by the means of tutorial programs adult illiteracy can be remediated and its effects lessened and (b) success, however, depends on a number of variables including IQ.

REFERENCES

Dahl, G. (1986). *Handbuch zum reduzierten Wechsler Intelligenztest; Anwendung, Auswertung, statistische Analysen, Normwerte.* Königstein/Ts.: Hain Verlag.

Fagan, W.T. (1987). A comparison of the reading processes of adult iliterates and four groups of school age readers. *The Alberta Journal of Educational Research, 33,* 123-136.

Goldmann-Eisler, F. (1968). *Psycholinguistics.* London: Academic Press.

Grissemann H. (1986). *Bestimmungs-und Abgrenzungsprobleme des funktionalen Analphabetismus.* Tagung "Funktionaler Analphabetismus: Fakten - Zahlen Strategien". Nationale Schweizerische UNESCO-Kommission, Bern.

Hamadache, A. (1984). Les analphabètes du "Quart Monde". *Le courrier de l'UNESCO, 2,* 22-25.

Jaffe, J. & Feldstein, S. (1970). *Rhythms of Dialogue.* New York: Academic Press.

Linder, M. & Grissemann, H. (1981). *Zürcher Lesetest.* Bern: Huber Verlag.

Oswald, M.L. & Müller, H.M. (1982). *Deutschsprachige Analphabeten..* Stuttgart: Klett Verlag.

Schlatter, T. & Hänni, R. (1989). Zur Steuerung des Sprecherwechsels in einem Arzt-Patient Gespräch. Eine Einzelfalluntersuchung. *Archiv für Psychologie, 141,* 273-286.

Siegman, A.W. & Feldstein, S. (1979). *Of Speech and Time. Temporal Speech Patterns in Interpersonal Contexts.* Hillsdale, N.J.: Erlbaum.

Smith, M C. (1990). Reading Habits and Attitudes of Adults at Different Levels of Education and Occupation. *Reading Research and Instruction, 30,* 50-58.

Zaugg, W. (1986). *Gesprächsanalysator. Benutzerhandbuch.* Unveröffentlichtes Manuskript. Psychologisches Institut, Universität Bern.

APPENDICES

Wortlesetest 1 (Single Word Reading Test 1)

in	*in*	die	*the*	wieder	*again*	acht	*eight*
an	*on*	sei	*was*	weiter	*further*	blau	*blue*
so	*so*	sie	*she*	Lieder	*songs*	auf	*on*
er	*he*	der	*the*	leider	*unfortunately*	faul	*lazy*
zu	*closed*	nie	*never*	lügen	*lie*	laufen	*run*
da	*there*	nein	*no*	Liebe	*love*	Säbel	*sabre*
ab	*off*	dein	*your*	beide	*both*	Orgel	*organ*
im	*in*	beim	*at*	dabei	*thereby*	Sport	*sport*

Wortlesetest 2 (Single Word Reading Test 2)

oben	*above*	Abend	*evening*	heute	*today*
droben	*above*	baden	*bathe*	teuer	*expensive*
Boden	*floor*	braten	*roast*	einmal	*once*
Borsten	*boistles*	Bart	*beard*	schreien	*cry*
Oberst	*colonel*	aber	*but*	endlich	*finally*
Körbe	*baskets*	bald	*soon*	langsam	*slowly*
Krone	*crown*	bleiben	*stay*	springen	*jump*
Kurt	*Kurt*	glauben	*believe*	sprudeln	*bubble*

Leseabschnitt 3 ("Dackeltext")

Schnell ging Fridolin, der kleine Dackel, den Weg zurück, den er gekommen war. Doch umsonst suchte er in allen Gassen und Strassen. Umsonst lief er den Bahnhof auf und ab. Es war kein Halsband zu sehen. "Vielleicht ist mir das Halsband abgefallen, wie ich aus dem Zug gesprungen bin!" sagte er sich. Er ging den Weg zurück bis zum Wassergraben, in den er gekollert war, und kletterte den Bahndamm hinauf. Dann lief er den Schienen entlang, bis er wieder beim Bahnhof ankam. Aber nichts, gar nichts war zu finden.

He hurried back the way he had come. In vain he looked in all the alleyways and streets. In vain he ran up and down the station. There was no sign of the collar. Perhaps the collar fell off when I jumped out of the train, thought Fridolin. He went back to the ditch he had rolled into and clambered up to the top of the embankment. Then he run along the track until he got back to the station again. There was no sign of the collar anywhere.

Facets of Dyslexia and its Remediation
S.F. Wright and R. Groner (Editors)
© 1993 Elsevier Science Publishers B.V. All rights reserved.

READING ACQUISITION IN ANALPHABETIC ADULTS

U. Tymister,

University of Freiburg, Germany

INTRODUCTION

Illiteracy is a problem that has been discussed by the German public for the last ten years. It affects German adolescents and adults who in spite of having attended school have not acquired sufficient skills which enable them to perform specified actions in connection with written language, such as writing or reading letters, filling out forms or writing job applications. Since then, it has been mainly the German 'Volkshochschulen', schools focusing on adult education, that have taken up the challenge of compensatory teaching in this field. By now more than 18,000 learners are taking reading and writing lessons in over 400 adult education facilities.

DIAGNOSIS

While planning the lessons, an first important step was to assess the underlying skills necessary for the acquirement of reading and writing skills. In the following, two possible means of diagnosis will be discussed.

1. Basic Skills

The elementary approach postulates the following skills as prerequisites for acquiring written language skills:
- the ability to hear and distinguish phonetic differences
- the ability to write and read letters of the alphabet
- the ability to memorize sequences of written characters and similar processes.

Further capabilities such as spatial orientation, logical deduction etc. were tested and the learners were categorized accordingly. This was similar to the differentiation test of Breuer & Weuffen (1986).

2. Developmental Approach

A further means of categorizing the learners is based on their written output. According to this approach, sequences of written characters have a logographic, an alphabetic and an orthographic level. In this concept one distinguishes three levels of learning. On the *logographic level* the learner tries to reproduce the sequence of written symbols as a whole. A person using the *alphabetic strategy* proceeds letter by letter, pronouncing or reading every single one. At the *orthographic level*, finally, the learner attempts to apply spelling rules. (Börner, 1989)

TEACHING METHODS

There are 3 teaching methods applied in the instruction process with illiterates. These will be briefly described; for further information see Tymister (1990).

1. Elementary Approach
A special teaching method based on the assessment of the learners' basic skills was found: e.g. if a learner was not able to differentiate the letters 'b' and 'd' when hearing them, the teacher resorted to the graphic forms of the letters. The circle in 'b' comes after the vertical line drawn downwards whereas 'd' is the other way round.

2. Language Experience Approach
The language experience approach was also frequently applied. Following this approach, oral statements made by the learners themselves and dealing with their personal experience were put into writing either by the teacher or by an advanced learner. This approach aims at helping the learners lose their fear of writing.

3. Morpheme Method
Finally, a linguistic method one might call the building block method remains to be mentioned. A word is divided into its morphemic constituents, i.e. morphemes marking the beginning and end of a word, intermediate morphemes and main morphemes (see below).

	WORD		
BEGINNING	MAIN/BASIC	INTERMEDIATE	END MORPHEME
un-	condition-	al-	ly
	wait-		ed
	dog-house		

A learner can find it difficult to write and segment words which do not consist of two main morphemes. A word like "Hochhaus" (German for "skyscraper") which can be divided into the two basic building blocks "high" and "house" - cf. "doghouse" in English - is much easier to handle. (Füssenich/Gläss, 1986)

These various approaches are frequently applied selectively and therefore, only reach people with particular types of illiteracy who respond to one or the other of the above mentioned approaches.

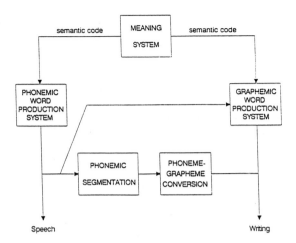

FIG. 1. Model of the Writing Process. (Reprinted with permission, Eigler, 1990).

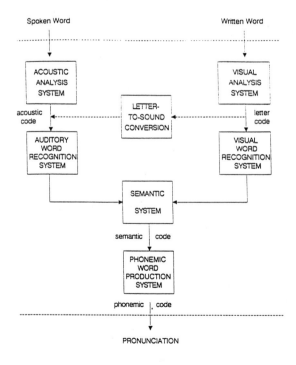

FIG. 2. Model of the Reading Process (Reprinted with permission, Eigler, 1990)

THE LEARNING PROCESS

The Cognitive Model

The theoretical assumptions about the learning process underlying these methods are partly rooted in the ideas of Ellis (1984), who developed a cognitive structural model valid for the process of reading as well as writing. His model makes it possible to make assumptions about how an adult reads or writes a word. However, he does not state how an adult might go about acquiring these skills. Figs 1 and 2 show a model of the writing and reading (words) processes, respectively, adapted from Ellis (1984).

Developmental Model

Frith (1980) worked out a developmental model that describes how children acquire writing skills. Her model was adopted in Germany and applied to adults. Frith's considerations are based on the idea that the acquirement of reading and writing abilities - like for Piaget the entire cognitive development - takes place step by step, each step being a prerequisite for the following one. No single step can be left out, nor is it possible to resume a former step. Beginning at the logographic level and continuing through the alphabetic level, the learner reaches the level of knowledge termed orthographic. In this process, reading and writing take turns in bringing about the next higher level.

Top-down/Bottom-up Model

Stanovich (1980) developed a model that took into account top-down as well as bottom-up processes. His ideas can be situated in a line of dispute incited by Gough's (1972) claim that all the letters within a reader's scope of vision must be seen before the reader is able to allocate a certain meaning to a specific series of letters.

TEXT COMPREHENSION
enables ⇅ facilitates
SENTENCE COMPREHENSION
enables ⇅ facilitates
SEMANTIC-SYNTACTIC STRUCTURES
enables ⇅ facilitates
WORD IDENTIFICATION
enables ⇅ facilitates
IDENTIFICATION OF WRITTEN AND ORAL PATTERNS
enables ⇅ facilitates
KNOWLEDGE / IDENTIFICATION OF LETTERS

Other models viewed the process the other way round, starting with an expectation of meaning followed by the deciphering of individual letters. Stanovich sees his model as an interactive and compensatory approach: shortcomings on one level can be compensated for by processes on other levels. His ideas are illustrated in the flow chart on the previous page (cf. Urban &Vanecek, 1990).

CONCLUSION

The reading/writing process consists of two processes: the momentary process of reconstructing a certain word and a developmental process of skill reinforcement. Ellis' model focuses on the aspect of momentary performance. His model is based on the idea that the performer, when writing a word, is either able to transcribe it directly from his mind's image of the word or that he needs to pronounce the word in order to be able to use phonological criteria as a guideline for writing. Thus, according to Ellis, there are two ways to write a word. However, this appears to be too limited when the importance of individual learning strategies is stressed.

Frith, on the other hand, describes the development of the skills in question, at the same time outlining a strict sequence of steps.

Stanovich's model, which can be viewed as a more general type of model, combines both components of the reading/writing process. In the course of both momentary performance as well as skill reinforcement, the different levels of text (re)production interact - from top to bottom or vice versa - ultimately leading to automatisms. And each individual attains them his or her own way.

The aspect of individual learning is a very important part of the following approach based on Stanovich.

THE FREIBURG INTEGRAL APPROACH

Based on Stanovich's ideas, the present author's "Freiburg integral approach" aims at individually supporting and activating the participants of literacy courses. For this purpose it is essential to assess the individual levels of knowledge, which can differ very greatly among functional illiterates. Writing and reading texts with contents taken from the everyday down-to-earth life of the participants is an adequate method to stimulate their own writing and reading activities. Furthermore, skills are not only trained and improved in this way, but self-confidence is heightened. The following four levels of learning can serve as a guideline for assessing the individual level of knowledge. These levels of learning are further differentiated by an individual hierarchy of learning objectives in accordance with Gagné:

Beginners who do not know the alphabet

These persons can only read single letters and copy words letter by letter, if at all. They know all the different letter forms, but are not necessarily able to associate them with sounds.

Beginners who know the alphabet

These persons can read and write groups of letters and short words.

Advanced learners with slight knowledge of rules

These persons know simple rules. Sentences are still read word for word, and rules for spelling are based on intuition.

Advanced learners with knowledge of rules

These persons are able to compose texts by themselves, sometimes almost without mistakes. Spelling is checked with the help of simple rules.

Since the group of learners is very heterogeneous, these four levels of learning are of great help to the teacher wishing to utilize a differentiated approach. The individual learner gets his or her own set of problems to solve, exercises, possibilities of feed-back and transfer, all based on a text that was chosen as a common working basis for the whole group. Thus the feeling created by learning together in a group is retained, while at the same time allowing individual learning progress. When applying this teaching method it is important to take into account as much as possible the individual learning capacities, learning speeds, learning methods and achievements.

The hierarchy of learning objectives is based on the consideration that knowledge of basic skills enables a positive transfer towards acquisition of higher skills. At the same time the higher skills interact with the underlying ones. By taking the participants' preliminary knowledge into account, their deficiencies are concentrated on and systematically dealt with instead of merely going through the whole hierarchy from bottom to top.

Checking the spelling of written text, writing text, writing sentences, writing words

Identifying e / ä	Identifying groups of signals	Special problems: capital letters
Identifying morpheme units Long i	Identifying groups of letters	Identifying syllabic units
Understanding the connection between long vowels and orthographic lengthening	Working out problems in associating letters with sounds as ie/ei, g/k etc.	Understanding the connection between short vowels and orthographic reduplication

Understanding that there is no one to one equivalence of letters and sounds

Knowledge of the alphabet

The learning objectives are arranged in such a way that the acquisition of spelling skills corresponds to and correlates with the ability to write words, sentences and texts. This means that beginners first of all focus on writing words, whereas the advanced learners are occupied with writing sentences and texts and with checking their spelling. This can be illustrated with the following example.

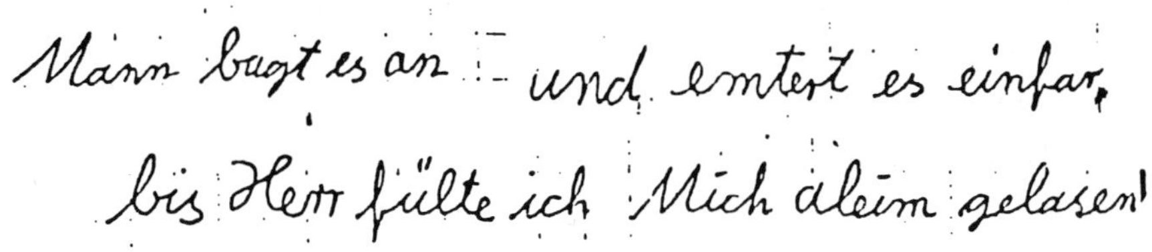

FIG. 3. Sample of text from an advanced learner. "Mann bagt es an und emtert es einfar. bis Herr fülte ich Mich aleim gelasen (Let's start changing things. Up till now I felt left alone with my problems).

This text was written by an advanced learner. His problem is that he mixes up several related letters and sounds such as p/b, g/k, d/t, n/m, r/ch. When to use capital letters and markers for lengthening of vowels are further problems he has to cope with. In accordance with this diagnosis, his individual hierarchy of learning objectives is depicted below.

Checking the spelling of written text, writing text, writing sentences, writing words

Identifying e / ä	Identifying groups of signals	*Special problems: capital letters*
Identifying morpheme units Long i	Identifying groups of letters	Identifying syllabic units
Understanding the connection between long vowels and orthographic lengthening	*Working out problems in associating letters with sounds as as ie/ei, g/k etc.*	*Understanding the connection between short vowels and orthographic reduplication*

Understanding that there is no one to one equivalence of letters and sounds

Knowledge of the alphabet

Key: Italics indicate skills taught in the lesson and non-italics existing skills.

Now the teacher can select the learning material especially for this learner. He will choose exercises concerning the phonemes like the connection between long/short vowels and orthographic lengthening/reduplication.

PERSPECTIVES

The integral approach has not only proven efficient in traditional courses. It has also been possible to apply this approach successfully in computer supported teaching, a useful complement to individual learning: The learners were able to choose their lessons according to their needs. This and the increased motivational impetus sparked by a new learning media leads to good learning results.

REFERENCES

Börner, A. (1989). Abschlu·bericht Projekt Alphabetisierung. Marburg.

Breuer, H. & Weuffen, M. (1986). Gut vorbereitet auf das Lesen und Schreibenlernen? Berlin 6.Aufl.

Eigler, G. (1990). Funktionaler Analphabetismus - auch ein psychologisch-erziehungs-wissenschaftliches Problem. *Unterrichtswissenschaft* 1990/2.

Ellis, A.W. (1984). *Reading, Writing and Dyslexia: A Cognitive Analysis*. London.

Gagné, M. (1970). The Conditions of Human Learning. New York: Holt, Rinehart & Winston, Inc.

Füssenich, I. & Gläss, B. (1984). Alphabetisierung und Morphemmethode. Kritik eines schriftdidaktisches Verfahrens aus linguistischer und psyholinguistischer Sicht. In OBSt 26, S.36-69.

Stanovich, K.E. (1980). Toward an interactive-compensatory model of individual differences in the development of reading fluency. *Reading Research Quarterly* 1980, 16, 32-71.

Tymister, U. (1990). Arbeiten mit Analphabeten in Volkshochschulen. In *Unterrichts-wissenschaft* , 2/1990, S.113-124

Urban, W. & Vanecek, E. (1990). Untersuchungen zur sprachlichen Minderleistung zum Funktionalen Analphabetisms bei Jugendlichen und jungen Erwachsenen in Österreich. Wein.

AUTHOR INDEX

SUBJECT INDEX

A

Accommodation, 125, 130-132, 134, 135, 181, 182, 186-188
Achievement measures, 103
Acuity, 125, 129, 131-135
Adult,
  analphabetic, 623
  dyslexics, 411, 413, 415, 419
  illiteracy, 607, 619
Alphabetic, 314, 315, 317
Asthenopia, 591, 599, 601
Attention, 229, 230, 357-359, 363, 365, 366, 367, 369, 370
Automaticity, 581, 582
Automatisation, 376, 381, 383, 387-391
Autonomous lexicon 393, 394, 403, 407

B

Behaviour, 427, 428
Bilingual, 248, 255
Binocular,
  fixation, 140, 141, 143, 149, 150, 152, 156-158
  full correction, 585, 591, 593, 594, 598, 599, 601
  problems, 588, 599
  vision, 181, 585-588, 591, 598, 601- 605

C

Cerebral dominance, 108, 119, 120, 122
Children, 3, 5, 7, 8, 10-12
Chinese, 245, 247, 248, 250, 253-256

Cluster analysis, 440-442, 445, 448, 449
Cognitive,
  measures, 101-103
  neuropsychology, 291, 292, 294, 299 301, 302
  factors, 163, 164
Computer, 210, 212, 218, 219, 248, 257 258, 349, 350, 551-553, 556, 561, 564, 567, 568, 570, 572, 576, 578, 610
  aided instruction, 581
Contrast sensitivity, 125, 131, 134, 139, 144-149, 158, 160, 161
Convergence, 125
Cooperative teaching, 517, 533
Curriculum differentiation, 514, 515, 519, 520, 524, 525, 531-535

D

Definition, 437, 438, 443, 444, 449, 457, 480
Developmental,
  models, 291, 292, 296, 301
  Test of Visual-Motor Integration, 169
Dunlop Test, 140-142, 144, 157, 162, 181, 182, 186-188
Dyseidetic, 314, 315
Dyslexia, 158, 179, 180, 182, 184, 185 187-190, 217, 230, 257, 358, 359-361 366, 367, 403, 405-407
Dysphonetic, 314, 315

E

Early screening, 483, 484